ANNUAL REVIEW OF EARTH AND PLANETARY SCIENCES

ANNUAL REVIEW OF EARTH AND PLANETARY SCIENCES

VOLUME 11, 1983

GEORGE W. WETHERILL, *Editor*
Carnegie Institution of Washington

ARDEN L. ALBEE, *Associate Editor*
California Institute of Technology

FRANCIS G. STEHLI, *Associate Editor*
University of Oklahoma

ANNUAL REVIEWS INC. 4139 EL CAMINO WAY PALO ALTO, CALIFORNIA 94306 USA

ANNUAL REVIEWS INC.
Palo Alto, California, USA

International Standard Serial Number: 0084-6597
International Standard Book Number: 0-8243-2011-5
Library of Congress Catalog Card Number: 72-82137

PRINTED AND BOUND IN THE UNITED STATES OF AMERICA

Annual Review of Earth and Planetary Sciences
Volume 11, 1983

CONTENTS

Special Announcement: Volume I of the *Annual Review of Immunology* (Editors: W. E. Paul, C. Garrison Fathman, and Henry Metzger) was published in April, 1983.

SOME RELATED ARTICLES IN OTHER *ANNUAL REVIEWS*

From the *Annual Review of Astronomy and Astrophysics*, Volume 20 (1982):

Planetary Magnetospheres, D. P. Stern and N. F. Ness

The Dynamics of Planetary Rings, Peter Goldreich and Scott Tremaine

The Satellites of Jupiter and Saturn, David Morrison

From the *Annual Review of Ecology and Systematics*, Volume 13 (1982):

The Evolutionary Effects of Mate Selection, William A. Searcy

Early Evolution of Reptiles, Robert L. Carroll

The Molecular Clock Hypothesis: Biochemical Evolution, Genetic Differentiation and Systematics, John P. Thorpe

Evolution and Classification of Beetles, John F. Lawrence and Alfred F. Newton, Jr.

Pheromones and Evolutionary Relationships of Tortricidae, Wendell L. Roelofs and Richard L. Brown

From the *Annual Review of Fluid Mechanics*, Volume 15 (1983):

Snow Avalanche Motion and Related Phenomena, E. J. Hopfinger

Breaking Waves on Beaches, D. H. Peregrine

Magneto-Atmospheric Waves, John H. Thomas

John Verhoogen

Ann. Rev. Earth Planet. Sci. 1983. 11: 1–9

PERSONAL NOTES AND SUNDRY COMMENTS

John Verhoogen
Department of Geology and Geophysics, University of California, Berkeley, California 94720

One may well wonder what made the Editors of *Annual Reviews* think that a prefatory chapter on my own experience could be of interest to anyone besides myself. I have no good stories to tell; it has all been very ordinary and straightforward. I was born in Brussels, Belgium, in 1912. Since early childhood I have been interested in rocks. At age 8 I dictated to my sister, whose penmanship was better than mine, a textbook on geology. It amounted to a statement that minerals and rocks come in great variety and many colors; this is still my view today. I graduated from high school in "classical humanities," which included Greek, Latin, and three modern languages besides my own, which is French. Thanks to a private tutor, who was the best teacher I ever had, I had enough mathematics (calculus, analytical geometry, descriptive geometry, etc) to pass the stiff entrance examination to the engineering school at the University of Brussels, where for five years I studied mining engineering. The choice of that subject was not random. My own thoughts were still on geology, or perhaps physics, but my father, who was an M.D., was a bit skeptical of the possibility of earning a living in either of these fields; mining engineering, which combined some geology with some physics and a salable skill, seemed a reasonable compromise. After finishing at Brussels in 1933, I spent a year at the University of Liège, where I took a degree in Geological Engineering. This had nothing to do with what is now called geological engineering; it consisted of a lot of paleontology, with some stratigraphy, structural geology, and petrology. As luck would have it, I then won a fellowship offered by the Commission for Relief in Belgium (CRB), now called the Belgian American Educational Foundation. The CRB was an outgrowth of Herbert Hoover's efforts in World War I to feed the Belgian population; funds left over after the war were used to rebuild university libraries and endow graduate fellowships in the United States. My choice of Stanford

0084–6597/83/0515–0001$02.00

University as my destination was determined by two factors. The first was an accidental meeting with H. Schenck, who taught paleontology at Stanford, thought highly of the place, and happened to be on sabbatical at Liège at that time. The second factor was that since the CRB paid transportation to wherever one wished to go, I naturally chose to go as far as possible.

When I arrived at Stanford in the fall of 1934, I had no precise idea of what I wanted to do, although I knew I was heading for an academic career. There never was much doubt about that: it was almost a family tradition. My father and his brother both taught medicine at the University of Brussels, an uncle on my mother's side taught in the school of engineering, and two of my sisters had earned doctorate degrees and were engaged in academic work. In 1934 I was much interested in the chemistry of sediments (the word geochemistry was not much used in those days), particularly coal. On the other hand, the "Institut des Parcs Nationaux du Congo Belge" was beginning to plan volcanological studies of the Virunga volcanoes that lie in what was then called the Albert National Park, north of Lake Kivu, in the Western Rift Valley of Africa, and I had been asked to have a look at the current state of volcanological research in the United States and Hawaii. Aaron Waters, then a young professor at Stanford, first convinced me and the CRB to extend my stay long enough to complete the requirements for a Ph.D. degree, and then sent me off to Berkeley to talk to Howel Williams, who suggested as a thesis topic a study of Mount St. Helens, a (then) dormant volcano in the Cascade Range.[1] After a summer of field work, I spent a few months at Berkeley doing petrography with Williams and learning elements of seismology from Perry Byerly. By an unusual arrangement, Stanford then granted me a degree for work done mostly at Berkeley. I also managed to spend a few weeks at the Kilauea Volcano Observatory. Both Kilauea and Mauna Loa were inactive at the time, but T. A. Jaggar was not. He ran the volcano observatory almost single-handedly and had a lot of volcanological lore to teach. I returned to Belgium in 1936, stopping on my way in Washington, D.C., to attend the annual meeting of the AGU. It is amusing to recall that in those days, attendance at the annual meeting was such that almost everyone could be seated around the table in the Board Room on the first floor of the National Academy of Sciences building.

During my university days, I used to spend a lot of time in the library, reading with interest and curiosity almost anything that came under my hand. Shortly after my arrival at Stanford, I thus discovered in the shelf H.

[1] I have recently reread my thesis and was disappointed not to find in it even the vaguest allusion to the possibility of a major eruption in 1980!

Jeffreys' *The Earth*, then in its second edition. Most of it I did not understand, but it nevertheless made a profound impression on me. So did N. L. Bowen's *The Evolution of Igneous Rocks*, which prompted me to start on a serious study of thermodynamics, and for the next twenty years placed petrology and geochemistry at the center of my interests. Then, in 1938–40, came an opportunity to observe at close range a spectacular eruption of the volcano Nyamuragira, also known as Nyamlagira, in the Virunga range. What came of these observations was, first and foremost, a sense of frustration at our inability to understand what goes on inside a volcano, and second, a conviction that the answer to such problems as magmatic activity should be sought in the thermodynamics of the Earth's deep interior.

The war years were spent in the Congo, working on the procurement of strategic minerals, while trying in my spare time to keep informed of scientific developments. There were not many scientific books or journals available in Africa in those days, but I did somehow manage to get hold of a copy of A. Eddington's *Internal Constitution of the Stars*. Here again was a source of frustration: if an astronomer can figure out to the last million degrees the temperature at the center of a body a million light years away, why can't we geologists determine the temperature in the Earth beneath our feet? But then there was also Eddington's exposition of von Zeipel's theorem on the condition for equilibrium in a rotating body with internal heat sources, which led me to believe that violation of von Zeipel's condition in the Earth makes for the likely occurrence of convection currents, as Holmes and others before him had been saying all along. As it turns out, von Zeipel's theorem is probably only a minor factor in triggering instability in our planet, but it did turn me into a firm believer in convection; this led, a few years later, to my suggestion that melting and formation of magma, a puzzling phenomenon in an otherwise solid mantle, could readily be expected in the rising limb of a convection cell, a "diapir" as they now say. All this to show how one thing leads to another, and how a train of thought may be started by a serendipitous event such as the turning up in central Africa, in the middle of a world war, of a copy of Eddington's book.

From 1947 on I was settled permanently in Berkeley, where I had the good luck to meet Frank Turner, who had arrived there a year before me, coming from New Zealand. If Frank and I were poles apart with regard to birthplaces, we soon found lots of things to talk about. He tried, rather unsuccessfully, to teach me some petrology, while I explained to him the mysteries of entropy and chemical potentials. Shortly afterwards, Bill Fyfe came to spend two years with us, bringing with him a rich new lore of quantum chemistry and a deep interest in geochemistry. At about the same time, Dick Doell had begun to build a spinner magnetometer and started us on a paleomagnetic spree that occupied me for the next fifteen years.

It is generally considered today that the major revolution in Earth Sciences of the second half of this century came in the 1960s, with the advent of plate tectonics. My own opinion is that the preceding decade, 1950–60, was probably even more significant. It saw such revolutionary discoveries as the approximate equality of heat flow in continents and oceans, which flatly contradicted all previous notions on the distribution of heat sources; the differences in apparent polar wander paths as seen from different continents, which left little doubt as to the reality of continental drift; and the development of Byerly's fault plane method, without which we probably would still not acknowledge many transform faults. In addition to adducing much evidence of heterogeneity (phase changes) in the upper mantle, Francis Birch's classic paper of 1952, "Elasticity and Constitution of the Earth's Interior," set the tone for all later research.

But perhaps it is not any specific discovery or hypothesis that best characterizes the advances of the decade. Perhaps the most significant component of the revolution of these years was a rather drastic change in the turn of mind, in the way to look at geological and, increasingly, geophysical problems. The turn was to a decidedly more quantitative, more analytical attitude. Physical and chemical theory was called upon with increasing frequency and sophistication. This evolution may perhaps best be described by a personal anecdote. In 1951 appeared the first edition of *Igneous and Metamorphic Petrology* by F. J. Turner and myself. Chapter 2 in that book is a brief introduction to chemical thermodynamics. An anonymous referee who had seen the manuscript had suggested deleting that chapter on the grounds that it was incomprehensible to geologists and useless to petrologists, but his advice was disregarded and the chapter stayed in. It was also left in the second edition (1960) in spite of the opinion of another referee, who thought that it should be omitted since all this stuff was already well known and familiar to most geologists.

The general and pervasive quantification of geology that occurred in those years was, of course, not entirely new. Arthur Holmes had been doing it for a long time. P. Niggli had written volumes on the application of thermodynamics (mostly of the Backhuis-Roozeboom type) to petrology, and the Geophysical Laboratory of the Carnegie Institution of Washington had been engaged for years in experimental work, mostly in what would now be called geochemistry. But it took a long time for this trend to seep down through the rank and file of the geological profession. A widely used introductory geology textbook published in the late forties still gave as proof of the spherical shape of the Earth the fact that innumerable ships had come safely into port by using maps drawn on the *assumption* that the Earth was round, as if no one had ever bothered to measure the radius of the Earth and its variation with latitude.

Another important change of the early 1950s was the realization by geologists that there was perhaps more to the Earth than just its crust. As late as 1950, the word "mantle" still meant to many geologists a very thin cover of broken and discolored products of rock weathering unevenly spread over the Earth's surface. What lies beneath an ill-defined crust was mostly unknown and generally considered to be uniform, remote, inert, and irrelevant. It took a minor revolution to convince geologists that most magmas do indeed rise from a very active mantle whose composition and behavior determine what happens to the crust. By the end of the 1950s it was generally conceded that the mantle might be worth looking at more closely, but then only its uppermost part; it took another decade for the notion to take hold that the whole mantle may be involved in surficial events. A logical extension of this trend to look deeper and deeper was the still more recent suggestion that perhaps the very core of the Earth may be involved in surficial geological events, through the agency of plumes rising from the core-mantle boundary.

In the early 1950s, the word "convection" also began to appear in geology textbooks, but then generally in connection with Griggs' experiments of the late 1930s on formation of mountains. The fundamental role that convection plays in heat transfer within the Earth was not recognized until many years later; by the middle 1960s geophysicists were still calculating geotherms from the equation that governs heat conduction. Curiously, it escaped everyone's (or almost everyone's) notice that these calculations were self-contradictory, insofar as the large horizontal temperature variations between subcontinental and suboceanic mantle they predicted were sufficient to cause the very convection that had been assumed not to take place.

Returning to the origin of mountains, the key word in the 1950s still was "geosyncline." The fact is that sedimentary thicknesses are far greater in fold-mountain ranges than in adjacent continental plates. Since much of the geosynclinal sediments bear evidence of deposition at or near sea level, the conclusion was drawn very early that in areas of developing mountains the crust must first be slowly depressed, at about the rate of sediment accumulation. The mechanism for this was indeed puzzling, and still is an unanswered question of plate tectonics. But who hears of geosynclines anymore? This is perhaps a good illustration of a tendency common to geologists: after long debate over a problem, and still with no solution on the horizon, we just start talking about something else.

Progress in Earth Sciences in my active days came mostly from developments in geochemistry and geophysics, particularly the latter. Plate tectonics, for instance, is a child of paleomagnetism and seismology; classical geological disciplines such as structural geology and stratigraphy have played only a minor role in its development. Most of the physical theory behind geophysics was already well known in the first quarter of this century; one can be a very good geophysicist today while still ignoring later developments in physics such as quantum mechanics. (An exception may be the theoretical advances of the last thirty years in the statistical treatment of observations, time series, power spectra, maximum entropy methods, etc, which are now widely used in geophysics.) On the whole, progress has resulted mostly from progress in instrumentation: computers, magnetometers, seismographs, mass spectrographs, and chemical analytical techniques allowing detection of incredibly small amounts of this or that in incredibly small samples (incredible, that is, to my generation, which occasionally still used a blowpipe). The mass spectograph, in particular, has been crucial in allowing precise geological dating.

Progress in instrumentation has been nowhere more spectacular than in space exploration; yet results of the last twenty years leave me slightly disappointed. I am still sorry that no little green men were found on Mars. I also remember being told long ago that exploration of the solar system would tell us how the Earth formed. This may be true eventually, but so far more questions have been raised than solved. We may know now a bit more about the rings of Saturn or the temperature on the surface of Venus, but we still don't know how the Moon was formed or where, or when it fell into orbit around the Earth. Satellites certainly have helped us to see features of the Earth's surface we hadn't seen before, but other features have tended to become more blurred. I am thinking in particular of the Earth's gravity field. I grant that our knowledge of this field has increased immensely in the last twenty years, but only with respect to the lower harmonics; wavelengths shorter than a few hundred kilometers have disappeared from our maps, and a broad, slightly positive, gravity anomaly has replaced the deep, sharp but narrow, negative anomalies associated with island arcs and trenches. These negative anomalies are perhaps *the* most spectacular feature of terrestrial gravity; yet because of their narrowness they have almost vanished from our gravity maps. More's the pity.

This may be an instance of a fundamental dichotomy—one might perhaps even say schizophrenia—that pervades the Earth Sciences. On the one hand, there is an obvious need to generalize, to simplify, to linearize, to look only at the first term in an expansion series; on the other hand, there is the fact that every geological situation is unique. No two rock samples are ever exactly alike, anymore than two human beings are ever identical; no two volcanoes erupt in the same manner; and no two mountain ranges have the same structure and history. Thus geophysicists construct models of a spherically symmetrical Earth that account for observed seismic travel times, period of free oscillations, etc, knowing full well that the radial density distribution predicted by their models may not actually exist anywhere in an Earth that is not spherically symmetrical at all levels. The model may be useful in an average sense, but what interests the geologist is precisely the departure from uniformity. If the Earth were spherically symmetrical all the way up to the surface, geologic maps, all tinted in one uniform color, would be very uninteresting indeed. Much of the occasional incomprehension and mistrust between geologists and geophysicists stems from this dichotomy between the particular and the general. When some years ago it was fashionable for seismologists to investigate the structure of the continental crust, they interpreted their observations in terms of superposed, horizontal, homogeneous, and isotropic layers for which geologists had no use at all; some accordingly resented spending good research money on what seemed to them a purposeless exercise.

Thus the urge to simplify, to unify, to generalize, which is the very stuff of science, finds itself pitted against nature's complexity, diversity, and uniqueness; as we stated before, no two planets, no two stones are identical. This same conflict also exists, of course, outside the Earth Sciences, but other sciences usually have enough subject matter to allow them to make valid generalizations based on good statistics. Geologists, poor chaps, have only one Earth, with only six continents (at the moment), and even fewer oceans. This is not a very strong statistical basis for a theory of the evolution of continents and oceans.

The concept of plate tectonics was hailed a decade ago for the simplification it brought into the description of geological structures and geological phenomena. All we need do, we were told, is to consider fewer than a dozen moving plates whose interactions would describe all that needed description. Note the intentional use of the word "description" rather than "explanation"; plate tectonics describes how plates move, but not why. But this great apparent simplification did not last very long; we soon found ourselves talking about ill-defined objects called microplates, no doubt paving the way for future picoplates. Continents are now seen to consist, at least in part, of bits and pieces of relatively small size that have drifted to and fro, and rotated this way and that, before becoming attached (for how long?) to a larger unit. Contours of plates presently well defined become more and more blurred as one looks back into the past. We are beginning to suspect that the theory may be applicable only to a small fraction of the Earth's history. Here again, a unifying concept founders on the diversity of Nature.

Much of the evidence for plate tectonics, microplates, etc, rests on a paleomagnetic basis; it is from the direction of the remanent magnetization of rocks that we infer their relative displacements. When paleomagnetism started in earnest some thirty years ago, not much was known of the magnetic properties of rocks. In particular, the mechanism by which igneous rocks acquire by cooling in the Earth's weak field a relatively strong and stable magnetization was not well understood; it still isn't. In the meantime, a wealth of information has been obtained, all of it pointing to the extraordinary complexity and variability of magnetic behavior, to the multiplicity of factors involved in rock magnetism and to the numerous processes by which the original magnetization of a rock could be altered. Much of the stable remanence in rocks seems to be carried by grains so small that they can hardly be seen under the optical microscope; neither their abundance, nor their size, shape, and composition can be easily ascertained. Nor can we in many instances determine precisely when they formed, as they may be products of chemical reactions (dehydration, oxidation, exsolution) occurring late in the history of the rock. It is amusing to reflect that if the pioneering paleomagnetists of the fifties had known, or even suspected, the full complexity of the magnetic properties of rocks, they would probably have thrown up their hands, declared rocks inherently unreliable, and turned to lesser things. Ignorance, it would seem, can sometimes be a blessing.

No doubt, though, that the success of paleomagnetism was largely a matter of luck. For there was an element of luck in the fact that among the first geological formations to be sampled were some folded sedimentary beds, such as the Rosehill shale studied by J. W. Graham, which showed great consistency in the direction of magnetic remanence when corrected for dip; this proved that the magnetization was stable and antedated folding. Good agreement was then found of "virtual" poles calculated from different formations of approximately the same age, even when sampled at far distant points on the same continent. Finally, apparent polar wander paths from the several continents showed enough internal consistency to establish real differences between continents, thus demonstrating their relative displacements. Discrepancies caused by intracontinental rotations, drift of microplates, magnetic instability, secondary magnetization acquired at a later date, or rapid "excursions" of the magnetic poles, showed up only later when enough evidence had accumulated to establish the rules that exceptions could only confirm. It was probably just by luck that exceptions did not show up first. One shudders at the thought of the chaos that might otherwise have ensued, just as one wonders what could have happened to the science of geology if William Smith had first surveyed the chaotic Franciscan formation of California instead of the well-ordered geological strata of England.

Paleomagnetism offers many illustrations of the clever ways in which

Nature manages to cover up its tracks. Considering the multiplicity of ways in which it could do so, it is perhaps surprising that it has not done it more often, or more effectively. If signals are not always as unambiguous as the optimist thinks they are, neither are they as distorted as the pessimist fears they might be. The optimist in this instance is the one who, having analyzed just a few samples, thinks he can discern a simple pattern; the pessimist is the one who expects the next sample to invalidate the optimist's theory. As it turns out, the pessimist is usually, but not always, right: there do seem to be a few threads of sufficient consistency in geological affairs to support expectations of progress in the understanding of the Earth's evolution. No doubt that early events at the time of, or very shortly after, the formation of the Earth will remain speculative for many years to come, as also will the crucial matter of the abundance of radioactive elements in the mantle and the core. But we now do have these solid threads to hang on to, such as the recognition of convection as the dominant mode of heat transfer in the deep Earth. There may be hope for progress, and there certainly will continue to be fun.

Ann. Rev. Earth Planet. Sci. 1983. 11 : 11–43

STRAIN ACCUMULATION IN WESTERN UNITED STATES[1]

J. C. Savage
US Geological Survey, Menlo Park, California 94025

INTRODUCTION

This review is principally concerned with recent geodetic strain measurements in western United States undertaken by the US Geological Survey as part of the earthquake studies program and, as a consequence, is heavily biased toward the author's own publications. Most of the publications reporting crustal-strain measurements in western United States prior to about 1968 have been compiled in one volume (National Geodetic Survey 1973), and more recent work (with complete bibliographies) is summarized in three successive quadrennial reports to the International Union of Geodesy and Geophysics (Meade 1971, Savage 1975, Thatcher 1979b).

The following conventions are employed in this paper: Strain, a dimensionless quantity, is reported in units of 10^{-6}. Extension is taken as positive. To distinguish between engineering and tensor shear strain, we denote the former by γ and quote the units as μrad, whereas the latter is denoted by ε and given the units of μstrain. Uncertainties in all cases are quoted as $\pm$ one standard deviation.

At the present time, geodetic techniques furnish the only reliable measure of strain accumulation in the seismic regions of the world. The superiority of geodetic techniques derives principally from the long baseline of the measurements. Difficulties in coupling fiducial marks to the Earth have generally made short-baseline measurements of strain accumulation questionable (Wyatt 1982). However, long baselines introduce the disadvantage that the measured strain is not closely localized.

The current technique of measuring strain accumulation involves the

[1] The US Government has the right to retain a nonexclusive royalty-free license in and to any copyright covering this paper.

comparison of two or more surveys of a geodetic network such as the one shown in Figure 1. In a trilateration survey the distances between geodetic stations are measured, and in a triangulation survey the horizontal angles between stations are measured. Ideally, a geodetic network should be so constructed that sufficient stations are intervisible that measurements of horizontal distance between them determines the relative horizontal positions of all stations. However, this constraint is not necessary in measuring strain; all that is required is that a sufficient number of lengths or angles can be measured within an area over which strain is reasonably homogeneous. Large networks may be divided into subnetworks if the foregoing conditions (reasonable homogeneity of strain and sufficient

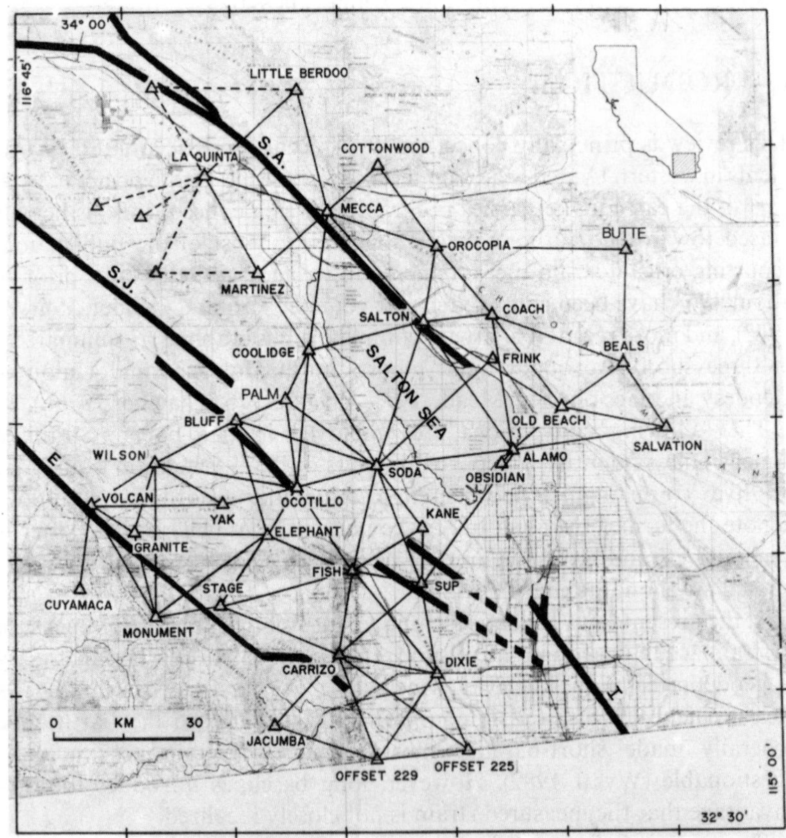

Figure 1 Map of the Salton trilateration network. The major faults, indicated by heavy lines trending northwest, are identified by initials as follows: S.A.—San Andreas, S.J.—San Jacinto, E—Elsinore, and I—Imperial. A location map of California is shown in the upper right corner.

observations) can be fulfilled within the subnetworks. In any case, what is calculated is the average strain field across the network chosen, and this average field may be of significance even when the strain field is quite inhomogeneous.

For trilateration surveys the elements for calculating strain are the observed extensions in the individual lines $e_i = \Delta L_i/L_i$, where L_i is the line length and ΔL_i is the change in length between surveys. The observed extensions are related to the surface strain tensor ε_{ij} by

$$e_i = \varepsilon_{11} \sin^2 \theta_i + 2\varepsilon_{12} \sin \theta_i \cos \theta_i + \varepsilon_{22} \cos^2 \theta_i, \tag{1}$$

where θ_i is the approximate azimuth of the line (measured clockwise from north), and the strain tensor is referred to a geographic coordinate system (1-axis directed east and 2-axis directed north). There will be one such equation for each line in the subnetwork, and if there are three or more lines, the average strain increments ε_{11}, ε_{12}, and ε_{22} can be determined. To the extent that the errors in measuring the extensions e_i are random with standard deviation σ, the standard deviations in ε_{11} and ε_{22} are about $(4/N)^{1/2}\sigma$ and the standard deviation in ε_{12} about $(2/N)^{1/2}\sigma$ for a network of N lines. There will in addition be a systematic error due to minor miscalibration of instruments that affects all values of e_i in any particular survey but is only randomly related to the systematic error in a subsequent survey (recalibration between surveys is presumed). This systematic error will increase the standard deviation in ε_{11} and ε_{22} but will not contribute to the standard deviation in ε_{12} or any other shear. For the trilateration surveys discussed in this paper, the estimated standard deviation in an increment of extension should approach 0.2 μstrain, whereas the standard deviation for tensor shear should approach $0.28/N^{1/2}$ μstrain. Inasmuch as strain rates will generally be reported, those estimates should be divided by the time interval between surveys. Savage et al (1981c) describe the errors in trilateration in more detail.

For triangulation surveys the elements for calculating the strain are the observed angle changes, and these are sufficient to calculate only shear strains. The change in the horizontal angle between stations at azimuths θ_i and θ_j is

$$\delta(\theta_j - \theta_i) = \gamma_1(\sin 2\theta_j - \sin 2\theta_i)/2 + \gamma_2(\cos 2\theta_j - \cos 2\theta_i)/2, \tag{2}$$

where $\gamma_1 = \varepsilon_{11} - \varepsilon_{22}$ and $\gamma_2 = 2\varepsilon_{12}$ are the shear components in a geographic coordinate system. The strain component γ_2 is equal to the decrease induced by strain in the right angle between northward- and eastward-directed lines, whereas γ_1 is equal to the increase in the angle between lines directed northwest and northeast. Thus, γ_1 measures right-lateral shear across a vertical plane striking N45°W (or left-lateral shear across a vertical plane striking N45°E), and γ_2 measures right-lateral shear across a vertical plane striking eastward (or left-lateral shear across a vertical plane striking northward). Notice that γ_1 and γ_2 (called *engineering shears*) are twice the corresponding tensor shear (e.g. $\gamma_2 = 2\varepsilon_{12}$). There will be one equation of the form (2) for each observed angle change, and two or more of those equations can be solved for γ_1 and γ_2. Given N observed angle changes, the standard deviation for γ_1 or γ_2 should be about $\sigma(8/N)^{1/2}$, where σ is the standard deviation for an observed angle ($0.8'' = 3.9$ μrad for first-order triangulation). A more detailed discussion of the analysis of triangulation data has been given by Prescott (1976).

As an example of a complete analysis of the deformation of a geodetic network, consider the Salton network (Figure 1), which spans the San Andreas, San Jacinto, and Elsinore faults just north of the Mexican border in southern California. By surveying those distances shown as solid lines in the figure, the relative positions of the individual geodetic stations (triangles in the figure) can be determined. (Three stations—Cuyamaca, Frink, and Obsidian—cannot be so determined.) A subsequent resurvey of the network would similarly determine the new relative positions of the monuments. The displacement generated in the time interval between the two surveys is simply the difference in the relative positions determined by the two surveys. Because only relative positions were determined in each survey, the displacement field so calculated is arbitrary to the extent that any displacement field generated by a rigid-body motion of the network as a whole can be added. This ambiguity is resolved by choosing the rigid-body displacement field such that the overall displacement field is as consistent as possible with the tectonic situation (e.g. displacements parallel to the trace of a strike-slip fault and perpendicular to the trace of a dip-slip fault). The velocity field (displacement divided by time interval) calculated for the Salton network is shown in Figure 2. Because the network is traversed by a set of subparallel right-lateral strike-slip faults trending about N40°W, the arbitrary rigid-body motion was selected to minimize the N50°E velocity components (Prescott 1981). The components of that velocity field are shown in Figure 3 as a function of distance perpendicular to fault strike (i.e. x' coordinate in Figure 2). The general pattern shown in Figure 3 is evident in Figure 2; velocities are directed N40°W and increase monotonically to the northeast. Figure 3, however, emphasizes that the N40°W velocity does not increase uniformly with distance perpendicular to the fault trend, but rather the velocity gradient flattens with distance from the fault.

It is clear from Figure 3 that the strain field is not uniform across the Salton network. The predominant strain is a right-lateral shear ($\gamma'_{xy} = \partial v'_y/\partial y'$) across a vertical plane striking N40°W. The shear-strain rate

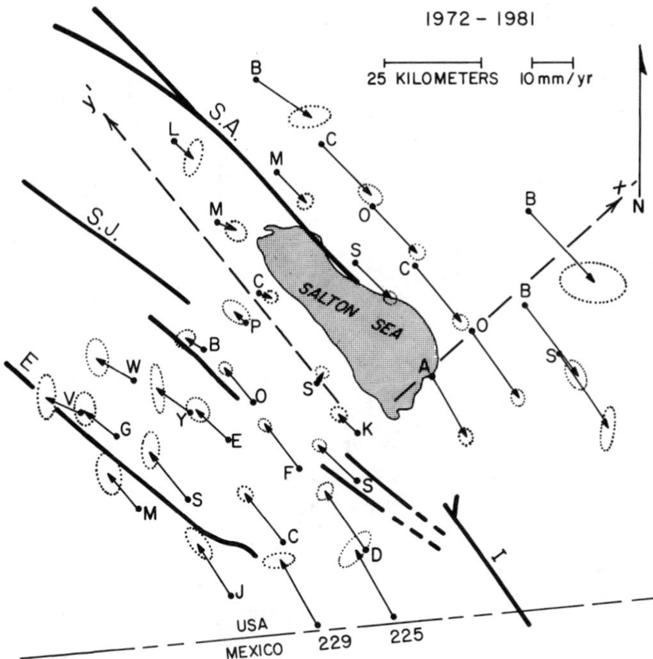

Figure 2 Average velocity field observed in the Salton network in the interval 1972–81. The faults are as identified in Figure 1. The dotted ellipse at the head of each velocity arrow indicates the 95-percent confidence limit.

(i.e. slope of the upper curve in Figure 3) shows a broad minimum value of about -0.5 μrad/yr near the center of the network but increases to near zero at the northeast and southwest extremes. Thus, describing the deformation of the Salton network by an average shear-strain rate (0.36 μrad/yr), as is done in this review, is quite a rough approximation. Most networks, however, are smaller than the Salton network, and the discrepancy between the maximum and average strain rates is not as extreme. The advantages of using the average strain rate to characterize the deformation of a network are that the strain rate can be expressed in three numbers (ε_{xx}, γ_{xy}, and ε_{yy}), each associated with a well-defined standard deviation, and that the calculation of the strain field is free from the arbitrary assignment of a rigid-body motion.

The conventional explanation for strain accumulation along faults involves continuous slip (fault creep) on the fault below some critical depth while above that depth the fault remains locked (Figure 4). At a large distance from the fault trace, the velocity on either block approaches a constant value such that the relative motion between two such distant points, one on either block, is simply the slip rate at depth on the fault. Eventually sufficient strain accumulates along the locked section of the fault to cause rupture, and slip occurs on the shallow part of the fault so that each side of the fault catches up with the remainder of its block.

Analytic expressions for strain accumulation for such a model follow from simple dislocation theory (Savage 1980). The fault is represented by a

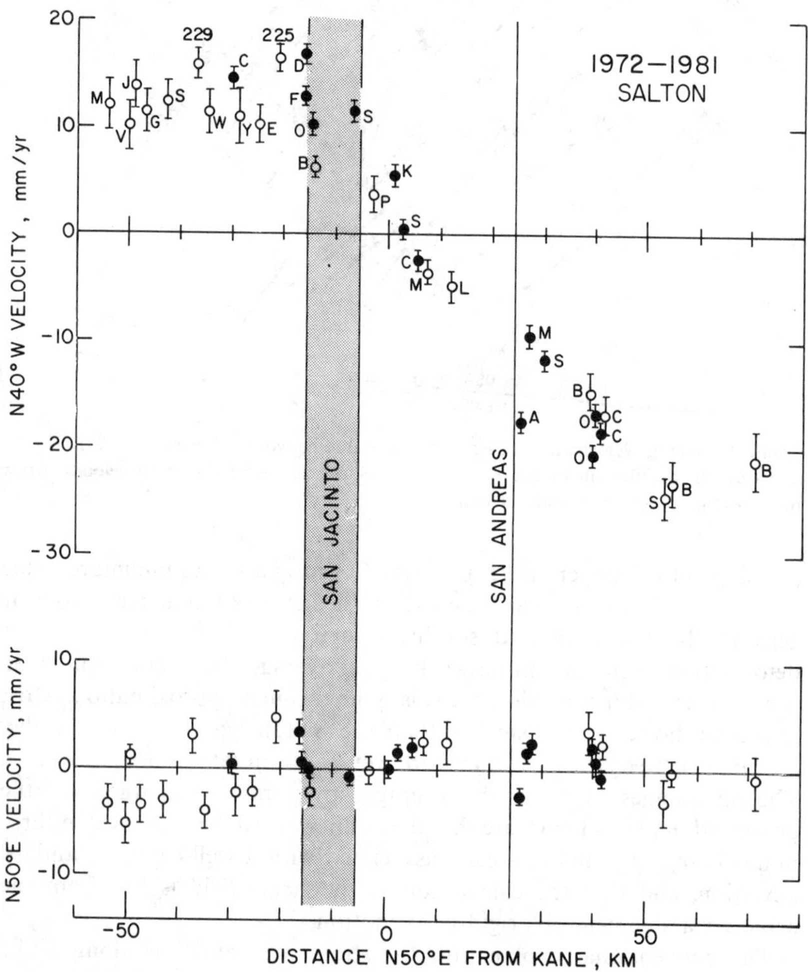

Figure 3 A plot of all y' and x' components of the average (1972–81) velocity field in the Salton network as a function of x' (see Figure 2 for the coordinate system). The locations of the San Andreas fault and the San Jacinto fault zone are indicated by the vertical lines.

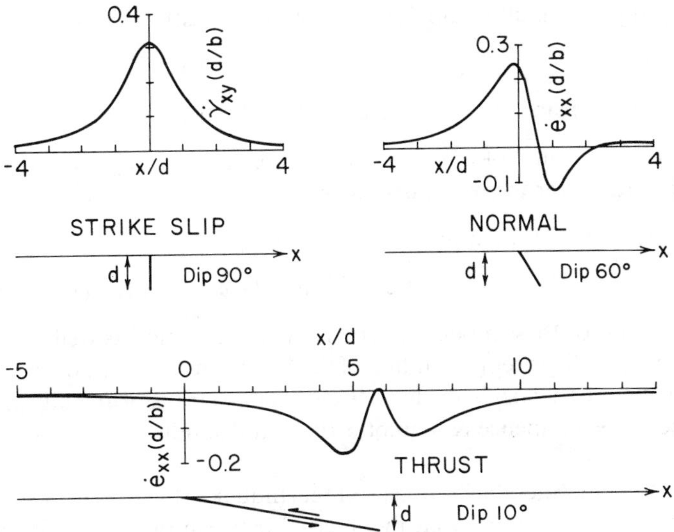

Figure 4 Plots of the strain accumulation as a function of distance from the fault for simple dislocation models of strike-slip, normal, and thrust faults. For each fault type, the diagram below the strain-accumulation plot shows the locked portion of the fault. The creep rate beneath the locked portion is taken to be b.

planar cut in an elastic halfspace, and the problem is reduced to two dimensions by assuming uniformity along fault strike. Below a depth d the fault is assumed to slip uniformly at the rate b. Then the deformation is equivalent to that produced by a dislocation with Burgers vector b located on the fault surface at depth d. In a right-handed coordinate system with the origin on the fault trace, x-axis along the horizontal component of fault dip, y-axis along fault strike, and z-axis vertically upward, strain-accumulation rates are as follows:

1. Strike slip on a vertical fault (b_y positive for left-lateral slip)

$$\dot{\varepsilon}_{xx} = \dot{\varepsilon}_{yy} = 0, \qquad \dot{\gamma}_{xy} = (b_y d/\pi)/(x^2 + d^2). \tag{3}$$

[The expressions for a dipping fault are obtained from (3) by replacing x with $x - d \text{ ctn } \delta$, where δ is the fault dip.] The average value of the shear-strain rate over the interval x_1 to x_2 is

$$\langle \dot{\gamma}_{xy} \rangle = (b_y/\pi)\,[\tan^{-1}(x/d)]_{x_1}^{x_2}/(x_2 - x_1), \tag{4}$$

where

$$[f(x)]_{x_1}^{x_2} = f(x_2) - f(x_1).$$

2. Dip slip on a fault of dip δ (b_t positive for thrust)

$$\dot{\varepsilon}_{yy} = \dot{\gamma}_{xy} = 0, \tag{5}$$

$$\dot{\varepsilon}_{xx} = (2b_t d/\pi) \sin^3 \delta (d - x \sin \delta \cos \delta)(x - d \operatorname{ctn} \delta)/D^4,$$

where $D^2 = x^2 \sin^2 \delta + d^2 - 2xd \sin \delta \cos \delta$. The average value of the extension rate over the interval x_1 to x_2 is

$$\langle \dot{\varepsilon}_{xx} \rangle = -(b_t/\pi)[(d \sin \delta)(d - x \sin \delta \cos \delta)/D^2 + \cos \delta \tan^{-1}(x/d - \operatorname{ctn} \delta)]_{x_1}^{x_2}/(x_2 - x_1). \tag{6}$$

Justifications for these models in terms of plate tectonics, as well as models with a more realistic representation of the Earth (an elastic lithosphere over a viscoelastic asthenosphere), have been given for transform faults (Savage & Prescott 1978, Spence & Turcotte 1979) and subduction zones (Savage 1982).

The "time-predictable" model of earthquake recurrence (Sykes & Quittmeyer 1981) implies that the interval between two successive major earthquakes at the same site should be simply the quotient of the strain drop at the time of the earlier earthquake and the average rate of accumulation of elastic strain in the interval between earthquakes. (The strain drop is the change in strain deduced from surveys just before and just after an earthquake. In measuring the strain drop, however, lines that cross the fault rupture must be excluded.) Thus, an estimate of the time to the next earthquake should be available where both strain drop in the preceding earthquake and strain accumulation have been measured by the same geodetic network. Such complete observations are seldom available. However, Rikitake (1976) has shown that shear-strain drops for many earthquakes, each measured by a different geodetic network (all of aperture 10 km or more), fall within a restricted range (35–170 μrad). Thus, a crude estimate of earthquake recurrence time should be given by the quotient of the average strain drop (47 μrad) and the *current* rate of strain accumulation. In the following, this quotient is used to estimate recurrence times even though the precision of the estimate is quite uncertain. Where possible this estimate is compared with geologic estimates of the same quantity. In western United States, where the historic record is relatively short, estimates of recurrence intervals, even when uncertain by a factor of two or more, may be of some use. A more sophisticated statistical treatment estimating the cumulative probability of an earthquake as a function of the elapsed time since the last earthquake has been based on the mean strain drop and the observed strain-accumulation rate (Rikitake 1976).

A less direct estimate of recurrence time can be made from a knowledge of the amount of slip that occurred in the preceding major earthquake and an observation of the present rate of strain accumulation. From Equations (4) or (6), the slip rate b at depth can be estimated from the observed strain rate and an independent estimate of d (usually 15 km for intraplate events and 15–30 km for interplate events). The recurrence time is then the slip in the preceding earthquake (a parameter that often can be obtained from the geologic record: scarp height, stream offset, etc) divided by the inferred creep rate b.

SAN ANDREAS FAULT SYSTEM

Strain accumulation in California is primarily in response to relative motion between the Pacific and North American plates. The relative motion of the Pacific plate with respect to the North American plate in California is thought to be 56 mm/yr N35°W (Minster & Jordan 1978). The plate boundary is generally taken to be the San Andreas fault, a major right-lateral transform fault running diagonally across California (Figure 5 *inset*), but there is good evidence that a substantial portion of the relative plate motion is accommodated by slip on subparallel faults or by continuous inelastic deformation across the plate margins (Minster & Jordan 1978). For example, geologic and geodetic evidence are in reasonable agreement that the long-term slip rate on the San Andreas fault in central California is about 38 mm/yr N41°W (Thatcher 1979c, Sieh 1982). The spreading

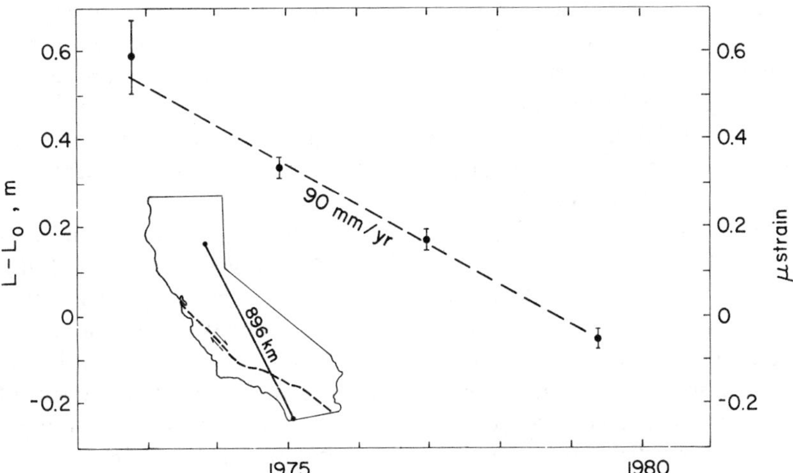

Figure 5 Observed length L less a nominal constant length L_0 as a function of time for the baseline between Otay Mountain and Quincy, California (see inset), measured by laser-ranging on a satellite (Smith 1980). The change in length is shown as a strain ($\Delta L/L$) on the scale at right. The dashed line on the inset map of California shows the San Andreas fault.

across the Basin and Range is less well known but may amount to 7 mm/yr N55°W (Thompson & Burke 1974). The vector sum of these two motions is 14 mm/yr N08°W less than the relative motion between the Pacific and North American plates. Thus, not even the combined deformation across the San Andreas fault and the entire Basin and Range province can account for the relative plate velocities. The remaining motion may be accommodated on the Hosgri-San Gregorio fault system off the coast of central California.

A remarkable long-baseline measurement (Figure 5) of strain from near San Diego to Quincy, California (80 km northwest of Lake Tahoe) has been made by laser-ranging upon a satellite (Smith et al 1979). The 896-km baseline appears to have shortened at a uniform rate $dL/dt = -90 \pm 20$ mm/yr in the interval between mid-1972 and mid-1979. This rate has been reduced to -80 ± 20 mm/yr in a recent reanalysis of the data (D. E. Smith, personal communication, 1982). If plate motion (56 mm/yr N35°W) alone were involved, the baseline (azimuth N20°W) should shorten at a rate not greater than 54 mm/yr, 1.2 standard deviations less than the observed value of 80 mm/yr.

Southern California

An unusual feature of the San Andreas fault system in southern California is the change from an average strike N40°W elsewhere to a strike near N65°W as the fault crosses the Transverse Ranges (Figure 6). Because the relative plate motion is about N35°W, a significant element of plate collision is introduced into the tectonics of southern California by this "big bend." Numerous thrust earthquakes in the Transverse Ranges attest to the existence of this collisional aspect.

Resurveys of triangulation networks (Figure 6) in southern California provide a measure of deformation along the San Andreas fault system since about 1930 (Thatcher 1979a). Strain rates measured in those networks during intervals in which no important earthquakes occurred near or within the network are shown in Table 1. The averages of the shear strains in Table 1 are $\langle \dot{\gamma}_1 \rangle = 0.26$ μrad/yr and $\langle \dot{\gamma}_2 \rangle = 0.12$ μrad/yr, equivalent to right-lateral shear accumulation at the rate of 0.28 μrad/yr across a vertical plane striking N33°W. The strike of the plane of maximum right-lateral shear is in excellent agreement with the direction of relative movement between the Pacific and North American plates (N35°W). The triangulation data indicate that there may be substantial variations in the strain rate as a function of time following a major earthquake. The best-documented case of such a variation occurred in the Imperial Valley network following the 1940 earthquake (magnitude 7.1): The strain rates in

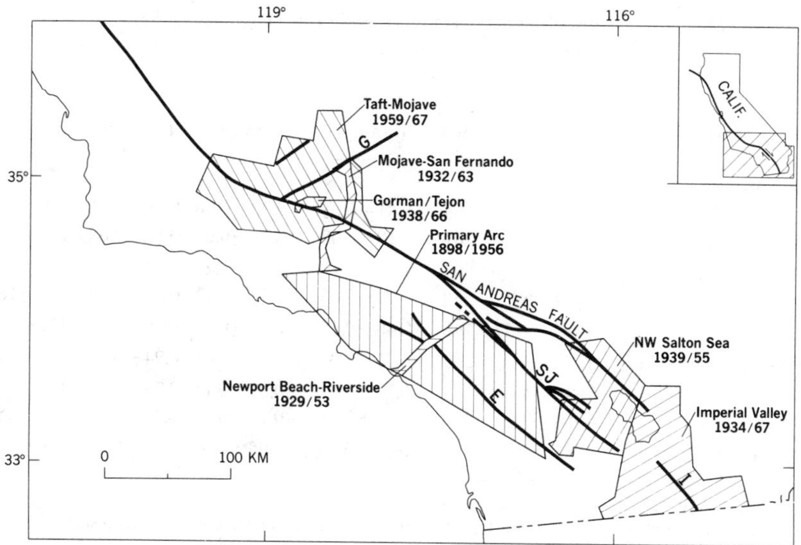

Figure 6 Location of triangulation networks in southern California for which strain-accumulation solutions are available (Thatcher 1979a). The time interval covered by the triangulation surveys is shown below the network name. The faults identified by initials are G—Garlock, S.J.—San Jacinto, E—Elsinore, and I—Imperial.

the 1941–54 interval are significantly larger than in 1954–67 interval (Table 1). The triangulation data also indicate that strain accumulation is not concentrated close to the faults (as, for example, is shown in Figure 3 and predicted in Figure 4) but rather appears to remain almost constant out to distances of 50 km or more from the fault (Thatcher 1979a). The preferred explanation of this broad distribution of strain accumulation is that it results from the superposition of strain accumulation from several subparallel faults (e.g. the Elsinore, San Jacinto, and San Andreas faults south of latitude 34° in Figure 6), each contributing strain concentrated close to its own trace. However, the possibility that inelastic strain has accumulated over a broad zone cannot be discounted.

The principal strain rates observed in the period 1971–81 for trilateration networks in southern California are shown in Figure 7. Except for the Garlock network, the principal strain rates are consistent with pure shear (i.e. $\dot{\varepsilon}_1 + \dot{\varepsilon}_2$ does not differ appreciably from zero), and the plane of maximum right-lateral shear (the bisector of the northwest quadrant of the principal strain axes) coincides very closely with the local strike of the northwest-trending faults. (Notice how the principal axes rotate counter-

Table 1 Shear strain rates observed in southern California (Thatcher 1979a)

Network	Time	Number of angles	γ_1 (μrad/yr)	γ_2 (μrad/yr)	γ^a (μrad/yr)	ψ^b (degrees)
Mojave-San Fernando	1932–52	199	0.31 ± 0.05	0.20 ± 0.06	0.38 ± 0.06	119 ± 6
	1952–63	201	0.10 ± 0.08	-0.14 ± 0.08	0.18 ± 0.08	162 ± 45
Gorman-Tejon	1938–49	93	0.27 ± 0.21	0.66 ± 0.25	0.72 ± 0.24	101 ± 14
	1949–66	80	0.09 ± 0.15	0.42 ± 0.17	0.42 ± 0.16	96 ± 18
Taft-Mojave	1959/60–67	146	0.40 ± 0.18	0.19 ± 0.18	0.46 ± 0.18	122 ± 19
Newport-Riverside	1933–53	255	0.23 ± 0.07	0.00 ± 0.06	0.24 ± 0.06	135 ± 9
Imperial Valley	1941–54	184	0.78 ± 0.09	-0.07 ± 0.11	0.78 ± 0.10	138 ± 4
	1954–67	123	0.09 ± 0.13	-0.17 ± 0.14	0.20 ± 0.14	166 ± 45
Primary Arc	1898–56	36	0.09 ± 0.04	-0.05 ± 0.05	0.10 ± 0.04	149 ± 19

[a] $\gamma = (\gamma_1^2 + \gamma_2^2)^{1/2}$ is the maximum shear strain.
[b] Strike, measured clockwise from north, of the vertical plane with maximum right-lateral shear.

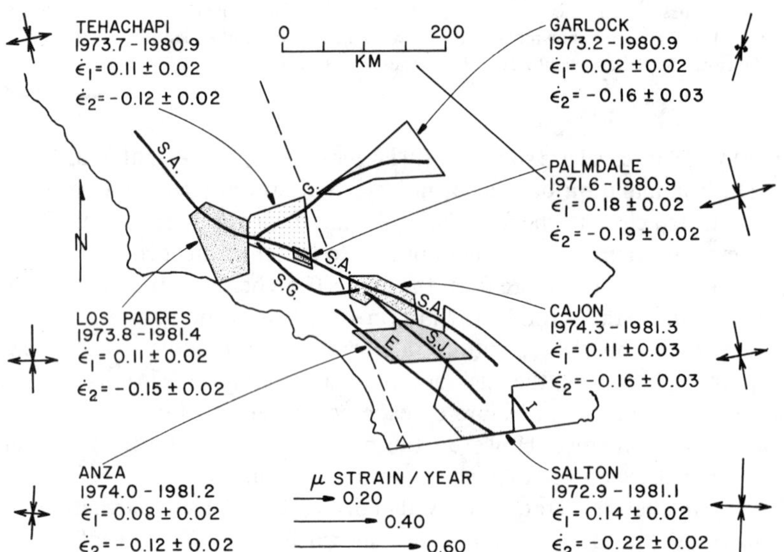

Figure 7 The principal trilateration networks in southern California along with the principal strain rates measured in the interval indicated beneath the network name. The principal strain rates with standard deviations are quoted in units of μstrain/yr. The portion of the satellite laser-ranging baseline (Figure 5) in southern California is shown as a dashed line extending north-northwest from the southwestern corner of California.

clockwise and then clockwise as one proceeds north through the "big bend" of the San Andreas fault.)

The observed rates of shear-strain accumulation for the six networks along the San Andreas fault are compared in Table 2 with the rates predicted by the simple strike-slip model in Figure 4. In the table we have taken b to be 56 mm/yr, the relative velocity between the Pacific and North American plates, despite the earlier arguments that not all of the relative plate motion was accommodated on the San Andreas fault, and estimated the expected strain rates from (4) for two different values of d (15 and 30 km). The estimated strain rates are directly proportional to b and can easily be corrected for other values. For example, the estimates for $b = 38$ mm/yr (the average slip rate on the San Andreas fault in central California) are 2/3 of those in Table 2. The cause of the variation in the calculated strain rates for a given d in Table 2 is the different values of x_1 and x_2 used, that is, the breadth and placement of the network relative to the fault. The Palmdale network, closely centered on the fault, should detect the maximum strain, whereas the Anza network, situated to one side of the fault, should detect appreciably less strain. The fit to the southern two networks (Salton and Anza) is satisfactory for either model ($d = 15$ km or 30 km), suggesting that the entire plate motion is accommodated on the San Andreas fault system there. The Cajon network apparently requires an even larger value of d and a smaller value of b. The Palmdale and Tehachapi networks can be fit using $b = 38$ mm/yr (the long-term average slip in central California), rather than 56 mm/yr and a depth $d = 30$ km. The overall picture then is that the full plate motion is accommodated on the San Andreas fault south of the "big bend," but that the obstruction of the "big bend" causes some of the plate motion to be distributed across the plate margins and apparently locks the

Table 2 Comparison of observed and calculated shear strain rates. γ

Network	Observed γ (μrad/yr)	Calculated γ $d = 15$ km (μrad/yr)	Calculated γ $d = 30$ km (μrad/yr)
Salton	0.36 ± 0.01	0.38	0.32
Anza	0.20 ± 0.02	0.17	0.21
Cajon	0.26 ± 0.03	1.02	0.56
Palmdale	0.37 ± 0.02	1.12	0.57
Tehachapi	0.23 ± 0.02	0.53	0.40
Los Padres	0.26 ± 0.02	0.38	0.32

San Andreas fault to greater depths in the "big bend." Finally, at the north end of the "big bend" the locked section becomes shallower. Thus, the simple model of Figure 4 affords a plausible explanation of the observed strain rates.

Contrary to the inference drawn from the triangulation data (Thatcher 1979a), the trilateration data indicate that strain accumulation is concentrated close to the fault trace. For example, the shear-strain rate for the Palmdale network, a small network close to the fault trace, is more than 1.6 times the strain rate for the larger Tehachapi network, which includes the Palmdale network within it. Concentration of strain close to the fault trace is also shown in Figure 3 for the Salton network. In that example it appears that both the San Jacinto and San Andreas faults are contributing to the strain accumulation.

A rough estimate of the recurrence time for major earthquakes on various segments of the San Andreas fault can be obtained from the observed strain rates and the average earthquake strain drop. For example, the average of the shear-strain rates for the three networks (Los Padres, Tehachapi, and Palmdale) along the "big bend" segment of the fault (the site of the 1857 rupture) is 0.29 μrad/yr, and the implied recurrence interval is 160 yr using the average earthquake strain drop, 47 μrad. This estimate is in excellent agreement with the average interval (160 yr) between major earthquakes on that segment of the fault over the past 1500 yr as determined from geologic evidence (Sieh 1981a). South of the "big bend," strain measurements span both the San Jacinto and San Andreas faults, and it is not clear exactly how the strain rates should be apportioned between the two faults (Figure 3). Simply dividing the average earthquake strain drop (47 μrad) by the strain rate (0.36 μrad/yr) observed across the Salton network would imply a recurrence time of 130 yr. The San Andreas fault within the Salton network has not ruptured within the past 600 yr (Sieh 1981b), whereas on the San Jacinto fault the recurrence time is about 100 yr (Wallace 1981).

The Garlock network spans the left-lateral strike-slip Garlock fault. The strain there appears to be a uniaxial north-northeast compression oblique to the fault strike so that the plane of maximum left-lateral shear coincides quite well with the average local strike of the Garlock fault. An estimate of the recurrence time on the Garlock fault based on the average shear-strain drop (47 μrad) is 260 yr. A tentative geologic estimate of the recurrence interval is 300–500 yr and an upper limit is 900–1800 yr (Sieh 1981a).

A question of major importance is whether strain accumulation in southern California is reasonably uniform from year to year. The trilateration data indicate that the accumulation of shear strain has been reasonably uniform over the 1973–81 interval (Figure 8), but real irregularities seem to occur in the extension perpendicular to the fault (Savage et al 1981a). On a longer term the situation is not completely clear. The triangulation data (Thatcher 1979a) certainly indicate that the strain rate may be anomalous in the years immediately following a major earthquake, but the rate presumably settles down to a more or less constant value subsequently. Good data to establish this latter presumption are not available. The shear-strain rates deduced from triangulation prior to 1967 (Table 1) are consistent with the strain rates deduced from trilateration subsequent to 1970 where similar areas can be compared. The average rate of shear-strain accumulation is 0.28 μrad/yr for both triangulation (Table 1) and trilateration (Figure 7), but the directions of maximum right-lateral shear differ: N33°W (in good agreement with the direction of relative plate motion) for triangulation and N51°W (in good agreement with the average strike of the San Andreas fault) for trilateration.

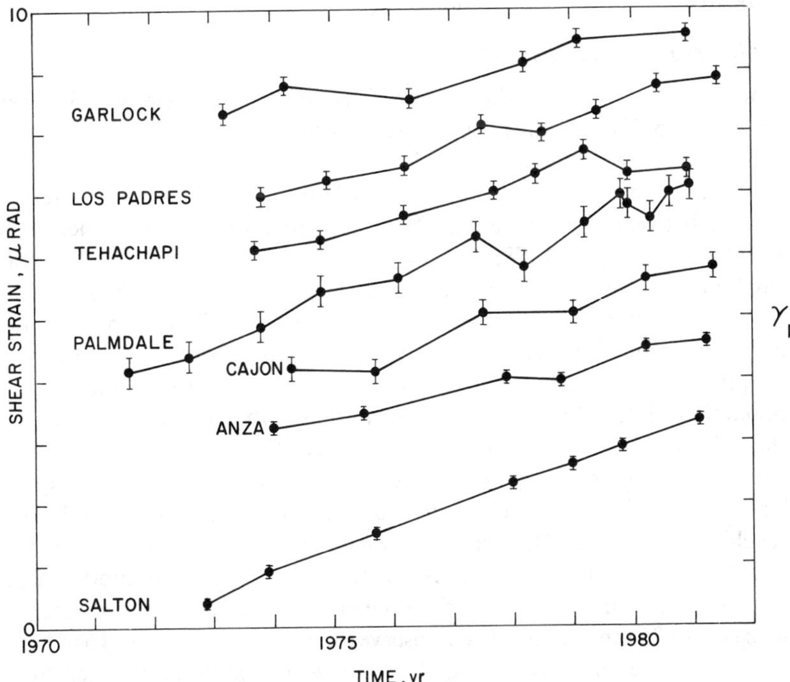

Figure 8 The shear-strain component $\gamma_1 = \varepsilon_{11} - \varepsilon_{22}$ as a function of time for the southern California trilateration networks (Figure 7). The initial level of shear strain for each network is arbitrary.

The satellite laser-ranging (SLR) baseline shown in Figure 5 indicates that the average strain over the 900-km length in the N23°W direction is -0.09 μstrain/yr. The average strain in the seven southern California trilateration networks ($\dot{\varepsilon}_{11} = 0.10$ μstrain/yr, $\dot{\varepsilon}_{22} = -0.15$ μstrain/yr, $\dot{\gamma}_2 = 0.04$ μrad/yr) would imply a strain rate of -0.14 μstrain/yr in the direction of the SLR baseline. Because the trilateration networks are concentrated along the principal faults where the strain rate is high, the fact that the SLR baseline strain is almost as large as the average network strain is surprising and suggests that strain accumulation is not too closely concentrated along the major faults. (See Figure 7 for the position of the SLR baseline relative to the trilateration networks.) Even if this average trilateration strain was the correct strain everywhere in southern California south of the Garlock fault, it would account for only 42 mm/yr of shortening in the SLR baseline. The remaining 38 ± 20 mm/yr would have to be taken up in the northern two-thirds of the baseline, which lies in the Sierra Nevada batholith, a relatively aseismic block that one would expect to be free of strain accumulation.

California Coast Ranges

North of the "big bend" the San Andreas fault system extends northwestward approximately parallel to the California coast for 750 km to the Mendocino triple junction (Figure 9). In this sector the San Andreas fault system apparently accommodates about 38 mm/yr of the 56 mm/yr relative motion between the Pacific and North American plates (Thatcher 1979c, Minster & Jordan 1978, Prescott et al 1981, Sieh 1982). The 1857 earthquake (magnitude 8+) ruptured the southernmost 150 km of this sector as well as about 200 km of the "big bend" section. The 1906 earthquake (magnitude 8.3) ruptured the San Andreas fault from the vicinity of the Pajaro network to the Mendocino triple junction. The remaining central 200 km of the San Andreas fault (shown as a solid rather than dashed line in Figure 9) is the locus of many small to moderate earthquakes but is not thought to have been the site of a major earthquake. This section does exhibit an unusual type of deformation—fault creep reaching the surface: The two blocks on either side of the fault slip past one another more or less continuously with the motion accommodated in a narrow (< 100 m) zone centered on the fault trace. Relative block motions in excess of 25 mm/yr have been observed in a span as narrow as 5 m across the fault trace (Burford & Harsh 1980). Appreciable fault creep at the surface has been observed outside of central California only on the Imperial fault in southern California (Figure 1) and the Anatolian fault in Turkey.

The southernmost 150 km of this sector of the San Andreas fault (shown as a dashed line in Figure 9) is at the present time aseismic, but it is known to

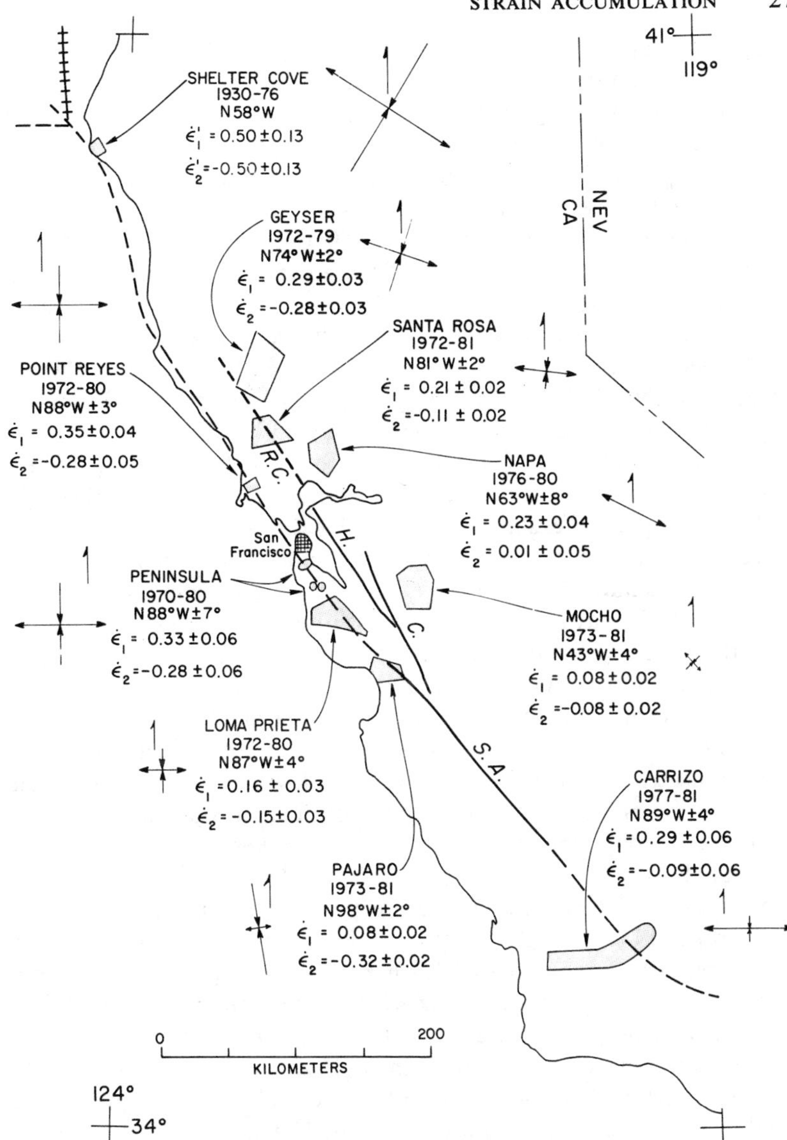

Figure 9 Strain rates measured at Geodolite networks in central and northern California. The time period covered, the azimuth of maximum extension, and the two horizontal principal strain rates in μstrain/yr are shown for each network. Also shown are the deviatoric strain rates measured at a triangulation network at Shelter Cove (Snay & Cline 1980). The major strike-slip faults (S.A.—San Andreas, C—Calaveras, H—Hayward, R.C.—Rodgers Creek) are shown by solid lines where fault creep on the surface is important and by dashed lines elsewhere. The Mendocino triple junction (junction of dashed and hachured lines) is shown at upper left and the west end of the "big bend" of the San Andreas fault at the lower right.

rupture in major earthquakes with recurrence time of about 300 yr (Sieh 1982). The strain rate measured at the southern end of the straight section (Carrizo network in Figure 9) in the interval 1977–81 was more nearly an east-west extension than pure shear. Presumably, this anomaly is due to the nearby "big bend." The azimuth of maximum right-lateral shear does coincide quite closely with the local strike of the San Andreas fault. An earthquake recurrence time of about 120 yr would be implied by the observed shear-strain rate (0.38 ± 0.04 μrad/yr) and the average shear-strain drop (47 μrad).

The next sector, 200 km farther north on the San Andreas fault (shown as a solid line in Figure 9), exhibits fault creep at the surface. Strain networks across this sector indicate that little, if any, long-term deformation occurs within the plates on either side of the fault, but rather that the deformation is concentrated along the fault (Savage & Burford 1970, Thatcher 1979c). Apparently the fault creep observed at the surface continues to depth on the fault, and the two plates slip past one another almost as rigid blocks. Minor hang-ups at asperities are relieved by limited rupture, i.e. small-to-moderate (magnitude <7) earthquakes. Thus, local, temporary accumulations of strain may be expected near asperities and steady accumulation near the endpoints of the creeping section of the fault (e.g. the Pajaro network in Figure 9). A detailed analysis of an extensive network near the north end of the creeping section shows shear-strain rates as high as 0.63 μrad/yr in a portion of the network, but the pattern of strain accumulation is complicated (Savage et al 1979). The low rate of secular strain accumulation along the creeping section of the San Andreas fault would imply a very long time between major earthquakes, if such earthquakes occur at all.

Near the north end of the creeping section of the San Andreas fault (i.e. the south end of the Pajaro network in Figure 9), the San Andreas fault system bifurcates, the eastern branch being made up of the Calaveras, Hayward, and Rodgers Creek faults, whereas the western branch is the San Andreas fault proper. Fault creep at the surface seems to be transferred from the creeping segment of the San Andreas fault farther south to the eastern branch with a decrease in surface creep along that branch to the north. (Creep rates at the surface are about 15 mm/yr on the Calaveras, 7 mm/yr on the Hayward, and 0 mm/yr on the Rodgers Creek fault.) Detailed analysis of an extensive Geodolite network spanning the San Francisco Bay region suggests that this decrease in the surface-creep rate to the north is accomplished by broadening the zone of deformation from a few meters on the southern Calaveras to discrete creep on the Hayward fault plus a few-kilometer-wide zone of deformation running parallel to the fault (Prescott et al 1981). It is not clear whether the strain accumulated in this zone represents elastic or inelastic deformation. In any case, the predominant strain accumulation between the latitude of San Francisco and the latitude of the Pajaro network is due to slip at depth on the San Andreas fault and not to motion on the Hayward-Calaveras fault system (e.g. note the low strain rate at the Mocho network). Moreover, strain accumulation on the San Andreas fault appears to be concentrated close to the fault as shown by the high rate of strain accumulation on the three small Peninsula networks as contrasted with low rates observed at the broader Loma Prieta and Pajaro networks. To explain these data, a fault depth $d = 6.5$ km and a slip rate at depth $b = 12$ mm/yr (Prescott et al 1981) are required in the strike-slip model of Figure 4. Independent geologic evidence corroborates the 12 mm/yr value as the long-term (last 1000 yr) slip rate on the San Andreas fault on the Peninsula south of San Francisco (Hall et al 1982). Prescott et al (1981) find that 32 ± 7 mm/yr of the relative motion between the Pacific and North American plates can be accounted for within the Geodolite networks in the San Francisco Bay area.

The recurrence time for earthquakes on the San Andreas fault immediately south of San Francisco, based on the strain rate observed in the Loma Prieta network, is about 150 yr. (It would be improper to use the Peninsula networks for this estimate as the aperture of these networks is much smaller than the networks used by Rikitake in deriving the 47-μrad average strain drop.) The geologic measurement of the recurrence time is 225 yr (Hall et al 1982).

Fault creep at the surface has not been observed on either branch (San Andreas fault proper on the west and Rodgers Creek fault on the east) of the San Andreas fault system north of San Francisco. The strain accumulation observed there is similar to that observed on the San Andreas fault proper just south of San Francisco (Figure 9). For example, the strain accumulation at Point Reyes is quite similar to that found in the small-aperture Peninsula networks, strain accumulation in the Santa Rosa network is similar to that observed at the Loma Prieta network, and strain accumulation in the Napa network, which does not span a through-going fault, contrasts with that observed in the Mocho network, which lies beside the creeping Calaveras-Hayward fault system. The strain rate at the Geyser network is unexpectedly high considering the position and breadth of the network. A producing geothermal area lies within the Geyser network.

Figure 10 shows the accumulation of the shear component γ_1 as a function of time for some of the Geodolite networks in Figure 9. This shear component measures right-lateral shear strain across a vertical plane striking N45°W. The San Andreas fault strikes about N40°W across these networks, and γ_1 is therefore a good measure of the shear strain resolved across the fault. It is clear from the figure that the accumulation of shear strain at each of the networks is reasonably uniform in central and northern California in the interval 1972–81. Only the Pajaro network shows a significant nonlinearity.

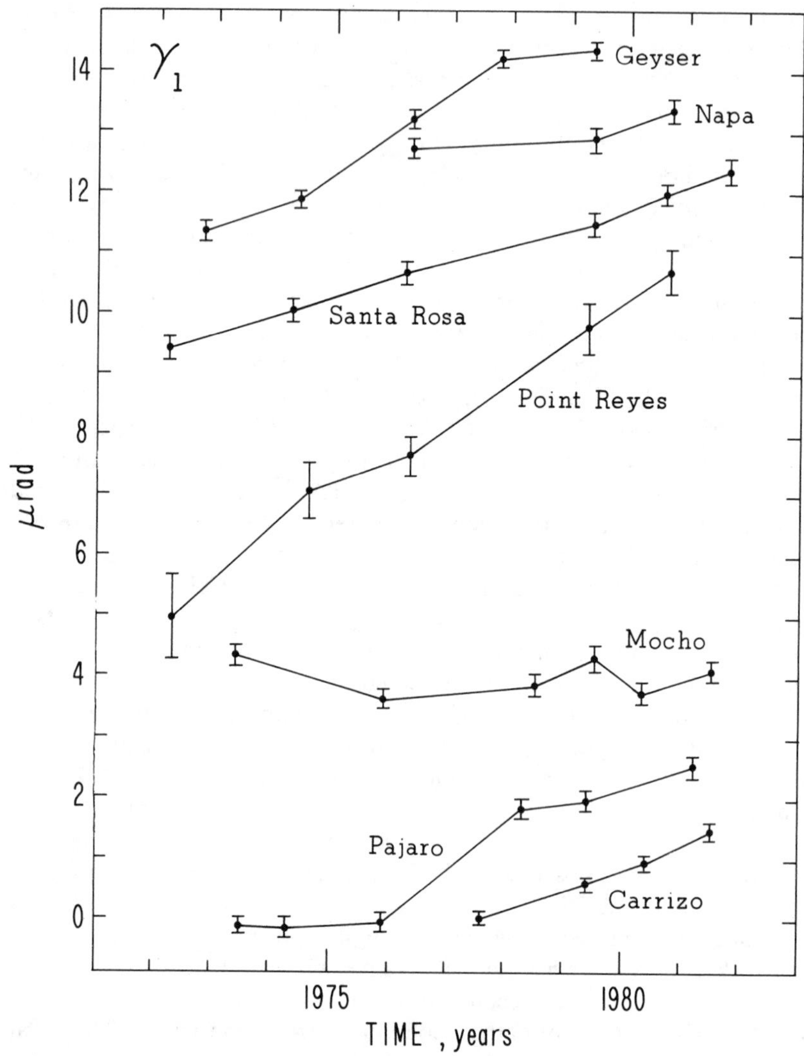

Figure 10 Accumulation of the shear component γ_1 as a function of time for some of the Geodolite networks shown in Figure 9. The initial level of shear strain for each network is arbitrary.

Strain rates have been deduced for those triangulation networks in central and northern California that have been surveyed two or more times (Thatcher 1975a,b, 1979c, Snay & Cline 1980). Of particular interest are the networks surveyed three times since the 1906 earthquake, for in those cases strain rates for two different epochs can be compared. The locations of such networks are shown in Figure 11 and the shear-strain rates deduced for them are shown in Table 3. Where comparison is possible, there is a tendency for the strain rate in the earlier epoch to be larger. This is suggested by the data at Fort Ross where the 1906–30 rate appears to be seven times larger than the 1930–69 rate. However, that comparison is questionable due to the large uncertainties in both rates. A better indication of the effect may be to compare the post-1930 measurements at Point Reyes with pre-1930 measurements at a similar network at Point Arena. The shear-strain rate (0.54 ± 0.17 μrad) measured at Point Reyes for the 1938–61 interval (Table 3) is quite comparable to the rate (0.63 ± 0.07) measured there in the interval 1972–80 (Figure 9). However, the strain rates measured at Point Arena for the 1906–25 and 1906–30 intervals, although consistent with each other, seem very high with respect to more recent strain-rate measurements at Point Reyes. Presumably, high strain rates occurred as a

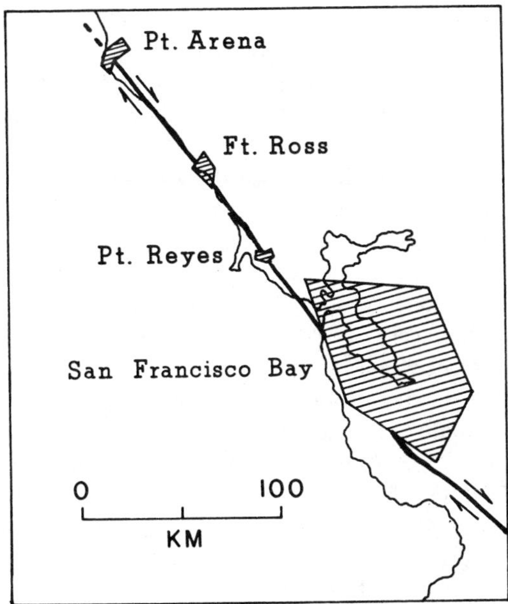

Figure 11 Location of triangulation networks from which strain rates have been deduced for two different epochs (see Table 3).

Table 3 Average shear-strain rates for various epochs at four strain networks (Figure 11) along the San Andreas fault (Thatcher 1975a)

Network	Epoch	γ_1 (μrad/yr)	γ_2 (μrad/yr)
Point Arena	1906–30	1.69 ± 0.67	-2.67 ± 0.67
	1906–25	1.32 ± 0.47	-1.86 ± 0.47
Fort Ross	1906–69	0.92 ± 0.08	-0.20 ± 0.08
	1906–30	2.26 ± 0.75	-1.00 ± 0.75
	1930–69	0.32 ± 1.05	-0.04 ± 1.05
Point Reyes	1930–61	0.94 ± 0.19	0.13 ± 0.19
	1938–61	0.54 ± 0.17	-0.14 ± 0.17
San Francisco Bay	1906–22	0.79 ± 0.25	-0.11 ± 0.25
	1922–47	0.43 ± 0.25	-0.14 ± 0.25

postearthquake effect in the years immediately following the 1906 earthquake (Thatcher 1975a).

Strain accumulation has been estimated from a small triangulation network at Shelter Cove (Figure 9) at the north-end of the San Andreas fault near the Mendocino triple junction (Snay & Cline 1980). The average engineering shear rate for the 1930–76 interval was 1.01 ± 0.18 μrad/yr, a rather high value even for a small-aperture network.

The triangulation network at Fort Ross affords an unusually complete set of data for estimating recurrence times. This network was surveyed in 1874, 1906 (postearthquake), 1930, and 1969, although the same angles were not measured in all surveys. Nason (1979) noted that as of 1969 only 69 μrad of shear had accumulated since the 1906 earthquake, whereas the strain drop between 1874 and 1906 (postearthquake) was 200 μrad. These data would suggest that a recurrence of the 1906 earthquake is not imminent. There is, moreover, the possibility that the strain rate is a decreasing function of time since the 1906 earthquake (Table 3). Thus, a recurrence time in excess of 180 yr would be indicated. Thatcher (1981) estimated a recurrence time of 225 yr for the San Andreas fault in northern California, based on the strain-accumulation rate 0.76 μrad/yr at Point Reyes and an estimate of 170 μrad (Rikitake 1976) for the strain drop in the 1906 earthquake. The 170-μrad strain-drop estimate is based on quite inhomogeneous data (Chinnery 1961, Figure 12). The 29-bar stress-drop estimate (Sykes & Quittmeyer 1981) for the 1906 earthquake would translate into a strain drop of 50 μrad, a value more consistent with the average strain drop (47 μrad). The recurrence time would then be 70 yr. This discrepancy illustrates the large uncertainties associated with estimates of recurrence time based on strain-accumulation rates.

BASIN AND RANGE PROVINCE

The Basin and Range province is dominated by north-trending oblique-slip (right-lateral, normal) faults. The overall spreading rate across the entire province is about 7 mm/yr in an east-west to southeast-northwest direction (Thompson & Burke 1974). The Salton and Garlock networks (Figure 7) strictly lie within the Basin and Range province but have already been discussed in connection with the San Andreas fault system. The largest historic earthquakes in the province have been concentrated in a zone along the 118th meridian from Owens Valley in California to Winnemucca, Nevada.

Relevant to studies of spreading in the Basin and Range province is a transcontinental strain measurement (Figure 12) made by very-long-baseline interferometry (VLBI) between radio observatories near Boston, Massachusetts, and Bishop, California (Allenby 1979). This 3930-km baseline has been measured repeatedly over a five-year period, and no significant change in length can be demonstrated (rms deviation from the average length is about 30 mm). With the inclusion of more recent data not shown in Figure 12, the average rate of change of line length for the interval September 1976 to June 1981 was found to be $dL/dt = -3.5 \pm 2.7$ mm/yr

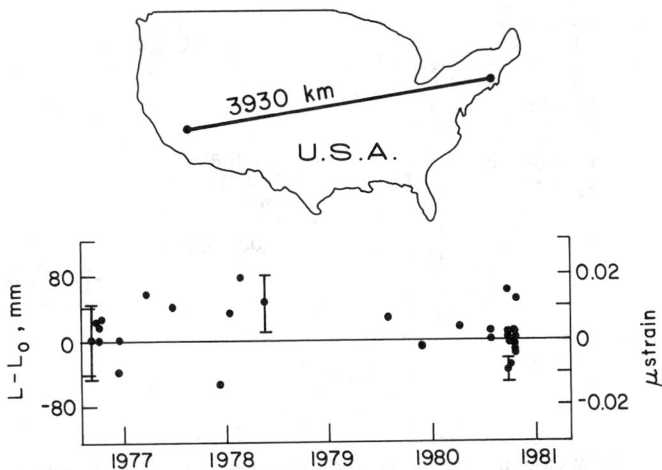

Figure 12 Observed length L less a constant nominal length L_0 as a function of time for a transcontinental baseline between Haystack and Owens Valley Radio Observatories measured by VLBI (National Research Council 1981). The scale on the right expresses the length change in terms of strain $\Delta L/L$. The error bars, representing one standard deviation, are shown for typical measurements; precision was substantially improved after 1978. (Data courtesy of Dr. C. Ma.)

(Ryan & Schupler 1982). The VLBI baseline crosses the Basin and Range province at an azimuth of N70°E, whereas spreading across the province is in a direction N55°W to N75°W. Thus, the extension rate on the VLBI line should be at least 0.6 of the spreading rate. The observed contraction of the VLBI line is not easily reconciled with Basin and Range spreading unless there is a compensating contraction along the baseline east of the Basin and Range province.

118th-Meridian Seismic Zone

The Owens Valley network (Figure 13) spans the rupture of the 1872 (magnitude 8+) Owens Valley earthquake (Savage & Lisowski 1980). The

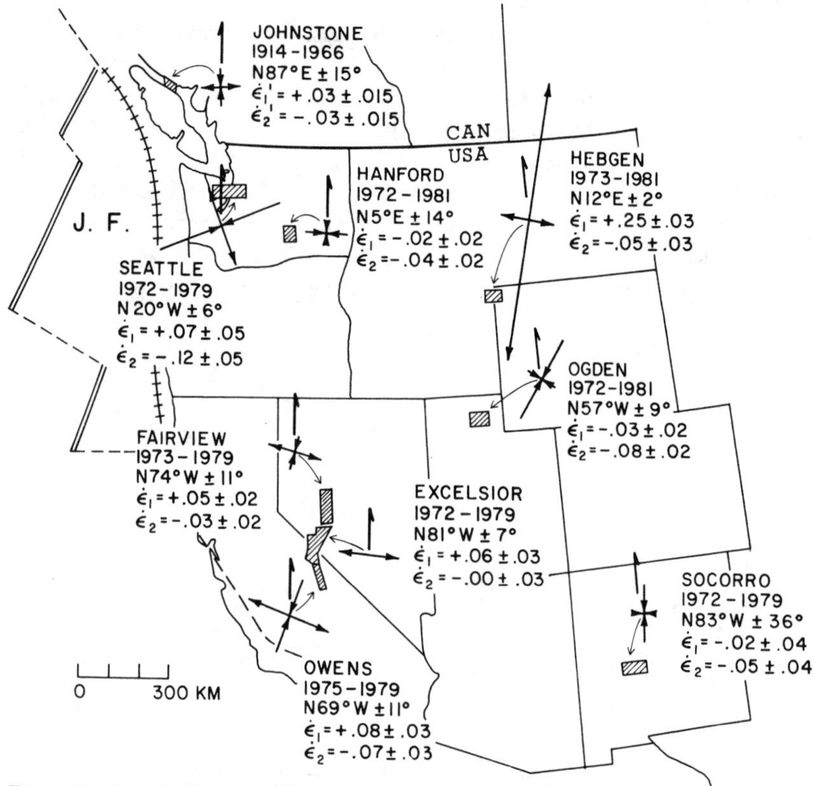

Figure 13 Location of trilateration networks in western United States and the principal strain rates measured in each. Shown beneath the network name are the time interval covered, the direction of maximum extension, and the principal strain rates (in μstrain/yr). Also shown in the southwest corner of Canada is one triangulation network with the principal deviatoric strain rates measured there. The plate boundaries along the Pacific coast are indicated by the usual symbols (ridge by double line, trench by hachured line, and transform by dashed line). The Juan de Fuca plate is identified by initials J.F.

western terminus of the transcontinental VLBI baseline (Figure 12) lies at the north end of this network. The strain rate inferred from trilateration surveys in 1974 and 1979 (Figure 13) is equivalent to a right-lateral shear rate of 0.16 ± 0.06 μrad/yr and an extension rate of 0.01 ± 0.03 μstrain/yr across the 1872 rupture (strike N20°W). Although these results might suggest pure strike slip at depth on the 1872 fault plane, they are not inconsistent with a ratio of two parts strike slip to one part dip slip as occurred in 1872. This can be demonstrated from the models in Figure 4 by first solving for b_y in (4) and substituting half of that value into (6) for b_t to determine $\langle \dot{\varepsilon}_{xx} \rangle$. For $d = 15$ km the predicted strike-slip creep at depth is 11 mm/yr and the predicted value of $\langle \dot{\varepsilon}_{xx} \rangle$ is 0.04 μstrain/yr, which is within the quoted uncertainties for the observed value (0.01 ± 0.03). A crude estimate of the recurrence time for the 1872 earthquake is 300 yr based on a strain drop of 47 μrad. Another estimate of the recurrence time is 550 yr based upon the time required to accumulate 6 m of strike slip (estimated strike slip in 1872) at the creep rate of 11 mm/yr found above. The geologic estimate for the recurrence time is 700–1000 yr (Sieh 1981a).

The Fairview Peak network spans the rupture associated with 1954 Fairview Peak earthquake (magnitude 7.1), as well as the rupture of the 1932 Cedar Mountain earthquake (magnitude 7.2) and the southern part of the 1954 Dixie Valley earthquake (magnitude 6.9) rupture. The network was surveyed in 1973, 1974, 1976, and 1979. The inferred strain rates (Figure 13) suggest that the faulting mechanism on the Cedar Mountain, Fairview Peak, and Dixie Valley faults should be normal slip with a minor right-lateral component. (The strain rates observed for Owens Valley and Fairview Peak do not differ significantly, but the fault strikes in the two networks do. The Owens Valley fault strikes N20°W, whereas the Cedar Mountain, Fairview Peak, and Dixie Valley faults all strike slightly east of north.) The estimated recurrence time for faults in the Fairview Peak network is about 600 yr if the standard 47-μrad strain drop is used. For the Fairview Peak earthquake the strain drop was 62μrad as observed by triangulation surveys in progress at the time of the earthquake (Rikitake 1976), and a better recurrence time might be 800 yr. Geologic estimates of recurrence time for major earthquakes somewhat farther north on the 118th-meridian seismic zone are 5000–6000 yr (Wallace 1981).

The possible seismic gap between the Owens Valley and Fairview Peak networks is spanned by the Excelsior network (Figure 13). Strain accumulation in the southern half of this network has been described by Savage et al (1981d). The Excelsior network was surveyed by trilateration in 1972, 1973, 1976, 1979, 1980, and 1982. However, the 1980 and 1982 surveys have been omitted in calculating the strain rates shown in Figure 13 to avoid coseismic effects introduced by the 1980 Mammoth Lakes earthquakes, a sequence of four magnitude 6 events that occurred on the west edge of the network. The observed 1972–79 strain rate is almost a pure east-west extension.

The spreading rate across the 118th-meridian seismic zone can be estimated from the strain rates measured for the Owens Valley, Excelsior, and Fairview Peak networks. These strain rates do not differ significantly (i.e. two standard deviations) from one another, and an average east-west strain rate is about 0.06 μstrain/yr. Because the breadth of the networks averages about 40 km, the minimum east-west spreading rate is about 2.5 mm/yr, somewhat in excess of the geologic estimates (Thompson & Burke 1974: 1 mm/yr over the past 12,000 yr and 0.4 mm/yr average over the past 15 Myr). Clearly, such a high east-west strain rate cannot persist across the entire Basin and Range province (600-km breadth) for that would imply a spreading rate of 36 mm/yr across the province, a rate that obviously exceeds the constraint imposed by the transcontinental strain measurement (Figure 12).

Eastern Basin and Range

The Socorro network (Figure 13) spans the Rio Grande rift at the eastern end of the Basin and Range province (Savage et al 1980). The network was surveyed by trilateration in 1972, 1973, 1976, and 1979, and no significant strain accumulation was detected. The measurements place an upper bound of about 1 mm/yr on the average east-west spreading across the 60-km broad rift in the interval 1972–79. The shear-strain rate in the Socorro network of $\dot{\gamma} = 0.01 \pm 0.03$ μrad/yr is consistent with the geologic estimates (Wallace 1981) that the recurrence interval exceeds 5000 yr.

The Ogden network spans the Wasatch fault near Ogden, Utah. The network was surveyed by trilateration in 1972, 1973, 1976, 1978, and 1981, and the average strain rate was found to be predominantly a N33°E compression of 0.08 μstrain/yr. The Wasatch fault, a major west-dipping normal fault, trends N10°W through the network and the expected strain rate was an east-west extension. Plots of strain accumulation as a function of time are not inconsistent with a linear accumulation of strain, although there may have been a reversal from east-west contraction to east-west extension in the interval 1978–81. The strain appears to be reasonably homogeneous across the network. I have no explanation for this observed anomalous strain (see, however, Zandt & Richins 1981). It does appear to be a valid observation of strain accumulation.

NORTHERN ROCKY MOUNTAINS

The Hebgen network lies on the east edge of Yellowstone National Park and spans the rupture of the 1959 Hebgen Lake earthquake (magnitude

7.1). The network was surveyed by trilateration in 1973, 1974, 1976, 1978, and 1981, and the strain accumulation, essentially a uniaxial N12°E extension at the rate of 0.25 μstrain/yr (Figure 13), appears to be quite uniform from survey to survey. The extension is perpendicular to the trend of the 1959 rupture and is consistent with accumulation of strain leading to further normal faulting along that rupture. Strain accumulating at the observed rate would attain 47 μstrain, the average earthquake strain drop, in only 200 yr. Whether this estimate of the recurrence interval is realistic is not known. Certainly, the observed rate of strain accumulation is surprisingly high considering that the average strain across a normal fault must include the effects of a band of compressive strain (Figure 4) that tends to diminish the overall average. If the depth of faulting on the Hebgen fault is 15 km, Equation (6) would require $\langle \dot{\varepsilon}_{xx} \rangle = -0.037\,(b_t/d)$. The slip rate at depth would then be $b_t = -100$ mm/yr, a rather high rate. In fact, that rate would imply that the 10-m slip associated with the 1959 earthquake could be recovered in only 100 yr. Although the strain rate in the Hebgen network showed no tendency to decrease in the 1973–81 interval, the high rate observed could be a temporary postearthquake effect. However, in view of the proximity of the Yellowstone hot spot to the Hebgen network, a high rate of strain accumulation need not be particularly surprising.

PACIFIC NORTHWEST

The primary source of strain accumulation in the Pacific Northwest is presumably the subduction of the Juan de Fuca plate (Figure 13) beneath the North American plate, which occurs at a rate of about 40 mm/yr in the direction N50°E. This interaction should affect primarily the coastal regions, but back-arc processes may occur farther inland (i.e. east of the Cascade Mountains).

A trilateration network across the Puget Sound lowlands near Seattle has been surveyed in 1972, 1973, 1974, 1976, and 1979 (Savage et al 1981a). The measured strain rate (Figure 13) is principally a N70°E compression of 0.12 μstrain/yr. This strain is reasonably consistent with what would be predicted by the model of Figure 4 for subduction of the Juan de Fuca plate: Equation (6) predicts $\langle \dot{\varepsilon}_{xx} \rangle = -0.12$ μstrain/yr N50°E for $b = 40$ mm/yr, $d = 35$ km, $x_1/d = 4.25$, and $x_2/d = 7.25$. The estimated recurrence time between great thrust earthquakes near Seattle would be about 250 yr using the 47-μrad average strain drop. No great thrust earthquakes have occurred off the Washington coast in the period for which records are available (about 200 yr). Some seismologists believe that subduction off the Washington-Oregon coast is accomplished aseismically, but the accumulation of significant northeasterly contraction would seem to contradict that presumption.

The shear-strain rates across Johnstone Strait, British Columbia, (Slawson & Savage 1979) have been determined from triangulation surveys in 1914 and 1966 by the Geodetic Survey of Canada. The principal deviatoric strain rates so determined are shown in Figure 13. These strain rates, which are only marginally above possible errors in measurement, suggest right-lateral shear parallel to the plate boundary rather than underthrusting normal to the plate boundary. Such transverse motion need not be inconsistent with the underthrusting of the Juan de Fuca plate in view of the location of the Johnstone Strait network (Figure 13) near the triple junction at the north end of the Juan de Fuca plate, where the influence of the Queen Charlotte transform fault may be felt (Savage et al 1981a).

An extensive trilateration network (Figure 13) was established at the Hanford Site near Richland, southcentral Washington, in the Columbia Plateaus geologic province to investigate crustal stability near a nuclear-waste-disposal site (Savage et al 1981a). The network was surveyed in 1972, 1973, 1975, 1978, 1979, and 1981, and the measured strain rate, though only marginally significant, indicates an east-west contraction, possibly related to the subduction of the Juan de Fuca plate. The observation is inconsistent with back-arc spreading. Geologic evidence in the Hanford area indicates a north-south compression active at least as recently as Pliocene time and possibly continuing to the present time. The focal mechanisms of the minor earthquakes in the area also suggest north-south compression. Notice that the strain rates observed at Hanford (Figure 13) suggest a small amount of north-south contraction; the principal contraction, however, is east-west.

ALASKA AND HAWAII

Interaction of the Pacific and North American plates is the primary source of strain accumulation in Alaska. In southeastern Alaska the plate boundary is the Fairweather fault (Figure 14), a right-lateral transform fault, and the relative plate velocity across the boundary is 58 mm/yr approximately parallel to the fault. Along the south coast of Alaska, southwest from the north end of the Fairweather fault, the relative plate motion is accommodated by subduction of the Pacific plate at a rate increasing westward from 58 mm/yr N20°W near the Fairweather fault and reaching a rate of 77 mm/yr N30°W at the end of the Alaska peninsula (Shumagin Islands in Figure 14).

The most extensive strain measurements in Alaska are in Prince William Sound (Figure 14) around the epicenter of the 1964 Alaska earthquake (magnitude 8.4). The shear-strain change deduced from a comparison of triangulation completed after the earthquake with triangulation completed

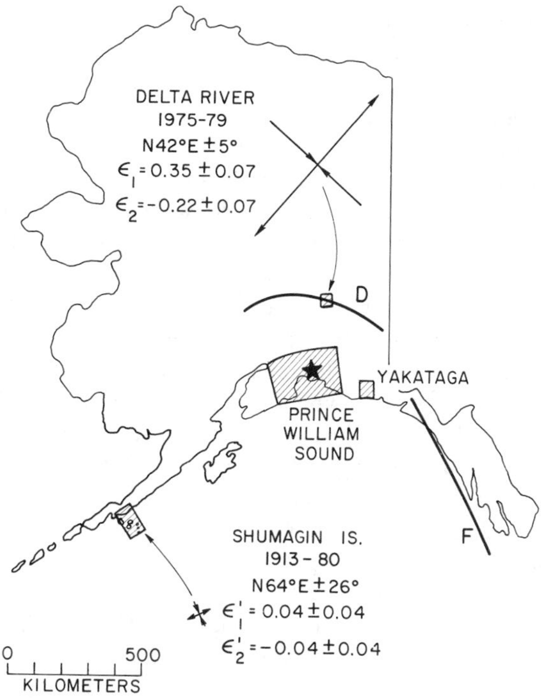

Figure 14 Map of Alaska showing the location of the principal geodetic strain networks. The Fairweather (F) and Denali (D) faults are shown as heavy lines. The strain accumulation measured at the Delta River and Shumagin Islands is indicated as follows: First line, network name; second line, dates of survey; third line, azimuth of extension axis; fourth and fifth lines, principal strain rates (deviatoric strain rates only for the Shumagin Islands network) in μstrain/yr. The star indicates the epicenter of the 1964 earthquake.

principally in the period 1941–48 in the Prince William Sound area was 80 ± 10 μrad with the axis of greatest extension oriented N20°W (Pope 1972). This extension is consistent with rebound from the N20°W compression implied by the relative plate movement, and the 80-μrad shear-strain drop is within the range of values used to derive the 47-μrad strain-drop average for all earthquakes (Rikitake 1976). Geologic evidence (uplifted marine terraces at Middleton Island) suggests that a similar earthquake occurred in Prince William Sound about 1350 yr earlier (Sieh 1981a). Thus, the average shear-strain rate there would be about $80/1350 = 0.06$ μrad/yr. For the thrust model in Figure 4 a somewhat higher rate of strain accumulation would be expected: Using $b = 60$ mm/yr, $d = 35$ km, $x_1 = 105$ km, and $x_2 = 280$ km in Equation (6), one finds $\langle \dot{\varepsilon}_{xx} \rangle = -0.18$ μstrain/yr, indicating a shear-strain rate three times the value (-0.06 μrad/yr) inferred from the observed strain drop and the geologic recurrence interval. A possible explanation of this discrepancy is that the 1350-yr interval might include a major earthquake that did not produce a terrace at Middleton Island.

A trilateration network in the Shumagin Islands (Figure 14) near the tip of the Alaska peninsula was surveyed in 1980 and 1981. The network is of particular interest because it spans a major seismic gap along the Alaska-Aleutian arc (Davies et al 1981). The Shumagin segment of the arc apparently last ruptured in 1903 or earlier and is considered ripe for a major earthquake. The amount of strain that accumulated in the one-year interval 1980–81 was less than the uncertainty in measurement, and all one can conclude is that the shear-strain rate is less than 0.5 μrad/yr. However, the 1980 trilateration network coincides with a portion of a triangulation network surveyed in 1913. The average shear-strain rates for the 1913–80 interval are $\dot{\gamma}_1 = 0.04 \pm 0.06$ μrad/yr and $\dot{\gamma}_2 = 0.06 \pm 0.05$ μrad/yr. This implies an orientation of the compressive axis N26°W$\pm$26°, quite consistent with the direction of relative motion between the Pacific and North American plates. The strain rate predicted from the thrust model in Figure 4 (average strain over the interval $x/d = 6$ to 7.5) is $\dot{\varepsilon}_{xx} = -0.23$ μstrain/yr, with $b = 77$ mm/yr (relative plate motion) and $d = 35$ km. The predicted strain rate is equivalent to $\dot{\gamma}_1 = 0.12$ μrad/yr and $\dot{\gamma}_2 = 0.20$ μrad/yr, about a factor of three higher than the observed values.

The other major seismic gap along the Alaska-Aleutian arc is located near Cape Yakataga (McCann et al 1980). A trilateration network was established there in 1979 and extended in 1980 (Figure 14). Only five measurements were common to both the 1979 and 1980 surveys, and the strain rates calculated for the one-year changes in those lines were not greater than the possible errors in measurement. The 1979 survey included eight lines from an electronic distance traverse measured in 1959. The changes between the two surveys suggested that the western edge of the Yakataga network had been displaced 3.4 m S40°E with respect to the eastern edge. The breadth of the network is about 40 km so that shear strains of about 85 μrad accumulated in the 20-yr interval. I presume this strain occurred as a coseismic effect of the 1964 Alaska earthquake and suggest that the 1964 rupture may have extended into the Yakataga area diminishing the significance of that seismic gap.

A triangulation network spanning the Denali fault at the crossing of the Delta River (Figure 14) was first surveyed by triangulation in 1941–42 and later surveyed by trilateration in 1970, 1975, and 1979 (Savage et al 1981b). The strain change in the interval 1941–42 to 1970 was consistent with 23 ± 4 μstrain extension approximately perpendicular to the Denali fault. Apparently this strain was a coseismic effect of the 1964 Alaska earthquake and not directly related to the Denali fault. The 1975 and 1979 surveys indicate that the strain accumulation along the Denali fault is essentially pure right-lateral shear at the rate of 0.6 ± 0.1 μrad/yr across a vertical plane striking N87°E. This plane of maximum shear is rotated about 30° counterclockwise from the local strike of the Denali fault but closely coincides with the strike of a major segment of the fault 50 km farther west. On this basis Savage et al (1981b) have suggested that the Denali fault behaves as a leaky transform fault with the southern block moving obliquely away from the northern block at a rate of about 20 mm/yr (17 mm/yr strike slip and 10 mm/yr spreading). It is possible that the apparent spreading is in fact due to postearthquake relaxation on the megathrust that ruptured in 1964, however. The geologic estimate of the strike-slip component of relative motion is 8–14 mm/yr (Sieh 1981a). The recurrence time estimated from the observed shear-strain rate and the average strain drop is only 80 yr. This is clearly too short. The geologic estimate of the recurrence time is 540–1880 yr, and there is good evidence that no major earthquake has occurred in the past 150 yr (Sieh 1981a).

A three-line strain network spanning the Denali fault at the crossing of the Nenana River (about 170 km west of the Delta River network) was surveyed in 1942 (triangulation) and 1970 (trilateration). The measured strain was a 15-μstrain extension approximately perpendicular to the Denali fault (Page 1972), presumably a coseismic effect of the 1964 Alaska earthquake.

Strain accumulation in Hawaii is dominated by volcanic processes that produce quite inhomogeneous strain; as a consequence, geodetic deformation data there has not generally been analyzed in terms of strain. Extensive trilateration networks for measuring deformation are located on the island of Maui (Berg et al 1981) and around Kilauea volcano on the island of Hawaii (Dvorak et al 1982).

Literature Cited

Allenby, R. J. 1979. Implications of very long baseline interferometry measurements on North American intra-plate crustal deformation. *Tectonophysics* 60:T27–35

Berg, E., Harris, D., Fisher, C. A. 1981. High precision single color laser distances, techniques and procedures to approach 0.3 ppm, Maui, Lanai, and Molokai network. *Rep. on NASA Grant NSG-7179*, Hawaii Inst. Geophys., Univ. Hawaii, Honolulu. 209 pp.

Burford, R. O., Harsh, P. W. 1980. Slip on the San Andreas fault in central California from alinement array surveys. *Bull. Seismol. Soc. Am.* 70:1233–61

Chinnery, M. A. 1961. The deformation of the ground around surface faults. *Bull. Seismol. Soc. Am.* 51:355–72

Davies, J., Sykes, L., House, L., Jacob, K. 1981. Shumagin seismic gap, Alaska peninsula: History of great earthquakes, tectonic setting, and evidence for high seismic potential. *J. Geophys. Res.* 86:3821–55

Dvorak, J., Dieterich, J. H., Okamura, A. 1982. Analysis of surface deformation data, Kilauea Volcano, Hawaii: October 1966 to September 1980. *J. Geophys. Res.* In press

Hall, N. T., Nelson, E. A., Fowler, D. R. 1982. *Holocene activity on the San Andreas fault between Crystal Springs Reservoir and San Andreas dam, San Mateo County, California.* Presented at Ann. Meet. Cordilleran Sect., Geol. Soc. Am., 78th, Anaheim, Calif.

McCann, W. R., Perez, O. J., Sykes, L. R. 1980. Yakataga gap, Alaska: Seismic history and earthquake potential. *Science* 207:1309–14

Meade, B. K. 1971. Crustal movement investigations. *EOS, Trans. Am. Geophys. Union* 52:IUGG7–9

Minster, J. B., Jordan, T. H. 1978. Present-day plate motions. *J. Geophys. Res.* 83:5331–54

Nason, R. 1979. Earthquakes and fault creep on the northern San Andreas fault. *Tectonophysics* 52:604 (Abstr.)

National Geodetic Survey. 1973. *Reports on Geodetic Measurements of Crustal Movement 1906–1971*, ed. B. K. Meade, C. A. Whitten. Washington DC: NOAA, US Dept. Commerce. 640 pp.

National Research Council. 1981. *Geodetic Monitoring of Tectonic Deformation—Towards a Strategy.* Washington DC: Natl. Acad. Press. 109 pp.

Page, R. A. 1972. Crustal deformation on the Denali fault, Alaska, 1942–1970. *J. Geophys. Res.* 77:1528–33

Pope, A. J. 1972. Strain analysis of horizontal crustal movements in Alaska based on triangulation surveys before and after the earthquake. In *The Great Alaska Earthquake of 1964: Seismology and Geodesy*, pp. 435–47. Washington DC: Natl. Acad. Sci. 596 pp.

Prescott, W. H. 1976. An extension of Frank's method for obtaining crustal shear strains from survey data. *Bull. Seismol. Soc. Am.* 66:1847–53

Prescott, W. H. 1981. The determination of displacement fields from geodetic data along a strike-slip fault. *J. Geophys. Res.* 86:6067–72

Prescott, W. H., Lisowski, M., Savage, J. C. 1981. Geodetic measurement of crustal deformation across the San Andreas, Hayward and Calaveras faults near San Francisco, California. *J. Geophys. Res.* 86:10853–69

Rikitake, T. 1976. *Earthquake Prediction.* Amsterdam: Elsevier. 357 pp.

Ryan, J., Schupler, B. R. 1982. Geodesy by radio interferometry: Baselines and earth rotation from Mark-III VLBI. *EOS, Trans. Am. Geophys. Union* 63:301 (Abstr.)

Savage, J. C. 1975. Crustal movement investigations. *Rev. Geophys. Space Phys.* 13:263–65

Savage, J. C. 1980. Dislocations in semismology. In *Dislocations in Solids*, ed. F. R. N. Nabarro, 3:251–339. Amsterdam: North Holland. 353 pp.

Savage, J. C. 1982. A dislocation model of strain accumulation and release at a subduction zone. *J. Geophys. Res.* In press

Savage, J. C., Burford, R. L. 1970. Accumulation of tectonic strain in California. *Bull. Seismol. Soc. Am.* 60:1877–96

Savage, J. C., Lisowski, M. 1980. Deformation in Owens Valley, California. *Bull. Seismol. Soc. Am.* 70:1225–32

Savage, J. C., Prescott, W. H. 1978. Asthenosphere readjustment and the earthquake cycle. *J. Geophys. Res.* 83:3369–76

Savage, J. C., Prescott, W. H., Lisowski, M., King, N. 1979. Geodolite measurements of deformation near Hollister, California. *J. Geophys. Res.* 84:7599–7615

Savage, J. C., Lisowski, M., Prescott, W. H., Sanford, A. R. 1980. Geodetic measurement of horizontal deformation across the Rio Grande rift near Socorro, New Mexico. *J. Geophys. Res.* 85:7215–20

Savage, J. C., Lisowski, M., Prescott, W. H. 1981a. Geodetic strain measurements in Washington. *J. Geophys. Res.* 86:4929–40

Savage, J. C., Lisowski, M., Prescott, W. H. 1981b. Strain accumulation across the Denali fault in the Delta River canyon, Alaska. *J. Geophys. Res.* 86:1005–14

Savage, J. C., Prescott, W. H., Lisowski, M., King, N. E. 1981c. Strain accumulation in southern California, 1973–1980. *J. Geophys. Res.* 86:6991–7001

Savage, J. C., Lisowski, M., Prescott, W. H., King, N. E. 1981d. Strain accumulation near the epicenters of the 1978 Bishop and 1980 Mammoth Lakes, California, earthquakes. *Bull. Seismol. Soc. Am.* 71:465–76

Sieh, K. E. 1981a. A review of geological evidence for recurrence times of large earthquakes. In *Earthquake Prediction*, ed. D. W. Simpson, P. G. Richards, pp. 181–207. Washington DC: Am. Geophys. Union. 680 pp.

Sieh, K. E. 1981b. Seismic potential of the dormant southern 200 km of the San Andreas fault. *EOS, Trans. Am. Geophys. Union* 62:1048 (Abstr.)

Sieh, K. E. 1982. *Seismotectonic studies of the San Andreas fault in southern California and implications concerning future large earthquakes.* Presented at US Geol. Surv. Semin., Apr. 13, Menlo Park, Calif.

Slawson, W. F., Savage, J. C. 1979. Geodetic deformation associated with the 1946 Vancouver Island, Canada, earthquake. *Bull. Seismol. Soc. Am.* 69:1487–96

Smith, D. E. 1980. Crustal motion measurements in California (SAFE). *NASA Tech. Memo. 80642*, pp. 3-40–41

Smith, D. E., Kolenkiewicz, R., Dunn, P. J., Torrence, M. H. 1979. The measurement of fault motion by satellite laser ranging. *Tectonophysics* 52:59–67

Snay, R. A., Cline, M. W. 1980. Geodetically derived strain at Shelter Cove, California. *Bull. Seismol. Soc. Am.* 70:893–901

Spence, D. A., Turcotte, D. L. 1979. Viscoelastic relaxation of cyclic displacements on the San Andreas fault. *Proc. R. Soc. London Ser. A* 365: 121–44

Sykes, L. R., Quittmeyer, R. C. 1981. Repeat times of great earthquakes along simple plate boundaries. In *Earthquake Prediction*, ed. D. W. Simpson, P. G. Richards, pp. 217–47. Washington DC: Am. Geophys. Union. 680 pp.

Thatcher, W. 1975a. Strain accumulation and release mechanism of the 1906 San Francisco earthquake. *J. Geophys. Res.* 80:4862–72

Thatcher, W. 1975b. Strain accumulation on the northern San Andreas fault zone since 1906. *J. Geophys. Res.* 80:4873–80

Thatcher, W. 1979a. Horizontal crustal deformation from historic geodetic measurements in southern California. *J. Geophys. Res.* 84:2351–70

Thatcher, W. 1979b. Crustal movements and earthquake-related deformation. *Rev. Geophys. Space Phys.* 17:1403–11

Thatcher, W. 1979c. Systematic inversion of geodetic data in central California. *J. Geophys. Res.* 84:2283–95

Thatcher, W. 1981. Crustal deformation studies and earthquake prediction research. In *Earthquake Prediction*, ed. D. W. Simpson, P. G. Richards, pp. 394–410. Washington DC: Am. Geophys. Union. 680 pp.

Thompson, G. A., Burke, D. B. 1974. Regional geophysics of the Basin and Range province. *Ann. Rev. Earth Planet. Sci.* 2:213–38

Wallace, R. E. 1981. Active faults, paleoseismology, and earthquake hazards in western United States. In *Earthquake Prediction*, ed. D. W. Simpson, P. G. Richards, pp. 209–16. Washington DC: Am. Geophys. Union. 680 pp.

Wyatt, F. 1982. Displacement of surface monuments: Horizontal motion. *J. Geophys. Res.* 87:979–89

Zandt, G., Richins, W. D. 1981. Interaction of high and low angle normal faults along the eastern Basin and Range, northern Utah. *EOS, Trans. Am. Geophys. Union.* 62:960 (Abstr.)

Ann. Rev. Earth Planet. Sci. 1983. 15: 45–73

ACCRETIONARY TECTONICS OF THE NORTH AMERICAN CORDILLERA

Jason B. Saleeby
Division of Geological and Planetary Sciences,
California Institute of Technology, Pasadena, California 91025

INTRODUCTION

Continental geology stands on the threshold of a change that is likely to be as fundamental as plate-tectonic theory was for marine geology. Ongoing seismic-reflection investigations into the deep crustal structure of North America are verifying that orogenic zones are underlain by low-angle faults of regional extent (Brown et al 1981). The growing body of regional field relations is likewise delineating numerous orogenic sutures that bound discrete crustal fragments. Paleomagnetic and paleobiogeographic studies are revealing major latitudinal shifts and rotations within and between suture-bounded fragments, particularly within the North American Cordillera. Such interdisciplinary studies are leading to a consensus that the Cordillera has been built by progressive tectonic addition of crustal fragments along the continent edge in Mesozoic and early Cenozoic time. Such crustal growth is referred to as accretionary tectonics. In this paper, we review some of the important concepts in accretionary tectonics, discuss the nature of the materials accreted between central Alaska and southern California in Jurassic and Cretaceous time, and consider the general relations between Cordilleran accretion and the movement of lithospheric plates.

The concept of continents growing by peripheral accretion through geologic time has long been a topic of great interest. With the advent of plate tectonics a number of different mechanisms for crustal accretion have arisen, along with mechanisms for crustal attrition. Accretion mechanisms include the growth of imbricated sedimentary prisms along inner-trench walls, slicing off of submarine topographic irregularities within subducting

0084–6597/83/0515–0045$02.00

plates, and collision of continents and volcanic arcs by ocean-basin closure. Tectonic attrition mechanisms include rifting, transform faulting, and strike-slip or underthrust removal of inner-trench wall materials coincident with or in place of accretionary prism growth. Growth of intraorogenic ocean basins by seafloor spreading is an additional important mechanism for creating accretionary materials as well as displacing crustal fragments. An important implication of plate kinematic theory is the likelihood for accretionary and attritionary mechanics to operate in series both in time and space along continental margins. Since attrition by nature leaves little material evidence of having operated, one of the major problems confronting Cordilleran geologists lies in the recognition of such attrition within the ancient record, particularly when interspersed with accretionary events.

The spectrum of accretion and attrition mechanisms viewed at cm yr^{-1} plate-transport rates over time scales of 100 m.y. leads one to suspect a highly mobile history for continental-margin orogens. The serial arrangement of subducting, transform, and rifting links along the modern Cordillera plate-juncture system and both serial and parallel arrangements in the western Pacific systems show the complex interplay of such mechanisms through space. Similar arrangements overprinted through time are suggested by the rock assemblages and structural patterns within the Cordillera, which presently resemble a collage of crustal fragments (Davis et al 1978). Recognition of the structural state of this collage by geologic field mapping and geophysical investigations will bring about a new level of understanding in the growth of continental crust, and the reading of stratigraphic records within the fragments and future palinspastic restorations will lead to a new level of understanding in paleogeography and Earth history. The first problem to be considered is the recognition of native North American crust from exotic fragments that have been accreted to its edge.

NATIVE AND EXOTIC ELEMENTS OF THE NORTH AMERICAN CORDILLERA

The North American Cordillera contains two regional paleotectonic belts of distinctly different character (Figure 1). The inner miogeoclinal belt consists of regionally continuous sedimentary sequences deposited on the North American sialic margin, which is in contrast to the outer eugeo-

---

Figure 1 Map showing generalized distribution of outer-belt tectonostratigraphic terranes and the miogeocline of northern Cordillera (modified after Coney et al 1980). Most terranes shown are actually composite. Also shown are some large strike-slip faults that are restored in the palinspastic base of Figure 4*b*. Pz = Paleozoic, Mz = Mesozoic.

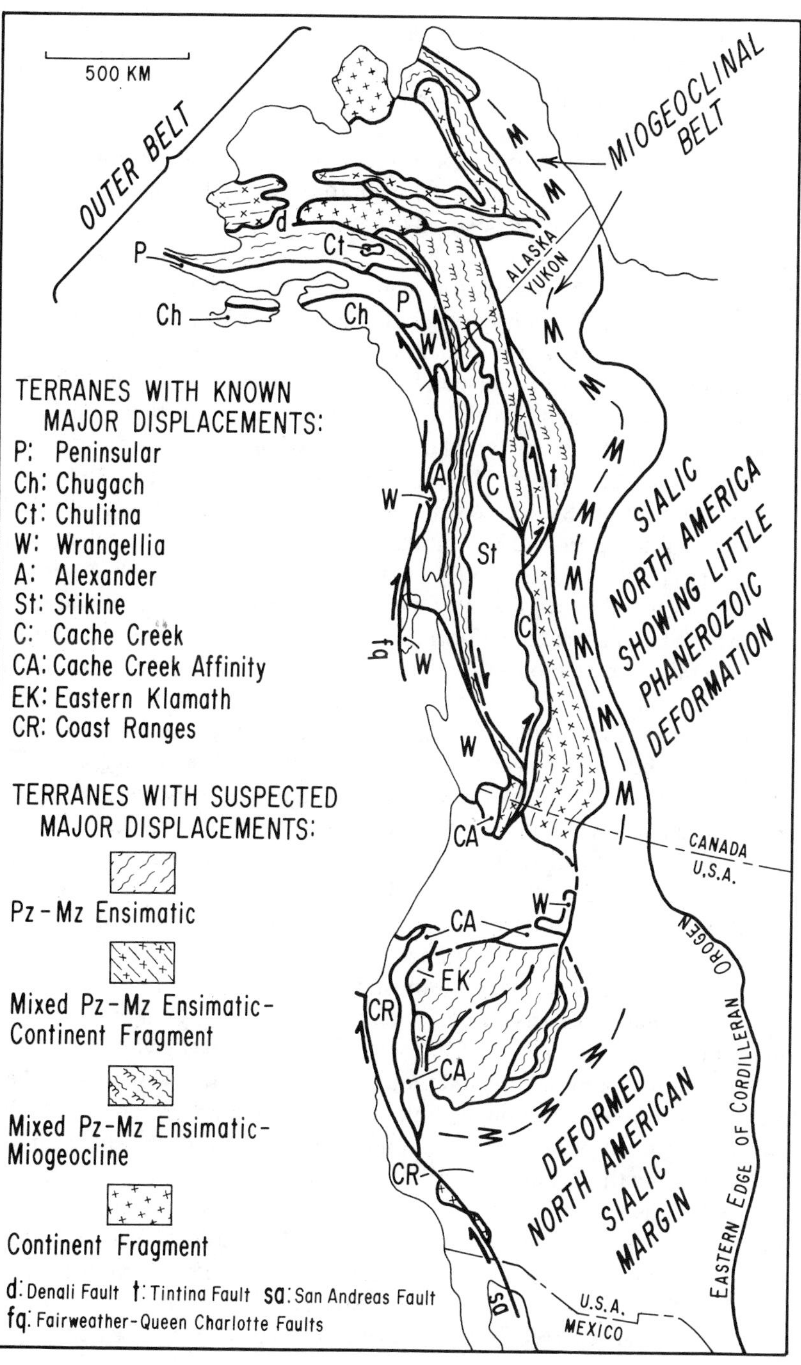
500 KM
OUTER BELT
MIOGEOCLINAL BELT
ALASKA
YUKON
SIALIC NORTH AMERICA SHOWING LITTLE PHANEROZOIC DEFORMATION
CANADA
U.S.A.
EASTERN EDGE OF CORDILLERAN OROGEN
DEFORMED NORTH AMERICAN SIALIC MARGIN
U.S.A.
MEXICO
TERRANES WITH KNOWN MAJOR DISPLACEMENTS:
P: Peninsular
Ch: Chugach
Ct: Chulitna
W: Wrangellia
A: Alexander
St: Stikine
C: Cache Creek
CA: Cache Creek Affinity
EK: Eastern Klamath
CR: Coast Ranges
TERRANES WITH SUSPECTED MAJOR DISPLACEMENTS:
Pz-Mz Ensimatic
Mixed Pz-Mz Ensimatic-Continent Fragment
Mixed Pz-Mz Ensimatic-Miogeocline
Continent Fragment
d: Denali Fault t: Tintina Fault sa: San Andreas Fault
fq: Fairweather-Queen Charlotte Faults

synclinal belt characterized by diverse ensimatic crustal fragments. The miogeocline is important with respect to understanding Cordilleran accretionary tectonics, for it represents the western edge of intact pre-Mesozoic North America. The miogeocline owes its origin to late Precambrian rifting with probable continental separation yielding a new passive margin, and subsequent sedimentation over a long period of drift (Stewart 1972). Such a depositional setting is shown by regional unconformable relations with a diverse Precambrian basement, and basal tholeiitic and active basinal clastic sequences overlain by thick stable-shelf sequences. The shelf sequences originally thickened substantially westward over a distance of perhaps 500 km, but their western limit has been structurally obscured by outer-belt tectonics. Stable miogeoclinal sedimentation persisted through mid-Paleozoic time until the first fragments of the outer belt were thrust eastward upon it in Mississippian and again in Permo-Triassic time (Speed 1979). Major outer-belt accretion then progressed throughout Mesozoic time, leaving the miogeocline sandwiched into its inner-belt position. The miogeocline lies within a marginal domain characterized by Mesozoic and younger deformation of North American sial. East of this domain lies cratonic North America, which for the most part escaped Phanerozoic deformation. Deformation of the sialic margin has been related to accretion of outer-belt crustal fragments (Coney 1981).

The regional internal structure of the outer belt is shown in Figure 1 (after Coney et al 1980). Numerous discrete and often unrelated fragments are recognized and shown as "terranes." A number of small terranes are grouped because they cannot be adequately shown at the scale of Figure 1. Many of the terranes are ensimatic assemblages with their oldest elements of Paleozoic age. Which of the terranes are exotic with respect to the ancient continental margin, and which formed along the fringes of the margin is one of the fundamental questions confronting Cordilleran geologists. The recognition of a number of terranes with significant latitudinal displacements has led to a new highly mobilistic view of Cordilleran tectonics.

The first major observational breakthrough directly supporting this highly mobilistic view was made by Monger & Ross (1971), who recognized regional suture-bounded disjunct faunal belts in British Columbia. They drew special attention to upper Paleozoic fusulinacean provinces and the existence of a belt of exotic verbeekinids that much more closely resemble Tethyan forms than coeval North American forms. Subsequent work has shown that the verbeekinids are an equatorial Tethyan-proto-Pacific form that almost always occurs intermixed with accreted oceanic crust, seamount, and abyssal plain assemblages (Yancy 1979, Danner 1977, Monger 1977a, Nestel 1980). The most expansive tract of these assemblages occurs in the Cache Creek terrane of British Columbia. Similar assemblages occur in northern Washington, eastern Oregon, and the western Sierra Nevada–Klamath Mountains belt; in each of these areas Cache Creek-type rocks are intermixed with other diverse assemblages. Such mixtures are shown as Cache Creek–affinity terranes in Figure 1. A number of other biostratigraphic studies have revealed displaced faunal assemblages in outer-belt terranes (Nichols & Silberling 1979, Nestel 1980, D. L. Jones, personal communication, 1982); however, the Cache Creek fauna stands out as the most acute problem in tectonic transport and paleobiogeography.

The observational basis for mobilistic Cordilleran accretion grew substantially through the 1970s. Northward translations in the range of hundreds to thousands of kilometers along with significant rotations within and between terranes are shown by numerous paleomagnetic studies. One of the most stunning studies was done on Wrangellia, a large terrane of upper Paleozoic to Triassic rocks occurring as large fragments in western British Columbia, southern Alaska, and eastern Oregon. Paleomagnetic studies indicate that each of the Wrangellian fragments originated within 15° of the paleoequator and thus the entire terrane has undergone at least 3000 km of northward translation since Triassic time (Hillhouse 1977, Jones et al 1977, Hillhouse & Gromme 1982). Other outer-belt terranes known to have been significantly displaced are labeled in Figure 1. Each of these is discussed below. Terranes that are suspected to have undergone displacement, but which presently lack definitive data, are shown in patterns denoting general compositional character. Those shown as "ensimatic" consist primarily of basinal sedimentary rocks, intermediate to mafic volcanic rocks, and masses of metamorphic tectonites that lack obvious continental basement protoliths. Such ensimatic assemblages are also mixed at scales ranging from small blocks to large nappes with fragments of Precambrian basement and sedimentary sequences that appear to have formed on or adjacent to continental basement. Similar tectonic mixtures occur between ensimatic assemblages and fringes of the miogeocline, and large isolated terranes of continental crust with uncertain heritage are also embedded in the outer belt.

The basis for recognizing the internal structure of the outer belt and for defining and resolving the major problems lies in structural-stratigraphic observations. Only with the mapping of terrane boundaries and an understanding of their internal stratigraphic and structural sequences can paleomagnetic, geochronological, paleobiogeographic, and deep crustal geophysical data be related directly to terrane accretion. With the advent of plate tectonics, field geology became more speculative and model-oriented. However, this field is now changing again dramatically. Attempts to refine and support popular plate-tectonic models of the 1970s by additional and more refined data are revealing misconceptions that are in essence as fundamental as the differences between geosynclinal theory and "uniformitarian" plate-tectonic analysis. One of the first major signs of a new conceptual framework in the field aspects of accretionary orogens is the concept of terrane analysis.

TERRANE ANALYSIS AND PALEOTECTONIC INFORMATION

The concept of terrane analysis is set forth in Coney et al (1980) and Jones et al (1983). The basic geologic field units are tectonostratigraphic terranes (Figure 1), which are fault-bounded entities often of regional extent, each characterized by a unique geological history. The cumbersome term "tectonostratigraphic" portrays an important aspect of both the terranes and their analysis. Tectonic processes account for their size, shape, and structural-geographic distribution, whereas internal stratigraphic records uniquely define them. Conventional stratigraphic records may be completely disrupted by tectonic or sedimentological reworking processes, or obscured by regional metamorphic overprints. Such modification phenomena may uniquely define terranes in a fashion similar to conventional stratigraphic or igneous sequences.

A critical aspect of terrane analysis is a built-in skepticism toward plate-tectonic modeling between groups of terranes. Genetic links must be physically demonstrated between adjacent terranes before paleogeographic and tectonic models can be constructed using them as building blocks. Until unequivocal links are established, the terranes are considered "suspect" to one another in that their formational stages may have been genetically and spatially unrelated, and they are now by chance juxtaposed by subsequent tectonic transport. Outside of physical continuity and facies gradation, different terranes can be linked by the detritus of one occurring in another or by adjacent terranes sharing common crosscutting plutons or overlap depositional sequences. Physical continuity and facies relations are the strongest signs of tectonogenetic linkages. The other criteria may only mark the "amalgamation" (Figure 2) of two or more unrelated terranes into a composite terrane, whereby related histories may be assumed following amalgamation. The amalgamation of two or more terranes into a composite terrane may or may not coincide with accretion to continental masses. Terrane accretion is used strictly in regard to firm juxtaposition with cratonic crust or cratonic-cored accretionary masses. Amalgamation of terranes may occur at great distances from their continental accretion sites. Examples are given below.

In their review of Cordilleran suspect terranes, Coney et al (1980) point out that over 70% of the orogen is composed of more than 50 terranes. Only the largest of these terranes and groups of terranes are shown in Figure 1. Nearly all of the terranes are elongate parallel to the north and northwest grain of the orogen. This pattern of fault-bounded interdigitated terranes is referred to as an "orogenic collage." The collage concept was introduced by Helwig (1974) in order to elucidate the composite nature of eugeosynclinal belts. An important implication of this concept is that accretionary orogens do not represent coherent bodies of paleotectonic information, but rather a fragmental array of clues. These clues are commonly referred to as "petrotectonic assemblages." Such rock assemblages are defined as distinct lithologic associations indicative of a particular plate-tectonic regime (Dickinson 1972). For example, the Cordilleran miogeocline is taken as the record of continental rift and subsequent passive drift. Examples of important outer-belt assemblages are thick sequences of volcaniclastic strata and plutonic belts of overall calc-alkaline nature taken as the remnants of subduction-related arcs, or ophiolite sequences taken as remnants of crust generated by seafloor spreading. Considering the concepts of petrotectonic assemblages and orogenic collage together brings us to a view of the Cordilleran orogen as a collection of plate-tectonic history fragments taken randomly from geologic time and space. The focus of the following sections is to merge the concepts of terrane analysis and petrotectonic assemblages in order to roughly characterize some of the better preserved fragments of plate-tectonic history.

FRAGMENTS OF PLATE TECTONIC HISTORY

For a number of the terranes shown in Figure 1, stratigraphic records are well enough preserved so that geological histories can be interpreted within a petrotectonic framework. Which segments of history relate directly to the North American margin and which relate to distant regions is a fundamental question in Cordilleran paleogeography. Some of the better-understood terranes are viewed below in terms of both their plate-tectonic and continental-margin accretion histories. Attention is focused first on some of the largest terranes of the Cordillera situated in British Columbia and Alaska. This is followed by an analysis of petrotectonic assemblages for a narrow yet important time interval in California history. The largest terranes of the Cordillera are Wrangellia and Stikine, both of which were amalgamated with other large terranes prior to accretion. Such amalgamated terrane groups along with their overlap assemblages are referred to as super-terranes. Discussion of the Wrangellian and Stikinian super-terranes centers around Figure 2, which shows highly diagrammatic stratigraphic columns and time relations in amalgamation and accretion.

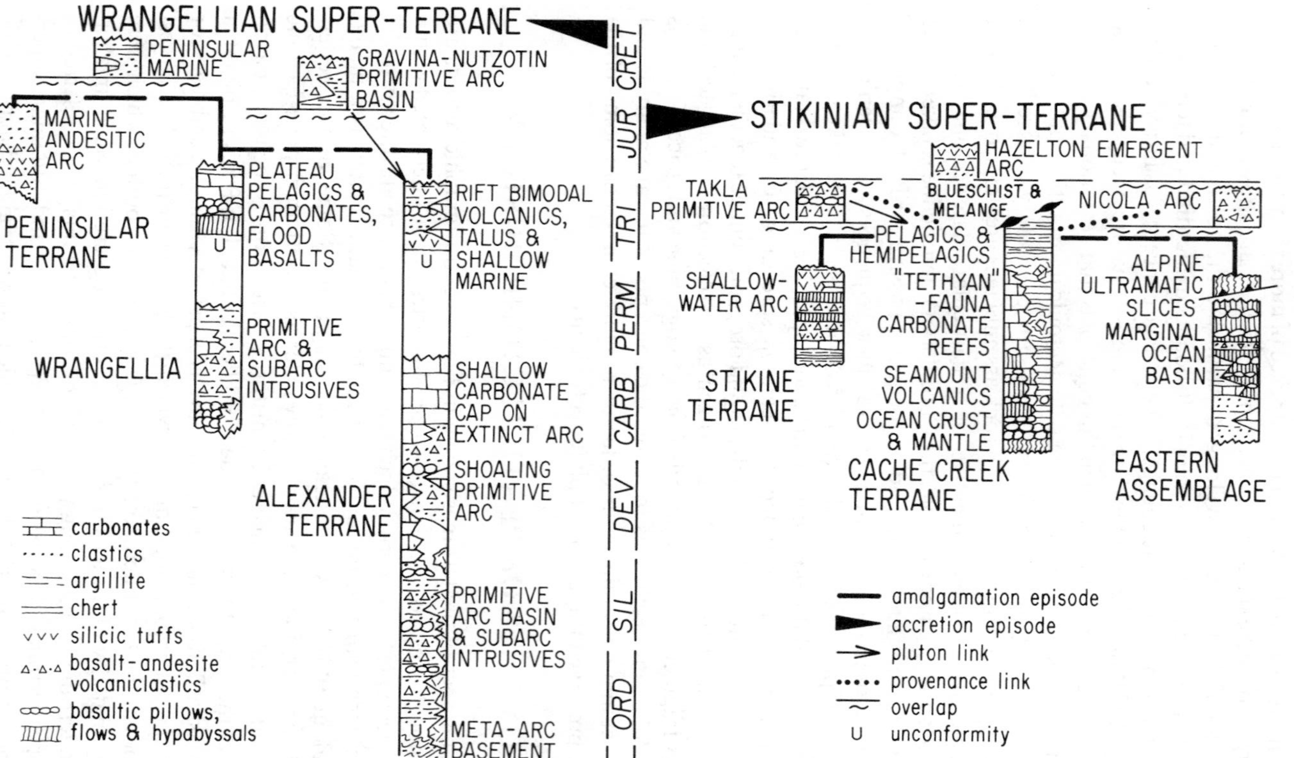

Figure 2 Diagram showing the petrotectonic, amalgamation, and accretion histories of the Wrangellian and Stikinian super-terranes of British Columbia and southern Alaska. References given in text.

Wrangellian Super-Terrane

The Wrangellian super-terrane consists of Wrangellia, the Alexander and Peninsular terranes, and Jura-Cretaceous overlap sequences and crosscutting plutons. The Alexander terrane contains within it one of the longest and most complete geologic histories in the outer belt. Nonmetamorphosed strata and shallow-level intrusives as old as Lower Ordovician are widespread (Churkin & Eberlain 1977, Saleeby et al 1983). Pre-Ordovician rocks lie beneath these strata and are cut by the intrusives. Ordovician through Silurian time is represented by basinal facies volcanic arc deposits. Volcanic units composed of basalt and basaltic-andesite pillows, pyroclastic flow breccias, and aquagene tuffs lie within volcanolithic turbidites and deep-water graptolitic shale. Widespread subvolcanic intrusives consist of complex dioritic, gabbroic, and trondhjemitic assemblages with local ultramafic members (Berg 1972, 1973, Saleeby et al 1983). Mafic dike swarms are widespread, and in places they resemble sheeted complexes of ophiolites. Such relations along with the basinal character of the stratified rocks are suggestive of an extensional or rifted volcanic arc. Disruption of the arc-basinal framework is shown in Devonian time by local unconformities that expose the subarc intrusives and cause local but significant gaps in the stratigraphic record. Dúring and following basin disruption, the arc sequence began shoaling with the intercalation of shallow-water carbonates. In Carboniferous time, arc activity ceased with prolonged deposition of shallow-water carbonates.

In contrast, the adjacent Wrangellia began an active basinal and then possible shoaling primitive arc history in Carboniferous through Middle Permian time (Jones & Silberling 1979). Older Paleozoic rock assemblages probably lie within the Wrangellian basement (Muller et al 1974, Yorath & Chase 1981), but their stratigraphic settings are as yet unclear. Wrangellia is most notable for its Triassic and earliest Jurassic oceanic plateau history (Jones et al 1977, Ben-Avraham et al 1981). After an Upper Permian–Lower Triassic hiatus, up to 6000 m of subaerial and shallow marine flood basalt capped the extinct arc. Following volcanism a thick carbonate sequence was deposited commencing with inner-platform limestone and dolomite and ending with basinal pelagic limestone, siliceous argillite, and carbonaceous shale. Such a stratigraphic sequence suggests submergence and rifting separation of Wrangellia followed by open ocean drift. The aberrant paleomagnetic poles are derived from the Triassic basaltic pile, which further substantiates the subsequent drift phase.

Triassic strata of the Alexander terrane show basaltic-rhyolitic bimodal volcanic sequences and basement-derived talus breccias intercalated with shallow marine strata. This apparent rift assemblage formed at significantly higher paleolatitudes than the roughly coeval flood basalts of Wrangellia (Hillhouse & Gromme 1980). New comparative paleobiogeographic data along with an earlier displacement history suggests that these higher paleolatitudes may have been in the southern Hemisphere (D. L. Jones, personal communication, 1982, Van der Voo et al 1980).

In early to mid-Jurassic time, the Wrangellia and Alexander terranes were amalgamated into a large composite terrane. Much of the original terrane boundary has been modified and reactivated by Cenozoic strike-slip faulting and rifting (Coney et al 1980, Yorath & Chase 1981). However, the early amalgamation episode is shown by overlap relations with the Upper Jurassic–Lower Cretaceous Gravina-Nutzotin belt, a primitive arc assemblage composed of basalt and basaltic-andesite volcaniclastic rocks and deep-basin turbidites (Berg et al 1972). Peridotitic to dioritic intrusive complexes thought to be the roots of Gravina-Nutzotin arc intrude the Alexander terrane (Berg et al 1972, Murray 1972). Further amalgamation of the composite Alexander-Wrangellia terrane and its Gravina-Nutzotin overlap with the diverse Peninsular terrane occurred prior to continental-margin accretion. This relation is shown by the overlap of mid-Cretaceous clastic sequences across both Wrangellia and Jurassic and Triassic basinal and arc-type strata of the Peninsular terrane (Packer & Stone 1974, Jones & Silberling 1979). The amount of northward translation that the amalgamated Peninsular, Wrangellia, and Alexander terranes underwent together is unclear inasmuch as the highly aberrant paleomagnetic pole positions were determined on pre-Cretaceous rocks. Poorly constrained paleopole aberrations in Cretaceous and Eocene strata of the Peninsular terrane (Stone & Packer 1977) indicate some postaccretion northward movement of the super-terrane. Such late-stage movement may have occurred along with accretion and northward transport of the outboard Chugach terrane (Gromme & Hillhouse 1981).

Continental-margin accretion of the Wrangellian super-terrane occurred in mid- to late Cretaceous time as shown by the age of the youngest basinal strata deformed during the collisional suturing event, and by the age of postcollisional intrusives (Coney 1981, Cowan & Brandon 1982, G. E. Gehrels & J. Saleeby, unpublished age data). In southeastern Alaska, the Gravina-Nutzotin flysch basin was tightly closed as it and its Alexander basement were obliquely underthrust along narrow, structurally complex terranes to the east (Figure 1). In southern Alaska the Peninsular and Wrangellian rocks are separated from inboard terranes by a partially closed Cretaceous flysch basin. A number of small terranes are structurally encased within the flysch assemblage, most notably the Chulitna terrane, whose fauna and Upper Triassic redbed sequence indicate an origin at substantially lower latitudes (Jones & Silberling 1979). As in southeastern

Alaska and British Columbia, accretion of the Wrangellian super-terrane in southern Alaska appears to have occurred by convergence, with significant dextral movement.

Numerous fragments of plate-tectonic history are well preserved in the Wrangellian super-terrane. Most notable are a series of separate and superimposed ensimatic volcanic arcs and arc basins scattered in time from Lower Ordovician to Lower Cretaceous. Other significant assemblages include the Triassic rift sequences of Wrangellia and the Alexander terrane, and the Wrangellia oceanic-plateau sequence. The final collisional suturing of the Wrangellian super-terrane to North America also represents an important plate-tectonic event. However, none of these history fragments can at present be related to a comprehensive plate model. Rock sequences of suitable age and petrotectonic heritage that could be forearc basin–subduction zone assemblages genetically related to any of the arc sequences have yet to be recognized. Furthermore, a subduction-magmatic couplet related to the closed ocean basin marked by the collisional suture has not been recognized.

Stikinian Super-Terrane

The Stikinian super-terrane consists of the Stikine and Cache Creek terranes, an eastern assemblage of Paleozoic strata that probably formed adjacent to the miogeocline, and a complex association of postamalgamation volcanic arcs and subarc intrusives of early Mesozoic age. Paleomagnetic data showing significant displacement are only available for postamalgamation arc rocks (Monger & Irving 1980), although paleobiogeographic data show earlier displacements between each of the major Paleozoic assemblages (Monger & Ross 1971). The Cache Creek terrane represents one of the clearly exotic elements of the Cordilleran collage distinguished by Carboniferous to Permian verbeekinids in shallow-water limestones that were deposited on subsiding seamounts (Monger 1977a, Souther 1977, Ben-Avraham et al 1981). Paleobiogeographic and carbonate petrographic features indicate an oceanic equatorial origin for the limestones in atoll-type environments (Monger 1975, 1977a, Danner 1977, Yancy 1979). Such buildups persisted locally into Lower Triassic time. Coeval deposition of pelagic sediments is shown by interbedding and intermixing relations with Carboniferous through Triassic radiolarian chert (Monger 1975, 1977a, Ben-Avraham et al 1981). Large slide-blocks of shallow-water limestone encased within younger chert are common, probably representing fringes of reefs that broke loose and moved downslope. Cache Creek basement rocks consist of large basaltic, commonly alkalic, volcanic piles presumed to be seamounts, and fragments of upper Paleozoic mafic and ultramafic rocks derived from oceanic crust (Monger 1977a, Souther 1977). Formation of melange and mid- to late Triassic blueschist facies (high pressure) tectonites coincides in time with amalgamation to the Stikine terrane and the eastern assemblage (Davis et al 1978, Monger et al 1982).

The Stikine terrane represents the largest individual crustal fragment in the Cordilleran outer belt. Unlike the Cache Creek terrane, the upper Paleozoic history of the Stikine terrane is dominated by shallow-water volcanic-arc activity with the eruption of andesitic to rhyolitic volcaniclastic rocks and basaltic flows (Monger 1977a). Permian fusulinaceans from expansive limestone units are of the schwagerinid family in contrast to the neighboring Cache Creek verbeekinids (Monger & Ross 1971). The Stikine schwagerinids are unlike coeval North American schwagerinids, but they closely resemble those present in the Permian McCloud limestone from the eastern Klamath terrane of California. The Stikinian and eastern Klamath schwagerinids may represent North American–derived branches that evolved in partial isolation along fringing arc systems in temperate climates (Monger & Ross 1971, Nestel 1980). The Stikine section shown in Figure 2 represents stratified sequences located along the east side of the terrane; a large portion of its interior is overlain by postaccretion overlap strata. Similar Paleozoic sequences lie along the west margin of the terrane, where shallow-water arc activity is evident back into early Carboniferous time (Monger 1977a).

The Stikine and Cache Creek terranes abut against an eastern assemblage of mixed ensimatic and fragmented continental edge material (Figure 1). In addition to tectonic slices of Precambrian and lower Paleozoic continent-affinity rocks and Mesozoic volcanogenic assemblages, upper Paleozoic sequences showing a history distinct from both the Stikine and Cache Creek terranes are widespread. These sequences consist of Carboniferous through Permian basinal sediments and great thicknesses of basaltic pillows, flows, and hypabyssals (Monger 1977a). Fault slices of Alpine-type peridotite are also common which along with widespread basaltic rocks suggest crustal fragments of ophiolitic character. The basinal volcanic-sedimentary association of this assemblage and its structural and spatial association with fragmented continental edge and miogeocline rocks suggest that it is the inboard remnants of a marginal ocean basin.

Amalgamation of the Stikine and Cache Creek terranes and eastern assemblage rocks is constrained to Middle or Upper Triassic time. Widespread primitive arc activity of the Takla and Nicola sequences overlap Paleozoic strata of Stikine and the eastern assemblage (Monger 1977a,b, Davis et al 1978). Melange-mixing and blueschist facies metamorphism in the Cache Creek terrane is thought to be related to Takla-Nicola subduction (Travers 1978, Monger et al 1982). The Cache Creek terrane can more definitely be linked to Stikine and eastern assemblage rocks by the intermixing of Takla-Nicola arc and Cache Creek detritus, and by 200 m.y. age Takla-Nicola affinity arc intrusives that crosscut Cache Creek rocks (Monger 1977a,b, Davis et al 1978, Tipper 1978, Monger et al 1982). Finally the Lower to Middle Jurassic Hazelton arc overlaps both the Takla and Nicola arcs and their underlying Paleozoic terranes (Davis et al 1978). The Hazelton overlap defines the uppermost stratigraphic sequence of the Stikinian super-terrane. Imbrication of the super-terrane rocks in mid to late Jurassic time is taken as the accretion episode. Paleomagnetic data on Mesozoic overlap sequences and on crosscutting plutons indicates that amalgamation and perhaps accretion occurred at California paleolatitudes, and subsequently the accreted Stikinian super-terrane slid northward into its present position by dextral strike-slip (Monger & Irving 1980).

The Stikinian super-terrane exhibits a wide variety of petrotectonic assemblages, but it is difficult to arrange them into a simple plate-tectonic model. If the Stikine terrane evolved along the fringes of the North American margin, then a complex pattern of interterrane movement is required for insertion of the Cache Creek terrane between Stikine and marginal ocean-basin rocks of the eastern assemblage. Furthermore, if Cache Creek rocks are subduction-zone relatives to the Takla-Nicola arc they occur in a somewhat peculiar axial position relative to the arc loci. Insight into some of the more subtle tectonic processes that may have influenced the map distribution of Stikinia's constituent terranes and overlap arcs may be derived from Stikinia-type rocks of California.

Stikinian Tectonics Viewed from Califnornia

Pre-Cretaceous rocks of the western Sierra Nevada–Klamath Mountains metamorphic belt and the northeastern Coast Ranges contain a distinct association of mid- to late Jurassic island arc and related ophiolitic rocks. Such arc rocks in the Sierran-Klamath belt were constructed on a basement that resembles the Stikinian super-terrane. Subsequently the arc assemblage, its basement, and the arc-related ophiolites were imbricated during the late Jurassic Nevadan orogeny (Davis et al 1978, Suppe & Foland 1978). The nature and evolution of the arc association and the timing of the Nevadan orogeny may hold important clues to the complex distribution of terranes and to the onset of major accretion throughout a large portion of the Cordillera. Attention is drawn first to the pre-Nevadan basement of the arc.

The pre-Nevadan basement contains analogues to the Cache Creek and Stikine terranes, the eastern assemblage, and the Takla-Nicola arc (Figure 3). Cache Creek elements consist of Permo-Carboniferous ophiolite and seamount fragments, verbeekinid-bearing limestone blocks, Permian to

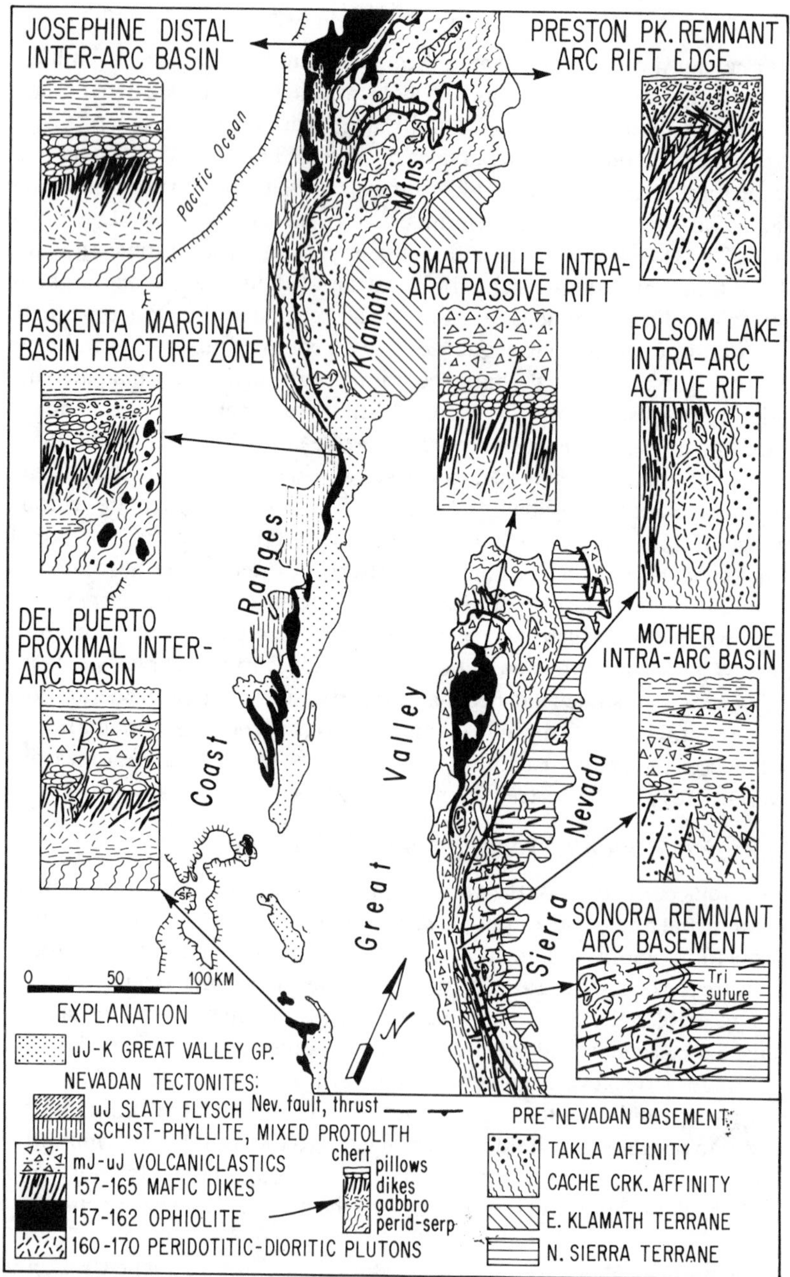

Figure 3 Generalized geologic map of northwest California showing distribution of pre-Nevadan basement rocks, and mid- to Upper Jurassic arc and related ophiolitic assemblages. Diagrammatic sections show critical primary relations in selected mid- to Upper Jurassic ophiolitic and arc associations. References given in text.

lower Mesozoic chert and argillite, and amphibolite-blueschist tectonites, which along with widespread melange record mid- to late Triassic deformation and metamorphism (Douglass 1967, Hotz et al 1977, Schweickert et al 1977, Behrman 1978, Davis et al 1978, Irwin et al 1978, Saleeby 1981, 1982, Wright 1982, Ando et al 1983). Pre-Nevadan basement rocks of the eastern Klamath and northern Sierra terranes bound Cache Creek–affinity rocks along regional tectonic contacts of pre-Nevadan age. The eastern Klamath terrane is distinguished by early Ordovician ophiolite sitting structurally above Devonian metamorphic tectonites (Lanphere et al 1968, Mattinson & Hopson 1972). Above the ophiolite rests Ordovician through Permian mafic to silicic arc rocks, continent-derived clastics, and shallow-water limestone (Irwin 1977, Potter et al 1977). Here lies the Permian McCloud limestone, which contains schwagerinids similar to those found in the Stikine terrane (Skinner & Wilde 1965, Monger 1977a, Nestel 1980). Paleozoic strata of the northern Sierra are somewhat different from those of the eastern Klamaths, although correlations of some specific units have been suggested (Schweickert & Snyder 1981). The most extensive unit in the northern Sierra is a continental rise-and-slope sequence of Ordovician-Silurian age, which along with possible correlatives in central Nevada is thought to have been deposited off the miogeocline edge (Bond & DeVay 1980, Schweickert & Snyder 1981). Above these strata along a regional unconformity lie Devonian through Permian strata dominated by arc deposits (D'Allura et al 1977, Varga & Moores 1981). A Cordilleran rise-slope depositional setting for the Ordovician-Silurian strata ties the northern Sierra terrane to an inboard position relative to most outer-belt Paleozoic terranes. Such a position is analogous to the eastern assemblage of the Stikinian super-terrane. The eastern Klamath and northern Sierra terranes were amalgamated by mid-Triassic time as shown by correlative Middle Triassic strata indicating an overlap relation (Albers & Robertson 1961, Skinner & Wilde 1965, D'Allura et al 1977).

Pre-Nevadan basement rocks of Jura-Triassic age along the western Sierran-Klamath belt have affinities to the Takla-Nicola arc assemblages. Most notable are large accumulations of basaltic and basaltic-andesite volcaniclastic rocks with local associated wherlitic-to-dioritic shallow-level intrusives. Many of the Sierran-Klamath sequences are coarsely clinopyroxene-phyric and often olivine-bearing or normative (Sharp & Wright 1981, Saleeby 1982), as are Takla-Nicola rocks (Monger 1977b, Preto 1977). In the southern and central Sierra, Takla-affinity rocks were erupted across an ophiolitic melange basement of Cache Creek–affinity (Saleeby 1981, 1982). Further north along the western Sierra, Takla-like rocks are strongly intermixed with Cache Creek–affinity rocks. Slices of early Mesozoic strata within this complex zone contain slide-blocks of verbeekinid limestone and, locally, cobbles of McCloud-type schwagerinid limestone (Behrman & Parkison 1978, Saleeby, unpublished data). In the northernmost Sierra and southwest Klamath Mountains, Takla-affinity rocks occur tectonically intermixed with rocks of Cache Creek–affinity and as a large tectonic slice (Irwin 1972, Davis et al 1978, Sharp & Wright 1981, Wright 1982).

The gross lithologic associations and the amalgamation history of the pre-Nevadan basement resemble those of the Stikinian super-terrane in too many respects to be a coincidence. Most notable is the joining of verbeekinid and schwagerinid-bearing terranes near or at the time of a distinct metamorphic-deformation event in Cache Creek rocks, and then the ensuing and perhaps partly coincident growth of a primitive arc sequence. The early Mesozoic amalgamation with Paleozoic strata that were deposited off the miogeocline edge is an additional important parallel. The plate-tectonic history of pre-Nevadan basement amalgamation is highly enigmatic as is that of the Stikinian super-terrane, but the relations outlined above indicate that the mid- to late Jurassic arc assemblage and the ensuing Nevadan orogeny were superimposed over a preexisting mosaic of terranes of Stikinian character.

Plutonic rocks of the superimposed arc consist of a distinct 170 to 160 m.y. age group (Figure 3). A critical aspect of this group is that it crosscuts basement rocks of Cache Creek and Takla-affinity as well as those of the northern Sierra and eastern Klamath terranes (Saleeby 1982, Wright & Sharp 1982). This plutonic belt is characterized by a compositional suite varying from clinopyroxene-rich ultramafics to dioritic and local granitoid end-members (Snoke et al 1982). Thick piles of basaltic-andesite and locally andesitic and dacitic volcaniclastic strata represent extrusive members of the arc. Age constraints on these strata are Callovian to possibly early Kimmeridgian (164 to 155 m.y. B.P., Armstrong 1978) in the central Sierra (Clark 1964, Behrman & Parkison 1978), and 162 to 159 m.y. B.P. in the northern Sierra (J. Saleeby & E. M. Moores, unpublished age data). Similar volcaniclastic rocks and age constraints occur in the western Klamaths just north of the area shown in Figure 3 (Wells & Walker 1953, Garcia 1979). Late Oxfordian to early Kimmeridgian flysch derived from first-cycle arc sources and from uplifted Cache Creek and craton-like sources is locally interbedded with, and for the most part above, the volcaniclastic strata in both the Sierras and Klamaths (Wells & Walker 1953, Clark 1964, Behrman 1978, Behrman & Parkison 1978, Harper 1980). Such flysch sequences constitute the main protolith for the Sierra-Klamath slate belt.

Mid- to Upper Jurassic ophiolitic rocks constitute a critical element in the petrotectonic analysis of the arc assemblage. Modern fringing arc systems of the western Pacific commonly contain a number of elements which include active arc segments with intra-arc basins, and remnant arc segments separated from the active arc by interarc basins formed by seafloor spreading (Karig 1970, 1971, 1972). Mid- to late Jurassic arc and ophiolitic assemblages of central and northern California together contain all of these elements. Some of the critical relations are represented by diagrammatic sections in Figure 3.

The process of intra-arc rifting is recorded in the Smartville block and Folsom Lake areas of the northern Sierra. The Smartville block possesses the intrusive and volcanic members of an ideal ophiolite, but with its pillow sequence interbedded with and giving way upward to arc volcaniclastic rocks (Xenophontos & Bond 1978). The great expanse of sheeted dikes suggests a zone of passive rifting within the arc with the generation of oceanic-type crust. The Smartville block is apparently in thrust contact above Cache Creek–Takla affinity basement as well as 160 m.y. age ophiolitic rocks of the Folsom Lake area (E. M. Moores, personal communication, 1980). The section in Figure 3 for Folsom Lake is a diagrammatic map view. Here Smartville-age sheeted dikes and periodotitic-to-dioritic arc intrusives were strongly deformed under solidus to hot subsolidus conditions along with polymetamorphic wall rocks of Cache Creek–Takla affinity (Springer 1980, Saleeby 1982). Such protoclastic deformation marks a zone of tectonically active intra-arc rifting; this is perhaps a transform segment or a zone of convergence that was superimposed on the rift immediately after it formed. In the Mother Lode belt east and south of the Folsom Lake complex, intra-arc basinal strata lie depositionally above the wall rock–basement complex (Clark 1964, Behrman 1978, Behrman & Parkison 1978, Saleeby 1982).

Age relations and petrotectonic assemblages suggesting that the arc split into interarc-basin and remnant-arc segments are preserved in the west-central Klamaths, central Sierras, and perhaps along the eastern edge of the Coast Ranges. A subtle but regional expression of this event is the abrupt termination of the arc plutons at about 160 m.y. B.P. and the regional overlap of late Oxfordian to early Kimmeridgian basement-derived flysch across the major volcaniclastic units. A more direct expression of this event occurs in the central Sierra, where widespread 160 m.y. age mafic dikes crosscut the extinct plutonic belt. The Sonora section of Figure 3 shows this critical relation in diagrammatic map view (Sharp & Saleeby 1979, Sharp 1980, Schweickert & Snyder 1981). The sequence of concentrated arc plutonism abruptly terminated and then followed by regional diking contemporaneous with ophiolite formation is taken as a sign of remnant-arc formation. Higher stratigraphic levels of the same or a similar age remnant-arc segment are preserved in the Preston Peak ophiolite of the west-central Klamaths (Snoke 1977, Saleeby et al 1982). Here

160 m.y. age basaltic dikes crosscut the arc basement and then coalesce upward into a intrusive complex that lies beneath but also crosscuts faulted diabase and basalt clast breccia. Silicious argillite in turn caps the mafic breccia.

West of the Sierran-Klamath arc lie the vestiges of 160 m.y. age interarc basin ophiolites. The Josephine ophiolite represents a complete oceanic-crust and upper-mantle sequence that is conformally overlain by flysch of the slate belt (Harper 1980, Saleeby et al 1982). This critical stratigraphic relation ties the Josephine ophiolite to the arc terrane as a juvenile basement "facies," which in some manner passed laterally into the older preexisting basement. It is suggested that the remnants of such a transition are preserved in the Preston Peak ophiolite (Saleeby et al 1982). Remnants of 160 m.y. age interarc-basin and marginal-basin ophiolites constitute the Coast Range belt (Bailey & Blake 1974, Blake & Jones 1981, Hopson et al 1981). Interarc-basin ophiolite formed proximal to the locus of arc activity is well preserved in the Del Puerto section, where abundant shallow-level silicic intrusives and proximal arc volcaniclastics form an integral and coeval part of the ophiolite section (Evarts 1977, Sharp & Evarts 1982). An additional important assemblage occurs along the northern segment of the Coast Range belt and is represented by the Paskenta section. Here the ophiolitic basin floor was fractured, uplifted, and eroded to its plutonic levels during its igneous activity, as shown by plutonic-clast talus breccias interbedded with pillows and local chert. Amphibolitic metamorphic tectonites also formed contemporaneous with ophiolite genesis, along with serpentinite melange zones that remained diapirically active into Cretaceous time (Hopson et al 1981, M. C. Blake & J. Saleeby, unpublished field and age data). Such relations are suggestive of a basin-floor fracture zone of probable transform origin (Bonatti & Honnorez 1976, Saleeby 1981). Geophysical modeling and basement-core data indicate that the California Great Valley is underlain by mainly mafic crust with a thickness of about 25 km (Cady 1975). A number of workers have suggested that the Valley floor is imbricated Jurassic ocean-type crust (Cady 1975, Suppe 1979). Different geometries for the Great Valley as the vestiges of inter- or intra-arc basin floor are modeled in Schweickert & Cowan (1975) and Saleeby (1981). An important unknown is what major structures may lie concealed beneath the Valley fill.

Nevadan structures in the western Sierra-Klamath belt consist of regional fault zones that run parallel to the trend of the belt. The main expression of the Nevadan orogeny in terms of rock deformation is a regional slaty cleavage best developed in the late Jurassic flysch. Such cleavage and related tight folds mark substantial crustal shortening, which is also indicated by known or inferred thrust components along a number of faults (Figure 3). The apparent regional pattern in Nevadan thrusting was westward-directed upper-plate movement of remnant-arc segments relative to the main locus of interarc-basin ophiolites. In the northern Sierra, eastward-directed overthrusting was widespread and perhaps led to the obduction of the Smartville block (McMath 1966, E. M. Moores, personal communication, 1980). High rank Nevadan tectonites appear to be spatially related to some westward-directed thrusts. Protoliths of the tectonites include arc-basement elements as well as arc-related rocks. Such tectonites are best displayed in the Klamath Mountains and in schists of the northern Coast Ranges (Coleman & Lanphere 1971, Klein 1977, Suppe & Foland 1978, Kays & Ferns 1980). Complex metamorphic paragenetic sequences between upper greenschist, amphibolite, and blueschist facies assemblages characterize these lower-plate tectonites. Radiometric and stratigraphic age constraints bracket the main tectonite-forming event to between 160 and 150 m.y. B.P., with local Cretaceous textural overprints and isotopic disturbances (Lanphere et al 1968, 1978, Coleman & Lanphere 1971, Suppe & Armstrong 1972, Suppe & Foland 1978, Mattinson 1981, Saleeby et al 1982, unpublished data). Northern Coast Range and western Klamath tectonites were in places structurally shuffled with Cretaceous rocks during subsequent Franciscan deformation.

Relations outlined above show a compressed sequence of events leading to the Nevadan orogeny. Primitive arc activity was concentrated between 170 and 160 m.y. B.P. At about 160 m.y. B.P., the arc ceased its main activity and interarc-basin ophiolites formed. Between 160 and 150 m.y. B.P., the entire assemblage was imbricated. Plate-tectonic explanations for the Nevadan orogeny have been viewed from several different perspectives. Moores (1970) and Schweickert & Cowan (1975) emphasize the oceanic character of the arc and the existence of major thrust structures, and thereby conclude the arc is an exotic fragment that collided with North America at Nevadan time. Behrman & Parkison (1978) and Davis et al (1978) emphasize the amalgamation history of the pre-Nevadan basement and thereby tie the arc to a position proximal to North America, making Nevadan deformation intraplate. An intermediate view is that an array of small plates consisting of interarc-basin-active arc and remnant-arc segments was created by 160 m.y. B.P. rifting, and along with the arc basement this microplate array was imbricated at Nevadan time (Saleeby 1981, 1982, Saleeby et al 1982). An additional aspect of this view is that significant transform links may have paralleled the margin during the 160 m.y. B.P. rifting event; such structures may have partly controlled the loci of Nevadan imbrication.

The parallels between the pre-Nevadan basement and the Stikinian super-terrane in conjunction with the roughly Nevadan-age accretion of Stikinia at California paleolatitudes point to the Cordillera-wide significance of the Nevadan event. Such significance is also suggested by the onset of major Cordilleran foreland folding and thrusting in late Jurassic time (Coney 1981). It appears that a large portion of the outer belt was accreted with Stikinia. Jurassic arc rocks and ophiolites and upper Paleozoic assemblages—both verbeekinid and McCloud-type schwagerinid-bearing—which are dispersed through northwest Nevada, eastern Oregon, and northern Washington (Danner 1977, Davis et al 1978, Brookes & Vallier 1978, Dickinson & Thayer 1978, Whetten et al 1980, Ketner & Wardlaw 1981) may be the vestiges of an expansive Stikinian basement regime with superimposed early Mesozoic arcs. Relationships in California suggest that rifting and interarc-basin-forming events, interspersed with amalgamation and accretion events, may account for the complex distribution of basement terranes and overlap arcs. In addition to the distinct 160 m.y. B.P. rifting event, there is evidence in the southern Sierra for at least local Takla-age interarc-basin formation and associated transform faulting (Saleeby 1982). This suggests that perhaps different ages of interarc basins formed within and adjacent to the Stikinian assemblage. In the next section, we briefly consider the petrotectonic and accretionary history of the outer belt in relation to these and other plate-tectonic phenomena.

PLATE TECTONIC ACCRETION OF THE CORDILLERAN OUTER BELT

Subduction of vast expanses of eastern Pacific ocean floor beneath western North America in late Mesozoic time led Hamilton (1969) to suggest that west coast eugeosynclinal belts were accreted island-arc and ocean-floor assemblages. Such a view is adopted here, but with some uniformitarian constraints. First, modern island-arc systems are located on the fringes of continents. There are no modern examples of active arcs migrating across major ocean basins, and thus accreted Cordilleran arcs are considered to have been parts of a fringing system. Second, far-traveled crustal blocks that are likely objects for tectonic accretion are oceanic plateaus and seamount chains. Plateaus may contain within them the remnants of distant extinct arc terranes. Such constraints and their effects on accretionary geometry are demonstrated in Figure 4*a*, a reversed-image map showing the plate-tectonic elements of the Melanesian Re-entrant (after Karig 1972, Hamilton 1979, Doutch et al 1981). The reversed image was made to facilitate direct comparison with Cordilleran geometry. Major crustal blocks of the Melanesian system that are likely to become accreted terranes are active and remnant-arc segments, rifted continental fragments such as the Lord Howe Rise, and oceanic plateaus and seamount chains. Also present is a complex family of small ocean basins, some still actively spreading. Such basins are likely sites of orogenic collapse and related tectonic accretion by collision. Young basin closure sutures are well demonstrated in eastern New Guinea and between the Ontong-Java plateau and the Solomon arc. The former probably represents closure of a small basin resulting in an arc-arc collision, the latter a larger-scale closure of part of the Pacific plate. Collision of the Ontong-Java plateau apparently caused a reversal in the polarity of the Solomon arc with the collisional suture lying along the former trench. Numerous collisions may be predicted in the near geologic future along the trace of the active trench.

The regional distribution of some analogous paleotectonic elements of the Cordillera is shown at the same scale on a Paleogene palinspastic base (Figure 4*b*). Preaccretion overlap arcs of Mesozoic age are differentiated from Paleozoic arc terranes and from early Mesozoic arc rocks superimposed over North American sial. Also shown are diagrammatic sutures that represent closed basins. The mark of such sutures are ophiolite belts, or highly deformed flysch or deep-basin sediments commonly lying in lower-thrust-plate structural positions. Closed basins represented by the Cache Creek terrane and the inboard edge of Wrangellia may have been of vast extent. Sutures marked by arc-related ophiolites and basinal clastic sediments are more likely the remains of smaller interarc or marginal basins. A close analogy is suggested between 160 m.y. B.P. ophiolites of California (Figure 3) and western Washington (Whetten et al 1978), and the family of active basins forming along the Melanesian system.

It is suggested that a complex fringing-arc system, like the Melanesian Re-entrant was collapsed into the Cordilleran outer belt in late Jurassic to mid-Cretaceous time. Two possible collapse mechanisms are considered. (1) The arrival of Wrangellia, like the Ontong-Java plateau, resulted in a major polarity change along the perimeter of the fringing system. Such a polarity change led to consumption of small internal basin floors, resulting in the accretion of basin-bounding fragments against the miogeoclinal buttress. (2) Absolute motion of North America accelerated in northwest and then west directions in late Middle and late Jurassic time in conjunction with a major opening phase in the Atlantic Ocean (Coney 1981, Gordan et al 1981). A similar age transform system has been suggested between the equatorial Atlantic and the southern Cordillera partly represented by the Mojave-Sonora megashear (Silver & Anderson 1974, personal communication, 1982). Perhaps the northwestward movement of North America resulted in a punctuated decoupling event between the Cordillera and the east Pacific realm, manifest by the megashear and a family of largely oblique interarc rift basins of 160 m.y. age. Subsequent rapid westward movement of North America at Nevadan time then led to basin closures perhaps concentrated along transform margins. The view that I tentatively

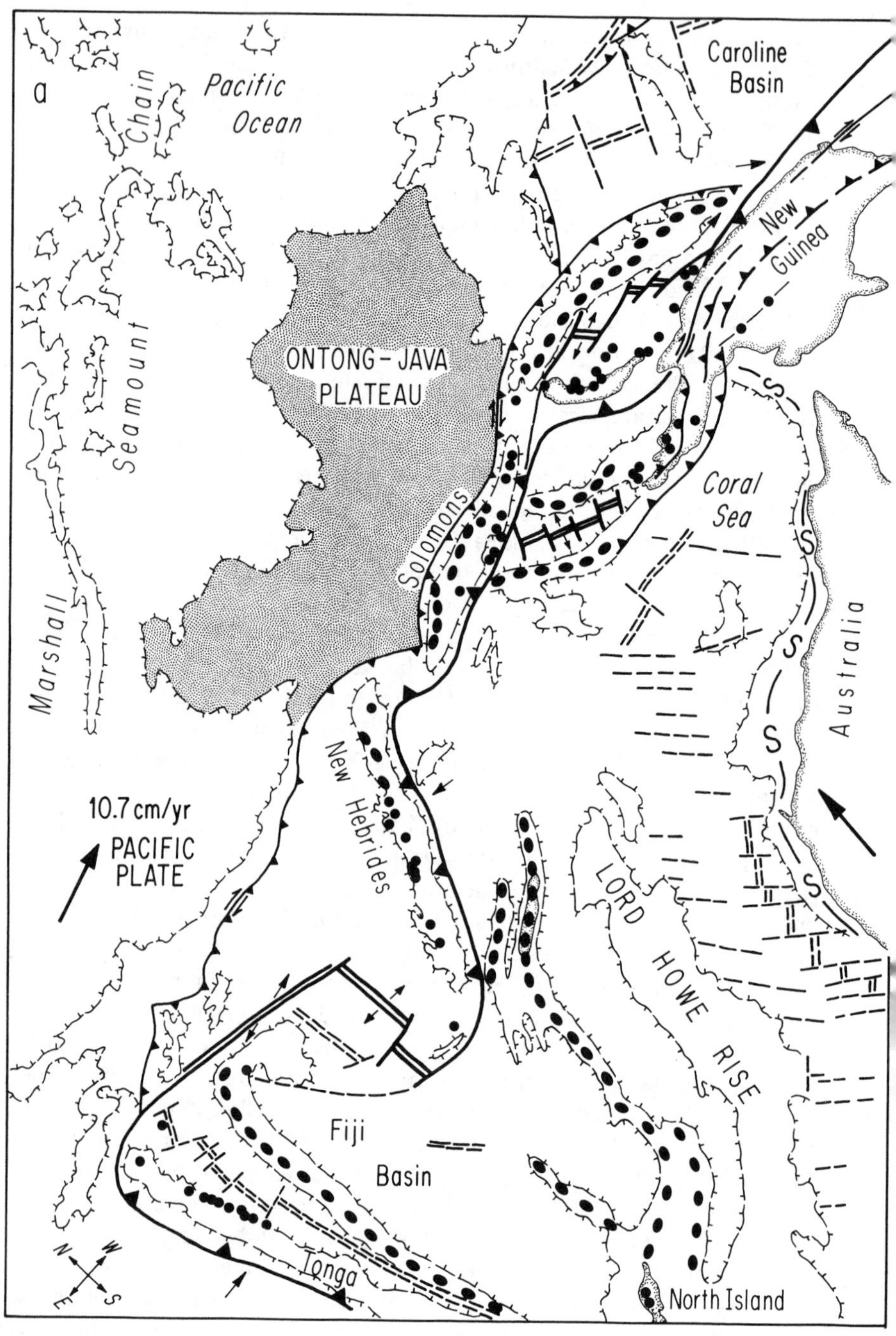
a
Pacific
Ocean
Caroline
Basin
Chain
Seamount
Marshall
ONTONG-JAVA
PLATEAU
New
Guinea
Solomons
Coral
Sea
Australia
10.7 cm/yr
PACIFIC
PLATE
New
Hebrides
LORD
HOWE
RISE
Fiji
Basin
Tonga
North Island
N
W
E
S

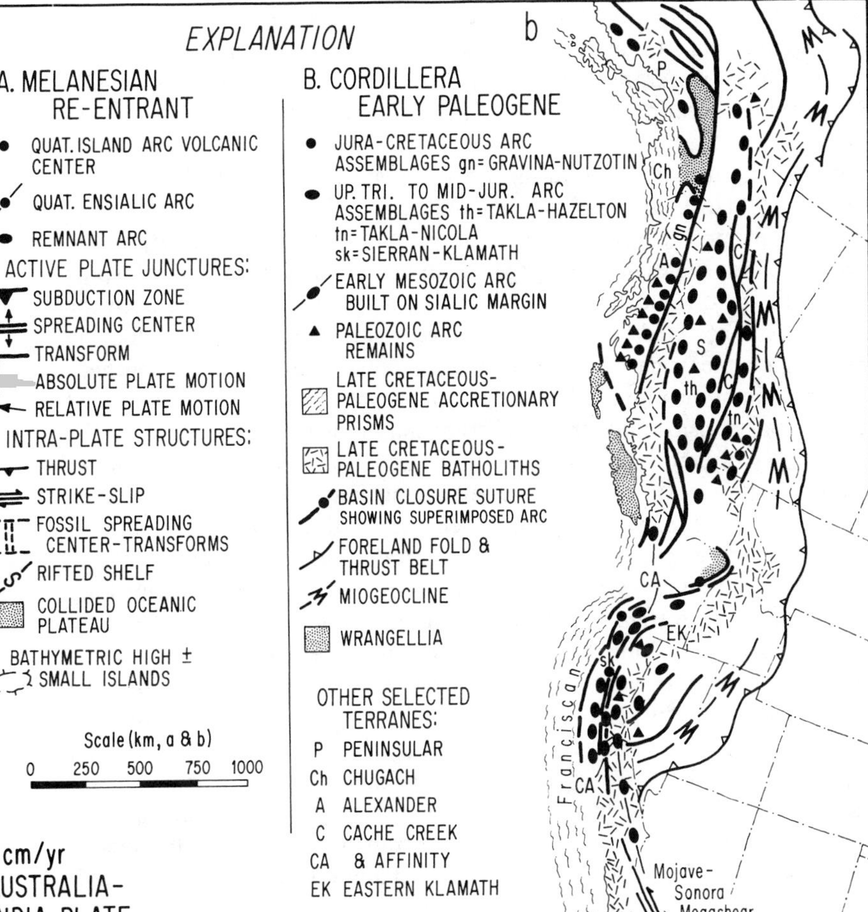

Figure 4 Comparative plate tectonic and paleotectonic maps for (*a*) the Melanesian Re-entrant, and (*b*) the Paleogene of the northern Cordillera. Melanesian Re-entrant shown in reverse image for comparison with Cordilleran geometry. Offset of major Cenozoic dextral faults and effects of Basin and Range extension in Cordilleran palinspastic base restored after Davis et al (1978), Dickinson & Thayer (1978), and Tempelman-Kluit (1979). References for arc assemblages and sutures given in text.

favor is that both mechanisms operated, perhaps in series considering Wrangellia's mid- to late Cretaceous accretion age. Critical questions are when and where Wrangellia made its initial impact, and did the Gravina-Nutzotin arc form on the fringes of the Cordillera following Wrangellia's arrival. Note the position of the Solomon arc, which is amalgamated onto the leading edge of the Ontong-Java plateau (Figure 4*a*). Perhaps the Gravina-Nutzotin arc formed in an analagous position.

A major problem in resolving important details in the outer-belt accretion history lies in the strong syn- and postaccretion dextral transport within the belt. Important results of this transport pattern have been the slicing-off of accreted terranes down to very narrow belts in the California region and at least part of the truncation pattern exhibited at the southern termination of the miogeocline. Furthermore, batholithic belts and related accretionary prisms that formed after the major terrane collisions show the effects of syngenetic northward transport. Examination of Figure 4*b* shows the localization of such batholithic belts along or adjacent to major sutures. Monger et al (1982) suggest that such localization is a direct result of terrane-suturing in British Columbia. Intrabatholithic structures of the Sierra Nevada suggest syngenetic dextral transport of around 200 km along preexisting suture trends that the batholith invaded. It is suggested that batholithic belts along with their wall rock terranes may undergo substantial lateral transport along preexisting suture trends during oblique subduction (Saleeby 1981, Beck et al 1983). Syn-batholithic oblique subduction is substantiated by several other lines of evidence. The Coast Range terrane of California contains Franciscan accretionary prism elements, mid-Jurassic ophiolite fragments, and slices of transported batholithic rocks, all showing northward transport in late Cretaceous as well as Cenozoic time (Alvarez et al 1980, Champion et al 1980, Luyendyk & Hornafius 1982). Prolonged late Mesozoic and younger dextral shear along the continental margin is also consistent with presumed relative motion patterns between oceanic plates of the east Pacific and the North American plate. Since early Jurassic time, the Pacific, Farallon, and Kula plates have undergone moderate to extreme northward movements relative to North America in conjunction with various underthrust components (Larsen & Chase 1972, Engebretson et al 1981). Thus a general pattern of oceanic plates moving northward and beneath North America, in conjunction with North America's movement pattern tied to the opening of the Atlantic, has assembled and laterally sheared-out the Cordilleran outer belt. A similar relative motion pattern is seen in the Melanesian system. With the arrival of the Ontong-Java plateau, perhaps in the near geologic future northeastern Australia will be armored with a Cordilleran-type orogen.

ACKNOWLEDGMENTS

Conversations and written communication with G. E. Gehrels, H. C. Berg, D. L. Jones and J. W. H. Monger were helpful during the writing of this review. Earlier interactions with these and numerous other earth scientists helped mold its content as well as my ideas. Field and laboratory support from the National Science Foundation, the United States Geological Survey, and the Alfred P. Sloan Foundation is gratefully acknowledged.

Literature Cited

Albers, J. P., Robertson, J. F. 1961. Geology and arc deposits of East Shasta copper-zinc district, Shasta County, California. *US Geol. Surv. Prof. Pap. 338*. 107 pp.

Alvarez, W., Kent, D. V., Silva, I. P., Schweickert, R. A., Larson, R. A. 1980. Franciscan complex limestone deposited at 17° south paleolatitude. *Geol. Soc. Am. Bull.* 91:476–84

Ando, C. D., Irwin, W. P., Jones, D. L., Saleeby, J. B. 1983. The ophiolitic North Fork Terrane in the Salmon River region, central Klamath Mountains, California. *Geol. Soc. Am. Bull.* In press

Armstrong, R. L. 1978. The pre-Cenozoic Phanerozoic time scale—A computer file of critical dates and consequences of new and in-progress decay constant revisions. *Am. Assoc. Pet. Geol. Stud. Geol.* 6:73–91

Bailey, E. H., Blake, M. C. 1974. Major chemical characteristics of Mesozoic Coast Range ophiolite in California. *US Geol. Surv. J. Res.* 2:637–56

Beck, M. E. Jr., Burmester, R. F., Engebretson, D. C., Schoonover, R. 1983. Northward translation of Mesozoic batholiths, western North America, paleomagnetic evidence and significance. *Geofis. Int.* In press

Behrman, P. G. 1978. Pre-Callovian rocks west of the Melones Fault Zone, central Sierra Nevada foothills. In *Mesozoic Paleogeography of the Western United States, Paleogeogr. Symp. 2*, ed. D. E. Howell, K. A. McDougall, pp. 337–48. Los Angeles, Calif: Pac. Sect. Soc. Econ. Paleontol. Mineral

Behrman, P. G., Parkison, G. A. 1978. Paleogeographic significance of the Callovian to Kimmeridgian strata, central Sierra Nevada foothills, California. In *Mesozoic Paleogeography of the Western United States, Paleogeogr. Symp. 2*, ed. D. E. Howell, K. A. McDougall, pp. 349–60. Los Angeles, Calif: Pac. Sect. Soc. Econ. Paleontol. Mineral.

Ben-Avraham, Z., Nur, A., Jones, D., Cox, A. 1981. Continental accretion: From oceanic plateaus to allochthonous terranes. *Science* 213:47–54

Berg, H. C. 1972. Geologic map of Annette Island, Alaska. *US Geol. Surv. Misc. Geol. Invest. Map I-684*

Berg, H. C. 1973. Geology of Gravina Island, Alaska. *US Geol. Surv. Bull. 1373*. 41 pp.

Berg, H. C., Jones, D. L., Richter, D. H. 1972. Gravina-Nutzotin Belt—Tectonic significance of an upper Mesozoic sedimentary and volcanic sequence in southern and southeastern Alaska. *US Geol. Surv. Prof. Pap. 800-D*, pp. D1–24

Blake, M. C. Jr., Jones, D. L. 1981. The Franciscan assemblage and related rocks in northern California: a reinterpretation. In *The Geotectonic Development of California, Rubey Vol. 1*, ed. W. G. Ernst, pp. 307–28. Englewood Cliffs, NJ: Prentice-Hall. 706 pp.

Bonatti, E., Honnorez, J. 1976. Sections through the earth's crust in the equatorial Atlantic. *J. Geophys. Res.* 81:4104–16

Bond, G. D., DeVay, J. C. 1980. Pre-Upper Devonian quartzose sandstone in the Shoo Fly Formation, northern California—Petrology, provenance, and implications for regional tectonics. *J. Geol.* 88:285–308

Brooks, H. C., Vallier, T. L. 1978. Mesozoic rocks and tectonic evolution of eastern Oregon and western Idaho. In *Mesozoic Paleogeography of the Western United States, Paleogeogr. Symp. 2*, ed. D. G. Howell, K. A. McDougall, pp. 133–145. Los Angeles, Calif: Pac. Sect. Soc. Econ. Paleontol. Mineral

Brown, L. D., Oliver, J. E., Kaufman, S., Brewer, J. A., Cook, F. A., Schilt, F. S., Albaugh, D. S., Long, G. H. 1981. Deep crustal structure: implications for continental evolution. *Am. Geophys. Union, Geodyn. Ser.* 5:38–52

Cady, J. W. 1975. Magnetic and gravity anomalies in the Great Valley and western

Sierra Nevada metamorphic belt, California. *Geol. Soc. Am. Spec. Pap. No. 168*. 56 pp.

Champion, D., Gromme, S., Howell, D. G. 1980. Paleomagnetism of the Cretaceous Pigeon Point Formation and the inferred northward displacement of 2500 km for the Salinian block, California. *EOS, Trans. Am. Geophys. Union* 46:948 (Abstr.)

Churkin, M. Jr., Eberlein, G. D. 1977. Ancient borderland terranes of the North American Cordillera: Correlation and microplate tectonics. *Geol. Soc. Am. Bull.* 88:769–86

Clark, L. D. 1964. Stratigraphy and structure of part of the western Sierra Nevada metamorphic belt, California. *US Geol. Surv. Prof. Pap. 410*. 70 pp.

Coleman, R. G., Lanphere, M. A. 1971. Distribution and age of high-grade blueschists, associated eclogites, and amphibolites from Oregon and California. *Geol. Soc. Am. Bull.* 82:2397–412

Coney, P. J. 1981. Accretionary tectonics in western North America. In *Relations of Tectonics to Ore Deposits in the Southern Cordillera*, ed. W. R. Dickinson, W. D. Payne, 14:23–38. *Ariz. Geol. Soc. Dig.*

Coney, P. J., Jones, D. L., Monger, J. W. H. 1980. Cordilleran suspect terranes. *Nature* 288:329–33

Cowan, D. S., Brandon, M. T. 1982. The geology and regional setting of the San Juan Islands. *Geol. Soc. Am. Abstr. with Programs* 14:157 (Abstr.)

D'Allura, J. A., Moores, E. M., Robinson, L. 1977. Paleozoic rocks of the northern Sierra Nevada: Their structural and paleogeographic significance. In *Paleozoic Paleogeography of the Western United States, Paleogeogr. Symp. 1*, ed. J. H. Stewart, C. H. Stevens, A. E. Fritsche, pp. 394–408. Los Angeles, Calif: Pac. Sect., Soc. Econ. Paleontol. Mineral

Danner, W. R. 1977. Paleozoic rocks of northwest Washington and adjacent parts of British Columbia. In *Paleozoic Paleogeography of the Western United States, Paleogeogr. Symp. 1*, ed. J. H. Stewart, C. H. Stevens, A. E. Fritsche, pp. 481–502. Los Angeles, Calif: Pac. Sect., Soc. Econ. Paleontol. Mineral

Davis, G. A., Monger, J. W. H., Burchfiel, B. C. 1978. Mesozoic construction of the Cordilleran "collage," central British Columbia to central California. In *Mesozoic Paleogeography of the Western United States, Paleogeogr. Symp. 2*, ed. D. E. Howell, K. A. McDougall, pp. 1–32. Los Angeles, Calif: Pac. Sect., Soc. Econ. Paleontol. Mineral

Dickinson, W. R. 1972. Evidence for plate-tectonic regimes in the rock record. *Am. J. Sci.* 272:551–76

Dickinson, W. R., Thayer, T. P. 1978. Paleogeographic and paleotectonic implications of Mesozoic stratigraphy and structure in the John Day inlier of central Oregon. In *Mesozoic Paleogeography of the Western United States, Paleogeogr. Symp. 2*, ed. D. G. Howell, K. A. McDougall, pp. 147–61. Los Angeles, Calif: Pac. Sect. Soc. Econ. Paleontol. Mineral

Douglass, R. C. 1967. Permian Tethyan fusulinids from California. *US Geol. Surv. Prof. Pap. 583-A*, pp. 7–43

Doutch, H. F., Packham, G. H., Rinehart, W. A., Simkin, T., Siebert, L., Moore, G. W., Golovchenko, X., Larson, R. L., Pitman, W. C. III, 1981. Plate tectonic map of the circum-Pacific region, southwest quadrant, scale 1:10,000,000. Tulsa, Okla: Am. Assoc. Pet. Geol.

Engebretson, D., Cox, A., Gordan, R. G. 1981. Relative motions between oceanic and continental plates in the northern Pacific basin since the Early Jurassic. *EOS, Trans. Am. Geophys. Union* 62:1034 (Abstr.)

Evarts, R. C. 1977. The geology and petrology of the Del Puerto ophiolite, Diablo Range, central California Coast Ranges. In *North American Ophiolites, Bull. 95*, ed. R. G. Coleman, W. P. Irwin, pp. 121–39. Portland: Oreg. Dept. Geol. Mineral Ind.

Garcia, M. O. 1979. Petrology of the Rogue and Galice Formation, Klamath Mountains, Oregon: Identification of a Jurassic island arc sequence. *J. Geol.* 86:29–41

Gordan, R. G., Cox, A., O'Hare, W. S. 1981. Paleomagnetic Euler poles for the absolute motion of North America during the Mesozoic and late Paleozoic. *EOS, Trans. Am. Geophys. Union* 62:853 (Abstr.)

Gromme, S., Hillhouse, J. W. 1981. Paleomagnetic evidence for northward movement of the Chugach terrane, southern and southeastern Alaska. In *The US Geological Survey in Alaska, Accomplishments During 1979*, ed. N. R. D. Albert, T. Hudson, pp. B70–72

Hamilton, W. 1969. Mesozoic California and the underflow of Pacific mantle. *Geol. Soc. Am. Bull.* 80:2409–30

Hamilton, W. 1979. Tectonics of the Indonesian region. *US Geol. Surv. Prof. Pap. 1078*. 345 pp.

Harper, G. D. 1980. *Structure and petrology of the Josephine ophiolite and overlying metasedimentary rocks, northwestern California*. PhD thesis. Univ. Calif., Berkeley. 259 pp.

Helwig, J. 1974. Eugeosynclinal basement and a collage concept of orogenic belts. *Soc. Econ. Paleontol. Mineral. Spec. Publ.* 19:359–76

Hillhouse, J. W. 1977. Paleomagnetism of the Triassic Nikolai greenstone, south-central Alaska. *Can. J. Earth Sci.* 14:2578–92

Hillhouse, J. W., Gromme, C. S. 1980. Paleomagnetism of the Triassic Hound Island Volcanics, Alexander Terrane, Southeastern Alaska. *J. Geophys. Res.* 85:2594–602

Hillhouse, J. W., Gromme, C. S. 1982. Paleomagnetism and Mesozoic tectonics of the Seven Devils volcanic arc in northeastern Oregon. *J. Geophys. Res.* 87:3777–94

Hopson, C. A., Mattinson, J. M., Pessagno, E. A. Jr. 1981. Coast range ophiolite, western California. In *The Geotectonic Development of California, Rubey Vol. 1*, ed. W. G. Ernst, pp. 307–28. Englewood Cliffs, NJ: Prentice-Hall. 706 pp.

Hotz, P. E., Lanphere, M. A., Swanson, D. A. 1977. Triassic blueschist from northern California and north-central Oregon. *Geology* 5:659–63

Irwin, W. P. 1972. Terranes of the western Paleozoic and Triassic belt in the southern Klamath Mountains, California. *US Geol. Surv. Prof. Pap. 800-C*, pp. 103–11

Irwin, W. D. 1977. Review of Paleozoic rocks of the Klamath Mountains. In *Paleozoic Paleogeography of the Western United States, Paleogeogr. Symp. 1*, ed. J. H. Stewart, C. H. Stevens, A. E. Fritsche, pp. 441–54. Los Angeles, Calif: Pac. Sect., Soc. Econ. Paleontol. Mineral

Irwin, W. D., Jones, D. L., Kaplan, T. A. 1978. Radiolarian from pre-Nevadan rocks of the Klamath Mountains, California and Oregon. In *Mesozoic Paleogeography of the Western United States, Paleogeogr. Symp. 2*, ed. D. E. Howell, K. A. McDougall, pp. 303–10. Los Angeles, Calif: Pac. Sect., Soc. Econ. Paleontol. Mineral

Jones, D. L., Silberling, N. J. 1979. Mesozoic stratigraphy: the key to tectonic analysis of southern and central Alaska. *US Geol. Surv. Open-File Rep. 79-1200*. 37 pp.

Jones, D. L., Silberling, N. J., Hillhouse, J. 1977. Wrangellia—A displaced terrane in northwestern North America. *Can. J. Earth Sci.* 14:2565–77

Jones, D. L., Howell, D. G., Coney, P. J., Monger, J. W. H. 1983. Recognition, character, and analysis of tectono-stratigraphic terranes in western North America. In *Oji Seminar Volume*, Cent. Acad. Publ. Jpn. In press

Karig, D. E. 1970. Ridges and basins of the Tonga-Kermadec island arc system. *J. Geophys. Res.* 75:239–54

Karig, D. E. 1971. Origin and development of marginal basins in the western Pacific. *J. Geophys. Res.* 76:2542–61

Karig, D. E. 1972. Remnant arcs. *Geol. Soc. Am. Bull.* 83:1057–68

Kays, M. A., Ferns, M. L. 1980. Geologic field trip guide through the north-central Klamath Mountains. *Oreg. Geol.* 42:23–35

Ketner, K. B., Wardlaw, B. R. 1981. Permian and Triassic rocks near Quinn River Crossing, Humboldt County, Nevada. *Geology* 9:123–26

Klein, C. W. 1977. Thrust plates of the north-central Klamath Mountains near Happy Camp, California. *Spec. Publ. Calif. Div. Mines Geol.* 28:23–26

Lanphere, M. A., Blake, M. C. Jr., Irwin, W. P. 1978. Early Cretaceous metamorphic age of the South Fork Mountain Schist in the northern Coast Ranges of California. *Am. J. Sci.* 278:798–815

Lanphere, M. A., Irwin, W. P., Hotz, P. E. 1968. Isotopic age of the Nevadan Orogeny and older plutonic and metamorphic events in the Klamath Mountains, California. *Geol. Soc. Am. Bull.* 79:1027–52

Larsen, R. L., Chase, C. G. 1972. Late Mesozoic evolution of the Pacific Ocean. *Geol. Soc. Am. Bull.* 83:3627–44

Luyendyk, B. P., Hornafius, J. S. 1982. Paleolatitude of the Point Sal ophiolite. *Geol. Soc. Am. Abstr. with Programs* 14:182 (Abstr.)

Mattinson, J. M. 1981. U-Pb systematics and geochronology of blueschists: Preliminary results. *EOS, Trans. Am. Geophys. Union* 62:1059 (Abstr.)

Mattinson, J. M., Hopson, C. A. 1972. Paleozoic ophiolitic complex in Washington and northern California. *Carnegie Inst. Washington Yearb.* 71:578–83

McMath, V. E. 1966. Geology of the Taylorsville area, northern Sierra Nevada, California. *Calif. Div. Mines Geol. Bull.* 190:173–83

Monger, J. W. H. 1975. Correlation of eugeosynclinal tectono-stratigraphic belts in the North American Cordillera. *Geosci. Can.* 2:4–9

Monger, J. W. H. 1977a. Upper Paleozoic rocks of the Western Canadian Cordillera and their bearing on Cordilleran evolution. *Can. J. Earth Sci.* 14:1832–59

Monger, J. W. H. 1977b. The Triassic Takla group in McConnell Creek map-area, north-central British Columbia. *Geol. Surv. Can. Pap. 76-29*. 45 pp.

Monger, J. W. H., Irving, E. 1980. Northward displacement of north-central British Columbia. *Nature* 285: 289–94

Monger, J. W. H., Ross, C. A. 1971. Distribution of fusulinaceans in the western Canadian Cordillera. *Can. J. Earth Sci.* 8: 259–78

Monger, J. W. H., Price, R. A., Tempelman-Kluit, D. J. 1982. Tectonic accretion and the origin of the two major metamorphic and plutonic welts in the Canadian Cordillera. *Geology* 10: 70–75

Moores, E. M. 1970. Ultramafic and orogeny, with models of the U.S. Cordillera and the Tethys. *Nature* 228: 837–42

Muller, J. E., Northcote, K. E., Carlisle, D. 1974. Geology and mineral deposits of Alert–Cape Scott map-area, Vancouver Island, British Columbia. *Geol. Surv. Can. Pap. 74-8*

Murray, C. G. 1972. Zoned ultramafic complexes of the Alaskan type: Feeder pipes of andesitic volcanoes. *Geol. Soc. Am. Mem.* 132: 313–35

Nestel, M. K. 1980. Permian fusulinacean provinces in the Pacific northwest are tectonic juxtapositions of ecologically distinct faunas. *Geol. Soc. Am. Abstr. with Programs* 12: 144 (Abstr.)

Nichols, K. M., Silberling, N. J. 1979. Early Triassic (Smithian) ammonites of paleoequatorial affinity from the Chulitna terrane, south-central Alaska. *US Geol. Surv. Prof. Pap. 1121-B*

Packer, D. R., Stone, D. B. 1974. Paleomagnetism of Jurassic rocks from southern Alaska and the tectonic implications. *Can. J. Earth Sci.* 11: 976–97

Potter, A. W., Hotz, P. E., Rohr, D. M. 1977. Stratigraphy and inferred tectonic framework of Lower Paleozoic rocks in the eastern Klamath Mountains, northern California. In *Paleozoic Paleogeography of the Western United States, Paleogeogr. Symp. 1*, ed. J. H. Stewart, C. H. Stevens, A. E. Fritsche, pp. 421–40. Los Angeles, Calif: Pac. Sect., Soc. Econ. Paleontol. Mineral

Preto, V. A. 1977. The Nicola Group: Mesozoic volcanism related to rifting in southern British Columbia. *Geol. Assoc. Can. Spec. Pap.* 16: 39–57

Saleeby, J. B. 1981. Ocean floor accretion and volcano-plutonic arc evolution of the Mesozoic Sierra Nevada. In *The Geotectonic Development of California, Rubey Vol. 1*, ed. W. G. Ernst, pp. 132–81. Englewood Cliffs, NJ: Prentice-Hall. 706 pp.

Saleeby, J. B. 1982. Polygenetic ophiolite belt of the California Sierra Nevada—geochronological and tectonostratigraphic development. *J. Geophys. Res.* 87: 1803–24

Saleeby, J. B., Gehrels, G. E., Eberling, G. D., Berg, H. C. 1983. Progress in Pb/U zircon studies of lower Paleozoic rocks of the southern Alexander terrane. In *The US Geological Survey in Alaska, Accomplishments During 1981*. In press

Saleeby, J. B., Harper, G. D., Snoke, A. W., Sharp, W. D. 1982. Time relations and structural-stratigraphic patterns in ophiolite accretion, west-central Klamath Mountains, California. *J. Geophys. Res.* 87: 3831–48

Schweickert, R. A., Cowan, D. S. 1975. Early Mesozoic tectonic evolution of the western Sierra Nevada, California. *Geol. Soc. Am. Bull.* 86: 1329–36

Schweickert, R. A., Snyder, W. 1981. Paleozoic plate tectonics of the Sierra Nevada and adjacent regions. In *The Geotectonic Evolution of California, Rubey Vol. 1*, ed. W. G. Ernst, pp. 182–202. Englewood Cliffs, NJ: Prentice-Hall. 706 pp.

Schweickert, R. A., Saleeby, J. B., Tobisch, O. T., Wright, W. H. III. 1977. Paleotectonic and paleogeographic significance of the Calaveras Complex, western Sierra Nevada, California. In *Paleozoic Paleogeography of the Western United States, Paleogeogr. Symp. 1*, ed. J. H. Stewart, C. H. Stevens, A. E. Fritsche, pp. 381–94. Los Angeles, Calif: Pac. Sect., Soc. Econ. Paleontol. Mineral

Sharp, W. D. 1980. Ophiolite accretion in the northern Sierra. *EOS, Trans. Am. Geophys. Union* 61: 1122 (Abstr.)

Sharp, W. D., Evarts, R. C. 1982. New constraints on the environment of formation of the Coast Range ophiolite at Del Puerto Canyon, California. *Geol. Soc. Am. Abstr. with Programs* 14: 233 (Abstr.)

Sharp, W. D., Saleeby, J. B. 1979. The Calaveras Formation and syntectonic mid-Jurassic plutons between the Stanislaus and Tuolumne rivers, California. *Geol. Soc. Am. Abstr. with Programs* 11: 127 (Abstr.)

Sharp, W. D., Wright, J. E. 1981. Jurassic metavolcanic sequences in the western Sierra-Klamath provinces: Character and tectonic significance. *Geol. Soc. Am. Abstr. with Programs* 13: 105 (Abstr.)

Silver, L. T., Anderson, T. H. 1974. Possible left-lateral Early to Middle Mesozoic disruption of the southwestern North American craton margin. *Geol. Soc. Am. Abstr. with Programs* 6: 955 (Abstr.)

Skinner, J. W., Wilde, G. L. 1965. Permian biostratigraphy and fusulinid faunas of the Shasta Lake area, northern California.

Kans. Univ., Harold N. Fiske Mem. Pap. Paleont. Contr., Protozoa, Art. 6. 90 pp.

Snoke, A. W. 1977. A thrust plate of ophiolitic rocks in the Preston Peak area, Klamath Mountains, California. *Geol. Soc. Am. Bull.* 88:1641–59

Snoke, A. W., Sharp, W. D., Wright, J. E., Saleeby, J. B. 1982. Significance of mid-Mesozoic peridotitic to dioritic intrusive complexes, Klamath Mountains–Western Sierra Nevada, California. *Geology* 10: 160–66

Souther, J. G. 1977. Volcanism and tectonic environments in the Canadian Cordillera —a second look. *Geol. Assoc. Can. Spec. Pap.* 16:3–24

Speed, R. C. 1979. Collided Paleozoic microplate in the western United States. *J. Geol.* 87:279–92

Springer, R. K. 1980. Geology of the Pine Hill intrusive complex, a layered gabbroic body in the western Sierra Nevada foothills, California, Summary, 1. *Geol. Soc. Am. Bull.* 91:381–85

Stewart, J. H. 1972. Initial deposits in the Cordilleran geosyncline: Evidence of a late Precambrian (<850 m.y.) continental separation. *Geol. Soc. Am. Bull.* 83:1345–60

Stone, D. B., Packer, D. R. 1977. Tectonic implications of Alaska Peninsula paleomagnetic data. *Tectonophysics* 37:183–201

Suppe, J. 1979. Structural interpretation of the southern part of the northern Coast Ranges and Sacramento Valley, California: Summary. *Geol. Soc. Am. Bull.* 90: 327–30

Suppe, J., Armstrong, R. L. 1972. Potassium-argon dating of Franciscan metamorphic rocks. *Am. J. Sci.* 272:217–33

Suppe, J., Foland, K. A. 1978. The Goat Mountain schists and Pacific Ridge complex: A redeformed but still intact late Mesozoic schupper complex. In *Mesozoic Paleogeography of the Western United States, Paleogeogr. Symp. 2*, ed. D. G. Howell, K. A. McDougall, pp. 431–51. Los Angeles, Calif: Pac. Sect., Soc. Econ. Paleontol. Mineral

Tempelman-Kluit, D. J. 1979. Transported cataclasite, ophiolite and granodiorite in Yukon: Evidence of arc-continent collision. *Geol. Surv. Can. Pap. 79-14.* 27 pp.

Tipper, H. W. 1978. Northeastern part of Quesnel (93B) map-area, British Columbia. *Geol. Surv. Can. Paper 78-1A*, pp. 67–68

Travers, W. B. 1978. Overturned Nicola and Ashcoft strata and their relation to the Cache Creek Group, southwestern Intermontane Belt, British Columbia. *Can. J. Earth Sci.* 15:99–116

Van der Voo, R., Jones, M., Gromme, C. S., Eberlein, G. D., Churkin, M. Jr. 1980. Paleozoic paleomagnetism and the northward drift of the Alexander Terrane, SE Alaska. *J. Geophys. Res.* 85:5281–96

Varga, R. J., Moores, E. M. 1981. Age, origin, and significance of an unconformity that predates island-arc volcanism in the northern Sierra Nevada. *Geology* 9:512–18

Wells, F. G., Walker, G. W. 1953. Geology of the Galice quadrangle, Oregon. *Geol. Quadrangle Map GO-25, scale 1–62,500.* Reston, Va: US Geol. Surv.

Whetten, J. T., Jones, D. L., Cowan, D. S., Zartman, R. E. 1978. Ages of Mesozoic terranes in the San Juan Islands, Washington. In *Mesozoic Paleogeography of the Western United States, Paleogeogr. Symp. 2*, ed. D. G. Howell, K. A. McDougall, pp. 117–32. Los Angeles, Calif: Pac. Sect., Soc. Econ. Paleontol. Mineral

Whetten, J. T., Zartman, R. E., Blakely, R. J., Jones, D. L. 1980. Allocthonous Jurassic ophiolite in northwest Washington. *Geol. Soc. Am. Bull.* 91:359–68

Wright, J. E. 1982. Permo-Triassic accretionary subduction complex, southwestern Klamath Mountains, northern California. *J. Geophys. Res.* 87:3805–18

Wright, J. E., Sharp, W. D. 1982. Mafic-ultramafic intrusive complexes of the Klamath-Sierra region, California: remnants of a middle Jurassic arc complex. *Geol. Soc. Am. Abstr. with Programs* 74:245–46 (Abstr.)

Xenophontos, C., Bond, G. C. 1978. Petrology, sedimentation, and paleogeography of the Smartville terrane (Jurassic)—Bearing on the genesis of the Smartville ophiolite. In *Mesozoic Paleogeography of the Western United States, Paleogeogr. Symp. 2*, ed. D. E. Howell, K. A. McDougall, pp. 291–302. Los Angeles, Calif: Pac. Sect., Soc. Econ. Paleontol. Mineral

Yancy, T. E. 1979. Permian positions of the northern hemisphere continents as determined from marine biotic provinces. In *Historical Geogeography, Plate Tectonics and The Changing Environment*, ed. A. J. Boucot, pp. 239–47. Corvallis: Oreg. State Univ. Press

Yorath, C. J., Chase, R. L. 1981. Tectonic history of the Queen Charlotte Islands and adjacent areas—a model. *Can. J. Earth Sci.* 18:1717–39

Ann. Rev. Earth Planet. Sci. 1983. 11: 75–97

THE STRUCTURE OF SILICATE MELTS

Bjørn O. Mysen
Geophysical Laboratory, Carnegie Institution of Washington, Washington, D.C. 20008

INTRODUCTION

The structure of silicate melts provides a basis for understanding the relations between the structure and the physical, chemical, and thermal properties of the melts. These data are required to determine the conditions of formation and evolution of magma in the Earth and terrestrial planets. In igneous processes, phase equilibria in melt-mineral-vapor systems, diffusion in melts, and thermodynamic, electrical, and rheological properties of magma systems are of particular importance. The structures of silicate melts relevant to igneous processes are therefore a subject of intensive research. As structural data become available and are correlated with the experimental determination of appropriate properties, a structural basis for the prediction of important properties of magmatic liquids and liquid-crystal systems may be developed.

STRUCTURE

The metal cations in silicate melts may be divided into network formers and network modifiers (Bottinga & Weill 1972). The network-forming cations occur in tetrahedral coordination in various polymers or units in the melt, whereas the network modifiers connect such units together. To date, direct structural studies have emphasized the network structure.

Melt Versus Glass

Most experimental data have been obtained on the quenched equivalents of melts (glass) on the assumption that the principal features of the structure of a melt are retained as the sample is quenched to a glass. Sweet & White (1969), with the aid of infrared spectroscopic data on compositions in the

0084–6597/83/0515–0075$02.00

system Na_2O-SiO_2, substantiated the assumption of structural similarity between silicate melts and their glasses. This conclusion has subsequently been supported by Raman spectroscopic data on glasses and melts in the same system and in the systems Na_2O-Al_2O_3-SiO_2 and GeO_2 (Seifert et al 1981). Taylor et al (1980) showed that X-ray radial distribution functions of glass and supercooled liquid of $NaAlSi_3O_8$ composition were similar, and suggested that melts and their glasses are structurally similar. It is, however, known that a small but significant heat effect is associated with the glass transition (Navrotsky et al 1980). This heat effect (of the order of 1 kcal mole^{-1}) is insufficient to account for reconstructive transformation as a silicate melt is transformed to a glass. These thermochemical data do indicate, however, that there are differences between melts and their glasses, although the differences (as yet unidentified) are small. In the light of such information, research on melt structures at 1 atm pressure has concentrated on silicate glass with the explicit or implicit assumption that the glass structure is a suitable approximation to that of the melt at the temperature from which the sample was quenched (quenching rates typically of the order of 500°C s^{-1}).

The limited structural data obtained on melts quenched at pressures corresponding to those of the upper mantle are based on the assumption that temperature-quenching at high pressure (10–40 kbar) results in a glass that retains the structural features induced by increased pressure. Structural changes of silicate liquids resulting from high pressure involve reduction in volume. Temperature-quenching under isobaric conditions, therefore, is not likely to result in a transition into a structural state of larger volume.

Structure of SiO_2 Glass

Compositionally, SiO_2 is the simplest chemical compound relevant to the structure of magmatic liquids. The silica content of most magmas exceeds 40 wt %. It is clear, therefore, that the structure of molten and glassy SiO_2 is important for the evaluation of the structure of magmatic liquids.

Structural information has been obtained with the aid of molecular-dynamics calculations (Soules 1979, Mitra 1982, Bell & Dean 1972, Gaskell & Tarrant 1980, Konnert & Karle 1973). The results of these calculations indicate a three-dimensionally interconnected structure, sometimes referred to as a "stuffed tridymite" structure (Navrotsky et al 1980), of vitreous SiO_2. In this structure, nearly all oxygen atoms act as bridges between two silicon atoms that are each in tetrahedral coordination with oxygen. Oxygen in such a structural position is referred to as "bridging oxygen" (BO).

A recent refinement by Gaskell & Tarrant (1980) indicates an asymmetric distribution of Si-O-Si angles (Figure 1). This asymmetry may result from at

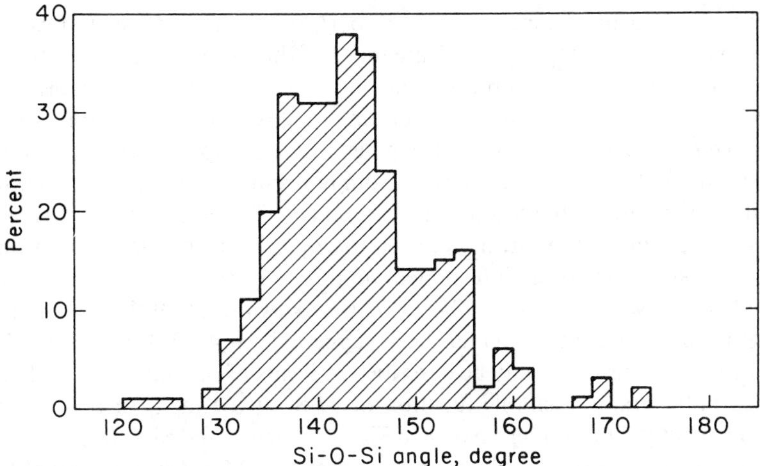

Figure 1 Calculated Si-O-Si angle distribution in vitreous SiO_2 (data from Gaskell & Tarrant 1980).

least two maxima in the distribution of Si-O-Si angles in vitreous SiO_2, as also suggested by Mikkelsen & Galeener (1980). Results from transmission electron microscopy indicate that two or more different structures may coexist in vitreous SiO_2 (Gaskell & Mistry 1979, Bando & Ishizuka 1979).

Data on local structural features in vitreous silica have been acquired with vibrational spectroscopy. Calculation of such spectra on the basis of a single structure (Bell & Dean 1972) is not fully consistent with the experimental data. These measured spectra can, however, be interpreted on the basis of two three-dimensional structures that differ in Si-O bond length and Si-O-Si angle (Mammone et al 1981, Seifert et al 1982). A 5–10° difference in angle has been calculated (Seifert et al 1982), compared with 8° as suggested by Vukcevich (1972) on the basis of physical properties of vitreous SiO_2.

Binary Metal Oxide-Silica Melts

The addition of metal oxide to silica melt results in the formation of nonbridging oxygen atoms (NBO). Such oxygens are bonded to both network-modifying metal cations and tetrahedrally coordinated cations in the silicate network. Metal cations (*M* cations) act, therefore, as linkage between various anionic structural units with tetrahedrally coordinated cations that occur in the melt. The degree of polymerization of a silicate melt is often expressed as the ratio of nonbridging oxygens per tetrahedrally coordinated cations (NBO/T). Thus, a three-dimensionally interconnected structure has $NBO/T = 0$, a sheet has $NBO/T = 1$, a chain or a single ring has $NBO/T = 2$, and so on. The types of anionic structural units and their relative proportions depend on the ratio M/Si and on the type of M cation.

It has been suggested that silicate melts of this type may be described in terms of relatively few simple anionic units (Bockris et al 1955, MacKenzie 1960), a suggestion originally made on the basis of systematic relations between coefficients of thermal expansion, activation energy of viscous flow, and M/Si of the melts. Results from molecular-dynamics calculations of melt structure on the compositional-join Na_2O-SiO_2 (Soules 1979) led to similar conclusions. Vibrational spectroscopic data on silicate melts and glasses have also been interpreted in this way (Brawer 1975, Verweij 1979a,b, Brawer & White 1975, 1977, Furukawa et al 1981, Mysen et al 1980). An alternative model has been suggested (Hess 1971, Masson 1977) in which silicate melts may be described in terms of continuously evolving silica polymers as a simple function of M/Si of the melt. In this model, separate SiO_4^{4-} monomers evolve to $Si_2O_7^{6-}$ dimers, and $Si_3O_{10}^{8-}$ trimers to infinite chains, branched chains, and so forth, as M/Si (or NBO/Si) of the melt decreases. Chromatography of trimethylsilyl derivatives of silicate glasses, in principle, supports this model (see Hess 1971 for summary). Such a simple polymer model is, however, difficult to reconcile with observed physical properties of silicate melts. For example, a melt property such as the activation energy of viscous flow ($E\eta$) to a first approximation reflects the strength of Si-O and Al-O bonds that must be broken in a melt in order to form the flow unit during viscous flow (Bockris & Reddy 1970, Taylor & Rindone 1970). Discontinuities in $E\eta$ versus bulk composition of binary metal oxide-silica melts (Figure 2) would therefore suggest distinct, discontinuous changes in the melt structure as a function of bulk composition. These rheological data cannot be readily explained with models in which the structure of melts is assumed to evolve continuously as a function of M/Si (Mysen et al 1980, Bottinga et al 1981). It should also be noted that the primary experimental supporting evidence for the latter models, chromatography of trimethylsilyl derivatives of glass, recently has been found unreliable (Kuroda & Kato 1979). The main problem is the silylation process itself. Kuroda & Kato (1979) found that the types and proportions of the trimethylsilyl derivatives depend strongly on the silylating agent used. No silylating agent was discovered that could reproduce the structure of known minerals. Comparative structural data from vibrational spectroscopy and chromatography (Verweij & Konijnendijk 1976, Smart & Glasser 1978), as well as chemical mass-balance calculations, consistently indicate that the structural analysis from chromatographic data results in overestimation of the degree of polymerization of the melt.

Vibrational spectroscopy is probably the most commonly employed method for studying the structure of silicate melts and glasses. Although rigorous theoretical treatment of the spectra of a disordered solid such as a melt or a glass may not be conducted owing to lack of long-range order, it has been shown (Etchepare 1972, Brawer 1975) that the vibration displacements of units of disordered silicate melt frequently resemble those found in relevant crystal structures. As a result of Raman (and to a lesser extent infrared) studies of a range of metal oxide-silica melts (metal cations Pb, Mg, Ca, Ba, K, Na, and Li) with bulk melt ratio of nonbridging oxygens per silicon [NBO/Si = (2 × O − 4 × Si)/Si] between 4 (orthosilicate) and 0 (tectosilicate), it has been concluded (Brawer & White 1975, 1977, Furukawa et al 1981, Verweij 1979a,b, Mysen et al 1980, 1982a) that the anionic structure of metal oxide-silica melts may be described in terms of

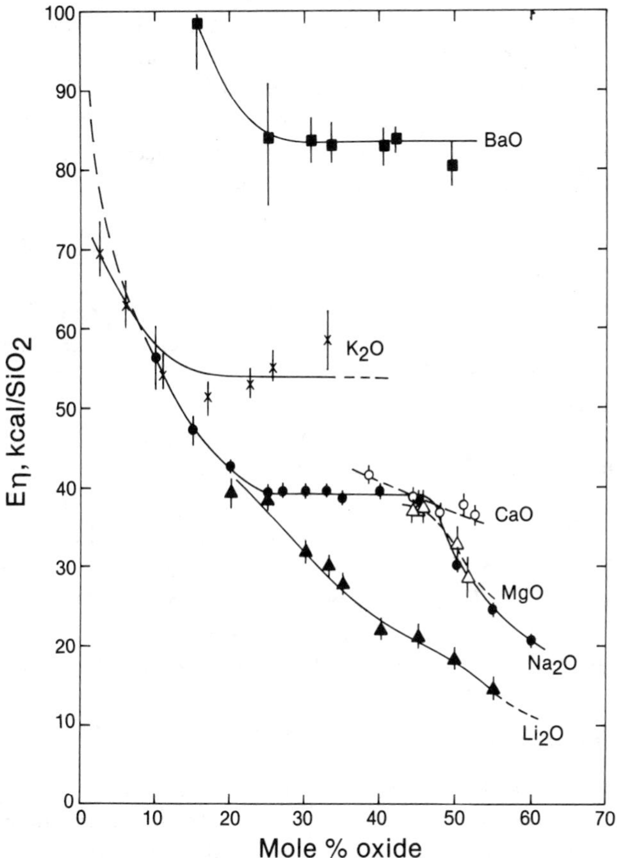

Figure 2 Activation energy of viscous flow ($E\eta$) of melts on binary metal oxide-silica joins (data compiled by Mysen et al 1982a).

individual anionic structural units such as monomers (NBO/Si = 4), dimers (NBO/Si = 3), chains (NBO/Si = 2), sheets (NBO/Si = 1), and three-dimensional network units (NBO/Si = 0). The identification of such structural units in metal oxide-silica melts has been based partly on coordinate analysis of silicate anions (Brawer 1975, Furukawa et al 1981) and partly on comparative studies of spectra of melts and crystals. Several of these units coexist in metal oxide-silica melts. The relative abundance of these units varies systematically with M/Si of the melt and with the type of M-cation (Figure 3). On any metal oxide-silica join, structural units with an average NBO/Si = 2 and NBO/Si = 1 show the maximum abundance in melts with bulk compositions near metasilicate and disilicate, respectively. In no melt is only one type of anionic unit observed (Furukawa et al 1981, Mysen et al 1982a). Furthermore, it can be seen from the data in Figure 3 that the abundance ratio of anionic units with average NBO/Si = 2 and 1 ($Si_2O_6^{4-}/Si_2O_5^{2-}$) decreases with increasing ionization potential of the metal cation. This decrease correlates with the increasing chemical potential of nonbridging oxygen, as defined and discussed by Hess (1980), as the ionization potential of the metal cation increases. Free oxygen, O^{2-} (bonded only to nontetrahedrally coordinated metal cations), occurs in silicate melt with bulk NBO/Si > 3 (Figure 3).

Although there is little direct information on the role of network-modifying metal cations in such melts, it may be inferred from variations in abundance of the structural units as a function of the type of M cation at constant bulk melt NBO/Si that small, highly charged cations such as Mg^{2+} (and probably Fe^{2+}) show a distinctive preference for anionic units with large NBO/Si. The stoichiometry of a magnesium disilicate melt ($MgSi_2O_5$), for example, is balanced by three-dimensional units and chains, dimers, and monomers. The presence of large, more electropositive cations such as Na^+, Ba^{2+}, or K^+ instead of Mg^{2+} or Fe^{2+} tends to enhance the stability of structural units such as sheets instead of chains and three-dimensional network units in a disilicate melt (Mysen et al 1982a). The interpretation of available spectroscopic data on silicate melts and glasses indicates, therefore, that many of the principles that govern crystal chemistry apply to silicate melts as well [see Dent Glasser (1979) and Liebau (1981) for recent reviews of pertinent data].

Aluminum in Silicate Melts

Aluminum is tetrahedrally coordinated in silicate melts at 1 atm pressure, provided that cations such as alkali metals or alkaline earths are present in sufficient amounts to charge-compensate Al^{3+} in order to create an effective 4+ electrical charge for the tetrahedrally coordinated cation. Riebling (1964, 1966) suggested that Al^{3+} would become a network

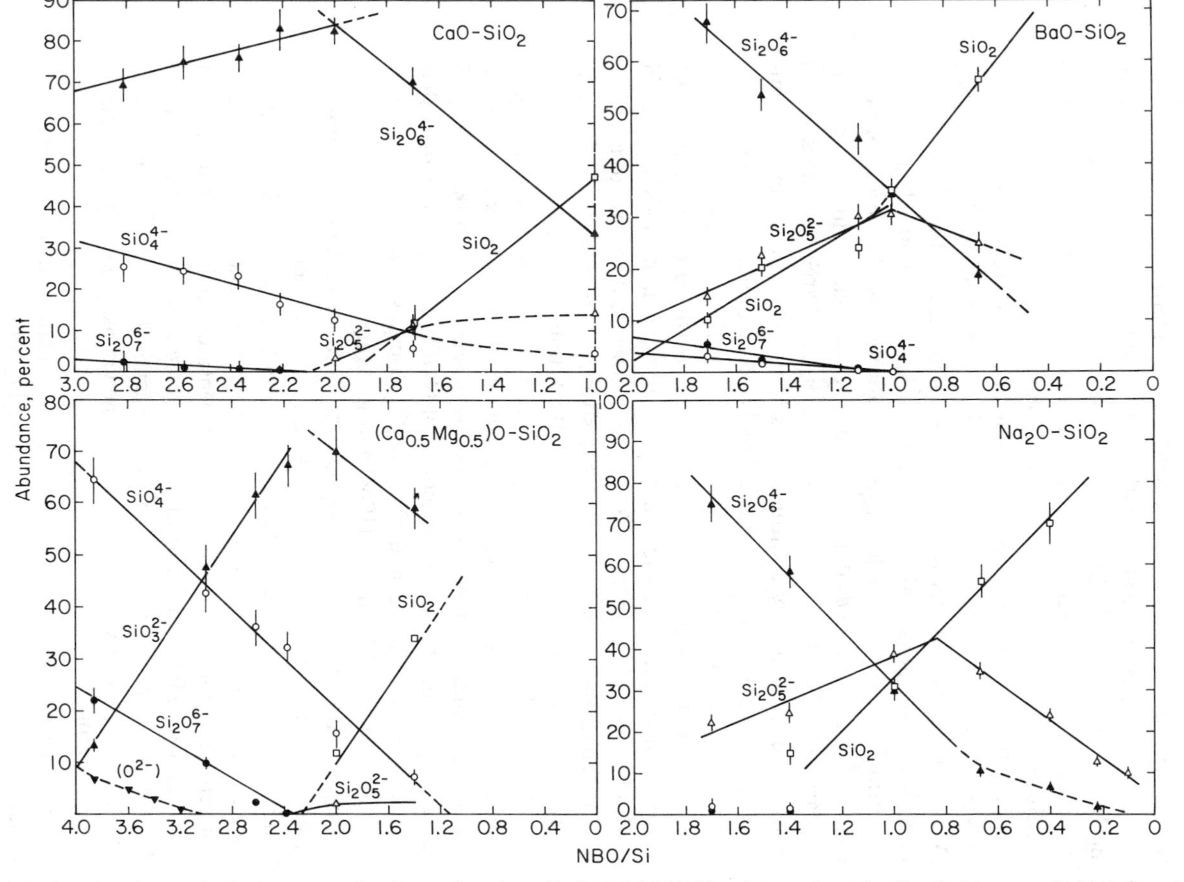

Figure 3 Relative abundance of anionic structural units as a function of bulk melt NBO/Si and type of metal cation for binary melts (data from Mysen et al 1982a).

modifier without such charge-compensation. Lacey (1963), on the other hand, suggested that under these circumstances Al-triclusters would be formed. In such triclusters, Al^{3+} would remain in tetrahedral coordination without charge-compensation with other metal cations. MacDowell & Beall (1969) suggested that such a structure would result in greater densification with decreasing M/Al than observed. Bottinga et al (1982) concluded, however, that whether Al^{3+} occurred in triclusters or as network modifiers, the effect on density would be approximately the same. The only available structural information is in the system Na_2O-Al_2O_3-SiO_2 (Mysen et al 1980). In this system, there is an increase in the abundance of nonbridging oxygens with decreasing Na/Al at constant silica content. Nonbridging oxygens could have been formed as the result of Al^{3+} changing from tetrahedral to octahedral coordination in the melt. It appears, therefore, that Al^{3+} requires electrical charge-compensation in order to be tetrahedrally coordinated, as is also the case for most aluminosilicate minerals. For magmatic liquids this requirement is met in all but peraluminous compositions [$Al^{3+} > Na^{+} + K^{+} + 0.5(Ca^{2+} + Mg^{2+})$]. Generally, both alkali metals and alkaline earths are involved in the charge-compensation in most natural magmatic liquids (Mysen et al 1981a) because $Al^{3+} > (Na^{+} + K^{+})$. It is necessary, therefore, to evaluate how different metal cations affect the structural role of Al^{3+} in tetrahedral coordination. Navrotsky et al (1980) found that there is a small ($\sim$1 kcal mole^{-1}) negative heat of mixing of melts on the join $NaAlSi_3O_8$-$CaAl_2Si_2O_8$. Cranmer & Uhlmann (1981) observed that the viscosity and activation energy of viscous flow of melts on the same join also change as discontinuous functions of Na/Ca, a result indicating that the melts of $NaAlSi_3O_8$ and $CaAl_2Si_2O_8$ composition are structurally different [see also Rossin et al (1964) and Riebling (1966)].

Taylor & Brown (1979a,b) reported data on X-ray radial distribution for melts of end-member feldspar compositions and for quenched melts on the join SiO_2-$NaAlSiO_4$. Their data were consistent with six-membered, three-dimensionally interconnected rings in melts with Na^{+} or K^{+} charge-compensation for Al^{3+}. For Ca^{2+}-compensated melts, four-membered rings of the type found in anorthite crystals were suggested. Such structural differences are consistent with the thermochemical data of Navrotsky et al (1980) and Henry et al (1982), as well as the viscosity data by Cranmer & Uhlmann (1981).

Seifert et al (1982) carried the problem further in a study of melt structures at 1 atm on the joins $MgAl_2O_4$-SiO_2, $CaAl_2O_4$-SiO_2, and $NaAlO_2$-SiO_2. On the basis of Raman spectra it was suggested that in melts with Mg^{2+} or Ca^{2+} for charge-compensation of Al^{3+}, most or all of the

Al^{3+} of the melt occurs in three-dimensional structural units similar to those suggested by Taylor & Brown (1979a,b). With bulk melt Al/Si $\neq 1$, three-dimensional structural units that resemble the structure of vitreous SiO_2 (for bulk melt Si/Al > 1) or AlO_2^- (for bulk melt Si/Al < 1) occur in the Ca^{2+}- and Mg^{2+}-aluminosilicate melts. Alkali metal for charge-compensation of Al^{3+} results in nearly random substitution of Al^{3+} for Si^{4+} in structures similar to vitreous SiO_2. The average T-O-T angles of the individual units decrease as a continuous function of Al/(Al + Si) of the melt.

The Structural Role of Aluminum in Melts with Nonbridging Oxygen

Natural magmatic liquids contain a significant proportion of nonbridging oxygens. [NBO/T is generally between 1.0 and 0.05 according to Mysen et al (1981a).] Inasmuch as such melts consist of coexisting anionic units with different NBO/T, it is necessary to determine the structural position of Al^{3+} in these coexisting units. Only then may the importance of Al^{3+} in magmatic liquids be properly evaluated.

Mysen et al (1981a), on the basis of Raman spectroscopic data, found that Al^{3+} in depolymerized melts (bulk melt NBO/T > 0) would show a preference for the unit in the melt with the smallest value of NBO/T. On the basis of bulk-melt-dependent Al/(Al + Si) frequency shifts of (Si, Al)-coupled antisymmetric vibrations of nonbridging oxygen-to-tetrahedral cation bonds, Mysen et al (1981a) estimated the distribution coefficients of Al^{3+} between the coexisting structural units in the melts (Table 1).

Table 1 Estimated aluminum partition coefficients between melt units at 1 atm, 1450°C[a]

	Al_2O_3		
	0–10 mole %	10–20 mole %	20–40 mole %
		$Na_2Si_2O_5$-$NaAlO_2$	
3D/sheet	—	2	2.1
3D/chain	—	—	3.0
Sheet/chain	—	—	1.4
		$Na_2Si_2O_5$-$CaAl_2O_4$	
3D/sheet	—	2.4	2.5
3D/chain	—	—	4.2
Sheet/chain	—	—	2.5

[a] Data from Mysen et al (1981a).

Other Tetrahedrally Coordinated Cations

Ferric iron, titanium, and phosphorus occur in magmatic liquids in abundances up to several weight percent. Despite their relatively low abundance, the presence of these elements profoundly affects the chemical properties of silicate melts (Kushiro 1975, Hess 1977, 1980). For example, the limits of liquid immiscibility are extended as Ti^{4+} or P^{5+} is added (Visser & Van Groos 1979, Ryerson & Hess 1980).

Vibrational spectroscopic data on glass in the system SiO_2-P_2O_5 indicate that P^{5+} is tetrahedrally coordinated and apparently forms sheet-like structural units in which one oxygen in the PO_4 tetrahedron is double-bonded to phosphorus (Wong 1976, Galeener & Mikkelsen 1979). In melts with nonbridging oxygens, Al^{3+}, or both, the double bond is no longer in evidence. Only *M*-O-P, Al-O-P, and P-O-P bonds are identified (Wong 1976, Nelson & Exharos 1979, Mysen et al 1981b). These observations are consistent with the existence of aluminum phosphate and metal phosphate complexes in the melts. The formation of *M*-O-P and Al-O-P bonds by adding P_2O_5 to compositions with bulk melt $NBO/T > 0$ results in a decrease in NBO/T (i.e. the melts becomes more polymerized).

Titanium in silicate melts probably occurs at least partly in tetrahedral coordination (Furukawa & White 1979, Tobin & Baak 1968). In fact, the structural role of Ti^{4+} may be similar to that of Si^{4+}. It appears, however, that Ti^{4+} tends to form separate titanate clusters rather than substituting for Si^{4+} (Mysen et al 1982a).

Iron is probably the only major element in magmatic rocks that can occur in several oxidation states under physical conditions relevant to igneous processes in the Earth. The structural roles of ferric and ferrous iron in silicate melts differ significantly because Fe^{3+} can be both a network former and a network modifier, whereas available data indicate that ferrous iron is generally a network modifier (Mao et al 1973, Brown et al 1978, Levy et al 1976).

Inasmuch as principles similar to those governing crystal chemistry also play an important role in controlling melt structure (Mysen et al 1980, Liebau 1981), it is likely that Fe^{3+} will be a network former in the presence of sufficient metal cations for charge-compensation (e.g. alkali metals or alkaline earths). This suggestion is supported by data on $NaFe^{3+}Si_2O_6$ melt (Brown et al 1978, Seifert et al 1979) and melts in the systems Na_2O-SiO_2-Fe-O (Virgo et al 1981, 1982, Fox et al 1982) and Na_2O-CaO-MgO-Al_2O_3-SiO_2-Fe-O (F. A. Seifert, personal communication, 1982). Although the available data are somewhat inconclusive, it appears that at least some tetrahedrally coordinated ferric iron exists in structural units in silicate melts in which $Si^{4+}/(Si^{4+}+Fe^{3+})$ is independent of bulk melt $Si^{4+}/(Si^{4+}$

$+Fe^{3+}$). Only the relative abundance of such Fe^{3+}-bearing units varies as a function of the ferric-iron content of the melt (Virgo et al 1982). Spectroscopic data by Fox et al (1982) can be interpreted similarly. In this respect Fe^{3+} differs from Al^{3+} but is similar to Ti^{4+}. One might expect, therefore, that charge-compensated Fe^{3+} and Ti^{4+} similarly affect the structure of silicate melts.

Although there are few detailed data on ferrous iron in silicate melts, its influence on liquid immiscibility (Roedder 1951, Visser & Van Groos 1979) and its similarity in size and radius to Mg^{2+} may indicate similar structural roles of Mg^{2+} and Fe^{2+} in silicate melts.

Effect of Pressure on the Structure of Melts

Many magmatic processes take place at pressures greater than 1 atm, where important melt properties differ from those at 1 atm. For example, the viscosity of many silicate melts decreases rapidly with increasing pressure (Figure 4). It appears, however, that for simple binary metal oxide-silica melts, the pressure dependence becomes less pronounced the greater the ratio M/Si (or NBO/Si) (Scarfe et al 1979). In aluminosilicate melts, the type of charge-compensating cation affects the pressure dependence of the viscosity as well as the compressibility of the melt (Figure 4; see also Kushiro 1980, 1981).

Crystal-liquid phase equilibria also are affected by pressure. For example, enstatite ($MgSiO_3$) melts incongruently to forsterite and liquid at pressure less than about 3 kbar, whereas at higher pressure, the melting behavior is congruent (Boyd et al 1964). In more complex systems such as basalt, olivine is typically a liquidus phase at crustal pressures, whereas pyroxene becomes a liquidus phase under pressure conditions of the upper mantle. Other properties showing petrologically important changes in the pressure range of magma formation and evolution include trace-element crystal-liquid partition coefficients (Shimizu 1974, Mysen & Kushiro 1979), cation diffusion (Watson 1979), and redox ratios of iron (Mysen & Virgo 1978, Mo et al 1981).

Despite the obvious need to understand such pressure effects in terms of melt structure, data for silicate melts at high pressures are scarce compared with the availability of data for 1 atm.

Results from molecular-dynamics calculations (Mitra 1982, Woodcock et al 1976) indicate that a pressure increase to about 5 kbar at 1500 K in vitreous SiO_2 results in a small (3–4°) reduction in Si-O-Si angle and an approximately 15% reduction in cavity volume. Evidently the change in cavity volume is reflected in a more than 10% compaction of vitreous SiO_2 under these physical conditions (Arndt & Stoffler 1969). The often dramatic change with pressure of properties (density and viscosity) that may be related to NBO/T of aluminosilicate melts has led to suggestions that Al^{3+} may undergo transformations from four-fold coordination at 1 atm pressure to six-fold coordination at upper-mantle pressures (i.e. Waff 1975). Such transformations have their analogues in equilibria such as

$$\text{albite} \rightleftharpoons \text{jadeite} + \text{quartz} \tag{1}$$

and

$$\text{anorthite} \rightleftharpoons \text{grossular} + \text{corundum} + \text{quartz} \tag{2}$$

in the pressure range of the upper mantle. It has been suggested on the basis of such equilibria that because one may calculate decreased activities of

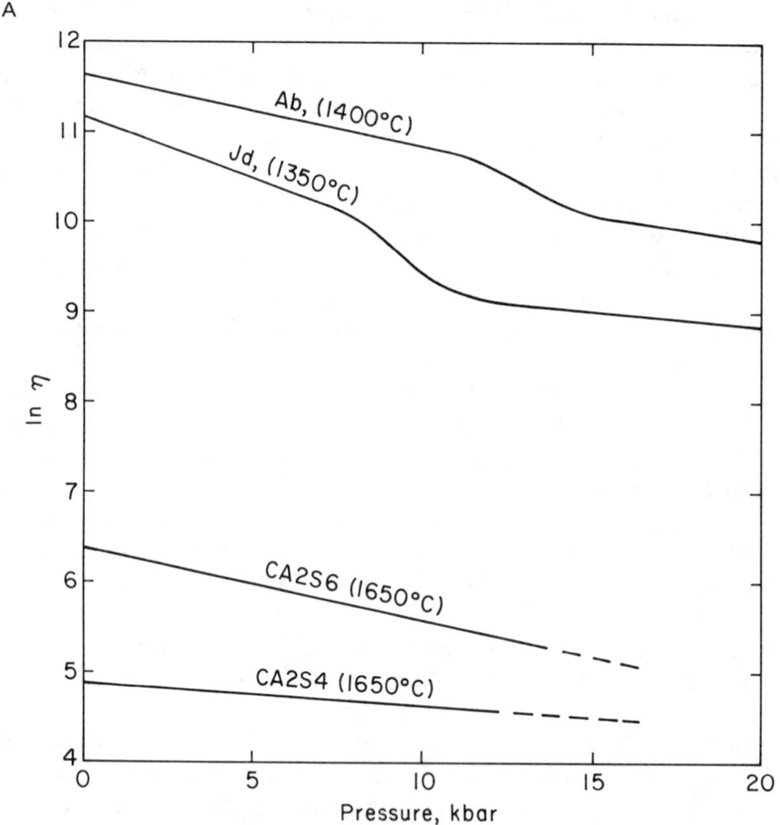

Figure 4 Pressure dependence of viscosity of melts in (*A*) the system Na_2O-CaO-Al_2O_3-SiO_2 (Jd, $NaAlSi_2O_6$; Ab, $NaAlSi_3O_8$; CA2S6, $CaAl_2Si_6O_{16}$; CA2S4, $CaAl_2Si_4O_{12}$) and (*B*) melts with NBO/Si > 0 (data from Kushiro 1981 and Scarfe et al 1979).

aluminosilicate components in melts with Al^{3+} in four-fold coordination with increasing pressure (Burnham 1981, Boettcher et al 1982), it is possible that analogous coordination transformations of Al may occur in a melt.

Although there are no experimental data on the structure of aluminosilicate melts at high pressure and temperature, Raman spectra of melts quenched under isobaric conditions at pressures above the metastable extensions of the relevant univariant equilibrium curves do not indicate reconstructive transformations of this kind (Sharma et al 1979, Mysen et al 1982b) within the sensitivity of the method (1–2% of the amount of Al present in melts of $NaAlSi_3O_8$ and $NaAlSi_2O_6$ composition). It has been suggested, however, that the intertetrahedral angles in the three-

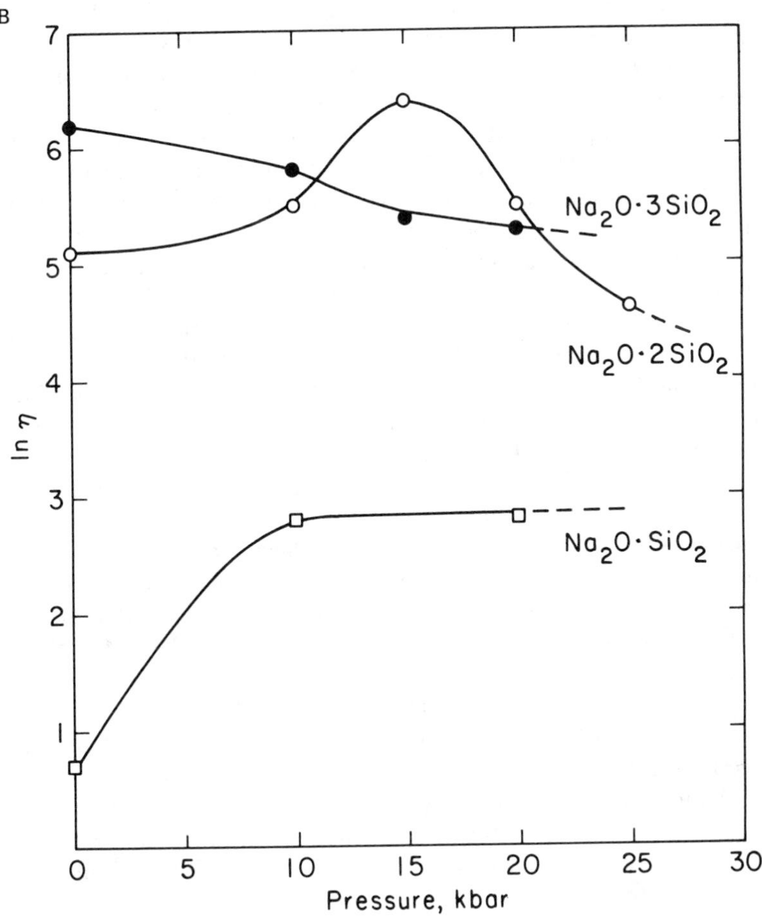

dimensional network decrease by several percent in this pressure range (Mysen et al 1982b), as is also suggested for vitreous SiO_2 (Woodcock et al 1976, Mitra 1982). Significant reduction in cavity volume (Mitra 1982) most likely will affect other physical properties in addition to density (diffusion, viscosity, and so on; see Angell et al 1976).

Volatiles (H_2O and CO_2) in Silicate Melts

Because of the openness of silicate melt structure, one might expect volatile components such as CO_2 and H_2O to dissolve in a molecular form, at least to some extent. The existence of molecular H_2O in Mg_2SiO_4 melt and molecular CO_2 in $NaAlSi_3O_8$ melt was suggested by Hodges (1973) and Mysen (1976), respectively. Solution of molecular H_2O and CO_2 in melts most likely will not affect the anionic structure of the melt significantly. Physical and chemical properties of silicate melts are, however, radically affected by dissolved volatiles. These effects are consistent with chemical interaction between the volatile components and the silicate melt. For example, crystal-liquid phase equilibria suggest that even at a few kilobars pressure, solution of H_2O results in the formation of nonbridging oxygens, whereas CO_2 solution results in the formation of bridging oxygens. [See, for example, Eggler & Rosenhauer (1978) for a discussion.] Even at 1 atm pressure, CO_2 dissolves primarily to form CO_3^{2-} complexes (Pearce 1964) and H_2O forms OH complexes (Uys & King 1963).

Liquidus boundaries between silicate minerals of different degree of polymerization (different NBO/Si; e.g. enstatite and forsterite) shift toward more silica-rich compositions with increasing activity of H_2O (Kushiro 1969) and toward more silica-deficient parts of the relevant systems with increasing activity of CO_2 (Eggler 1978). This liquidus-boundary shift implies that the activity of silica in the melt, defined by the coexisting minerals on the liquidus boundary, decreases with increasing water content and increases with increasing CO_2 content of silicate melts.

Vibrational spectra of quenched hydrous aluminosilicate melts indicate that nonbridging oxygens are formed as H_2O is dissolved and that the OH groups are associated with Si^{4+}, but probably not with Al^{3+} (Mysen et al 1982a). Furthermore, it is likely that some Al^{3+} becomes octahedrally coordinated in hydrous aluminosilicate melts (Burnham 1981).

Carbon dioxide dissolves in silicate melts to form CO_3^{2-} complexes (Pearce 1964, Eggler & Rosenhauer 1978). This extra oxygen is obtained by transforming two nonbridging oxygens in the melt to a bridging oxygen and CO_3^{2-}, thus decreasing the NBO/T ratio of the melt. As a result, the observed enhanced activity of silicate components in CO_2-bearing melts may be understood.

THE STRUCTURE OF NATURAL MAGMATIC LIQUIDS

The anionic structure of magmatic liquids may be estimated from the available data on the structure of simple silicate melts. The major cations in tetrahedral coordination are Al^{3+} and Si^{4+}, where Al^{3+} will be charge-compensated predominantly with Na^{+}, K^{+}, Ca^{2+}, and Mg^{2+}. Cations such as P^{5+}, Ti^{4+}, and Fe^{3+} also occur in tetrahedral coordination. Their abundance is, however, relatively low (total $\lesssim 5$ wt % oxide), and will not significantly affect the overall anionic structure of natural magmatic liquids.

The structure calculation can be carried out as follows. An oxide analysis is recast to molar or atomic proportions, and tetrahedrally coordinated cations are assigned on the basis of the structural data. The NBO/T ratio of the magmatic liquid is then calculated from charge balance. The number of nonbridging oxygens is the difference between total positive and total negative electrical charge ($2 \times O - 4 \times T$), and NBO/T is easily calculated. A few values of NBO/T for typical igneous rocks are shown in Table 2.

The anionic structure of most magmatic liquids will consist of units that have average values of NBO/T equal to 2, 1, and 0. Their proportions in a liquid may be estimated from data such as those in Figure 3 combined with data on aluminum distribution indicated in Table 1. Details of the calculation have been given by Mysen et al (1981a, 1982b).

It can be seen from the summary in Table 2 that three-dimensional network units are the most common in natural magmatic liquids. These

Table 2 Distribution of cations in some magmatic melts by percent

	1[a]	2	3	4	5	6	7	8	9	10
NBO/T	0.08	0.13	0.34	0.65	0.93	1.01	0.86	0.75	0.06	0.44
Si^{4+}(3D)[b]	40.7	38.3	41.7	35.5	35.1	35.8	34.9	36.2	36.6	37.2
Si^{4+}(sh)	20.4	19.1	18.4	17.7	17.5	17.9	17.4	20.6	18.3	18.6
Si^{4+}(ch)	20.4	19.1	18.4	17.7	17.5	17.9	17.4	20.6	18.3	18.6
Al^{3+}(3D)	10.2	13.2	15.0	15.9	15.5	15.7	15.6	8.8	15.4	14.5
Al^{3+}(sh)	4.6	5.8	6.7	6.7	6.6	6.8	4.2	6.5	6.1	—
Al^{3+}(ch)	3.2	3.8	4.1	4.6	4.3	4.1	4.4	2.9	4.1	3.0
Ti^{4+}	0.3	0.5	0.8	1.7	2.9	2.0	3.2	0.2	0.7	1.1
P^{5+}	0.1	0.2	0.2	0.2	0.5	—	0.3	—	0.1	0.1

[a] 1, rhyolite; 2, dacite; 3, andesite; 4, subalkaline basalt; 5, alkaline basalt; 6, alkali basalt; 7, hawaiite; 8, commendite; 9, basaltic andesite; 10, tholeiite.

[b] 3D, three-dimensional network structure; sh, sheet; ch, chain.

units will also contain most of the aluminum. Consequently, the anionic structure of anhydrous magmatic liquids at or near 1 atm pressure may be described as a combination of nearly pure $Si_2O_6^{4-}$ (NBO/Si = 2) and $Si_2O_5^{2-}$ (NBO/Si = 1) units, together with three-dimensional units in which the aluminum is located. The types of three-dimensional units depend on the relative abundance of alkali metals and alkaline earths in the magma (see Seifert et al 1982).

The proportions of the anionic units, even with constant NBO/T of the liquid, are closely related to the proportions of monovalent and divalent cations in the melt. For example, the abundance ratios of units with average NBO/Si = 2 relative to those with average NBO/Si = 1 will be greater in typical olivine tholeiite than in alkali basalt. As a result, chemical and physical properties of magmatic liquids may be related not only to the proportions of metal cations but also to their relative abundance.

PROPERTIES AND MELT STRUCTURE

The viscous behavior of silicate melts may be correlated with the anionic structure of the liquid. The main contribution to the activation energy ($E\eta$) of viscous flow of silicate liquid probably is the breakage of bonds in order to form the flow units (Bockris & Reddy 1970, Taylor & Rindone 1970, Riebling 1966). Although there is another potentially important contribution to the activation energy from the volume that must be created in the melt in order for the units to flow (Glasstone et al 1941), it is commonly agreed that bond energies represent the main contribution to activation energies of viscous flow in silicate melts (Bockris & Reddy 1970). This assumption is supported by the observed similarity of activation energies of viscous flow and energies of relevant bonds in silicate melts (Bockris & Reddy 1970).

In the systems Na_2O-Al_2O_3-SiO_2, CaO-Al_2O_3-SiO_2, and MgO-Al_2O_3-SiO_2, the joins SiO_2-$NaAlO_2$, $CaAl_2O_4$-SiO_2, and $MgAl_2O_4$-SiO_2 define lines of maximum viscosity (Rossin et al 1964, Riebling 1964, 1966) as exemplified in Figure 5. It has been suggested that the lower viscosity on each side of the aluminate-silica joins is probably due to nonbridging oxygens (Riebling 1964, 1966). Structural data from such melt compositions also indicate that nonbridging oxygens are formed on each side of the aluminate-silica joins. On the Al^{3+}-rich side, NBO/T increases with increasing aluminum content because some Al^{3+} is no longer charge-compensated with alkali metals or alkaline earths. On the Al^{3+}-deficient side of the aluminate-silica joins, nonbridging oxygens are formed because of the excess metal cations over that needed to charge-compensate Al^{3+}.

On the aluminate-silicate joins themselves, both viscosity and activation

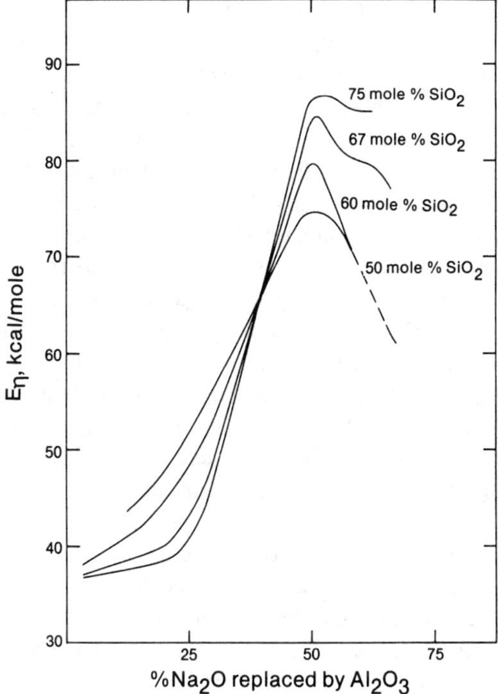

Figure 5 Activation energy of viscous flow of melts in the system Na_2O-Al_2O_3-SiO_2 (data from Riebling 1966).

energies of viscous flow decrease markedly with increasing Al/(Al + Si), although the rate of decrease is much slower on the Na join than on the Ca and Mg joins. In the system $NaAlO_2$-SiO_2 there is random (Al, Si) substitution, and the decrease in activation energy probably mainly reflects a weakening of the *T*-O bonds with increasing Al content (Taylor & Rindone 1970). In systems with divalent charge-compensation, partial (Al, Si) ordering results in a mixture of three-dimensionally interconnected SiO_2 and $Al_2Si_2O_8^{2-}$ units. Their proportions change with Al/(Al + Si). It is possible that this structural change has a smaller relative effect on the viscosity than the continuous increase of Al/Si in melts with essentially the same overall structure.

Crystal-Liquid Phase Equilibria and Melt Structure

The proportions of anionic units in a melt and the distribution of network-modifying cations between these anionic units will be reflected in the liquidus phase relations of a given chemical system. For example, the metasilicate/tectosilicate and orthosilicate/metasilicate liquidus boundaries in the system CaO-MgO-SiO_2 (Osborn & Muan 1960a), as summarized in Figure 6, show two important features. First, the orthosilicate/metasilicate boundary shifts to lower values of NBO/Si (lower M/Si) of the melt, and the metasilicate/tectosilicate boundary to higher values of bulk melt NBO/Si, with increasing Mg/(Mg + Si) of the system. The larnite (Ca_2SiO_4) liquidus field does not even extend to the Ca : Mg = 1 : 1 join, whereas the forsterite (Mg_2SiO_4) extends to Ca/Mg > 1. The abundance of monomers in binary melts increases rapidly with increasing Mg/(Mg + Ca) in melts of constant M/Si (Figure 3) at the expense of units with NBO/Si = 2 (and, to a lesser extent, units with NBO/Si = 1). This expansion of the orthosilicate mineral liquidus field with increasing Mg/(Mg + Ca) probably relates to the enhanced stability of monomers in the melt as Mg/(Mg + Ca) is increased. A comparison of the liquidus relations of fayalite (Fe_2SiO_4) and larnite (Ca_2SiO_4) in the system CaO-FeO-SiO_2 (Osborn & Muan 1960b) indicates that increasing $Fe^{2+}/(Fe^{2+} + Ca^{2+})$ has an effect on the melt structure similar to that of increasing Mg/(Mg + Ca).

The precipitation of forsterite from a melt of monticellite ($CaMgSiO_4$) composition indicates that in melts with both Ca^{2+} and Mg^{2+}, Mg^{2+} shows a preference for the most depolymerized anionic unit in the melt. In melt of $CaFeSiO_4$ composition, fayalite (Fe_2SiO_4) crystallizes on the liquidus, a result suggesting Fe^{2+} preference for monomers in Ca-Fe melts. This preference may also help explain why forsterite crystallizes from a melt of enstatite composition ($MgSiO_3$) and fayalite from a melt of ferrosilite ($FeSiO_3$) composition, whereas wollastonite ($CaSiO_3$) melts congruently. The enhanced preference of Mg^{2+} for monomers may also explain the observed metastable crystallization of forsterite from diopside ($CaMgSi_2O_6$) melt (Kirkpatrick et al 1981).

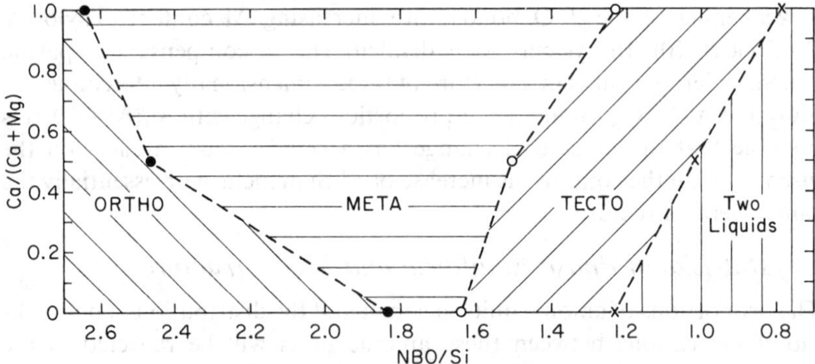

Figure 6 Relations between orthosilicate/metasilicate, metasilicate/tectosilicate, and liquid immiscibility in the system CaO-MgO-SiO_2 as a function of Ca/(Ca + Mg).

The crystal-liquid phase equilibria in the system CaO-MgO-SiO_2 also show that the liquidus field of tridymite or cristobalite (SiO_2) expands over that of metasilicate minerals (clinoenstatite or pseudowollastonite) with increasing Mg/(Mg + Ca) of a melt (Figure 6). The liquid immiscibility volume also expands with increasing Mg/(Mg + Ca). The data on proportions of anionic units in binary metal oxide melts (Figure 3) show that the abundance of three-dimensional network units in these melts also increases with increasing Mg/(Mg + Ca) of the melt. It would be anticipated that the same relations hold for increasing $Fe^{2+}/(Fe^{2+} + Ca)$ or $Fe^{2+}/(Fe^{2+} + M^{2+} + M^{+})$, and these results would explain the common occurrence of immiscible liquid in iron-rich magma.

SUMMARY

Experimental data on the structure of silicate melts and glasses show that only several types of anionic units coexist. The experimentally obtained structural data on melts confirm that the principles that have been established for crystal chemistry also apply to the structure of silicate melts. Early concepts of silicate-melt structure, based on crystal-chemical principles, to explain observed physical properties (e.g. Tomlinson 1953, Bockris et al 1955, MacKenzie 1960) generally have been confirmed with the more recently available structural data.

For bulk compositions relevant to igneous rocks, (Al + Si)-bearing, three-dimensional network units are the most abundant. Their overall stoichiometry is balanced by depolymerized units that, on average, have 2 and 1 nonbridging oxygens per tetrahedrally coordinated cation (mostly Si^{4+}). These anionic units are interconnected with monovalent and divalent metal cations. The divalent cations show preference for the anionic units with the largest values of NBO/T. Other cations in tetrahedral coordination (Fe^{3+}, Ti^{4+}, and P^{5+}) generally form separate clusters in the melts.

The structural data are consistent with results obtained with molecular-dynamics calculations, and can be used to understand physical and chemical properties of melts. Physical properties tend to depend primarily on the degree of polymerization of the melt and on the strength of the bridging-oxygen bonds. Chemical properties, such as crystallization behavior, also reflect the preference shown for various metal cations by the different types of anionic units in silicate melts.

Literature Cited

Angell, C. A., Finch, E. D., Woolf, L. A., Bach, D. 1976. Spin-echo diffusion coefficients of water to 2380 bar and −20°C. *J. Chem. Phys.* 65:3063–66

Arndt, J., Stoffler, D. 1969. Anomalous changes in some properties of silica glass densitified at very high pressures. *Phys. Chem. Glasses* 10:117–24

Bando, Y., Ishizuka, K. 1979. Study of the structure of silica glass by high-resolution microscopy. *J. Non-Cryst. Solids* 33:375–82

Bell, R. J., Dean, P. 1972. Localization of phonons in vitreous silica and related glasses. *Int. Conf. Phys. Non-Cryst. Solids, 3rd, Univ. Sheffield, 1970*, ed. R. W. Douglas, B. Ellis, pp. 443–52. London: Wiley-Interscience

Bockris, J. O'M., Reddy, A. K. N. 1970. *Modern Electrochemistry*. New York: Plenum. 622 pp.

Bockris, J. O'M., MacKenzie, J. D., Kitchner, J. A. 1955. Viscous flow in silica and binary silicate liquids. *Trans. Faraday Soc.* 51:1734–48

Boettcher, A. L., Burnham, C. W., Windom, K. E., Bohlen, S. R. 1982. Liquids, glasses and the melting of silicates to high pressures. *J. Geol.* 90:127–38

Bottinga, Y., Weill, D. F. 1972. The viscosity of magmatic silicate liquids: A model for calculation. *Am. J. Sci.* 272:438–75

Bottinga, Y., Weill, D. F., Richet, P. 1981. Thermodynamic modeling of silicate melts. In *Thermodynamics of Minerals and Melts*, eds. R. C. Newton, A. Navrotsky, B. J. Wood, pp. 207–45. New York: Springer

Bottinga, Y., Weill, D. F., Richet, P. 1982. Density calculations for silicate liquids. I. Revised method for aluminosilicate compositions. *Geochim. Cosmochim. Acta* 46:909–19

Boyd, F. R., England, J. L., Davis, B. C. T. 1964. Effect of pressure on the melting and polymorphism of enstatite, $MgSiO_3$. *J. Geophys. Res.* 69:2101–9

Brawer, S. A. 1975. Theory of the vibrational spectra of some network and molecular glasses. *Phys. Rev. B* 11:3173–94

Brawer, S. A., White, W. B. 1975. Raman spectroscopic investigation of the structure of silicate glasses. I. The binary silicate glasses. *J. Chem. Phys.* 63:2421–32

Brawer, S. A., White, W. B. 1977. Raman spectroscopic investigation of the structure of silicate glasses. II. Soda-alkaline earth-alumina ternary and quaternary glasses. *J. Non-Cryst. Solids* 23:261–78

Brown, G. E., Keefer, K. D., Fenn, P. M. 1978. Extended X-ray fine structure (XAFS) study of iron-bearing silicate glass. *Geol. Soc. Am. Abstr. with Programs* 10:373 (Abstr.)

Burnham, C. W. 1981. The nature of multicomponent silicate melts. *Phys. Chem. Earth* 13–14:197–227

Cranmer, D., Uhlmann, D. R. 1981. Viscosities in the system albite-anorthite. *J. Geophys. Res.* 86:7951–57

Dent Glasser, L. S. 1979. Non-existent silicates. *Z. Kristallogr. Kristallgeom. Kristallphys. Kristallchem.* 149:291–305

Eggler, D. H. 1978. The effect of CO_2 upon partial melting in the system Na_2O-CaO-MgO-Al_2O_3-SiO_2-CO_2 to 35 kb, with an analysis of melting in a peridotite-H_2O-CO_2 system. *Am. J. Sci.* 278:305–43

Eggler, D. H., Rosenhauer, M. 1978. Carbon dioxide in silicate melts. II. Solubilities of CO_2 and H_2O in $CaMgSi_2O_6$ (diopside) liquids and vapors at pressures to 40 kb. *Am. J. Sci.* 278:64–94

Etchepare, J. 1972. Study by Raman spectroscopy of crystalline and glassy diopside. In *Amorphous Materials*, ed. R. W. Douglas, B. Ellis, pp. 337–46. New York: Wiley

Fox, K. E., Furukawa, T., White, W. B. 1982. Transition metal ions in silicate melts. II. Iron in sodium silicate glasses. *Phys. Chem. Glasses*. In press

Furukawa, T., White, W. B. 1979. Structure and crystallization of glasses in the $Li_2Si_2O_5$-TiO_2 system determined by Raman spectroscopy. *Phys. Chem. Glasses* 20:69–80

Furukawa, T., Fox, K. E., White, W. B. 1981. Raman spectroscopic investigation of the structure of silicate glasses. III. Raman intensities and structural units in sodium silicate glasses. *J. Chem. Phys.* 75:3226–37

Galeener, F. L., Mikkelsen, J. C. 1979. The Raman spectra and structure of pure vitreous P_2O_5. *Solid State Commun.* 20:505–10

Gaskell, P. H., Mistry, A. B. 1979. High-resolution transmission spectroscopy of small amorphous silica particles. *Phil. Mag.* 39:245–57

Gaskell, P. H., Tarrant, I. D. 1980. Refinement of a random network model for vitreous silicon dioxide. *Philos. Mag. B* 42:265–86

Glasstone, S., Laidlen, K., Eyring, H. 1941. *The Theory of Rate Processes*. New York: McGraw-Hill

Henry, D. J., Navrotsky, A., Zimmermann, H. D. 1982. Thermodynamics of plagioclase-melt equilibria in the system albite-anorthite-diopside. *Geochim. Cosmochim. Acta* 46:381–93

Hess, P. C. 1971. Polymer models of silicate

melts. *Geochim. Cosmochim. Acta* 35:289–306

Hess, P. C. 1977. Structure of silicate melts. *Can. Mineral.* 15:162–78

Hess, P. C. 1980. Polymerization model for silicate melts. In *Physics of Magmatic Processes*, ed. R. B. Hargraves, pp. 3–49. Princeton, NJ: Princeton Univ. Press. 587 pp.

Hodges, F. N. 1973. Solubility of H_2O in forsterite melt at 20 kbar. *Carnegie Inst. Washington Yearb.* 72:495–97

Kirkpatrick, R. J., Kuo, L.-C., Melchior, J. 1981. Crystal-growth in incongruently-melting compositions: Programmed cooling experiments with diopside. *Am. Mineral.* 66:223–42

Konnert, J. H., Karle, J. 1973. The computation of radial distribution functions for glassy materials. *Acta Crystallogr. A* 29:702–10

Kuroda, K., Kato, C. 1979. Trimethylsilylation of hemimorphite. *J. Inorg. Nucl. Chem.* 41:947–51

Kushiro, I. 1969. The system forsterite-diopside-silica with and without water at high pressures. *Am. J. Sci.* 267-A:269–94

Kushiro, I. 1975. On the nature of silicate melt and its significance in magma genesis: Regularities in the shift of liquidus boundaries involving olivine, pyroxene and silica minerals. *Am. J. Sci.* 275:411–31

Kushiro, I. 1980. Viscosity, density, and structure of silicate melts at high pressures, and their petrological applications. In *Physics of Magmatic Processes*, ed. R. B. Hargraves, pp. 93–121. Princeton, NJ: Princeton Univ. Press. 587 pp.

Kushiro, I. 1981. Viscosity change with pressure of melts in the system CO-Al_2O_3-SiO_2. *Carnegie Inst. Washington Yearb.* 80:339–41

Lacey, E. D. 1963. Aluminum in glasses and melts. *Phys. Chem. Glasses* 4:234–38

Levy, R. A., Lupis, C. H. P., Flinn, P. A. 1976. Mossbauer analysis of the valence and coordination of iron cations in SiO_2-Na_2O-CaO glasses. *Phys. Chem. Glasses* 17:94–103

Liebau, F. 1981. The influence of cation properties on the conformation of silicate and phosphate anions. *Struct. Bonding Cryst.* 2:197–232

MacDowell, J. F., Beall, G. H. 1969. Immiscibility and crystallization in Al_2O_3-SiO_2 glasses. *J. Am. Ceram. Soc.* 52:17–25

MacKenzie, J. D. 1960. Structure of some inorganic glasses from high temperature studies. In *Modern Aspects of the Vitreous State*, ed. J. D. MacKenzie, pp. 188–218. Washington DC: Butterworths. 266 pp.

Mammone, J. F., Sharma, S. K., Nocil, M. F. 1981. Ring structures in silica glass—A Raman spectroscopic investigation. *EOS, Trans. Am. Geophys. Union* 62:425 (Abstr.)

Mao, H.-K., Virgo, D., Bell, P. M. 1973. Analytical study of the orange soil returned by the Apollo 17 astronauts. *Carnegie Inst. Washington Yearb.* 72:631–38

Masson, C. R. 1977. Anionic constitution of glass-forming melts. *J. Non-Cryst. Solids* 25:1–41

Mikkelsen, J. C., Galeener, F. L. 1980. Thermal equilibration of Raman active defects in vitreous silica. *J. Non-Cryst. Solids* 37:71–85

Mitra, S. K. 1982. Molecular dynamics simulation of silicon dioxide glass. *Philos. Mag. B* 45:529–48

Mo, X.-X., Stebbins, J. F., Carmichael, I. S. E. 1981. The partial molar volume of Fe_2O_3 in silicate liquids and the pressure dependence of oxygen fugacity. *EOS, Trans. Am. Geophys. Union* 62:1065 (Abstr.)

Mysen, B. O. 1976. The role of volatiles in silicate melts: Solubility of carbon dioxide and water in feldspar, pyroxene and feldspathoid melts to 30 kb and 1625°C. *Am. J. Sci.* 276:969–96

Mysen, B. O., Kushiro, I. 1979. The effect of pressure on the partitioning of nickel between olivine and aluminosilicate melt. *Earth Planet. Sci. Lett.* 42:383–89

Mysen, B. O., Virgo, D. 1978. Influence of pressure, temperature, and bulk composition on melt structure in the system $NaAlSi_2O_6$-$NaFe^{3+}Si_2O_6$. *Am. J. Sci.* 278:1307–22

Mysen, B. O., Virgo, D., Scarfe, C. M. 1980. Relations between the anionic structure and viscosity of silicate melts—a Raman spectroscopic study. *Am. Mineral.* 65:690–710

Mysen, B. O., Virgo, D., Kushiro, I. 1981a. The structural role of aluminum in silicate melts—a Raman spectroscopic study at 1 atmosphere. *Am. Mineral.* 66:678–701

Mysen, B. O., Ryerson, F. J., Virgo, D. 1981b. The structural role of phosphorus in silicate melts. *Am. Mineral.* 66:106–17

Mysen, B. O., Virgo, D., Seifert, F. A. 1982a. Structure of silicate melts: Implications for chemical and physical properties of natural magma. *Rev. Geophys.* 20:353–83

Mysen, B. O., Virgo, D., Danckwerth, P., Seifert, F. A., Kushiro, I. 1982b. Influence of pressure on the structure of melts on the joins $NaAlO_2$-SiO_2, $CaAl_2O_4$-SiO_2 and $MgAl_2O_4$-SiO_2. *Geochim. Cosmochim. Acta.* In press

Navrotsky, A., Hon, R., Weill, D. F., Henry, D. J. 1980. Thermochemistry of glasses and liquids in the systems $CaMgSi_2O_6$-$CaAl_2SiO_6$-$NaAlSi_3O_8$, SiO_2-$CaAl_2Si_2O_8$-$NaAlSi_3O_8$ and SiO_2-Al_2O_3-

CaO-Na_2O. *Geochim. Cosmochim. Acta* 44:1409–33

Nelson, B. O., Exharos, G. J. 1979. Vibrational spectroscopy of cation-site interactions in phosphate glasses. *J. Chem. Phys.* 71:2739–47

Osborn, E. F., Muan, A. 1960a. Plate 2, the system CaO-MgO-SiO_2. In *Phase Equilibrium Diagrams of Oxide Systems.* Columbus, Ohio: Am. Ceram. Soc.

Osborn, E. F., Muan, A. 1960b. Plate 7, the system CaO-"FeO"-SiO_2. In *Phase Equilibrium Diagrams of Oxide Systems.* Columbus, Ohio: Am. Ceram. Soc.

Pearce, M. L. 1964. Solubility of carbon dioxide and variation of oxygen ion activity in soda-silica melts. *J. Am. Ceram. Soc.* 47:342–47

Riebling, E. F. 1964. Structure of magnesium aluminosilicate liquids at 1700°C. *Can. J. Chem.* 42:2811–21

Riebling, E. F. 1966. Structure of sodium aluminosilicate melts containing at least 50 mole % SiO_2 at 1500°C. *J. Chem. Phys.* 44:2857–65

Roedder, E. 1951. Low-temperature liquid immiscibility in the system K_2O-FeO-Al_2O_3-SiO_2. *Am. Mineral.* 36:282–86

Rossin, R., Bersan, J., Urbain, G. 1964. Etude de la viscosite de laitiers liquides appartenant au systeme ternaire: SiO_2-Al_2O_3-CaO. *Rev. Hautes Temp. Refract.* 1:159–270

Ryerson, F. J., Hess, P. C. 1980. The role of P_2O_5 in silicate melts. *Geochim. Cosmochim. Acta* 44:611–25

Scarfe, C. M., Mysen, B. O., Virgo, D. 1979. Changes in viscosity and density of melts of sodium disilicate, sodium metasilicate, and diopside composition with pressure. *Carnegie Inst. Washington Yearb.* 78:547–51

Seifert, F. A., Virgo, D., Mysen, B. O. 1979. Melt structures and redox equilibria in the system Na_2O-FeO-Fe_2O_3-Al_2O_3-SiO_2. *Carnegie Inst. Washington Yearb.* 78:511–19

Seifert, F. A., Mysen, B. O., Virgo, D. 1981. Structural similarity of glasses and melts relevant to petrological processes. *Geochim. Cosmochim. Acta* 45:1879–84

Seifert, F. A., Mysen, B. O., Virgo, D. 1982. Three-dimensional network structure in the systems SiO_2-$NaAlO_2$, SiO_2-$CaAl_2O_3$ and SiO_2-$MgAl_2O_4$. *Am. Mineral.* 67:696–718

Sharma, S. K., Virgo, D., Mysen, B. O. 1979. Raman study of the coordination of aluminum in jadeite ($NaAlSi_2O_6$) melt as a function of pressure. *Am. Mineral.* 64:779–88

Shimizu, N. 1974. An experimental study of the partitioning of K, Rb, Cs, Sr and Ba between clinopyroxene and liquid at high pressures. *Geochim. Cosmochim. Acta* 38:1789–98

Smart, R. M., Glasser, F. P. 1978. Silicate anion constitution of lead silicate glasses and crystals. *Phys. Chem Glasses* 19:95–102

Soules, T. F. 1979. A molecular dynamics calculation of the structure of sodium silicate glasses. *J. Chem. Phys.* 71:4570–78

Sweet, J. R., White, W. B. 1969. Study of sodium silicate glasses and liquids by infrared spectroscopy. *Phys. Chem. Glasses* 10:246–51

Taylor, M., Brown, G. E. 1979a. Structure of mineral glasses. I. The feldspar glasses $KAlSi_3O_8$, $NaAlSi_3O_8$, $CaAl_2Si_2O_8$. *Geochim. Cosmochim. Acta* 43:61–77

Taylor, M., Brown, G. E. 1979b. Structure of mineral glasses. II. The SiO_2-$NaAlSiO_4$ join. *Geochim. Cosmochim. Acta* 43:1467–75

Taylor, M., Brown, G. E., Fenn, P. M. 1980. Structure of silicate mineral glasses. III. $NaAlSi_3O_8$ supercooled liquid at 805°C and the effects of thermal history. *Geochim. Cosmochim. Acta* 44:109–19

Taylor, T. D., Rindone, G. E. 1970. Properties of soda aluminosilicate glasses. V. Low-temperature viscosities. *J. Am. Ceram. Soc.* 53:692–95

Tobin, M. C., Baak, T. 1968. Raman spectra of some low-expansion glasses. *J. Opt. Soc. Am.* 58:1459–60

Tomlinson, J. W. 1953. Some aspects of the constitution of liquid oxides. In *Physical Chemistry of Melts*, pp. 22–33. London: Inst. Min. Metall.

Uys, J. M., King, T. B. 1963. The effects of basicity on the solubility of water in silicate melts. *Trans. Metall. Soc. AIME* 227:492–500

Verweij, H. 1979a. Raman study of alkali germanosilicate glass. I. Sodium and potassium metagermanosilicate glasses. *J. Non-Cryst. Solids* 33:41–53

Verweij, H. 1979b. Raman study of the structure of alkali germanosilicate glasses. II. Lithium, sodium and potassium digermanosilicate glasses. *J. Non-Cryst. Solids* 33:55–69

Verweij, H., Konijnendijk, W. L. 1976. Structural units in K_2O-PbO-SiO_2 glasses by Raman spectroscopy. *J. Am. Ceram. Soc.* 59:517–21

Virgo, D., Mysen, B. O., Seifert, F. A. 1981. Relationships between the oxidation state of iron and structure of silicate melts. *Carnegie Inst. Washington Yearb.* 80:308–11

Virgo, D., Mysen, B. O., Danckwerth, P., Seifert, F. A. 1982. The anionic structure of 1-atm melts in the system SiO_2-$NaFeO_2$.

Carnegie Inst. Washington Yearb. 81: In press
Visser, W., Van Groos, K. 1979. Effects of P_2O_5 and TiO_2 on the liquid-liquid equilibria in the system K_2O-FeO-Al_2O_3-SiO_2. *Am. J. Sci.* 279:970–88
Vukcevich, M. R. 1972. A new interpretation for the anomalous behavior of vitreous silicate. *J. Non-Cryst. Solids* 11:25–63
Waff, H. S. 1975. Pressure-induced coordination changes in magmatic liquids. *Geophys. Res. Lett.* 2:193–96
Watson, E. B. 1979. Calcium diffusion in a simple silicate melt to 30 kbar. *Geochim. Cosmochim. Acta* 43:313–23
Wong, J. 1976. Vibrational spectra of vapor-deposited binary phosphosilicate glasses *J. Non-Cryst. Solids* 20:83–100
Woodcock, L. V., Angell, C. A., Cheeseman, P. 1976. Molecular dynamics studies of the vitreous state: Simple ionic systems and silica. *J. Chem. Phys.* 65:1565–77

Ann. Rev. Earth Planet. Sci. 1983. 11: 99–132

CENOZOIC GLACIATION IN THE SOUTHERN HEMISPHERE[1]

J. H. Mercer
Institute of Polar Studies and Department of Geology and Mineralogy, Ohio State University, Columbus, Ohio 43210

INTRODUCTION

With the end of glaciation in Gondwana in early Permian time, a prolonged interval began when the Earth was without ice sheets, though not necessarily free of mountain glaciers. Not until sometime during the Tertiary Era did ice sheets build up once more, first in Antarctica and later in the Northern Hemisphere. The fundamental cause of the cooling that led to the formation of the Antarctic Ice Sheet and, eventually, to major glaciation in lower latitudes is generally held to have been the increasing thermal isolation of Antarctica as Australia, and later South America, separated from it, allowing zonal oceanic circulation to develop. Rifting between Australia and Antarctica began between 110 and 90 m.y. ago (Cande & Mutter 1982), but not until much later, probably sometime between 30 and 25 m.y. ago, had it progressed far enough for deep circulation to begin as the South Tasman Rise cleared north Victoria Land (Kennett et al 1974, Kennett 1977, Weissel et al 1977). However, Antarctica did not reach its present high level of thermal isolation until South America had separated from it, enabling the Antarctic Circumpolar Current to become established (Savin et al 1975). According to Barker & Burrell (1977, 1982), coherent spreading between the Antarctic Peninsula and Tierra del Fuego started about 29 m.y. ago, and deep circulation began 23.5 ± 2.5 m.y. ago, thereafter increasing in volume.

The initial ice fields and ice caps that later expanded to become the

[1] Contribution No. 445 of the Institute of Polar Studies, The Ohio State University.

0084–6597/83/0515–0099$02.00

Antarctic Ice Sheet must have formed when Antarctica was much warmer than it is today—perhaps as much as 20°C warmer according to Kvasov & Verbitsky (1981), who believe that the first ice accumulated on an upland plateau about 1800 m in elevation. On the other hand, the temperature of the tropical sea surface is thought to have been about the same as it is now throughout the Tertiary Era (Matthews & Poore 1980), or lower during the late Eocene, Oligocene, and early Miocene and the same during the rest of the Tertiary (Savin et al 1975). Consequently the equator-to-pole temperature gradient was much less steep at the start of Antarctic glaciation than it is today. Middle latitudes would then have been warmer than they are now (although by less than were the polar regions), confining glaciers to higher elevations than at present. Within the tropics, however, if highlands existed with elevations and precipitation levels similar to today's glaciated mountains, they may have been comparably ice-covered throughout the Tertiary Era.

The initial Antarctic ice fields would have been composed of temperate ice (at the pressure melting point). By the time that the ice-covered area had grown enough to be termed an ice sheet, the high, central part may already have become cold (below the pressure melting point). The marginal zone would have stayed temperate as the ice sheet was expanding, but after the ice had reached the coast and as temperatures continued to drop, the entire ice sheet would have become cold. It then reached full-bodied state as ice shelves developed, thickened, and grounded in sounds and embayments. At the same time as ice shelves formed around Antarctica, the extent of cold surface water increased sharply, and the Antarctic Convergence in its present form first appeared. These events, in conjunction with the climatic effects of orbital variations (Hays et al 1976), probably set the stage for the major late Cenozoic glacial-interglacial cycles in middle and low latitudes of the Southern Hemisphere, i.e. in the subantarctic islands, South America, New Zealand, New Guinea, and Africa, all of which contain glaciers today, and in Australia, which is now ice-free. Even at its greatest extent, however, the total volume of ice in middle and low southern latitudes was insignificant in global or hemispheric terms; the Antarctic Ice Sheet has, throughout its existence, contained more than 98% of the ice in the Southern Hemisphere.

Early in its history when it was still essentially a temperate glacier, the ice sheet in East Antarctica may have repeatedly undergone great changes in volume as it responded to orbitally controlled climatic fluctuations (Woodruff et al 1981), but once it had extended to the coast and had become cold down to sea level, it was immune from these influences. However, the marginal areas of the ice sheet are sensitive to changes in sea level, which largely determine the position of the ice sheet grounding line on the continental shelf: if sea level drops, the grounding line will advance (Hollin 1962). Thus, ever since the midlatitude ice sheets in the Northern Hemisphere first formed, the Antarctic Ice Sheet has varied in unison with them, as they have controlled glacio-eustatic sea level. The variations have been small in East Antarctica, where the continental shelf is narrow, but in West Antarctica major changes have occurred as the grounding lines in the Ross and Weddell embayments have migrated back and forth over hundreds of kilometers. However, these changes in the size of the Antarctic Ice Sheet—and, therefore, in the total volume of ice in the Southern

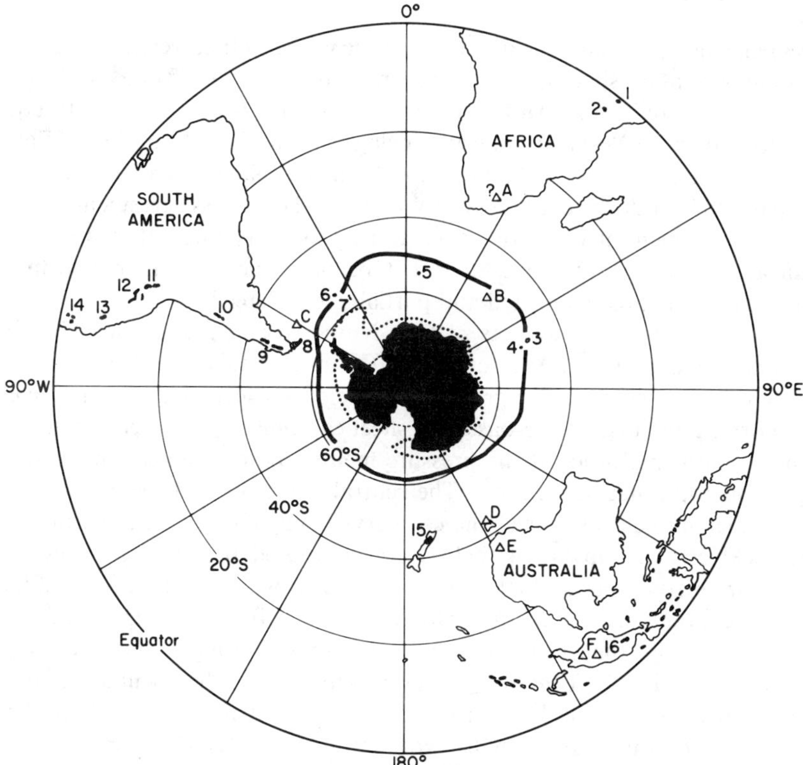

Figure 1 Present ice cover in the Southern Hemisphere (*solid black*): 1. Mount Kenya; 2. Kilimanjaro; 3. Kerguelen; 4. Heard Island; 5. Bouvet Island; 6. South Georgia; 7. South Sandwich Islands; 8. Tierra del Fuego; 9. Patagonian ice fields; 10. Santiago-Mendoza Andes; 11. Cordillera Real; 12. Cordillera Vilcanota and Quelccaya Ice Cap; 13. Cordillera Blanca; 14. Ecuadorian glaciers; 15. Southern Alps; 16. Irian Barat (New Guinea). Formerly glaciated areas (*triangles*): A. Drakensberg Mountains (?); B. Marion Island; C. Falkland Islands; D. Tasmania; E. Snowy Mountains; F. Papua-New Guinea. Heavy line around Antarctica is the extent of summer sea ice at last glacial maximum (Hays 1978); dotted line is the extent of summer sea ice today.

Hemisphere—have been minor compared to the repeated buildup followed by wastage to extinction of the ice sheets in middle latitudes of the Northern Hemisphere. The closest southern parallel to the growth and disappearance of the northern ice sheets has been the change in the extent of summer sea ice, which at the last glacial maximum covered ten times as much ocean as it does today, reaching almost to the present limit of winter sea ice (Hays 1978, Cooke & Hays 1982) (Figure 1).[2]

BIRTH AND GROWTH OF THE ANTARCTIC ICE SHEET

When, where, and how the initial Antarctic ice fields that were forerunners of the present ice sheet formed and grew is still in dispute. Today, if the ice cover were removed, Antarctica would consist of an East Antarctic continent and a West Antarctic archipelago (Figure 2*b*). The outline of East Antarctica is generally thought not to have changed greatly since it separated from Australia 110–90 m.y. ago, but there is less agreement about the development of West Antarctica during the same interval; has it been an archipelago throughout, or did it become one only after a former extensive land area had rifted and partially foundered?

The ice sheet now consists of two distinct parts: the ice sheet based mainly on land in East Antarctica, and the much smaller ice sheet in West Antarctica, most of which is grounded far below sea level. This distinction is important, because a land-based ice sheet can build up as a temperate ice mass, consisting of ice at the pressure melting point, whereas this is not possible for a marine ice sheet. The central dome of a marine ice sheet is grounded far below sea level and can survive only if buttressed by fringing ice shelves. All known ice shelves are composed of cold ice below its pressure melting point, floating in water that is within half a degree of its freezing point of $\sim -2°C$ (Robin & Adie 1964); this is apparently a requirement for their formation and persistence. During climatic cooling, therefore, an ice sheet that *originates* as a marine ice sheet will form later than an adjacent land-based ice sheet, because it needs a much colder climate. For this reason, many workers (e.g. Bentley & Ostenso 1961, Mercer 1973) conclude that the Antarctic Ice Sheet originated as a temperate ice sheet covering East Antarctica only, at about the same time that a heavy ice cover developed on what was then a mountainous West Antarctic archipelago. Later, after further climatic cooling, the ice became cold down to sea level, allowing ice shelves to form and then expand to cover the channels between the glaciated islands of West Antarctica, and

[2] *Note added in proof* Burckle et al 1982 (*Nature* 299:435–37) dispute this, and conclude that the ice-age limit of summer sea ice is not yet known.

between this archipelago and the mainland of East Antarctica. The growing ice shelves thickened in their inland portions until they became grounded on the sea floor in central West Antarctica, enabling an ice sheet dome to build up on the grounded portion. LeMasurier & Rex (1982), however, suggest a fundamentally different origin for the ice sheet in West Antarctica, believing that, like the ice sheet in East Antarctica, it started out as a temperate ice sheet on a land surface above sea level. Later, when much of this land surface began to sink below sea level as the result of block faulting, further cooling had rendered the ice sheet cold, and it was able to persist as a marine ice sheet buttressed by ice shelves. This mode of origin does not

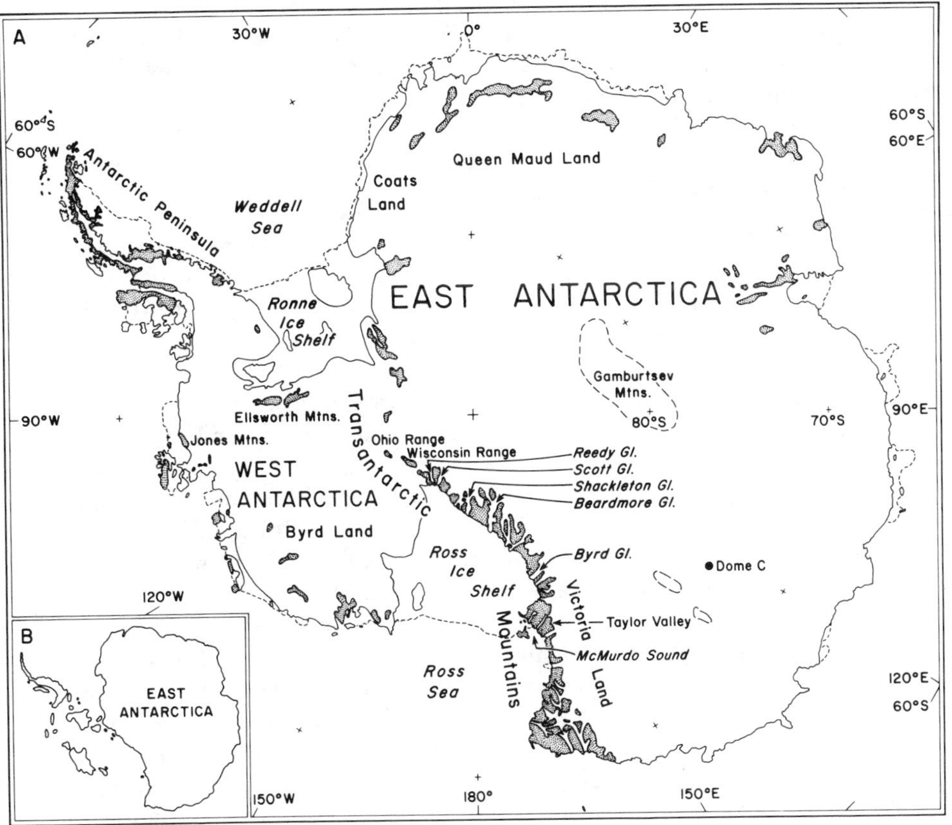

Figure 2 *A*. Location map of Antarctica. Stippled: mountains, partially ice-free (generalized). Dashed lines offshore: ice-shelf margins. Dashed lines within East Antarctica: subglacial mountains. *B*. Antarctic ice cover, if ice grounded below sea level in West Antarctica were removed (with no allowance for isostatic recovery). Ice cover may have resembled this during last interglacial and during middle Miocene time.

require the ice sheet in West Antarctica to be younger than that in East Antarctica.

The marine geological records of both the Antarctic seas and the world ocean, as well as the terrestrial geological record of Antarctica, contain information about the initiation and subsequent history of the Antarctic Ice Sheet. Considerable disagreement has arisen about the interpretation of this evidence, and opinions are constantly being modified as fresh data come from continuing field and laboratory investigations.

Marine Geological Record

Two lines of investigation in marine geology have been particularly important in reconstructing Antarctic glacial development: measurement of the oxygen isotopic composition of calcitic foraminiferal tests, and the history of ice rafting. Additional evidence has come from interpretations of changing microfaunal assemblages in terms of temperature, and from the identification and dating of sedimentary hiatuses, some of which are thought to have been produced by strong bottom currents during heightened Antarctic glaciation.

δ^{18}O OF SEA WATER, GLOBAL COOLING, AND ANTARCTIC GLACIAL DEVELOPMENT Sea water contains two relatively abundant isotopes of oxygen: ^{16}O and ^{18}O. The lighter molecule $H_2{}^{16}O$, being more volatile than $H_2{}^{18}O$, is preferentially removed from the ocean during evaporation. Conversely, during condensation the heavier molecule $H_2{}^{18}O$ preferentially condenses out, so that precipitation becomes richer in $H_2{}^{16}O$ as the temperature falls. Consequently, buildup of glaciers on land makes sea water richer in ^{18}O. This will be reflected in a change in the oxygen isotopic composition of marine microfossils, but not to a detectable extent until the ice has built up to the dimensions of a small ice sheet. However, an increase in δ^{18}O *calcite* can reflect a drop in temperature at the time the organism was secreting its test as well as an increase in the volume of global ice, and on an ice-free Earth the temperature effect would compose the entire signal. This introduces an inherent ambiguity into the isotopic record that leaves room for differences of opinion about how the Cenozoic isotopic curve should be interpreted; for example, which episode of ^{18}O enrichment records the initial formation of the Antarctic Ice Sheet? During later episodes of ^{18}O enrichment, how much of the signal represents further glacial buildup, how much represents change in the isotopic composition of the ice (as the temperature of the snowfall changes), and how much represents ocean cooling? Hudson (1977) and Savin (1977) give useful summaries of the assumptions, methodology, conclusions, and results of recent oxygen isotopic studies concerned with the elucidation of these problems.

As a rule, abyssal benthic foraminifera have been analyzed in preference to planktic when attempting to reconstruct changes in global ice volume. This preference is based on the assumption that when the temperature of deep bottom water, which is derived from surface water near the Antarctic coast, had dropped to within a few degrees of freezing point (as it would have once extensive sea ice had formed around Antarctica), any further enrichment in ^{18}O would have chiefly signaled an increase in ice volume (Shackleton 1967, Shackleton & Opdyke 1973). The curve of the oxygen isotopic composition of benthic foraminifera during the Cenozoic shows a major but gradual ^{18}O enrichment between 53 and 40 m.y. ago, and two episodes of rapid ^{18}O enrichment: one about 38 m.y. ago at the Eocene-Oligocene boundary, and the other between 16.5 and 13 m.y. ago during middle Miocene time (Savin 1977).

Traditional interpretation of the $\delta^{18}O$ curve The traditional interpretation of these two episodes of rapid ^{18}O enrichment is that they reflect, respectively, a major cooling of Antarctic surface water (and thus of deep bottom water worldwide) at the Eocene-Oligocene boundary, and the buildup of an ice sheet in Antarctica, plus further cooling, during the middle Miocene (Savin et al 1975, Shackleton & Kennett 1975, Kennett & Shackleton 1976, Savin 1977, Savin et al 1981, Woodruff et al 1981). Savin et al (1975) conclude that the temperature of deep bottom water in the tropical Pacific Ocean at the end of the Eocene-Oligocene cooling about 38 m.y. ago was ~7°C and, therefore, was still too high for the presence of Antarctic sea ice. Savin (1977) revises this figure to ~5°C, but believes this is still too high for sea ice to have been forming around Antarctica. On the other hand, Shackleton & Kennett (1975) estimate that the temperature of bottom water at ~1000 m depth at Deep-Sea Drilling Project (DSDP) Site 277 south of New Zealand had fallen to ~5°C by the beginning of the Oligocene, implying that conditions were severe enough for sea ice to be forming around Antarctica. However, they conclude that the isotopic evidence precludes the presence of an ice sheet at that time. They reason that, unless its absence is assumed, a correction must be applied to the calculation of ocean temperatures; when this correction is applied, the calculated bottom water temperature at Site 277 is ~9°C, which is much too high for the presence of a major ice sheet.

Kennett & Shackleton (1976) estimate that the event ~38 m.y. ago was completed in only about 100,000 years, but Shackleton (1979) believes that such a close estimate is probably not justified; nevertheless he agrees that it took place in well under 1 m.y. Kennett & Shackleton (1976) think that it represents the crossing of some critical threshold, not yet identified but in some way related to the gradual isolation of Antarctica; they point out that intense circumantarctic flow did not start until much later in the Oligocene.

Kvasov & Verbitsky (1981) have a different view, stating that deep circulation south of Tasmania began ~38 m.y. ago and caused the abrupt cooling; however, they cite no supporting evidence for this assumption.

The second episode of rapid ^{18}O enrichment in benthic foraminifera was between 16.5 and 13 m.y. ago, with the greatest change occurring between 14.8 and 14 m.y. ago (Woodruff et al 1981). Savin et al (1975) tentatively suggest that the thermal isolation of Antarctica resulting from the opening of Drake Passage between Tierra del Fuego and the Antarctic Peninsula triggered the inferred mid-Miocene glacial buildup. However, Barker & Burrell (1977) conclude that deep circulation through Drake Passage began about 23.5 m.y. ago, several million years before the isotopic excursion. They later suggest (Barker & Burrell 1982) that the ~7 m.y. difference relates to a minimal volume of Antarctic Circumpolar Current needed for effective disruption of meridional heat transport. However, Kennett (1977) points out that subantarctic water temperatures were rising 23.5–16 m.y. ago; he concedes that the cause of the inferred mid-Miocene glacial buildup is not yet clear, but suggests that an increase in Antarctic snowfall accompanying the warming may have contributed. Schnitker (1980) believes that the warming resulted from the upwelling of North Atlantic deep water in high southern latitudes that started after the Iceland-Faeroe Rise subsided.

Between 14.8 and 14 m.y. ago, when the mid-Miocene isotopic change was most rapid, frequent major oscillations in the $\delta^{18}O$ curve imply that the Antarctic Ice Sheet was very unstable. The variations were apparently cyclical, with periods of ~100,000 years, and this suggests to Woodruff et al (1981) that the ice sheet was perhaps responding to the same orbital variations that later are thought to have controlled the Northern Hemisphere ice sheets (Hays et al 1976). Moore et al (1982), however, after studying the variance of the late Miocene and Pleistocene carbonate records in a core (RC-11-209) from the equatorial Pacific Ocean, find that whereas a 100,000-year oscillation dominates the Pleistocene, a 400,000-year oscillation dominates the late Miocene; they conclude that this 400,000-year cycle may reflect the natural period of "glacial-interglacial" oscillation in Antarctica.

An alternative interpretation of the $\delta^{18}O$ curve Matthews & Poore (1980) have recently challenged the foregoing traditional interpretation of the oxygen isotopic record, pointing out that it starts by comparing the past isotopic composition of the ocean to that of the modern ocean, and that the resultant calculated tropical sea surface temperatures between the late Eocene and the middle Miocene are several degrees lower than today's (e.g. Savin et al 1975, Savin 1977). They prefer to compare Tertiary data to average late Pleistocene conditions, and to assume that tropical sea surface temperatures stayed constant. Tropical planktic $\delta^{18}O$ *calcite* is then constrained to vary only as a function of the ice-volume $\delta^{18}O$ *water* effect, whereas the benthic $\delta^{18}O$ *calcite* may vary as a function of either ice volume $\delta^{18}O$ *water* effect or bottom-water temperature variation. When benthic and tropical planktic $\delta^{18}O$ *calcite* vary together, as at the Eocene-Oligocene boundary about 38 m.y. ago, an ice-volume signal is implied, whereas when the benthic record alone shows a change, as between 16.5 and 13 m.y. ago (middle Miocene), this signifies a change in bottom-water temperature alone. Thus Matthews & Poore (1980) interpret the abrupt enrichment in ^{18}O of tropical surface water near the Eocene-Oligocene boundary as marking a major increase in global ice volume, representing a lowering of average glacio-eustatic sea level of at least 50 m (that is, only slightly less than the volume of the present ice sheet in East Antarctica). They do not rule out the presence of significant land ice before then, and they urge that ice be considered to have been present back into geologic time until proved otherwise.

The hypothesis of Matthews & Poore (1980) that significant ice was present in Antarctica during the Paleogene receives some support from Barron et al (1981), Corliss (1981), Kvasov & Verbitsky (1981), LeMasurier & Rex (1982), Denton et al (1982), and Webb (1982). Barron et al (1981) point out that a fundamental problem in paleoclimatology is how the globally ice-free climate generally inferred for Cretaceous and early Paleogene time could have been maintained; with a climate model based on the zonal energy balance, the necessary equator-to-pole temperature gradient is especially hard to effect with known mechanisms of plausible magnitude, even if the past major changes in geography are taken into account. They support Matthews & Poore (1980) in urging critical reexamination of the general acceptance of above-freezing polar temperatures during the Cretaceous Period and no significant accumulation of ice until middle Miocene time. Corliss (1981) finds no major change in the benthic fauna at the Eocene-Oligocene boundary at DSDP Site 277 south of New Zealand, despite an isotopic enrichment which, if interpreted in terms of temperature alone, implies a cooling of 3°C. He concludes that either the foraminifera possessed wide environmental tolerance, or the drop in temperature was less than 3°C, in which case some of the isotopic signal must represent accumulation of ice. As was mentioned earlier, Kvasov & Verbitsky (1981) state, without supporting evidence, that deep circulation between Antarctica and Australia began at the Eocene-Oligocene boundary 38 m.y. ago [whereas Kennett et al (1974) and Weissel et al (1977) believe it did not start until 30–25 m.y. ago] and the resultant cooling of Antarctica caused major ice fields to form on several mountain massifs in

East Antarctica. They believe that these glaciers would have coalesced and expanded to cover most of East Antarctica in about 100,000 years. Noting that this is the same as the duration of the rapid isotopic change, as estimated by Kennett & Shackleton (1976), Kvasov & Verbitsky (1981) conclude that the formation of the ice sheet intensified the cooling and accounts for its abruptness. They do not suggest that any of the rapid isotopic change relates directly to an increase in ice volume, but such a conclusion is a logical requirement of their glacial reconstruction.

LeMasurier & Rex (1982), noting the presence in West Antarctica of subglacially erupted hyaloclastite as much as 27 ± 1 m.y. old, conclude that an ice sheet was present in West Antarctica no later than the middle Oligocene. Denton et al (1982) believe that their interpretation of the glacial history of the Dry Valleys area, south Victoria Land, supports LeMasurier & Rex's (1982) chronology for the development of an ice sheet in West Antarctica, and is more compatible with Matthews & Poore's (1980) timetable for growth of the ice sheet in early Oligocene time than with the conventional view that it formed during the middle Miocene. Finally, in McMurdo Sound, Antarctica, Webb (1982) interprets the lower part of an ocean core, which according to Barrett & McKelvey (1981) consists throughout of glaciomarine sediments, as being of late Paleocene, early Eocene, and late Eocene age. He believes that five major hiatuses in the sequence represent intervals when grounded ice from the Transantarctic Mountains extended at least 100 km into the Ross Sea; at these times significant amounts of ice accumulated in East Antarctica also, contributing to global depression of sea level. The hiatuses are from early to middle Eocene, late Eocene to middle Oligocene, late Oligocene to early Miocene, middle Miocene to early Pliocene, and near the Pliocene-Pleistocene boundary. Webb (1982) also suggests that an earlier hiatus may reflect buildup of ice in Antarctica near the Cretaceous-Tertiary boundary.

ICE RAFTING Although Geitzenauer et al (1968) and Margolis & Kennett (1971) conclude, from the presence of quartz grains identified as being of glacial origin, that ice rafting was in progress far from the Antarctic coast during early and middle Eocene time, this view has not been upheld. Formerly it was supported by an age of 42 ± 9 m.y. (late Eocene) for hyaloclastite in West Antarctica, inferred to have been erupted subglacially (LeMasurier 1972), but this age has since been revised to 15 ± 5 m.y.; the oldest hyaloclastite is 27 ± 1 m.y. old (LeMasurier & Rex 1982). The indication that ice rafting was active during the Eocene near the coast of the Ross embayment in waters with a cool-temperate fauna (Webb 1982) does not favor long-distance ice rafting. The first unequivocal evidence for ice rafting beyond near-shore waters comes from DSDP Site 270 in the southern Ross Sea (lat 77° 30′S), where a thick glaciomarine sequence, whose basal glacial unit contains middle to late Oligocene dinoflagellates, covers a preglacial glauconitic sandstone dated at 26 ± 0.4 m.y. by K-Ar (Hayes & Frakes 1975).

After the late Oligocene, ice rafting steadily expanded its range northward, closely following the boundary between earlier calcareous and later siliceous sedimentation, until at the very end of the Miocene cold, ice-laden water abruptly expanded northward by about five degrees of latitude to near its present limits (Hayes & Frakes 1975, Plafker et al 1976). Mercer (1973) points out that ice rafting would have abruptly increased its range when ice shelves and floating ice tongues first formed, as the tabular bergs they calve may be two or three orders of magnitude larger than bergs from temperate outlet glaciers.

Ciesielski et al (1982) conclude from the history of ice rafting and of the erosion of sea-floor sediments by strong bottom currents in the southwestern South Atlantic Ocean that Antarctic Bottom Water as cold and saline as today's was first produced at the end of the Miocene, when temperature had dropped to the level at which ice shelves could form. They believe that the extensive late Miocene erosion of circumantarctic deep-sea sediments was the work of vigorous bottom currents as the production of Antarctic Bottom Water reached a maximum. Such a volume of Antarctic Bottom Water—well in excess of today's—was, the authors believe, produced beneath the vast ice shelf that would have extended over much of West Antarctica before its central part grounded to become the West Antarctic ice sheet. This reconstruction conflicts with that of LeMasurier & Rex (1982), who believe that the West Antarctic ice sheet has been in place since the end of the Oligocene (see below).

Terrestrial Geological Record

Four types of terrestrial study have provided information about Antarctic glacial history: (*a*) study of ancient landforms and deposits of glacial drift, (*b*) determination of the subglacial topography by radio-echo sounding, (*c*) dating of volcanic rocks identified as having been erupted subglacially, and (*d*) coring of sedimentary sequences. In East Antarctica most of the information comes from the Transantarctic Mountains, and in West Antarctica, most comes from the volcanic province of northern Byrd Land.

PRE-PLEISTOCENE GLACIATION IN EAST ANTARCTICA At a 2600-m elevation in the Wisconsin Range, Transantarctic Mountains (lat 86°S), Mercer (1968a) reports compact till with properties diagnostic of deposition beneath wet-based ice, covering two units thought to be the products of mass-wasting. He interprets this sequence as showing the very start of

Antarctic glaciation, as a local ice cap formed on the high plateau after an interval of periglacial conditions. With continued cooling, he suggests, valley glaciers descended the interior flank of the Transantarctic Mountains and eventually reached low ground where they formed piedmont lobes that thickened and became the nucleus of the East Antarctic ice sheet. Drewry (1975) has since detected by radio-echo sounding a network of glaciated valleys on the inland flank of the Transantarctic Mountains, and he points out that, according to a preliminary survey, they are present also in the northeastern part of the subglacial Gamburtsev Mountains and on smaller mountain massifs in central East Antarctica northeast of the Gamburtsev massif. Drewry (1975) believes that both the southern part of the Transantarctic Mountains and the northeastern part of the Gamburtsev Mountains were the centers of growth for the East Antarctic ice sheet, but that glaciation began first in the Transantarctic Mountains because of higher precipitation. Denton et al (1971) and Kvasov & Verbitsky (1981), however, think that the massif of the Gamburtsev Mountains, and perhaps also the mountains in Queen Maud Land, were the chief centers of growth (Figure 2).

An important factor in determining where the first ice accumulated is the position of the pole at that time. The movement of Antarctica after it separated from Australia is not known with total confidence. According to one reconstruction (e.g. Smith & Briden 1977, Weissel et al 1977), the late Eocene (40 m.y. ago) pole was in Coats Land on the east side of the Weddell Sea, about 12° of latitude from its present position. This would place north Victoria Land near lat 60°S at that time instead of its present 72°S, and under these circumstances a small, temperate ice sheet might well have been present in Queen Maud Land while *Nothofagus* (southern beech) forests flourished along the coast of the Ross embayment. However, a more recent reconstruction places the late Eocene pole considerably nearer its present position (Suarez & Molnar 1980). Virtually nothing is known about the glacial history of Queen Maud Land, and no offshore marine cores have yet been obtained.

Mercer (1968a) and Drewry (1975) note the importance of uplift in initiating mountain glaciation, and Drewry (1975) calculates that the Transantarctic Mountains have risen 1500–2000 m since the late Oligocene, when they were only 500–1500 m in elevation. Mayewski (1975) believes that all traces of former local glaciers were later obliterated when the ice sheet overrode the Transantarctic Mountains sometime before 4.2 m.y. ago, and Denton (1979), after working in the Byrd Glacier area (lat 80°S), concludes that large portions of the Transantarctic Mountains were uplifted through a pre-existing ice sheet. However, Denton et al (1982), after an extensive study of landforms and sediments in south Victoria Land (lat

78°S), have since modified these views, and now believe that the Transantarctic Mountains there were initially covered by a large local ice cap, whose outlet glaciers carved the valleys on both flanks. Later, a thick ice sheet from East Antarctica twice overrode the mountains, before and after the fiord stage in Taylor Valley, which, as the cores obtained by the Dry Valleys Drilling Program have shown, occurred no later than the late Miocene (McKelvey 1981). On both occasions the ice sheet moved from southwest to northeast at a sharp angle to the pre-existing valleys, and Denton et al (1982) believe that this direction of ice movement is hard to explain unless ice from East Antarctica was deflected by the presence of thick ice to the east in the Ross embayment. They therefore favor LeMasurier & Rex's (1982) contention that the West Antarctic ice sheet originated during the Oligocene, at about the same time as the East Antarctic ice sheet (see below).

Stump et al (1980) have obtained whole-rock K-Ar ages of 18.32 ± 0.35 and 15.86 ± 0.30 m.y. for basaltic lava flows overlying and interbedded with hyaloclastites at Sheridan Bluff and Mount Early in the Transantarctic Mountains at the head of Scott Glacier (lat 87°S). $^{40}Ar/^{39}Ar$ spectral analyses have been carried out and confirm the integrity of these samples. Stump et al (1980) reject the possibility that the hyaloclastites were erupted beneath a valley glacier and do not consider the possibility of eruption beneath a local ice cap; they thereby conclude that they were erupted under the East Antarctic ice sheet in early Miocene time.

PRE-PLEISTOCENE GLACIATION IN WEST ANTARCTICA The direct evidence for pre-Pleistocene glaciation in West Antarctica consists of till interbedded with volcanic rocks and of volcanic mountains partly or wholly composed of hyaloclastites, inferred to have been erupted subglacially rather than subaqueously. In the Jones Mountains, till resting on a striated surface of basement rocks is covered by basaltic lava flows. Radiometric ages of flows within 3 m of the till range from 9 m.y. up to 300 m.y., but higher up the ages are more consistent, ranging from 6.1 to 24 ± 12 m.y., with several clustering between 7 and 10 m.y. Rutford et al (1972) conclude that the West Antarctic ice sheet was present by about 7 m.y. ago.

LeMasurier & Rex (1982) have studied exposures of hyaloclastites of Oligocene to Pleistocene age in the volcanic province that extends 1400 km east-west and 350 km north-south in northern Byrd Land. They interpret these exposures as having been subglacially erupted, ruling out a submarine origin on account of the absence of marine interbeds, the great thickness of the deposits, and the improbably great amount of uplift that such an origin would imply. At four sites, whole-rock K-Ar ages of more than 14 m.y. have been obtained, the oldest being 27 ± 1 m.y., i.e. late Oligocene. LeMasurier

& Rex (1982) conclude that the West Antarctic ice sheet formed during middle Oligocene time or earlier on a flat, prevolcanic subaerial erosion surface that has since been fragmented by block faulting. On the assumption that the entire Ross Sea–Byrd Land region has tectonic unity, they suggest that the erosion surface formerly extended not only over the half-million km^2 or so of the present volcanic province, but also over the Byrd Subglacial Basin to the south. As evidence they point to an apparent near-absence of sediment fill in the basin, implying that it has been ice-filled since formation. However, a recent geophysical study by Jankowski & Drewry (1981), which suggests that the basin is floored by a sequence of interbedded sedimentary and volcanic strata of unknown age that overlies basement, throws some doubt on this assumption. Furthermore, the advanced state of development of glaciated landforms on the flank of the Transantarctic Mountains between the Reedy Glacier and the Ohio Range, now partially buried by the grounded West Antarctic ice sheet (Mercer 1968a), implies that a long interval of glacial erosion of the Transantarctic Mountains preceded emplacement of the West Antarctic ice sheet. Until the presence or absence of Oligocene and Miocene sediments in the Byrd Subglacial Basin has been determined, the hypothesis of LeMasurier & Rex (1982) that an ice sheet covered much of West Antarctica at the end of the Oligocene must remain in doubt. On the other hand, their investigations do provide strong evidence for a small ice sheet—a half-million km^2 or so—covering the volcanic province of northern Byrd Land at the times that the hyaloclastites were erupted. However, the age determinations must be treated with some skepticism until $^{40}Ar/^{39}Ar$ spectral analyses have been performed on the dated rocks.

GLACIATION IN MIDDLE AND LOW LATITUDES BEFORE 1 M.Y. AGO

The Antarctic climate is now severe enough for the presence of cold ice at sea level, and thus of ice shelves, which are necessary for the existence of the marine West Antarctic ice sheet. Clearly, then, this ice sheet must have been present when the scale of glaciation in southern middle latitudes was greater than it is today. Thus the earliest known glaciation in southern middle latitudes, which necessarily relates to a time of more extensive ice cover than today's, gives a minimal age for the formation of the West Antarctic ice sheet. Since ice shelves first formed around Antarctica, conditions in the surrounding seas, especially the extent of sea ice in summer, have greatly influenced the climate of middle latitudes. A close correspondence is therefore to be expected between variations of midlatitude glaciers and changes in the surface temperature of the Antarctic and subantarctic seas.

The most detailed record of pre-late-Pleistocene glaciations in southern middle latitudes comes from the high plains east of the Andean Cordillera in southern South America where, by a rare combination of circumstances, glacial sediments are excellently exposed interbedded with basaltic lava flows ranging in age from late Miocene to early Pleistocene. In Australia and New Zealand, only one glaciation before the late Pleistocene has been recognized; it is believed to be early Pleistocene in age. However, marine sediments in New Zealand of late Miocene and Pliocene age record several occasions when the climate was colder than it is today, and some of these events were of glacial-age severity. The apparent absence of pre-Pleistocene glacial sediments in New Zealand may relate to the recency of the uplift of the Southern Alps, which are thought to have reached most of their present height during the second half of the Pleistocene.

Late Miocene to Early Pleistocene Glaciations in South America

Today, the Andean Cordillera south of latitude 46°S has heavy ice cover, including two ice fields. During glaciations, ice extended eastward over the Patagonian plains where, as the result of episodic vulcanism during the Neogene, glacial sediments are now interbedded with basaltic lava flows. Exposures of interbedded basalt and till are known from three areas in Santa Cruz province, Argentina: south of Lago Buenos Aires near latitude 47°S, where they are late Miocene or earliest Pliocene in age; north of Lago Viedma near latitude 49°30′S, where they are mid-Pliocene in age; and south of Lago Argentino near latitude 50°30′S, where they are latest Pliocene and early Pleistocene in age (Figures 3 and 4). All previously published age determinations have, where necessary, been revised according to the time scale of LaBrecque et al (1977).

LATE MIOCENE–EARLIEST PLIOCENE GLACIATION NEAR LAGO BUENOS AIRES

South of Lago Buenos Aires, till lies between basaltic lava flows with average ages of 7.03 ± 0.11 m.y. and 4.63 ± 0.07 m.y. The site is over 100 km from the nearest large existing glacier in the central cordillera, and although the topography at the end of the Miocene undoubtedly differed from today's, the cordillera was already in place with its axis in its present location. In other words, glaciation of this site implies a more severe climate in latest Miocene or earliest Pliocene time than prevails today (Mercer & Sutter 1982).

PLIOCENE GLACIATION NEAR LAGO VIEDMA North of Lago Viedma, till at one site lies between basaltic lava flows with indistinguishable ages: 3.64 ± 0.07 m.y. above and 3.57 ± 0.09 m.y. below. At another site nearby, till is covered by flows 3.64 ± 0.07 m.y. and 3.59 ± 0.14 m.y. old, covers a flow 3.78 ± 0.03 m.y. old, and contains basaltic clasts 3.62 ± 0.05 and 3.57 ± 0.57 m.y.

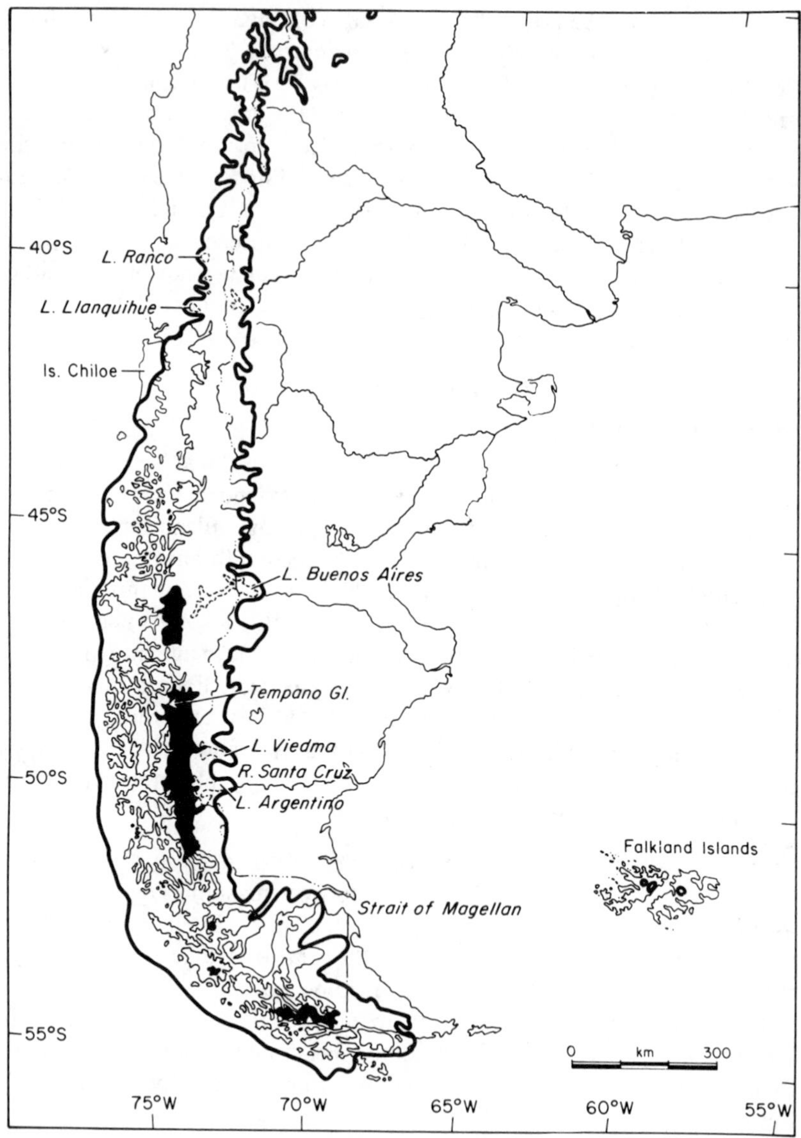

Figure 3 Southern South America. Solid black: present icefields. Thick line: ice margin at last glacial maximum (after Hollin & Schilling 1981).

old. These dates from the two sites show that the till interbedded with the basalt was deposited about 3.6 m.y. ago (Mercer 1976).

LATE PLIOCENE AND EARLY PLEISTOCENE GLACIATION NEAR LAGO ARGENTINO South of Lago Argentino, a sequence of interbedded tills and basaltic lava flows is more than one million years old. Six tills are present. Five are bracketed by ages of 1.05–1.51 m.y., 1.71–1.91 m.y. (two tills), 1.91–2.11 m.y., and 2.11–2.12 m.y. The sixth till is >2.11 m.y. old, but its maximum age is not known. At another site 25 km to the east, till lies between two lava flows with indistinguishable ages: 2.02 ± 0.09 m.y. above, and 2.01 ± 0.10 m.y. below. East of Lago Argentino in the valley of the Rio Santa Cruz, close to the outermost end moraine, coarse outwash gravel underlies basalt 2.73 ± 0.06 m.y. old, implying that glaciation in the mountains west of Lago Argentino had begun before then (Mercer 1976).

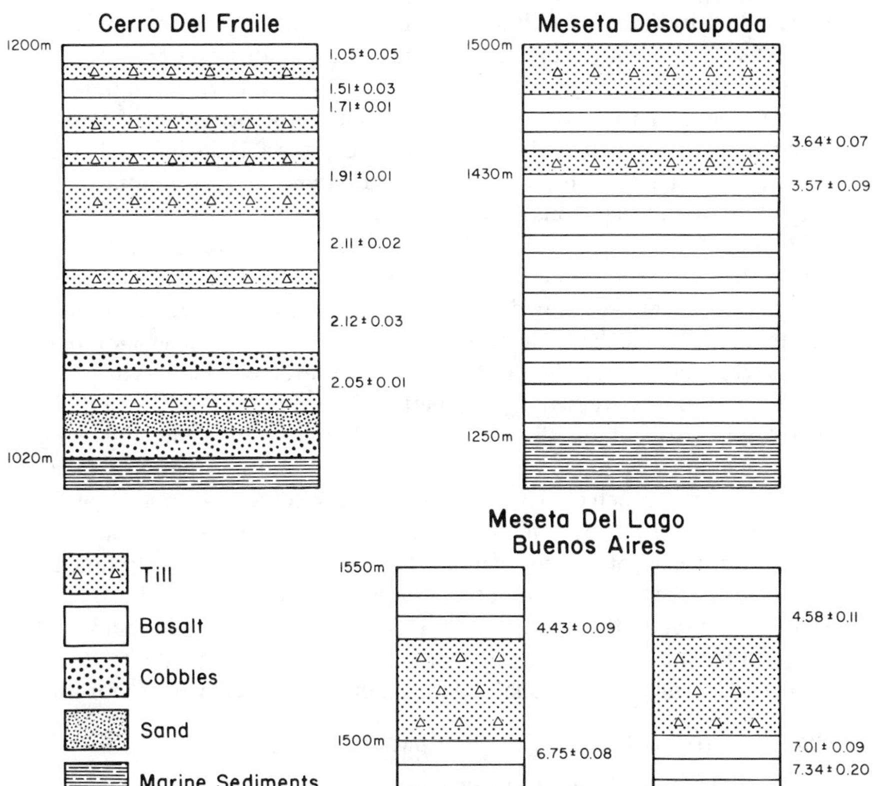

Figure 4 Diagrammatic sections of interbedded till and basalt in southern Argentina. Ages in millions of years. At a third site in the Meseta del Lago Buenos Aires, basalt 5.05 ± 0.07 m.y. old covers till.

PLIOCENE GLACIATION IN BOLIVIA The only record of pre-Pleistocene glaciation in low latitudes comes from near La Paz, Bolivia ($\sim 16°$S), where an ignimbrite dated by K-Ar on biotite at 3.27 ± 0.14 and 3.28 ± 0.13 m.y. covers till deposited by a glacier originating in the Cordillera Real. Clapperton (1979) thinks that this till may be the same age as that dated at 3.6 m.y. in southern Argentina (Mercer 1976).

Early Pleistocene Glaciation and Late Neogene Cold Spells in New Zealand

As mentioned previously, the scarcity of direct evidence for glaciation in New Zealand before the late Pleistocene is believed to be due mainly to the recency of major mountain uplift (Suggate 1978, Bowen 1978). The earliest recognized glaciation, the Ross, is known only from a restricted area on the west side of the Southern Alps, South Island. It is generally held to be earliest Pleistocene in age, but may be late Pliocene (Fleming 1973). If glaciers were present during earlier cold episodes, no direct evidence for them has yet come to light. However, cool or cold climates at the times of the pre-Pleistocene glaciations in southern South America (that is, between 7 and 4.6 m.y. ago and about 3.6 m.y. ago) have been inferred from biological changes in both New Zealand and the surrounding seas, suggesting that glacial climates were then circumpolar in extent.

In uplifted marine sediments in New Zealand, Kennett & Watkins (1974) find evidence for a severe cooling of Pleistocene intensity during the late Miocene Kapitean Age, dated by Loutit & Kennett (1979) at 6.2–5.3 m.y. ago, followed by much warmer conditions in the early Pliocene. Devereux et al (1970), who have studied Pliocene and Pleistocene sediments at one of the same outcrops studied by Kennett & Watkins (1974), interpret the microfaunal content as showing that throughout Pliocene and early Pleistocene times, surface water was never much warmer than it is today and at times was much colder. Severe cold occurred near the middle of the Pliocene, perhaps near the end of the Pliocene, and in the early Pleistocene. At another site, according to Fleming (1973), mainly nonmarine sediments of early Pleistocene age—fission track dates suggest an age of 1.1–1.3 m.y., which is younger than the Ross Glaciation—contain cold-temperate pollen. Fleming (1973) suggests that these sediments relate to three "glacials," but no known glacial sediments provide direct evidence for this.

Plio-Pleistocene Glaciation in Australia

In Tasmania (West Coast Range), Colhoun (1982) notes the presence of highly weathered till, deposited by an ice cap covering ~ 1000 km^2 (compared to ~ 100 km^2 at the last glacial maximum). At one locality this weathered till rests on a paleosol containing pollen of Tertiary age.

Colhoun (1982) thereby concludes that the glaciation is no younger than early Pleistocene, and may be older.

Miocene-Pliocene Midlatitude Glaciation and Southern Ocean Temperature

Known occurrences of glacial sediments interbedded with basaltic lava show that glaciations affected southern South America sometime between 7 and 4.6 m.y. ago, 3.6 m.y. ago, and repeatedly after about 2 m.y. ago. Does the rarity of known glaciations before 2 m.y. ago merely reflect the incompleteness of the record, or is it because the first glaciations were in fact rare events, separated by long nonglacial intervals? The marine geological record on the whole supports the second interpretation.

A widespread, moderate-to-severe cooling of surface waters in middle and high latitudes of both hemispheres at the end of the Miocene is well documented—in the Southern Hemisphere, for example, in and near New Zealand (Kennett & Watkins 1974, Kennett & Vella 1975, Loutit 1981) and in the southeast South Pacific (Bandy et al 1971). Hayes & Frakes (1975) point out that the abrupt climatic cooling and northward expansion of Antarctic surface water near the Miocene-Pliocene boundary must have strongly influenced the course of glaciation at lower latitudes. The age of the first known major glaciation in southern South America confirms their forecast.

The cold spell at the end of the Miocene gave way during the early Pliocene to a pronounced and rapid warming, followed by a brief return to cold in mid-Pliocene time; this cooling peaked 3.5–3.6 m.y. ago, at the same time that southern South America experienced major glaciation (Ciesielski 1975, Kennett & Vella 1975, Keany 1978). Although in the Northern Hemisphere the midlatitude ice sheets are believed to have first formed, on a small scale, about 3.2 m.y. ago (Shackleton & Opdyke 1977), many workers believe that in high southern latitudes the climate had warmed again by then, after the cold episode peaking about 3.5 m.y. ago. Bandy et al (1971) and Kennett & Vella (1975) believe that most of the Gauss Chron (3.4–2.5 m.y. ago) was comparatively warm; Barrett (1975) infers from the character of glaciomarine sedimentation in the Ross Sea that the Ross Ice Shelf disappeared at that time; and Shackleton & Cita (1979) conclude that the oxygen isotopic composition of deep water at DSDP Site 397 (near the Canary Islands) after 3.4 m.y. ago implies that either the Antarctic Ice Sheet shrank to half its present size, or temperatures in high southern latitudes rose enough to make deep water 1°C warmer than it is today. On the other hand, Ciesielski & Wise (1977) believe that cold conditions continued after 3.5 m.y. ago in high southern latitudes. Opposing trends in temperature in the Northern and Southern hemispheres would not be easy to account for.

Evidently, more investigations are needed to determine the behavior of glaciers and the course of ocean surface temperatures in both northern and southern middle latitudes between 3.5 and 2.5 m.y. ago.

LATE PLEISTOCENE GLACIATION IN THE SOUTHERN HEMISPHERE

During the late Pleistocene, and expecially during that part of the last glaciation that is within the reach of ^{14}C dating, the glacial record of the Southern Hemisphere becomes much clearer.

Late Pleistocene Glaciation in Antarctica

During the late Pleistocene and Holocene the volume of the ice sheet in East Antarctica has remained rather constant, as only the marginal zone has been appreciably affected by changes in the position of the grounding line on the continental shelf brought about by changes in eustatic sea level (Hollin 1962). The climate has remained too cold for much melting to occur. The ice sheet in West Antarctica, on the other hand, has been much more affected by changing sea level, which has brought about major movements of the grounding lines across the continental shelf. Also, on at least one occasion (the last interglacial) the ice sheet in West Antarctica may have been affected briefly but drastically by climatic change: warming above the critical level of about 0°C for midsummer may have destroyed the ice shelves and thereby caused the disappearance of all ice grounded below sea level. Local glaciers that do not reach the sea, however, have remained unaffected by changes in sea level, and because melting has been minimal, have fluctuated chiefly in response to changes in the local precipitation.

ANTARCTICA DURING THE LAST GLOBAL INTERGLACIAL (SANGAMON-EEM)

During the current global interglacial, melting plays an insignificant role in the budget of the Antarctic Ice Sheet, because average temperatures are below freezing point even in summer (except in the northwest of the Antarctic Peninsula). Shrinkage of the ice sheet since the last (global) glacial maximum has been almost entirely the result of rising glacio-eustatic sea level, caused by warmth in the Northern Hemisphere. However, ice shelves are now absent from those parts of the Antarctic Peninsula where midsummer temperatures are above freezing point, probably because such temperatures bring the temperature of the sea surface above -1.5°C (Robin 1958). A warming of about 5°C would bring the 0°C isotherm for midsummer from the northwestern Antarctic Peninsula southward to the fronts of the Ross and Ronne ice shelves, and with further warming these would start to recede. The end result would be the disappearance of all ice grounded below sea level and virtual deglaciation of West Antarctica. The melting of this ice would add an estimated 4–5 m thick layer of water to the present world ocean, and although Clark & Lingle (1977) show that the actual rise in sea level would not be globally uniform, the widespread occurrence, on supposedly more or less stable coasts (e.g. Bahamas, southern England, South Africa, Mauritius, western and southeastern Australia, Pacific islands), of a strandline of the last interglacial age ($\sim$125,000 years ago) a few meters above present sea level suggests that deglaciation of much of West Antarctica may in fact have occurred at that time (Mercer 1968b). Because much of the ice that melted had previously been displacing an equal mass of water, the oxygen isotopic change resulting from West Antarctic deglaciation would have been equivalent to considerably more than 5 m of sea-level rise—probably about 10 m. An isotopic change of this magnitude in marine carbonate is still hard to detect unequivocally above background noise, but the isotopic composition of both benthic foraminifera in a deep-sea core (Shackleton & Opdyke 1973) and corals and molluscs in a dated strandline (Fairbanks & Matthews 1978) suggests that during the last interglacial, the amount of water locked up in ice sheets was less than at present by an amount equivalent to 5–10 m of sea level. This is supporting, but not compelling, evidence for the hypothesis that the West Antarctic ice sheet was absent during the last interglacial.

ANTARCTICA DURING THE LAST GLOBAL GLACIATION During the present interglacial, average midsummer temperatures are below freezing point over the whole of Antarctica, except for the extreme northwest of the Antarctic Peninsula. Subaerial melting is a negligible factor in the mass balance of the ice sheet, and accumulation is offset almost entirely by calving of icebergs. During the last (global) glaciation, therefore, increased cold in itself would not have caused an increase in glaciation; on the contrary, the likely decrease in snowfall accompanying the lower air temperatures and the greater extent of sea ice in summer would be expected to result in glacial shrinkage. This has in fact been shown to be the case for local glaciers in south Victoria Land, which were smaller than they are today at the last (global) glacial maximum (Stuiver et al 1981). By contrast the ice sheet, most of whose periphery consists of either grounded ice cliffs or floating ice shelves and glacier tongues, will be affected much more by a change in sea level than by a change in accumulation rate. A drop in sea level produced by the buildup of ice sheets in the Northern Hemisphere will cause the grounding line of the ice sheet to advance seaward, whereas a rise in sea level will have the opposite effect. This prediction that the expansion and contraction of the ice sheet would be in unison with the variations of the ice sheets in middle latitudes of the Northern Hemisphere was discussed by

Hollin (1962), but only recently has it been firmly demonstrated by ^{14}C dating (Stuiver et al 1981).

In East Antarctica, where the continental shelf is narrow, the grounding line can advance only a short distance after a 100–150 m fall in eustatic sea level caused by buildup of Northern Hemisphere ice sheets; consequently, major changes will be confined chiefly to a narrow marginal zone, where ice will locally thicken by several hundred meters. Further inland, thickening will diminish rapidly with increasing distance from the coast, and will be negligible in the central parts of the ice sheet. In West Antarctica, however, a glacial-age drop in sea level will produce major changes, as the grounding line advances several hundred kilometers across the seafloor beneath the present Ross and Ronne ice shelves.

South Victoria Land In the Dry Valleys of south Victoria Land, Denton et al (1971) show that large lakes were formerly ponded against ice lobes that entered the valley mouths from grounded ice in McMurdo Sound and the Ross Sea. This has happened at least four times, all within the last 1.2 m.y. These Ross Sea glaciations, which were controlled by sea level, were out of phase with the fluctuations of alpine glaciers on the valley sides, which reflected local snowfall.

The most recent Ross Sea glaciation left prominent ice-cored moraines on the valley side at the mouth of Taylor Valley. Stuiver et al (1981) have obtained ^{14}C ages for algal mats in the uppermost stranded deltas ranging from about 17,000 to 21,200 BP (Before Present; that is, before A.D. 1950). They infer that lake levels were highest when grounded ice in the Ross Sea was thickest, and that this occurred when the ice-sheet grounding line reached its most northerly position as glacio-eustatic sea level dropped to its lowest point. The dates are similar to those for the maximum extent of the Laurentide Ice Sheet in the Middle West of the United States. Deltas at successively lower elevations record progressive thinning of the Ross Sea ice as sea level rose. Ice was still grounded in McMurdo Sound about 8300 years ago, and the grounding line probably did not reach its present position until after 6700 BP. Evidence for comparable changes in the West Antarctic ice during what is believed to have been the last glaciation (no ^{14}C dating is available) has been found in the Ellsworth Mountains (Rutford et al 1980), where the ice level is controlled by the position of the grounding line of the Ronne Ice Shelf.

Southern Transantarctic Mountains Observations in the southern Transantarctic Mountains support the reconstruction of Stuiver et al (1981) in the Ross Sea area. Alongside both the Reedy and Beardmore glaciers that now flow from the East Antarctic ice sheet through the Transantarctic Mountains to the Ross Ice Shelf, ice-cored lateral moraines, similar to those dated at about 20,000 BP in the McMurdo Sound area, are stranded far above the glacier surfaces at the ice-shelf ends; these moraines, however, progressively approach the ice surface up-glacier, implying major change in ice thickness in the Ross embayment and stability of the East Antarctic ice sheet during a glacial-interglacial hemicycle. Mercer (1968a, 1972) finds that the surface levels of the Reedy and Beardmore glaciers were 500–700 meters higher at their mouths than they are today when the ice-cored moraines were formed, and he infers that an extension of the grounded West Antarctic ice sheet then occupied much of the area now covered by the Ross Ice Shelf. On the other hand, Mayewski (1975), after investigations in the Shackleton and Beardmore glacier areas, deduces a much smaller expansion of the grounded West Antarctic ice sheet, and Whillans (1976) interprets the results of radio-echo sounding in central West Antarctica as showing that the elevation of the ice divide has changed little in the past 30,000 years.

Glaciation Beyond the Range of ^{14}C Dating in Middle and Low Latitudes

Between ~30,000 BP (the limit of reliable ^{14}C dating by conventional techniques) and ~1 m.y. ago—that is, during most of the last half of the Pleistocene—little is known about the timing of glacial maxima in southern middle and low latitudes, on account of both the inherent unreliability and the fortuitous scarcity of K-Ar dating in this age range. Several end-moraine belts and drift sheets relating to glaciations of successively lesser extent lie in southern South America on both the west (Mercer 1976, Porter 1981) and east sides (Caldenius 1932, Flint & Fidalgo 1964, 1969, Mercer 1976) of the Andes, and in New Zealand on the east and west sides of the Southern Alps (Suggate 1965, Fleming 1973). Any attempt at trans-Pacific correlation of these drift sheets is hazardous because of the very different histories of mountain uplift in the two areas; in fact for this reason, the oldest and outermost moraine belts in South America may well have no counterparts in New Zealand. Correlation is difficult even between the dry eastern and the wet western sides of the same mountain range, because of the resultant contrast in weathering rates.

Limiting ages for some glacial deposits have been obtained from southern South America and from Marion Island in the same latitude belt in the Indian Ocean. In southern South America, the glacier flowing along the Strait of Magellan reached its maximum late-Cenozoic extent near lat 52°S after 1.2 m.y. ago and before 170,000 ± 35,000 BP. Although inconclusive evidence that the glaciation was more than 1 m.y. ago (Mercer 1976) (which would have bracketed it rather closely between 1.2 and 1 m.y. ago) is now considered invalid, the glaciation must in fact be a great deal older than

170,000 years, because an end moraine of the Lago Buenos Aires glacier near lat 47°S, 30 km inside the glaciation limit and formerly considered to date from the last glaciation (Caldenius 1932), is covered by a lava flow 177,000 ± 57,000 years old (Mercer 1982a). On Marion Island (lat 46°54′S, long 37°45′E), which is now ice-free, glaciers calved into the sea sometime between 105,000 ± 25,000 and 276,000 ± 30,000 years ago (whole-rock K-Ar ages of lava flows) (Hall 1978); that is, probably during the glaciation that preceded the Sangamon-Eem interglacial 125,000 years ago. In Africa on Kilimanjaro (lat 3°S, 5895 m) two glaciations antedate volcanic activity radiometrically dated at about 400,000 years (Downie 1964).

The Last Glaciation in Middle Latitudes

In middle latitudes (between lat 30°S and 60°S), glaciers are now present in southern South America, in New Zealand, and on several subantarctic islands. During the last glaciation they were also present in Australia, perhaps in southern Africa, and on several islands that are now cold-temperate (Figure 1).

SOUTHERN SOUTH AMERICA During glacial maxima in southern South America, glaciers east of the mountains ended on the semiarid plains, and no chronology based on ^{14}C dating has been obtained. West of the mountains, the glaciers south of lat 43°S extended beyond the modern coastline, so that features of full-glacial age are now inaccessible. Only north of lat 43°S, on Isla Chiloé and in the Chilean lake region to the north (Figure 3), where the glaciers ended in a moist environment on what is now dry land, have full glacial events been dated by ^{14}C. This dating control shows that during the last glaciation, glaciers were largest at a time beyond the range of ^{14}C dating (Mercer 1976), perhaps at the same time as the first and smaller of two main peaks of global ice volume at 73,000 BP (Shackleton & Opdyke 1973), and that they reached a second major maximum extent after 19,450 BP (Mercer 1976) and before 18,900 BP (Porter 1981). Mercer (1976) and Porter (1981) interpret a rise in the level of Lago Llanquihue (Figure 3) ~13,000 BP as caused by a final glacial readvance beyond the mountains, but later field results (unpublished) cast some doubt on this interpretation: glacial sediments on the east side of Isla Chiloé cover wood ~14,355 years old, and the Cordillera at lat 47°30′S was extensively deglaciated by 12,800 BP.

By 11,000 BP, the Tempano Glacier (lat 48°S) was smaller than it is today, and remained so for several thousand years (Mercer 1970). Geological evidence implies that the glaciers did not readvance between 11,000 and 10,000 BP, when full glacial conditions briefly returned to northwest Europe during the Younger Dryas Stade. However, this pattern of behavior conflicts with the climatic reconstruction of Heusser & Streeter (1980), who conclude from palynological studies that a much colder and wetter interval, peaking at 10,000 BP, followed a brief warm episode at 11,300 BP.

NEW ZEALAND In New Zealand, glaciers reached their maxima after 18,600 BP on the west side of the mountains (Suggate & Moar 1970), and after 19,200 BP on the east side (Soons & Burrows 1978). A later and smaller readvance was in progress 14,100 BP (Suggate 1965). Evidence for a readvance 11–10,000 BP is equivocal. Moar (1973) detects no evidence in the pollen record for cooling at that time, whereas Burrows & Russell (1975)

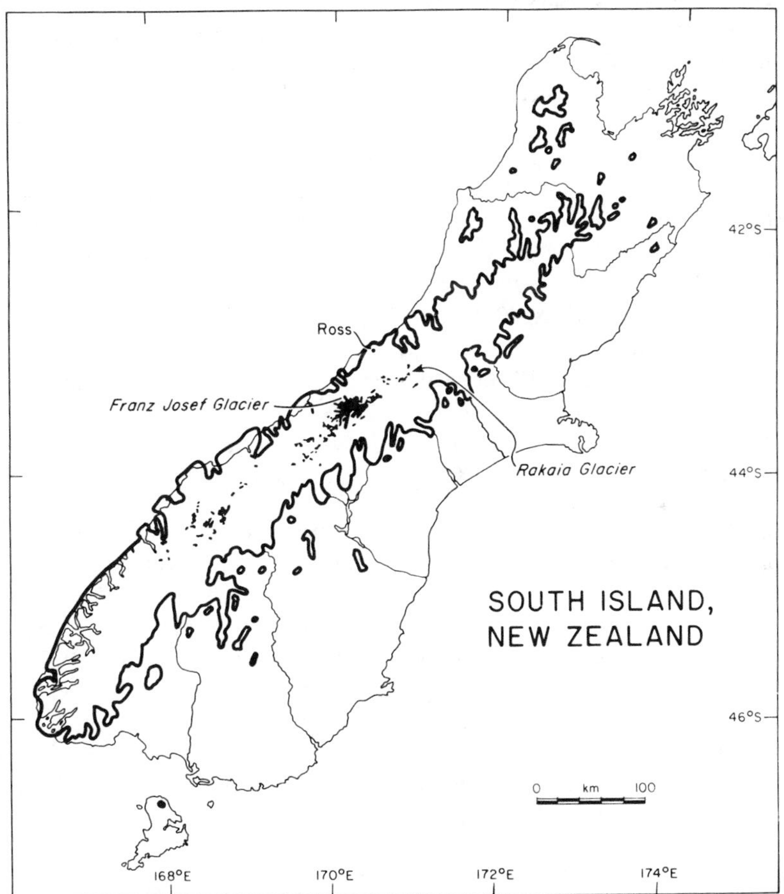

Figure 5 South Island, New Zealand. Solid black : present glaciers. Thick line : ice margin at last glacial maximum (after Hollin & Schilling 1981).

claim that the Rakaia Glacier then advanced 20 km beyond its present terminus, and Wardle (1978) reports that the Franz Josef Glacier was 7.5 km beyond its present terminus, and advancing, sometime between 11,650 and 10,750 BP (Figure 5).

SOUTHERN SOUTH AMERICA AND NEW ZEALAND COMPARED Climatically and topographically, southern South America and New Zealand (South Island) are very similar, so that close correspondence of glacial behavior during the last glaciation is to be expected. In both areas, the glacial maximum was between 20,000 and 18,000 BP, and was followed by a readvance, probably between 14,500 and 14,000 BP. In southern South America, a pollen study points to a marked cooling between 11,000 and 10,000 BP while the geological evidence implies less ice than at present, whereas in New Zealand no cooling has been detected in pollen studies while geological evidence suggests a marked glacial readvance.

Contrasting behavior of glaciers in the two areas seems inherently unlikely, so the conflicting evidence for events between 11,000 and 10,000 BP poses an unresolved problem. However, support for the South American geological evidence for no cooling comes from three independent lines of investigation in the Southern Hemisphere: (*a*) at Lago Ranco, Chile (lat 40°S), Hoganson & Ashworth (1981) detect no perceptible change in the beetle fauna between 12,500 BP and the present; (*b*) on Marion Island in the southwest Indian Ocean (lat 46°54′S), the climate warmed abruptly 14,000 BP and for the past 11–12,000 years has been similar to today's (Schalke & van Zinderen Bakker 1971); and (*c*) Lorius et al (1979) conclude from the oxygen isotopic composition of ice at Dome C in East Antarctica that the warmest part of the past 11,000 years there was the interval 11–8000 B.P.

AUSTRALIA Australia is now ice-free. During the last glaciation, cirque glaciers formed in New South Wales and ice caps and valley glaciers in Tasmania. In New South Wales glaciers were confined to the Snowy Mountains (lat 36°S), where basal organic sediments from between the two end moraines of a cirque glacier are ~20,200 years old (Costin 1972). In Tasmania (lat 42°S), Bowler et al (1976) report an ice cover of ~5000 km^2. Colhoun (1982), however, believes that less ice was present, although its limits are not yet fully known; the smaller and better-studied of two ice caps covered ~108 km^2 at its maximum extent 18–20,000 BP, compared with ~1000 km^2 during earlier glaciations.

SOUTHERN AFRICA In southern Africa, evidence for Pleistocene glaciation is equivocal. Harper (1969) calculates from the elevation of the lower limit of frost action that the higher peaks of the Drakensberg Mountains (lat 30°S, 3800 m) must have been above the snowline during the last glaciation, but he finds no physical evidence that glaciers actually formed. Hastenrath & Wilkinson (1973) note cirquelike forms at 2900–3100 m, but have no information on their age.

COLD-TEMPERATE AND SUBANTARCTIC ISLANDS On islands north of lat 60°S and south of the tree limit, glaciers are present today on Kerguelen, Heard and Bouvet islands, South Georgia, and the South Sandwich Islands. During the last glaciation, Marion Island to the south of Africa, MacQuarie Island and the Auckland Islands south of New Zealand, and the Falkland Islands east of southern South America also supported glaciers. The greatest glacial-interglacial changes in ice cover have occurred on those islands that are now north of the Antarctic Convergence but were south of it during the last glaciation. An example is Marion Island (lat 46°54′S, 1230 m), which is now ice-free but was largely ice-covered during the last glaciation, with glaciers 6–7 km long flowing to the sea from a central ice cap (Hall 1978). This ice cover disappeared rapidly after 14,000 BP (Schalke & van Zinderen Bakker 1971). By contrast the Falkland Islands (lat 51°–52°30′S), although lying 4–5° further south than Marion Island, remained north of the Antarctic Convergence during the last glaciation, and only cirque glaciers developed (Clapperton & Sugden 1976). South Georgia (lat 54°S), which has remained south of the Antarctic Convergence, is still heavily ice-covered; during the last glaciation, the ice probably extended beyond the present coastline to the present −120 m contour (Clapperton et al 1978).

The Last Glaciation in Low Latitudes

In low latitudes during the last glaciation, glaciers were present in the Andes of South America, in East Africa, and in New Guinea—all areas in which glaciers are found today.

SOUTH AMERICA In tropical South America the snowline is lowest in the equatorial regions, where precipitation and cloud cover are high year-round, and rises to the south as the dry season lengthens. The glaciers are extremely sensitive to any change in insolation and, therefore, in cloud cover, and the growing evidence for ice-age aridity in the South American tropics (Damuth & Fairbridge 1970) implies that the expansion of Andean glaciers at that time resulted from a considerable drop in temperature. Glaciers are present in Ecuador, Bolivia, and Peru; from Peru, enough ^{14}C dating is now available for a tentative chronology.

Mercer & Palacios (1977) report that the last major glacial maximum in the Cordillera Vilcanota (lat 13° 45′S, long 71°20′W) was sometime between 28,000 and 14,000 BP, and later investigations have shown that the maximum was at the very end of this interval, about 14,000 BP (Mercer

1982b). The glaciers were then much smaller than during an earlier advance that is beyond the range of ^{14}C dating but, judging from the fresh appearance of the moraines, is perhaps no older than the early part of the last glaciation. After 14,000 BP the ice receded rapidly, and by 12,200 BP the Quelccaya Ice Cap (lat 13°55′S, long 70°50′W) was close to its present size. A readvance began after 11,500 BP and probably culminated shortly after 11,000 BP. By 10,000 BP the ice had receded again to near its present margins.

AFRICA Today, glaciers are present south of the equator on Kilimanjaro (lat 3°S, 5895 m) and Mount Kenya (lat 0°10′S, 5200 m). Kilimanjaro has traces of four glaciations; the best preserved features are from the youngest glaciation, when ice covered 150 km^2 compared to 4 km^2 today (Downie 1964). Mount Kenya, whose glaciers today cover only 0.7 km^2, had 420 km^2 of ice during the last glaciation according to Baker (1969). No ^{14}C dating has been obtained from Kilimanjaro or Mount Kenya, but in the Ruwenzori (5110 m), situated about 800 km west of Mount Kenya and lying just north of the equator, glaciers receded from the youngest major moraines shortly before 14,700 BP (Livingstone 1962).

NEW GUINEA In the island of New Guinea, glaciers are now confined to the higher summits of Irian Barat, the western part of the island. In 1972 they covered 7 km^2, down from 13 km^2 in 1936 (Allison & Peterson 1976). At the last glacial maximum, glaciers with a total area of $\sim$2000 km^2 were widespread throughout the entire island between lat 4°S and 10°S. In Irian Barat 1400 km^2 were ice-covered, and in Papua-New Guinea, the eastern part of the island, about 600 km^2 of ice was distributed among 20 glaciated mountains (Galloway et al 1973). Hope & Peterson (1976) believe that the end moraines of the last glacial maximum in Irian Barat were abandoned shortly before 15,000 BP. Two readvances culminated soon after 13,000 BP and soon after 11,400 BP. Galloway et al (1973) report that in one valley in Irian Barat, glaciers reached their greatest extents shortly after 10,300 BP, the age of a log in a diamicton identified as till, but Hope & Peterson (1976) believe that the diamicton may be a mudflow.

Comparison of Glaciation in Low and Middle Latitudes

In the three tropical areas that have some ^{14}C-dating control—Peru, New Guinea, and equatorial Africa—the glaciers appear to have reached their maxima between about 15,500 and 14,000 BP, rather than 20–18,000 BP, when glaciers in middle latitudes of the Southern Hemisphere were largest and when global ice volume was greatest. This maximum extent of tropical glaciers occurred at about the same time that Northern Hemisphere ice sheets readvanced and some lobes reached their maxima, and probably about the same time as the final "late glacial" readvances in New Zealand and middle-latitude Chile. In both low and middle latitudes of the Southern Hemisphere the ice retreated rapidly after 14,000 BP, and deglaciation was far advanced by 12,500 BP. A minor readvance is believed to have taken place soon after 11,500 BP in New Zealand and Peru, but not in southern South America.

SUMMARY AND FUTURE WORK

During the past decade, the Deep-Sea Drilling Project has revealed a history of Antarctic glaciation longer than had previously been suspected. At first a measure of agreement emerged as studies of both ice-rafted detritus and oxygen isotopic ratios in the cores seemed to show that calving glaciers were present in the Ross embayment by the late Oligocene, that the ice sheet in East Antarctica built up during the middle Miocene, and that the ice sheet in West Antarctica formed during the late Miocene. Recently, however, this consensus has broken down and the picture has become less clear. Some workers have urged a reinterpretation of the marine oxygen isotopic record that would push the initial formation of ice sheets back into the Paleogene. Others have found evidence for unexpectedly early glaciation on Antarctica itself and in its coastal waters; calving glaciers are thought to have been present no later than the Eocene, an ice sheet present somewhere on the continent by the Eocene/Oligocene boundary, and the ice sheet in West Antarctica in place before the end of the Oligocene. Resolution of the problems of Antarctic glacial history will require further deep-sea drilling, particularly off Queen Maud Land, and drilling through the West Antarctic ice sheet to reach the stratified sediments that are thought to be present in the Byrd Subglacial Basin. Ocean cores from waters off Queen Maud Land should indicate whether glaciation in that part of Antarctica began significantly earlier than in areas that border the Ross embayment. The recovery of sediments from the Byrd Subglacial Basin should clarify the glacial history of the southern Transantarctic Mountains, and test the hypothesis that the West Antarctic ice sheet originated as a terrestrial glacier.

The start of major glaciation in middle latitudes has also been pushed back in recent years. It is now known to have begun by the end of the Miocene, and appears to have been followed first by a nonglacial interval, and then by renewed glaciation ~3.6 m.y. ago. Although these latest Miocene and mid-Pliocene glaciations were comparable in extent to those of the Pleistocene, they seem to have been unaccompanied by simultaneous buildup of ice sheets in the Northern Hemisphere; the generally accepted date for the initial formation of these is ~3.2 m.y. ago. There is no obvious reason for such a delay in the development of ice sheets in the Northern Hemisphere, particularly as tidewater glaciers are known to have been present in southeastern Alaska by the end of the Miocene. Further investigations are needed to confirm or refute a 3.2 m.y. date for the first Northern Hemisphere ice sheets.

During global late-Pliocene and Pleistocene glacial-interglacial cycles, the volume of the Antarctic Ice Sheet was controlled chiefly by sea level, which in turn was controlled by the volume of the Northern Hemisphere ice sheets. Thus the fluctuations of the Antarctic Ice Sheet give a proxy record of climatic change in the Northern Hemisphere, not the Southern. Because of the prevailing low temperatures, even during most interglacials, the ice sheet has been virtually unaffected by regional climatic fluctuations in Antarctica itself, except perhaps during exceptionally warm interglacials when the marine ice sheet in West Antarctica may have distintegrated.

Variations of southern midlatitude glaciers, in contrast to those of the Antarctic Ice Sheet, give proxy paleoclimatic records for the Southern Hemisphere. Major cold episodes appear to have peaked at about the same times in the two hemispheres, but there is some indication that postglacial warming may have been faster, and peaked earlier, in the Southern Hemisphere. If so, a likely explanation is the great contrast in the amount of heat needed to reduce the cryospheres of the two hemispheres from glacial to interglacial state, because of the absence of midlatitude ice sheets in the Southern Hemisphere. Thus accurate determination of the speed and timing of deglaciation in southern middle latitudes should give valuable information about causes and mechanisms of climatic change.

Acknowledgments

I thank D. H. Elliot and R. K. Matthews for reviewing parts of the manuscript. Field investigations by the author were supported by the US National Science Foundation.

Literature Cited

Allison, I., Peterson, J. A. 1976. Ice areas on Mt. Jaya: their extent and recent history. In *The Equatorial Glaciers of New Guinea*, ed. G. S. Hope, J. A. Peterson, I. Allison, U. Radok, 3:27–38. Rotterdam: Balkema. 244 pp.

Baker, B. H. 1969. Moraines, Mt. Kenya. In *Palaeoecology of Africa*, ed. E. M. van Zinderen Bakker, 4:62–63. Capetown: Balkema. 274 pp.

Bandy, O. L., Casey, R. E., Wright, R. C. 1971. Late Neogene planktonic zonation, magnetic reversals, and radiometric dates, Antarctic to the tropics. *Am. Geophys. Union Antarct. Res. Ser.* 15:1–26

Barker, P. F., Burrell, J. 1977. The opening of Drake Passage. *Mar. Geol.* 25:15–34

Barker, P. F., Burrell, J. 1982. The influence upon Southern Ocean circulation, sedimentation, and climate of the opening of Drake Passage. In *Antarctic Geoscience*, ed. C. Craddock, 43:377–85. Madison: Univ. Wis. Press. 1172 pp.

Barrett, P. J. 1975. Textural characteristics of Cenozoic preglacial and glacial sediments at Site 270, Ross Sea, Antarctica. In *Initial*

Reports of the Deep Sea Drilling Project, D. E. Hayes, L. A. Frakes, et al, 28:757–66. Washington DC: GPO

Barrett, P. J., McKelvey, B. C. 1981. Cenozoic glacial and tectonic history of the Transantarctic Mountains in the McMurdo Sound region: recent progress from drilling and related studies. *Polar Rec.* 20:543–48

Barron, E. J., Thompson, S. L., Schneider, S. H. 1981. An ice-free Cretaceous? Results from climate model simulations. *Science* 212:501–8

Bentley, C. R., Ostenso, N. A. 1961. Glacial and subglacial topography of West Antarctica. *J. Glaciol.* 3:882–911

Bowen, F. E. 1978. Stratigraphy: Lower Quaternary. Nelson and Westland. In *The Geology of New Zealand*, ed. R. P. Suggate, G. R. Stevens, M. T. Te Punga, pp. 567–70. Wellington: Gov. Printer. 820 pp.

Bowler, J. M., Hope, G. S., Jennings, J. N., Singh, G., Walker, D. 1976. Late Quaternary climates of Australia and New Guinea. *Quat. Res.* 6:359–74

Burrows, C. J. 1979. A chronology for cool-climate episodes in the Southern Hemisphere 12000–1000 Yr B.P. *Palaeogeogr. Palaeoclimatol. Palaeoecol.* 27:287–347

Burrows, C. J., Russell, J. B. 1975. Moraines of the upper Rakaia Valley. *R. Soc. N.Z. J.* 5:463–77

Caldenius, C. C. 1932. Las glaciaciones cuaternárias en la Patagonia y Tierra del Fuego. *Geogr. Ann.* 14:1–164

Cande, S. C., Mutter, J. C. 1982. A revised identification of the oldest sea-floor spreading anomalies between Australia and Antarctica. *Earth Planet. Sci. Lett.* 58:151–60

Ciesielski, P. F. 1975. Biostratigraphy and paleoecology of Neogene and Oligocene silicoflagellates from cores recovered during Antarctic Leg 28, Deep Sea Drilling Project. In *Initial Reports of the Deep Sea Drilling Project*, D. E. Hayes, L. A. Frakes, et al, 28:625–64. Washington DC: GPO

Ciesielski, P. F., Wise, S. W. 1977. Geologic history of the Maurice Ewing Bank of the Falkland Plateau (Southwest Atlantic sector of the Southern Ocean) based on piston and drill cores. *Mar. Geol.* 25:175–207

Ciesielski, P. F., Ledbetter, M. T., Ellwood, B. B. 1982. The development of Antarctic glaciation and the Neogene paleoenvironment of the Maurice Ewing Bank. *Mar. Geol.* 46:1–51

Clapperton, C. M. 1979. Glaciation in Bolivia before 3.27 Myr. *Nature* 277:375–77

Clapperton, C. M., Sugden, D. E. 1976. Maximum extent of glaciers in part of West Falkland. *J. Glaciol.* 17:73–77

Clapperton, C. M., Sugden, D. E., Birnie, R. V., Hanson, J. D., Thom, G. 1978. Glacier fluctuations in South Georgia and comparison with other island groups in the Scotia Sea. In *Antarctic Glacial History and World Palaeoenvironments*, ed. E. M. van Zinderen Bakker, 8:95–104. Rotterdam: Balkema. 172 pp.

Clark, J. A., Lingle, C. S. 1977. Future sea-level changes due to West Antarctic ice sheet fluctuations. *Nature* 269:206–9

Colhoun, E. A. 1982. The glaciations of the West Coast Range, Tasmania. In press

Cooke, D. W., Hays, J. D. 1982. Estimates of Antarctic Ocean seasonal sea-ice cover during glacial intervals. In *Antarctic Geoscience*, ed. C. Craddock, 131:1017–25. Madison: Univ. Wis. Press. 1172 pp.

Corliss, B. H. 1981. Deep-sea benthonic foraminiferal faunal turnover near the Eocene/Oligocene boundary. *Mar. Micropaleontol.* 6:367–84

Costin, A. B. 1972. Carbon-14 dates from the Snowy Mountains area, southeastern Australia, and their interpretation. *Quat. Res.* 2:579–90

Damuth, J. E., Fairbridge, R. W. 1970. Equatorial Atlantic deep-sea arkosic sands and ice-age aridity in tropical South America. *Geol. Soc. Am. Bull.* 81:189–206

Denton, G. H. 1979. Glacial history of the Byrd-Darwin Glacier area, Transantarctic Mountains. *Antarct. J. US* 14:57–58

Denton, G. H., Armstrong, R. L., Stuiver, M. 1971. The late Cenozoic glacial history of Antarctica. In *Late Cenozoic Glacial Ages*, ed. K. K. Turekian, 10:267–306. New Haven: Yale Univ. Press. 606 pp.

Denton, G. H., Prentice, M. L., Kellogg, D. E., Kellogg, T. B. 1982. Tertiary history of the Antarctic Ice Sheet: evidence from the Dry Valleys. *Nature*. In press

Devereux, I., Hendy, C. H., Vella, P. 1970. Pliocene and early Pleistocene sea temperature fluctuations, Mangaopari Stream, New Zealand. *Earth Planet. Sci. Lett.* 8:163–68

Downie, C. 1964. Glaciations of Mount Kilimanjaro, northeast Tanganyika. *Geol. Soc. Am. Bull.* 75:1–16

Drewry, D. J. 1975. Initiation and growth of the East Antarctic Ice Sheet. *J. Geol. Soc. London* 131:255–73

Fairbanks, R. G., Matthews, R. K. 1978. The marine oxygen isotope record in Pleistocene coral, Barbados, West Indies. *Quat. Res.* 10:181–96

Fleming, C. A. 1973. The Quaternary record of New Zealand and Australia. In

Quaternary Studies, ed. R. P. Suggate, M. M. Cresswell, pp. 155–62. Wellington: R. Soc. N.Z. 321 pp.

Flint, R. F., Fidalgo, F. 1964. Glacial geology of the east flank of the Argentine Andes between latitude 39°10′S and latitude 41°20′S. *Geol. Soc. Am. Bull.* 75:335–52

Flint, R. F., Fidalgo, F. 1969. Glacial drift in the eastern Argentine Andes between latitude 41°10′S and latitude 43°10′S. *Geol. Soc. Am. Bull.* 80:1043–52

Galloway, R. W., Hope, G. S., Löffler, E., Peterson, J. A. 1973. Late Quaternary glaciation and periglacial phenomena in Australia and New Guinea. In *Palaeoecology of Africa*, ed. E. M. van Zinderen Bakker, 8:125–38. Capetown: Balkema. 198 pp.

Geitzenauer, K. R., Margolis, S. V., Edwards, D. S. 1968. Evidence consistent with Eocene glaciation in a South Pacific deep sea sedimentary core. *Earth Planet. Sci. Lett.* 4:173–77

Hall, K. J. 1978. Evidence for Quaternary glaciation of Marion Island (sub-Antarctic) and some implications. In *Antarctic Glacial History and World Palaeoenvironments*, ed. E. M. van Zinderen Bakker, 12:137–47. Rotterdam: Balkema. 172 pp.

Harper, G. 1969. Periglacial evidence in southern Africa during the Pleistocene Epoch. In *Palaeoecology of Africa*, ed. E. M. van Zinderen Bakker, 4:71–101. Capetown: Balkema. 274 pp.

Hastenrath, S., Wilkinson, J. 1973. A contribution to the periglacial morphology of Lesotho, southern Africa. *Biul. Peryglac.* 22:157–67

Hayes, D. E., Frakes, L. A. 1975. General synthesis, Deep Sea Drilling Project, Leg 28. In *Initial Reports of the Deep Sea Drilling Project*, D. E. Hayes, L. A. Frakes, et al, 28:919–42. Washington DC: GPO

Hays, J. D. 1978. A review of the Late Quaternary climatic history of Antarctic seas. In *Antarctic Glacial History and World Palaeoenvironments*, ed. E. M. van Zinderen Bakker, 6:57–71. Rotterdam: Balkema. 172 pp.

Hays, J. D., Imbrie, J., Shackleton, N. J. 1976. Variations in the Earth's orbit: pacemaker of the Ice Ages. *Science* 194:1121–32

Heusser, C. J., Streeter, S. S. 1980. A temperature and precipitation record of the past 16,000 years in southern Chile. *Science* 210:1345–47

Hoganson, J. W., Ashworth, A. C. 1981. *Late glacial climatic history of southern Chile interpreted from Coleopteran (beetle) assemblages.* Presented at Ann. Meet. Geol. Soc. Am., 94th, Cincinnati

Hollin, J. T. 1962. On the glacial history of Antarctica. *J. Glaciol.* 4:173–95

Hollin, J. T., Schilling, D. H. 1981. Late Wisconsin-Weichselian mountain glaciers and small ice caps. In *The Last Great Ice Sheets*, ed. G. H. Denton, T. Hughes, 3:179–206. New York: Wiley-Interscience. 484 pp.

Hope, G. S., Peterson, J. A. 1976. Palaeoenvironments. In *The Equatorial Glaciers of New Guinea*, ed. G. S. Hope, J. A. Peterson, I. Allison, U. Radok, 9:173–205. Rotterdam: Balkema. 244 pp.

Hudson, J. D. 1977. Oxygen isotope studies on Cenozoic temperatures, oceans, and ice accumulation. *Scott. J. Geol.* 13:313–23

Jankowski, E. J., Drewry, D. J. 1981. The structure of West Antarctica from geophysical studies. *Nature* 291:17–21

Keany, J. 1978. Paleoclimatic trends in Early and Middle Pliocene deep-sea sediments of the Antarctic. *Mar. Micropaleontol.* 3:35–49

Kennett, J. P. 1977. Cenozoic evolution of Antarctic glaciation, the circum-Antarctic Ocean, and their impact on global paleoceanography. *J. Geophys. Res.* 82:3843–60

Kennett, J. P., Shackleton, N. J. 1976. Oxygen isotopic evidence for the development of the psychrosphere 38 MY ago. *Nature* 260:513–15

Kennett, J. P., Vella, P. 1975. Late Cenozoic planktonic foraminifera and paleoceanography at DSDP Site 284 in the cool subtropical South Pacific. In *Initial Reports of the Deep Sea Drilling Project*, J. P. Kennett, R. E. Houtz, et al, 29:769–82. Washington DC: GPO

Kennett, J. P., Watkins, N. D. 1974. Late Miocene-Early Pliocene paleomagnetic stratigraphy, paleoclimatology and biostratigraphy in New Zealand. *Geol. Soc. Am. Bull.* 85:1385–98

Kennett, J. P., Houtz, R. E., Andrews, P. B., Edwards, A. R., Gostin, V. A., Hajos, M., Hampton, M. A., Jenkins, D. G., Margolis, S. V., Ovenshine, A. T., Perch-Nielsen, K. 1974. Development of the circum-Antarctic current. *Science* 186:144–47

Kvasov, D. D., Verbitsky, M. Y. 1981. Causes of Antarctic glaciation in the Cenozoic. *Quat. Res.* 15:1–17

LaBrecque, J. L., Kent, D. V., Cande, S. C. 1977. Revised magnetic polarity time scale for Late Cretaceous and Cenozoic time. *Geology* 5:330–35

LeMasurier, W. E. 1972. Volcanic record of Antarctic glacial history: implications with regard to Cenozoic sea levels. *Inst. Br. Geogr. Spec. Publ.* 4:59–74

LeMasurier, W. E., Rex, D. C. 1982. Volcanic

record of Cenozoic glacial history in Marie Byrd Land and western Ellsworth Land: revised chronology and evaluation of tectonic factors. In *Antarctic Geoscience*, ed. C. Craddock, 89:725–34. Madison: Univ. Wis. Press. 1172 pp.

Livingstone, D. A. 1962. Age of deglaciation in the Ruwenzori Range, Uganda. *Nature* 194:859–60

Lorius, C., Merlivat, L., Jouzel, J., Pourchet, M. 1979. A 30,000-yr isotope climatic record from Antarctic ice. *Nature* 280:644–48

Loutit, T. S. 1981. Late Miocene paleoclimatology: subantarctic water mass, southwest Pacific. *Mar. Micropaleontol.* 6:1–27

Loutit, T. S., Kennett, J. P. 1979. Application of carbon isotope stratigraphy to Late Miocene shallow marine sediments, New Zealand. *Science* 204:1196–99

Margolis, S. V. Kennett, J. P., 1971. Cenozoic paleoglacial history of Antarctica recorded in sub-antarctic deep-sea cores. *Am. J. Sci.* 271:1–36

Matthews, R. K., Poore, R. Z. 1980. Tertiary $\delta^{18}O$ record and glacio-eustatic sea level fluctuations. *Geology* 8:500–14

Mayewski, P. A. 1975. Glacial geology and Late Cenozoic history of the Transantarctic Mountains, Antarctica. *Ohio State Univ., Inst. Polar Stud. Rep. 56.* 168 pp.

McKelvey, B. C. 1981. The lithologic logs of DVDP cores 10 and 11, eastern Taylor Valley. In *Dry Valley Drilling Project*, ed. L. D. McGinnis, 5:63–94. Washington DC: Am. Geophys. Union. 465 pp.

Mercer, J. H. 1968a. Glacial geology of the Reedy Glacier area, Antarctica. *Geol. Soc. Am. Bull.* 79:471–86

Mercer, J. H. 1968b. Antarctic ice and Sangamon sea level. *Int. Assoc. Sci. Hydrol., Gen. Assem. Berne, Publ.* 79:217–25

Mercer, J. H. 1970. Variations of some Patagonian glaciers since the Late-Glacial. II. *Am. J. Sci.* 269:1–25

Mercer, J. H. 1972. Some observations on the glacial geology of the Beardmore Glacier area. In *Antarctic Geology & Geophysics*, ed. R. J. Adie, pp. 427–33. Oslo: Universitetsforlaget. 876 pp.

Mercer, J. H. 1973. Cainozoic temperature trends in the southern hemisphere: Antarctic and Andean glacial evidence. In *Palaeoecology of Africa*, ed. E. M. van Zinderen Bakker, 8:85–114. Capetown: Balkema. 198 pp.

Mercer, J. H. 1976. Glacial history of southernmost South America. *Quat. Res.* 6:125–66

Mercer, J. H. 1982a. Holocene glacier variations in southern South America. *Striae* 18:35–40

Mercer, J. H. 1982b. The last glacial-deglacial hemicycle in Peru. *Am. Quat. Assoc. Conf., 7th, Abstr.*, p. 139

Mercer, J. H., Palacios, O. 1977. Radiocarbon dating of the last glaciation in Peru. *Geology* 5:600–4

Mercer, J. H., Sutter, J. F. 1982. Late Miocene-earliest Pliocene glaciation in southern Argentina: implications for global ice sheet history. *Palaeogeogr. Palaeoclimatol. Palaeoecol.* 38:185–206

Moar, N. T. 1973. Late Pleistocene vegetation and environment in southern New Zealand. In *Palaeoecology of Africa*, ed. E. M. van Zinderen Bakker, 8:179–98. Capetown: Balkema. 198 pp.

Moore, T. C., Pisias, N. G., Dunn, D. A. 1982. Carbonate time series of the Quaternary and late Miocene sediments of the Pacific Ocean: a spectral comparison. *Mar. Geol.* 46:217–33

Plafker, G., Bartsch-Winkler, S., Ovenshine, A. T. 1976. Paleoglacial implications of coarse detritus in DSDP Leg 36 cores. In *Initial Reports of the Deep Sea Drilling Project*, P. F. Barker, I. W. D. Dalziel, et al, 36:857–67. Washington DC: GPO

Porter, S. C. 1981. Pleistocene glaciation in the southern Lake District of Chile. *Quat. Res.* 16:263–92

Robin, G. de Q. 1958. *Glaciology III: Seismic Shooting and Related Investigations. Scientific Results, Norwegian-British-Swedish Antarctic Expedition, 1949–52*, Vol. 5. Oslo: Norsk Polarinstitutt. 134 pp.

Robin, G. de Q., Adie, R. J. 1964. The ice cover. In *Antarctic Research*, ed. R. Priestley, R. J. Adie, G. de Q. Robin, 8:100–17. London: Butterworths. 360 pp.

Rutford, R. H., Craddock, C., White, C. M., Armstrong, R. L. 1972. Tertiary glaciation in the Jones Mountains. In *Antarctic Geology & Geophysics*, ed. R. J. Adie, pp. 239–50. Oslo: Universitetsforlaget. 876 pp.

Rutford, R. H., Denton, G. H., Andersen, B. G. 1980. Glacial history of the Ellsworth Mountains. *Antarct. J. US* 15:56–57

Savin, S. M. 1977. The history of the Earth's surface temperature during the past 100 million years. *Ann. Rev. Earth Planet. Sci.* 5:319–55

Savin, S. M., Douglas, R. G., Stehli, F. G. 1975. Tertiary marine paleotemperatures. *Geol. Soc. Am. Bull.* 86:1499–1510

Savin, S. M., Douglas, R. G., Keller, G., Killingley, J. S., Shaughnessy, L., Sommer, M. A., Vincent, E., Woodruff, F. 1981. Miocene benthic foraminiferal isotope records: a synthesis. *Mar. Micropaleontol.* 6:423–50

Schalke, H. J. W. G., van Zinderen Bakker, E. M. 1971. History of the vegetation. In *Marion and Prince Edward Islands*, ed. E. M. van Zinderen Bakker, J. M. Winterbottom, R. A. Dyer, 7:89–97. Capetown: Balkema. 427 pp.

Schnitker, D. 1980. Global paleoceanography and its deep water linkage to the Antarctic glaciation. *Earth Sci. Rev.* 16:1–20

Shackleton, N. J. 1967. Oxygen isotope analyses and Pleistocene temperatures re-assessed. *Nature* 215:15–17

Shackleton, N. J. 1979. Evolution of the Earth's climate during the Tertiary Era. In *Evolution of Planetary Atmospheres and Climatology of the Earth*, pp. 49–69. Toulouse: Cent. Natl. d'Études Spatiales. 574 pp.

Shackleton, N. J., Cita, M. B. 1979. Oxygen and carbon isotope stratigraphy of benthic foraminifers at Site 397: detailed history of climatic change during the late Neogene. In *Initial Reports of the Deep Sea Drilling Project*, U. von Rad, W. B. F. Ryan, et al, 47(1):433–45. Washington DC: GPO

Shackleton, N. J., Kennett, J. P. 1975. Paleotemperature history of the Cenozoic and the initiation of Antarctic glaciation: oxygen and carbon isotope analyses in DSDP sites 277, 279 and 281. In *Initial Reports of the Deep Sea Drilling Project*, J. P. Kennett, R. E. Houtz, et al, 29: 743–55. Washington DC: GPO

Shackleton, N. J., Opdyke, N. D. 1973. Oxygen isotope and paleomagnetic stratigraphy of Equatorial Pacific core V28-238. *Quat. Res.* 3:39–55

Shackleton, N. J., Opdyke, N. D. 1977. Oxygen isotope and palaeomagnetic evidence for early Northern Hemisphere glaciation. *Nature* 270:216–19

Smith, A. G., Briden, J. C. 1977. *Mesozoic and Cenozoic Paleocontinental Maps*. Cambridge: Cambridge Univ. Press. 63 pp.

Soons, J. M., Burrows, C. J. 1978. Dates for Otiran deposits, including plant microfossils and macrofossils, from Rakaia Valley. *N.Z. J. Geol. Geophys.* 21:607–15

Stuiver, M., Denton, G. H., Hughes, T. J., Fastook, J. L. 1981. History of the marine ice sheet in West Antarctica during the last glaciation: a working hypothesis. In *The Last Great Ice Sheets*, ed. G. H. Denton, T. Hughes, 7:319–439. New York: Wiley-Interscience. 484 pp.

Stump, E., Sheridan, M. F., Borg, S. G., Sutter, J. F. 1980. Early Miocene subglacial basalts, the East Antarctic ice sheet, and uplift of the Transantarctic Mountains. *Science* 207:757–59

Suarez, G., Molnar, P. 1980. Paleomagnetic data and pelagic sediment facies, and the motion of the Pacific Plate relative to the spin axis since the late Cretaceous. *J. Geophys. Res.* 85:5257–80

Suggate, R. P. 1965. Late Pleistocene geology of the northern part of the South Island, New Zealand. *N.Z. Geol. Surv. Bull.* 77. 91 pp.

Suggate, R. P. 1978. The late mobile phase: Quaternary. Introduction. In *The Geology of New Zealand*, ed. R. P. Suggate, G. R. Stevens, M. T. Te Punga, 8:542–44. Wellington: Gov. Printer. 820 pp.

Suggate, R. P., Moar, N. T. 1970. Revision of the chronology of the Late Otiran Glacial. *N.Z. J. Geol. Geophys.* 13:742–46

Wardle, P. 1978. Further radiocarbon dates from Westland National Park and the Omoeroa River mouth, New Zealand. *N.Z. J. Bot.* 16:147–52

Webb, P-N. 1982. *Paleoclimatic, eustatic and biogeographic responses to Paleogene-Neogene glacial and interglacial events in the western Ross Sea.* Presented at Int. Symp. Antarct. Earth Sci., 4th, Adelaide

Weissel, J. K., Hayes, D. E., Herron, E. M. 1977. Plate tectonics synthesis: the displacement between Australia, New Zealand, and Antarctica since the Late Cretaceous. *Mar. Geol.* 25:231–77

Whillans, I. M. 1976. Radio-echo layers and the recent stability of the West Antarctic ice sheet. *Nature* 264:152–55

Woodruff, F., Savin, S. M., Douglas, R. G. 1981. Miocene stable isotope record: a detailed deep Pacific Ocean study and its paleoclimatic implications. *Science* 212:665–68

Ann. Rev. Earth Planet. Sci. 1983. 11: 133–63

RADIOACTIVE NUCLEAR WASTE STABILIZATION: Aspects of Solid-State Molecular Engineering and Applied Geochemistry

Stephen E. Haggerty
Department of Geology, University of Massachusetts, Amherst, Massachusetts 01003

INTRODUCTION

Considerable debate, both sociological and technological, currently surrounds the disposal of potentially high-risk radioactive nuclear waste (radwaste) products derived from reactors, nuclear weaponry, and clinical, as well as other, research (Lee 1980, Krugmann & von Hippel 1977, Angino 1977, de Marsily et al 1977, Hobbs 1979). Breaking the sociological barrier of acceptance is likely to be as difficult, or even more difficult, than attaining a technological solution. It is not only a matter of ethics and science but also a matter of survival and responsibility, as eloquently outlined in the practical and philosophical thesis presented by Rochlin (1977). In the face of rapidly diminishing energy resources and the ever-increasing demand for energy, modern technology must provide convincing and sound guidelines to resolve which energy alternative is the most productive, the most readily available, and the most versatile. Whether or not nuclear power continues to be adopted, there seems little doubt that public opinion will continue to effect the utilization of radioactive materials for all purposes. We may in fact be forced into the realization that there are no known alternatives comparable to the magnitude of the need. No matter what future social attitudes and no matter what scientific progress may be made, we can neither escape the problem nor relinquish the obligation to safely immobilize the by-products of nuclear-fission processes. These products can no longer be indiscriminately disseminated into the environment.

No internationally agreed-upon policy has yet been adopted for the safe

0084–6597/83/0515–0133$02.00

disposal of radioactive nuclear waste products, a fact that does not come as a surprise when one considers the following unresolved questions. For example, (*a*) should radwaste be placed in a retrievable surface storage facility until a well-developed and proven method of disposal is established? (*b*) Should the waste products be viewed as an economic resource of rare metals and hence placed in a respository conducive to present or future mining practice? (*c*) Can low-level and high-level wastes be treated in a similar manner? (*d*) Should particular emphasis be placed on separating the long-lived α-emitting uranic and transuranic elements from the shorter-lived β and γ emitters? (*e*) Is it technologically sound to have a single method of disposal, or should several systems be adopted in the event that one mechanism fails in long-term durability tests? (*f*) Should the waste products be diluted and dispersed, or should they be concentrated and contained? (*g*) If long-term immobilization is considered as a viable means of ultimate disposal, can geological stability be guaranteed? (*h*) Of the choices presently available for disposal, how do we set about ascertaining the reliability of each system or, alternatively, developing models that will assure effective containment?

In 1954, Congress amended the Atomic Energy Act to permit the use of nuclear fuel in privately owned facilities for commercial energy generation (National Academy of Sciences 1975). Among the stipulations was a requirement that high-level radioactive nuclear wastes, which are derived from the reprocessing of spent-fuel elements, be solidified and then "safely shipped to a federal repository within 10 years after reprocessing" (National Academy of Sciences 1975). The waste is aged in a highly diluted aqueous solution for a "cooling-off" period prior to the production of a fine-grained (0.1–100 microns) powdery material by either spray-drying or fluidized-bed calcination. This product is highly toxic, both radiogenically and carcinogenically, on inhalation or ingestion. Shipments of waste were expected to be transported to suitably sited repositories in 1983; however, there are many unresolved problems and no single solution to the safe fixation of radwaste has emerged. Considerable attention has been given to the problem, and immobilization in glass is currently employed in Europe (de Marsily et al 1977). Molecularly engineered ceramics have also been proposed (Roy 1977) and these, together with synthetic rocks (Ringwood 1978, Ringwood et al 1979b), are among the most likely containment methods for adoption in the United States. Glass containment is unsuitable because of its inherent instability. Unless it can be demonstrated that these immobilization concepts are effective on a geological time scale, the mode of an interim, retrievable surface storage facility should probably be adopted, notwithstanding the fact that much effort has already been devoted to the multiple-barrier concept in permanent geological repositories (National Academy of Sciences 1975, Westerman 1981).

The initial barrier must be chemically stable, but the prudent evaluation of interim-storage or permanent-disposal methods must also depend upon whether we can demonstrate that the first immobilization barrier is stable over protracted periods of time commensurate with the half-lives of the noxious elements that are being contained. It is this aspect of the problem that is addressed in this article, and to this end the following questions are discussed: (*a*) How does nature deal with the problem of stabilizing radwaste-equivalent elements? (*b*) For what periods of time can these elements be immobilized? (*c*) What are the effects and potential consequences of radionuclide decay on mineral structures, and do these consequences affect the long-term stability of such structures? (*d*) What constraints can be placed on the intensive parameters (e.g. temperature, pressure, oxygen partial pressure) of minerals in ceramics or in synthetic rocks that are designed for first-barrier containment, based on their stabilities in a geological setting? (*e*) What are the limiting factors on the acceptable concentrations of active radwaste products to the proportions of the inert crystalline barriers? (*f*) Are there other minerals in nature that have an equivalent or higher receptor capacity for radwaste elements than the minerals or structures that have been proposed in ceramic or synthetic-rock assemblages?

Concern over the disposal of radioactive nuclear waste has resulted in accelerated research programs worldwide. The literature is voluminous and only a selected aspect of the disposal problem is covered in this review. For those seeking a more detailed perspective, reference should be made to publications from the US Department of Energy (e.g. Assistant Secretary for Nuclear Energy 1980) and the volumes on Nuclear Waste Management (*Scientific Basis for Nuclear Waste Management* 1979–1981).

RADWASTE

It is estimated that liquefied radioactive military waste amounts to approximately 75 million gallons and that at least 2,500 metric tons of radioactive waste also exist in stockpiles from reactors in the United States (Krugmann & von Hippel 1977). The wastes differ from the standpoint that military wastes are volumetrically large but highly diluted, whereas civilian wastes are more concentrated and hence considerably more toxic. There are additional wastes from medical and related research, but the quoted volumes do not reflect these concentrations. By the year 2000, based on reactor capacity and present methods of waste processing, approximately 13,800 m^3 of solidified high-level radioactive nuclear waste will have accumulated, containing an anticipated 53,000 MCi of activity and a heat-production capacity of 195 MW, even after the initial 10-year decay period (National Academy of Sciences 1975).

High-level radwaste stems from the reprocessing of spent-fuel cartridges. The word "spent" is used in the broadest sense because the fuel cells are replaced annually when only 3% of the fissionable uranium has been activated. This is necessary in order to maintain a constant high level of efficiency, but it is also necessary to remove the accumulated fission products because they have large or larger neutron-capture cross sections than that of uranium. A 10-year isolation period in water follows the removal of the fuel cell and it is only after this time that the rods are disaggregated, dissolved in nitric acid, and recycled with the ensuing formation of the waste products.

In general terms, the wastes may be classified according to their periods of intense radionuclide activity and in terms of the categories of elements that are produced in the fission and in the decay cycle. The long-lived transuranic and α-emitting elements (i.e. those having a mass number greater than that of uranium), make up the actinide class of elements. They are formed by neutron capture by uranium and remain hazardous for extraordinarily long periods of time (400 to 2 million years). Rapid transmutation of these actinides (neptunium, plutonium, americium, and curium) can be achieved (for example, in a nuclear reactor into fissile products and then into shorter-lived fission products); however, this would involve separating these elements from the radwaste calcine, a process that has yet to be perfected. Clearly, once the calcined waste has been incorporated into the primary crystalline barrier it will be even more difficult to separate these elements. The following half-lives for members of the transuranic series illustrate the need for long-term assessments in evaluations of interim storage and permanent disposal: ^{241}Am, $t_{1/2} = 433$ yr; ^{240}Pu, $t_{1/2} = 6600$ yr; ^{243}Am, $t_{1/2} = 7370$ yr; ^{245}Cm, $t_{1/2} = 9320$ yr; ^{239}Pu, $t_{1/2} = 24{,}400$ yr; ^{237}Np, $t_{1/2} = 2.14$ million yr. The range of time for active concern over safe disposal is therefore as much as 1 million years, using an upper limit that is perhaps excessively optimistic. However, as Benedict (National Academy of Sciences 1975) and Angino (1977) note, the waste extraction of U and Pu is 99.5% efficient but if a 99.9% achievement level could be attained for U, Pu, and Np, and a 99% level for Am and Cm, the long-term activity would be reduced by a factor of 100 and the waste problem would then be one of concern for a period of only approximately 1000 years. That technology has not been developed, so the geological time-scale is a more realistic perspective. A 100,000-year repository ought to be considered the minimum requirement.

After the 10-year storage period the isotopes of strontium (^{90}Sr) and cesium (^{137}Cs) are the most deleterious, with half-lives ($t_{1/2}$) of 30 years and a hazard factor that will decrease by three orders of magnitude only after 350 years (National Academy of Sciences 1975). These are also the high heat-producing radioisotopes that emit β and γ particles. They are of relatively minor concern in the secondary containment barriers but the potential thermal effect that they may have on the stabilizing medium of the crystalline primary barriers should not be ignored; this feature may in fact be utilized to advantage (see the subsequent section on metamictization).

The remainder of the fission products fall into the same class as Sr and Cs. These are the rare-earth elements (REE: yttrium, cerium, neodymium, samarium, europium, gadolinium, terbium, dysprosium, holmium, erbium, thulium ytterbium, and lutetium), zirconium, molybdenum, ruthenium, rhodium, palladium, barium, rubidium, technetium, tellurium, and iodine. These elements are of intermediate atomic number ($Z = 37$–71) and, excluding ^{90}Sr, ^{137}Cs and ^{129}I, are relatively inert but constitute the major proportion of the radwaste calcine. Hence they are a significant factor in the design of the first isolation barrier.

The remaining elements are also relatively inert; they are the result of processing contaminants and include sodium, phosphorous, aluminum,

Periodic Table of the Elements

Groups / Periods	IA	IIA	IIIB	IVB	VB	VIB	VIIB	VIII	VIII	VIII	IB	IIB	IIIA	IVA	VA	VIA	VIIA	0
1	1 H 1·008																	2 He 4·003
2	3 Li 6·94	4 Be 9·01											5 B 10·81	6 C 12·01	7 N 14·007	8 O 15·999	9 F 18·998	10 Ne 20·18
3	11 Na 22·99	12 Mg 24·31											13 Al 26·98	14 Si 28·09	15 P 30·97	16 S 32·06	17 Cl 35·45	18 A 39·95
4	19 K 39·10	20 Ca 40·08	21 Sc 44·96	22 Ti 47·90	23 V 50·94	24 Cr 51·99	25 Mn 54·94	26 Fe 55·85	27 Co 58·93	28 Ni 58·71	29 Cu 63·54	30 Zn 65·37	31 Ga 69·72	32 Ge 72·59	33 As 74·92	34 Se 78·96	35 Br 79·90	36 Kr 83·80
5	37 Rb 85·47	38 Sr 87·62	39 Y 88·90	40 Zr 91·22	41 Nb 92·91	42 Mo 95·94	43 Tc 98·91	44 Ru 101·07	45 Rh 102·90	46 Pd 106·4	47 Ag 107·87	48 Cd 112·40	49 In 114·82	50 Sn 118·69	51 Sb 121·75	52 Te 127·60	53 I 126·90	54 Xe 131·30
6	55 Cs 132·90	56 Ba 137·34	57 La 138·91	72 Hf 178·5	73 Ta 180·95	74 W 183·85	75 Re 186·2	76 Os 190·2	77 Ir 192·2	78 Pt 195·09	79 Au 196·97	80 Hg 200·59	81 Tl 204·37	82 Pb 207·2	83 Bi 208·98	84 Po	85 At	86 Rn
7	87 Fr	88 Ra 226·03	89 Ac															

L	58 Ce 140·12	59 Pr 140·91	60 Nd 144·24	61 Pm	62 Sm 150·4	63 Eu 151·96	64 Gd 157·25	65 Tb 158·92	66 Dy 162·50	67 Ho 164·93	68 Er 167·26	69 Tm 168·93	70 Yb 173·04	71 Lu 174·97
A	90 Th 232·04	91 Pa 231·04	92 U 238·03	93 Np 237·05	94 Pu	95 Am	96 Cm	97 Bk	98 Cf	99 Es	100 Fm	101 Mv	102 No	

Legend: Processing Contaminants. Actinides. Fission Products.

Figure 1 Periodic table of the elements illustrating the variety of ions present in radioactive nuclear-waste products.

iron, nickel, and chrome. Silicon, copper, lithium, and boron are also likely to be present in relatively minor concentrations.

In summary, the classes of elements to be contended with in radwaste fixation are fission products, actinides, and processing contaminants (Figure 1). The fission products (Sr, Cs, REE, Zr, Mo, Ru, Rh, Pd, Ba, Rb, Tc, Te and I) are of intermediate atomic number ($Z = 37$–71) and make up about 84% of radwaste. Within this group, ^{90}Sr, ^{137}Cs, and ^{129}I are toxic, and ^{90}Sr and ^{137}Cs are relatively short-lived but are high heat generators. The actinides have $Z = 90$–96 and constitute approximately 2% of radwaste. The actinides Th, Pa, and U are long-lived but cool to acceptable levels for remote processing within 10 years, while Np, Pu, Am, and Cm have extremely long half-lives, are α-emitters, and are hazardous for times in excess of 1 million years. The inert processing contaminants are the lighter elements ($Z = 11$ to 29) and are in concentrations of $\sim$14%.

STABILIZATION SYSTEMS

In principle, the first barrier stabilizes the calcined radwaste into a monolithic form that is mechanically stable and chemically inert. Contained in this form it can be readily handled in remote operations for incorporation into canisters and into subsequent sequences of multiple barriers of carbon steel and concrete (National Academy of Sciences 1975, Assistant Secretary for Nuclear Energy 1980). A large number of first-barrier stabilization methods have been proposed, including cement, metallic glasses, alkali- or zinc-based borosilicate glasses, high-silica-content glasses, titanate and alumina ceramics, supercalcined ceramics, and synthetic rocks. These barriers fall into two categories: those that are amorphous and those that are crystalline.

Any form of glass has the advantage that the effect of natural partitioning coefficients of elements among a series of mineral phases can be ignored. Ideally the calcined waste is totally dissolved into the glass matrix. However, if dissolution is not possible because of supersaturation or solution characteristics, the insoluble components will be suspended as isolated particles within the glass. An additional advantage is that the production skills and technology for the industrial-scale manufacture of glass ingots is well established. However, there are also several major disadvantages to glass. Among these is its inherent thermodynamic instability over historical, let alone geological, time. The instability stems from the deterioration of glass by devitrification, in which the amorphous state is transformed into the crystalline state. In producing glass the high-temperature liquid state of nonbridging atoms or the development of interatomic bonds is achieved at lower temperatures by rapid quenching. If allowed to cool slowly or to age the crystalline state will result. Although the method of glass containment has been demonstrated to be a practical means of stabilization and is widely employed in West Germany, France, and the United Kingdom (Angino 1977, de Marsily et al 1977), it has failed both the crystalline test and accelerated leaching tests for some critical elements, notably Cs and Sr. In one specific demonstration, such a glass was shown to be partially crystalline after only 11 years of nuclear-waste confinement (Morris et al 1978). Because the atomic species are not bound into glass, their leachability is considerably higher than if the ions were tightly bonded into crystalline structures. It was not determined whether crystalline phases had in fact formed in the glass test cited above, but embrittlement and mosaic devitrification had nonetheless set in. The inversion of glass to the crystalline state increases the surface area, induces a loss of integrity by microfracturing, and greatly increases the probability of leaching. Various estimates of the centerline temperatures of the first-barrier canisters, induced by β and γ emitters in the natural decay process, are in the range of 100–500°C. These temperatures are likely to continue over periods of at least 100 years, a span that is more than adequate for annealing and hence crystallization. The initial objective of rapid quenching—to maintain the glass state—is therefore defeated. From the available data on leaching and devitrification, it would appear that glass is unacceptable as a first stabilization barrier (Morris et al 1978, Merritt 1976).

The concept of immobilization of calcine into crystalline materials was first conceived by Hatch (1953) and was demonstrated to be viable in the Materials Science Laboratories of Pennsylvania State University (Roy 1980, McCarthy 1976, McCarthy & Davidson 1975, 1976). In a lengthy study of the phase chemistries and phase formations of the wastes, and waste plus inert additives, a ceramic (Supercalcine) has been formulated in what Roy (1977) refers to as a "molecularly engineered product tailor-made specifically for radwaste immobilization." Considering the enormous range of elements (from $Z = 11$–96), the equally large range in ionic radii [from 0.3 Å (phosphorous) to 1.9 Å (cesium)], and the coordination polyhedra that are required for this elemental spectrum (IV, VI, VIII, XII), the solution results in an unexpectedly small number of crystalline phases. Using aluminum and strontium nitrates and SiO_2 as suspension additives, McCarthy (1976) and McCarthy & Davidson (1975, 1976) found that heat treatment of the calcine in this matrix at 1000°C for as little as 30 minutes led to an assemblage of crystalline phases and characteristic structure types. The structural types are pollucite (cubic), scheelite (tetragonal), fluorite (cubic), apatite (hexagonal), spinel (cubic), and corundum (hexagonal), with each structure exhibiting a preferred uptake of specific radwaste elements.

The perovskite ($CaTiO_3$) structure was also considered as a leach-resistant refractory phase in which Sr and Ru would partially or completely replace sites conventionally occupied by Ca and Ti. Leaching tests of the Supercalcine ceramic have shown that most elements are retained, except for Cs which is readily dissolved from the pollucite framework structure (McCarthy & Davidson 1976). Among the structural groups in this Supercalcine ceramic, the minerals that exhibit similar compositions in natural occurrences are pollucite ($CsAlSi_2O_6$), one of only two known cesium-bearing minerals in which Cs is a dominant cation (the other being avogadrite, [KCs] BF_4); the fluorite-structured baddeleyite (ZrO_2) and cerianite (CeO_2); and spinel (AB_2O_4, where $A = Fe^{2+}$,Mg,Mn,Ni,Zn and $B =$ Al,Cr,Fe^{3+},Ti). The synthetic scheelite, apatite, and corundum-structured phases bear only a faint resemblance to equivalent minerals in nature.

A similar approach to that of the Penn State group has been taken at the Sandia Laboratory in Albuquerque, New Mexico. A high-titanium ceramic has been proposed that yields an assemblage of phases having rutile, fluorite, and pollucite structures in association with metal, amorphous silica, and $Gd_2Ti_2O_7$ (Roy 1977). A modified form of the rutile-based ceramic, which is inert and highly refractory, has been successfully produced, and accelerated leach tests yield results that are highly encouraging (Dosch et al 1981). The component minerals essential to the partitioning of specific radwaste elements in this ceramic are similar to those discussed for SYNROC below.

High-alumina nuclear-waste ceramics developed by the Rockwell International Science Center, Thousand Oaks, California (Morgan et al 1981), are modeled on magnetoplumbite (ideally $PbFe_{12}O_{19}$), which can structurally incorporate Cs, Sr, Si, Na, Ca, Ba, La, Nd, Mn, Fe, Ce, K, and Ni. Associated minerals are uraninite, spinel, and corundum. This waste formulation is tailored specifically to the high Al_2O_3 Savannah River defense product and initial tests show that magnetoplumbite is leach-resistant and has a high thermal stability, specifically for Na, K, Rb, and Ca (Jantzen et al 1982, Morgan & Cirlin 1982). The composition of magnetoplumbite employed in this waste form differs markedly from the naturally occurring mineral, and hence its stability cannot be assessed in a geological context.

A synthetic rock (SYNROC) has also been suggested by Ringwood et al (1979a,b, Ringwood 1978) at the Australian National University, Canberra. They propose that radwaste immobilization can be achieved by an assemblage of minerals that are closer to naturally occurring phases, adopting the attitude that if a geological repository is to be the final means of disposal, then geological materials should constitute the first, as well as the final, barrier. Moreover, the stabilities of minerals and their element-retention capacities can be assessed in a geological setting over geological time. In the initial design of SYNROC, Ringwood et al (1979b) proposed an assemblage consisting of hollandite ($BaMn_8O_{16}$) modified compositionally to $BaAl_2Ti_6O_{16}$, perovskite ($CaTiO_3$), zirconolite ($CaZrTi_2O_7$), celsian ($BaAl_2Si_2O_8$), kalsilite ($KAlSiO_4$), and leucite ($KAlSi_2O_6$). In a refined version of this assemblage, SiO_2 was eliminated with the concomitant disappearance of celsian, kalsilite and leucite, an improvement that was necessary in order to force Cs into the hollandite structure, which proved to be particularly resilient to accelerated leaching tests. The revised SYNROC, therefore, is reduced to an oxide assemblage of $BaAl_2Ti_6O_{16}$, perovskite, and zirconolite, with effective partitioning of the radwaste elements as shown in Table 1. United States military waste is rich in iron (5–50 wt% Fe_2O_3), aluminum (5–80 wt% Al_2O_3), and sodium (3–6 wt% $Na_2O + Na_2SO_4$), and a SYNROC specific to these compositions has been proposed (Ringwood et al 1979a) in which spinel ($FeAl_2O_4$) and nepheline ($NaAlSiO_4$) are components in addition to hollandite, perovskite, and zirconolite. Although $BaAl_2Ti_6O_{16}$ is not recognized in nature, several closely related minerals are known, as discussed in a subsequent section and summarized in Table 1. Nepheline, zirconolite, and pervoskite, on the other hand, are restricted geologically but are not uncommon, and spinel solid solutions are widespread.

The proportion of waste that is proposed for incorporation into these first-barrier containment systems, excluding glass, is as follows: the Supercalcine ratio is 50% waste and 50% inert additives; the Sandia ceramic is 25% waste in 75% ceramic; the Rockwell product requires sufficient Al_2O_3 and REE to stabilize magnetoplumbite; and the SYNROC proportion may be varied from less than 5% to 20% waste. Ringwood et al (1979b) argue that if the concentration of waste is dominant, the waste will dominate the assemblage that develops. However, if the proportions and the structural flexibility of the inert additives are maintained at a high level, or varied within small degrees, and provided the resulting assemblage is in thermodynamic equilibrium and not supersaturated with radwaste elements, the end products will always be approximately similar, differing only in the factor of dilution. This is an important consideration if the composition of the waste product is changed with developing technology, or if several different types of waste forms are processed by the same technique.

In these four containment techniques the principle of element stabilization into crystalline solids is essentially similar and overcomes the drawbacks associated with glass. On a production scale, it is envisaged that the calcine and the inert matrix will be thoroughly mixed and hot-pressed at elevated temperatures (1000–1300°C) either into pellets or into relatively large blocks (e.g. 50 × 100 cm) suitable for incorporation into second-barrier metal canisters. None of these methods have yet been fully tested, and acceptance by the Nuclear Regulatory Commission must await a satisfactory demonstration that the containment system will meet the stringent criteria of immobilization with a large margin of safety.

Table 1 Principal high-level waste elements introduced into SYNROC[a]

"Hollandite" ($BaAl_2Ti_6O_{16}$)		Zirconolite ($CaZrTi_2O_7$)	Perovskite ($CaTiO_3$)
Cs^+	Mo^{4+}	U^{4+}	Sr^{2+}
K^+	Ru^{4+}	Zr^{4+}	(U^{4+})
(Na^+)	Rh^{3+}	Y^{3+}	(Y^{3+})
Ba^{2+}	Fe^{3+}	Gd^{3+}	(Gd^{3+})
	Cr^{3+}	La^{3+}	(La^{3+})
	Ni^{2+}	Na^+	
	Fe^{2+}		

Some Related Minerals in Nature

Freudenbergite	$Na_2FeTi_7O_{16}$
Priderite	$(K,Ba)(TiFe)_8O_{16}$
Loparite	$(Ce,Na,Ca)_2(Ti,Nb)_2O_6$
Kobeite	$(Y,U)(Ti,Zr,Fe,Nb)_2O_6$
Polymignite	$(Ca,La,Y,Th,Mn,Fe,Ce)(Zr,Ti,Nb,Ta)_2O_6$
Pyrochlore group	$A_{2-m}B_2O_6$
	(A = Na,Ca,K,Sr,Ba,Mn,REE,Pb,B,U,Th)
	(B = Nb,Ta,Ti),
	Pyrochlore Nb + Ta > 2Ti and Nb > Ta
	Microlite Nb + Ta > 2Ti and Ta ≥ Nb
	Betafite 2Ti ≥ Nb + Ta
Transitional armalcolites	$(FeMgCa)(TiZrCrAl)_2O_5$
Crichtonite group	$AM_{21}O_{38}$
	(A = Sr,Ca,Na,REE,Pb,Ba,K)
	(M = Ti,Zr,Cr,Al,Fe,Mg,Mn)
	Crichtonite—Sr
	Senaite—Pb
	Davidite—REE
	Loveringite—Ca
	Landauite—Na
	Phase X—Ba
	Phase Y—K

[a] The principal high-level waste elements introduced into SYNROC are those proposed by Ringwood et al (1979b). The list of related minerals in nature is discussed in the text and contains some of the alternatives suggested in this review.

IMPLICATIONS OF MINERAL GEOCHEMISTRY

Supercalcine and SYNROC contain minerals whose geochemistry can be evaluated on their ability to retain specific radwaste elements over geological time. Both products also contain synthetic phases not recognized in nature. Only a partial assessment, therefore, can be made of these phases but some valuable insights are nonetheless apparent as to the likely durability of the assemblages in interim repositories or under conditions of permanent disposal.

The natural minerals in Supercalcine are pollusite, baddeleyite, cerianite, and spinel; those in SYNROC are perovskite and zirconolite.

Pollucite ($CsAlSi_2O_6 \cdot nH_2O$) is an important mineral component because Cs is a major constituent. It is a relatively rare mineral in nature and is most commonly associated with granite pegmatites (Vilasov 1966). The thermal stability of pollucite has not been experimentally determined, but it is known to form at relatively low temperatures as demonstrated by McCarthy et al (1978), who reacted radwaste calcine with shale and basalt under hydrothermal conditions at $\sim$300 bars and 150–200°C for 7–30 days. Loss of the H_2O radical takes place at 720°C (Newnham 1967), a temperature that is above the centerline range of thermal activity within the canisters. It is clearly undesirable to have any hydrous phases within the first barrier because dehydration will alter the mechanical strength of the unit. However, the anhydrous analog of pollucite must be a stable phase according to the heat treatment (1000–1300°C) and X-ray data reported for Supercalcine (McCarthy & Davidson 1976). Cs is readily leached from pollucite in natural weathering conditions and in laboratory experiments (Vilasov 1966, McCarthy & Davidson 1975), but provided the repository is rich in aluminosilicates the recombination of Cs will very probably lead to the formation of second-generation pollucite in the event that the canister is breached (McCarthy et al 1978). Ringwood et al (1979b) add, furthermore, that although the framework silicate structures allow for the ready accommodation of large-radius cations such as cesium, these structures are equally conducive to the liberation of cations during leaching. It was this particular aspect that initially led to the removal of leucite, kalsilite, and celsian in the redesign of SYNROC (Ringwood et al 1979b).

Baddeleyite (ZrO_2) is a high-temperature, highly refractory, stable mineral that occurs in alkalic associations with carbonatites (Vilasov 1966, Van Wambeke 1971). It has also been recognized in meteorites (Ramdohr 1973), in lunar basalts (Haggerty 1973), in terrestrial gabbros (Keil & Fricker 1974), and in kimberlites (Kresten 1974, Raber & Haggerty 1979). Baddeleyite is typically a late-stage mineral in the paragenetic crystallization sequence of these rock types, a result of the incompatible nature of Zr, which does not easily enter the structures of common rock-forming silicates. Secondary baddeleyite is a widespread alteration product of kimberlitic zircons ($ZrSiO_4$) that is induced by desilicification according to $ZrSiO_4 = ZrO_2 + SiO_2$, a reaction that also models zircon decomposition in impactites (El Goresy 1956). Reactions that involve zircon and ilmenite, again in kimberlites, produce ZrO_2 in association with zirconolite ($CaZrTi_2O_7$) and diopside ($CaMgSi_2O_6$). It is conjectured that the reaction is triggered by the effects of an influx of magmatic calcite ($CaCO_3$) during late-stage carbonatitic events related to the injection of kimberlites (Raber & Haggerty 1979). Baddeleyite is highly resistant to weathering and in addition to accommodating Zr the mineral is also capable of absorbing Hf, Al, Ca, Th, Y and the rare-earth elements into its structure. Natural occurrences generally contain on the order of 0.1–3 wt% of these elements. High-temperature tetragonal baddeleyite inverts to cubic ZrO_2 at lower temperatures ($\sim$1200°C), and it is the latter polytype that is present in Supercalcine (Butterman & Foster 1967, McCarthy & Davidson 1975).

Cerianite (CeO_2) forms a solid-solution series with ZrO_2 at high temperatures; this property arises from the similarity in both the structure and the ionic radii of Ce and Zr. Cerianite is a relatively rare mineral found in carbonatites, pegmatites, and in the weathering rinds of some alkalic rocks (Vilasov 1966). Although the stability of cerianite is unknown, its major attribute as a phase in Supercalcine is that large concentrations of REE ($\sim$4.5 wt%) and ThO_2 (5.1 wt%) can be accommodated, based on the compositions of CeO_2 reported in natural occurrences.

Spinel ($MgAl_2O_4$ *sensu stricto*) forms extensive solid solutions with other members of this mineral group, most notably chromite ($FeCr_2O_4$), hercynite ($FeAl_2O_4$), picrochromite ($MgCr_2O_4$), magnesioferrite ($MgFe_2O_4$), ulvöspinel (Fe_2TiO_4), magnetite (Fe_3O_4), and trevorite ($NiFe_2O_4$). These minerals have a widespread distribution in igneous and metamorphic rocks (Haggerty 1976), commonly form at relatively high temperatures (500–1200°C), and are stable compounds in the context of immobilizing some of the processing contaminants of radwaste in Supercalcine.

Perovskite ($CaTiO_3$) has the extraordinary capacity of allowing the substitution of Sr, Ba, Na, REE, Y, and Cd for Ca, and Nb, Zr, Mo, Pu, Rh, U, Sn, Ru, Fe, Cr, and Al for Ti. The mineral is characteristic of carbonatites, kimberlites, and other closely related silica-undersaturated rocks, where it may be present as either a primary crystallization product or as a secondary reaction phase associated with ilmenite [$(Fe,Mg)TiO_3$]. Titanium has an extremely high affinity for Ca and correspondingly lower affinities for either Si or Fe (Verhoogen 1962), so perovskite will certainly occur in SYNROC and is a probable constituent of Supercalcine. Together with zirconium, titanium is an incompatible element that forms highly refractory and stable phases, features that are shared by most of the elements that may be substituted into the perovskite structure. In natural perovskite it is the uptake of the rare-earth elements that is most pronounced, and concentration levels of up to 13 wt% have been reported (Boctor & Boyd 1979). Smaller concentrations of U, Th, Zr, and Hf are also commonly present and an additional noteworthy substitution is niobium, which leads to the mineral lattrapite when Nb becomes a dominant cation (Nickel & McAdam 1963). Na and Ce replacement lead to loparite, high Th produces irinite, and Zr + Al results in uhligite. Perovskite is resistant to chemical alteration, even under the most rigorous of geological settings, and because of its range of possible substitutions, it is unquestionably an ideal choice for the immobilization of most of the fission products and actinides in radwaste calcine.

Zirconolite ($CaZrTi_2O_7$) has many of the properties of baddeleyite (ZrO_2) and perovskite ($CaTiO_3$) in terms of structure, stability, element substitution, and geological environments of formation (Vilasov 1966). The combined refractory and incompatible nature of Ti and Zr in zirconolite dictates that it is a late-stage product of mafic magma crystallization. Zirconolite acts in a spongelike manner, readily accommodating other refractory and large-radius cations retained in residual liquids. These elements are concentrated because of their inability to enter the rock-forming silicates. Concentrations of 4.6 wt% U_3O_8 and 8.3 wt% ThO_2 are reported for some zirconolites, whereas others have up to 6.2 wt% REE (Vilasov 1966). Other elements that may readily substitute for Ca are Na, K, Sr, Ba, Pb, Y, and Bi, and Ta, Nb, Fe, Mn and Sn may replace Zr and Ti either directly or in coupled substitution with Ca. Zirconolite is recognized in lunar-highland basalts, in terrestrial gabbros and pyroxenites, in carbonatites, and in reaction relationships between ilmenite and zircon in kimberlites (Haggerty 1973, Raber & Haggerty 1979, Borodin et al 1957, 1961, Busche et al 1972, Williams 1978, Meyer & Boctor 1974). It has been shown to be a relatively high-temperature mineral (Wark et al 1973), and its dominant role in radwaste immobilization is the ability to stabilize the actinides. In addition, zirconolite can also structurally facilitate the incorporation of considerably large concentrations of fission-produced elements.

If we now consider the stability of radwaste elements in their natural mineralogical settings, we first note that baddeleyite, zirconolite, and perovskite are common to kimberlites and carbonatites. The most active age span for the formation of these rock types is during the Cretaceous, with intrusion ages of 50–125 m.y. (Allsopp & Barrett 1975, Davis 1978). It has been demonstrated (Ringwood et al 1981) that significant proportions of the actinides and fission products are still retained in their respective structures, notwithstanding the fact that tectonic activity, erosion, and deeply lateritized weathering are common. Relatively high confidence levels may, therefore, be given to these minerals as potential radwaste immobilizers. Although conditions are markedly different in the lunar environment, zirconolite and baddeleyite have, nevertheless, survived for 3.8 to 4.3 b.y. Members of the spinel mineral group are equally stable and this phase may be useful in stabilizing a selection of the processing contaminants, chiefly Fe, Cr, Ni, Co, and Zn. Cerianite remains an unknown quantity; it may well be stabilized in the presence of Zr but the fact that it does form in the weathered rinds of alkali-rich rocks suggests that its high mobility would be a disadvantage as a containment mineral. For similar reasons one may conclude that pollucite is an unacceptable mineral receptor, specifically because Cs is readily mobilized from the framework structure. From a mineralogical standpoint it appears that a refractory oxide rather than a refractory silicate media is the preferred method of radwaste stabilization. As noted earlier, many of the phases in Supercalcine cannot be evaluated in a geological context, and this also applies to the $BaAl_2Ti_6O_{16}$ phase in SYNROC. For these phases it will be necessary to undertake severe leaching tests to demonstrate their stability.

ALTERNATE MINERALS

Although accelerated-leaching and stability tests may prove to be effective in establishing the resistance of synthetic phases under geological conditions, the time parameter cannot be accurately simulated and, hence, there will always be an element of uncertainty in projected extrapolations. One method of alleviating this situation is to examine the proposed stabilization systems, notably Supercalcine and SYNROC, in an attempt to substitute alternate minerals that have demonstrated stabilities on a geological time scale. In doing this one is forced to refine those criteria that constitute the minimum requirements for the first barrier, namely mechanical stability and chemical inertness. Compositional simplicity of the inert additive is one refinement, a second is that the number of mineral phases be as small as possible, and a third is that the barrier be highly inert, enhancing the probability of long-term endurance. Taking these factors into account, I have chosen to model a series of mineral alternatives on the basis of SYNROC. One form of SYNROC is silica-free, highly refractory, and has a simple additive chemistry. More significantly, it includes perovskite and zirconolite, two phases that have proven records of geological stability.

The choice of alternate minerals is limited to oxides having structures allowing extensive substitution of a wide variety of elements of variable valency state, ionic radii, and coordination number. Although there are many minerals that fulfill these criteria, there are only a few that have sufficient latitude in accepting high proportions of radwaste elements. In the following discussion, a concerted effort has been made to select only those minerals that will easily accommodate specific radwaste elements in relatively large concentrations, or alternatively those minerals or mineral series in which waste components are critical elements in the composition of the phase. An additional criterion has been to take the composition of the calcine itself into account. This restricts the choice of naturally occurring minerals even more, considering that radwaste is dominated by REE ($\sim$25 mole %) and Zr ($\sim$13 mole %). The very nature of radwaste implies that the minerals should also be actinide-bearing. Those present in nature can, of course, only be evaluated with respect to U and Th.

Within these constraints the only minerals that appear to be satisfactory alternatives are members of the pyrochlore, the crichtonite, and the euxenite-priorite (specifically, kobeite and polymignite) series (Table 1).

The pyrochlore ($A_{2-m}B_2O_6$) mineral series is cubic, where the A site contains Na, Ca, K, Sr, Ba, Mn, Pb, B, U, Th, light lanthanides La $\rightarrow$ Eu, and Y plus the heavier lanthanides Gd $\rightarrow$ Lu. The B site is dominated by Nb, Ta, and Ti, but may also contain Fe, Sb, Zr, Sn, and W. Moderate to extensive substitution is possible in both the A and B sites, and although the anion given in the general formula is oxygen, OH, F, Cl, and N may also be present. The pyrochlore series is divided into three subgroups based on the abundances of Nb, Ta, and Ti in the B site (Hogarth 1977). Pyrochlore has $Nb+Ta > 2Ti$ and $Nb > Ta$; microlite has $Nb+Ta > 2Ti$ and $Ta \geq Nb$; and betafite has $2Ti \geq Nb+Ta$. Compositions among members of the pyrochlore group can be broadly related to rock-type association and to temperatures of formation (Vilasov 1966, Hogarth 1977, Van Der Veen 1963, Gold 1966, Perrault 1967, Petruck & Owens 1975, Van Wambeke 1978, Cerny et al 1979, McMahon & Haggerty 1979). Alkalic host rocks have high CaO ($\sim$19 wt%) and TiO_2 (7–10 wt%) contents and a marked enrichment in the cerium rare earth element group (ΣCe). Pyrochlore from carbonatites are distinguished by high ZrO_2 contents ($\sim$10 wt%) and diminished ΣCe. Concentrations of TiO_2, CaO, REE, and U decrease with decreasing temperature, whereas the contents of Nb and Na increase. Some, but not all, members of the pyrochlore group undergo radiation metamict damage, and secondary processes produce fersmite (Ca,Ce,Na)$(Nb,Ti,Fe)_2(O,OH,F)_6$ and/or columbite-tantalite (Fe,Mn) $(Nb,Ta)_2O_6$. Other alteration products are encountered under conditions of equatorial weathering, but in temperate climates such as the Oka carbonatite complex in Quebec, Canada, pyrochlore has remained remarkably unaffected, considering that the age of the REE deposit is dated at 117 m.y. (Shafigullah et al 1969).

Crichtonite ($AM_{21}O_{38}$) has rhombohedral symmetry in which A is the dominant large cation, and M are the smaller cations Ti, Zr, Cr, Al, Fe, Mg, and Mn (Grey et al 1976). The large cations (Grey et al 1976, Grey & Gatehouse 1978, Grey & Lloyd 1976, Campbell & Kelly 1978, Gatehouse et al 1978, Kelly et al 1979) may be Sr (crichtonite), Pb (senaite), REE (davidite), Ca (loveringite), Na (landauite), or Ba and K (Haggerty 1975, Smyth et al 1978). These minerals are isostructural, with the large cations occupying one of the anion sites in a closely packed anionic stacking sequence and the smaller cations occupying both the tetrahedral and octahedral interstices in the lattice. Members of the crichtonite series are present as accessory minerals in mafic and alkalic rocks. Davidite has a wide distribution that ranges from pneumatolic veins to metamorphic rocks (Butler & Hall 1960), and the two unnamed Ba and K members (phases X and Y in Table 1) have so far only been recognized in upper-mantle-derived nodules in kimberlites (Haggerty 1975, Smyth et al 1978). There are no reported alteration effects on the mineral group apart from radiation damage, which is prevalent in davidite and loveringite. Phases compositionally related to the chrichtonite series but also to armalcolite [(FeMg)Ti_2O_5] are referred to here as transitional armalcolites, reported by Cameron (1978) from the Bushveld layered intrusion and by Raber & Haggerty (1979) from a kimberlite. The minerals have not been crystallographically characterized but are potentially useful alternatives because of their refractory high TiO_2 (62–68 wt%), REE, Ba, and Ca contents.

Kobeite (Y,U) $(Ti,Zr,Fe,Nb)_2O_6$, although classified as a member of the AB_2O_6 orthorhombic euxenite group is reported to be cubic (Vilasov 1966). This mineral is found in some granite pegmatites, is known to alter to anatase (TiO_2) plus a Nb-rich constituent, and is commonly metamict.

Polymignite (Ca,La,Y,Th,Mn,Fe,Ce) $(Zr,Ti,Nb,Ta)_2O_6$ is orthorhombic and occurs in granite pegmatites and syenites coexisting with pyrochlore and zircon. It is not reported to be metamict and its weathering or stability characteristics are unknown (Vilasov 1966).

In order to provide any reasonable assessment of the geological stability of the SYNROC $BaAl_2Ti_6O_{16}$ phase, closely approximating analogs must be considered since it does not occur naturally. The proposed hollandite-structured phase has the advantages that a broad range of elements can be substituted and that Cs can be stabilized. There are two naturally occurring minerals that are remarkably similar to $BaAl_2Ti_6O_{16}$: freudenbergite ($Na_2FeTi_7O_{16}$–$Na_2Fe_2Ti_6O_{16}$) and priderite (K,Ba) $(Ti,Fe)_8O_{16}$. Priderite does in fact have the tetragonal hollandite structure (Norrish 1951), and freudenbergite is monoclinic (McKie & Long 1970). In what is becoming a recurring theme in this analysis, we find that priderite and freudenbergite are only recognized in highly alkalic rocks. The former is present in leucite-bearing lamproites, and the latter in apatite-rich alkali syenites. An unusual example has also been found in a crustal-derived xenolith from a kimberlite in Liberia (Haggerty et al 1979), where freudenbergite is associated with perovskite, sphene, ilmenite, and rutile. In all reported instances these minerals are stable, crystalline products that have formed at magmatic temperatures ($\sim 1200°C$). Freudenbergite may be a key mineral in the crystal stabilization of radwaste Na in a silica-free system, and its incorporation should be seriously considered. Priderite is also an obvious choice, as are priderite-related synthetic phases. Bayer & Hoffman (1966) have shown that the open framework (*B* ions) and the channels (*A* ions) that are present in the hollandite structure are conducive to a considerable number of possible elemental substitutions, with the structure being stabilized by large alkali or alkaline earths (*A* ions) in the channels. The *A*-ion channel substitution of K and Rb has been experimentally verified (Bayer & Hoffman 1966) in combination with an extensive substitutional series (Al,Ti,Cr,Fe,Ga,Mg,Co,Ni,Cu,Zn,Sb,Sn,In) in the corresponding *B*-site framework. Ringwood et al (1979b), recognizing the potential of the structure, have synthesized the end members $K_2Al_2Ti_6O_{16}$, $SrAl_2Ti_6O_{16}$, and $Cs_2Al_2Ti_6O_{16}$, and have demonstrated the existence of extensive solid solutions with their idealized $BaAl_2Ti_6O_{16}$ component. Bayer & Hoffman (1966), however, were unsuccessful in producing a stoichiometric Cs-channeled compound in combination with the *B*-framework ions of Al, Ti, Cr, Fe, and Ga, and they conclude that cesium is present as an *incorporated* phase that is tentatively identified as a Cs-titanate.

Cesium is of considerable concern in radwaste stabilization but cannot be evaluated in hollandite or related minerals under geological conditions. Whether Cs is structurally bound or whether it is present as a discrete phase in hollandite needs to be established. With regard to other potential Cs-bearing phases, one possibility is the large-radius cation site known to be present in crichtonite structures, or those available in freudenbergite and priderite.

Representative chemical analyses of natural minerals are given in Tables 2 and 3 for perovskite and for zirconolite. Compositions of suggested alternate minerals (pyrochlore, crichtonite, kobeite, polymignite, freudenbergite, and priderite) are also tabulated; from these data it is apparent that large proportions of characteristic radwaste elements are readily accommodated in natural settings. Aside from the transuranic actinides (Np,Pu,Am,Cm), which do not appear in these analytical data, the prominent omissions are molybdenum and phosphorous. The concentration of Mo in the waste calcine is ~ 12 mole %, and for P is ~ 3 mole %. It

Table 2 Representative analyses of naturally occurring SYNROC minerals, SYNROC-related minerals, and SYNROC alternatives[a]

	1	2	3	4	5	6	7	8	9	10	11	12	13
SiO_2	1.5	0.35	0.15	0.45	2.03	0.02	0.02	—	0.78	1.79	0.45	1.81	3.30
TiO_2	29.66	27.78	53.29	10.05	63.62	70.21	80.10	70.6	—	26.11	18.90	4.15	5.43
ZrO_2	31.79	32.91	—	—	—	—	—	—	38.20	15.99	29.71	1.24	1.66
Al_2O_3	0.91	0.49	—	—	0.47	0.10	0.01	2.3	0.47	0.34	0.19	0.51	—
Cr_2O_3	—	0.19	0.15	—	—	—	0.08	—	—	—	—	—	—
Fe_2O_3	2.72	—	—	8.74	18.94	17.81	—	12.4	0.74	8.93	7.66	1.36	1.43
FeO	2.78	7.45	1.33	—	—	—	9.20	—	0.27	—	2.08	0.21	—
MgO	0.56	0.28	0.6	2.20	0.47	0.03	0.75	—	0.06	—	0.16	0.15	0.08
MnO	0.13	0.02	—	0.77	0.26	0.14	0.02	—	—	1.82	1.32	0.20	1.28
CaO	10.68	4.19	34.78	25.95	—	—	0.42	trace	5.29	0.63	6.98	10.05	18.1
NiO	—	—	—	—	—	—	—	—	—	—	—	—	—
ZnO	—	—	—	—	—	—	—	—	—	—	—	—	—
SnO	—	—	—	—	—	—	—	—	—	—	—	—	—
SrO	—	—	—	—	—	—	—	—	2.04	—	—	0.20	0.54
Na_2O	0.44	—	1.10	4.03	6.90	8.58	8.75	0.6	7.96	—	0.59	3.26	4.15
K_2O	0.24	—	—	0.03	1.33	0.04	0.06	5.6	0.34	—	0.77	0.22	—
Cs_2O	—	—	—	—	—	—	—	—	—	—	—	—	—
BaO	—	—	—	—	—	—	0.59	6.7	—	—	—	0.04	—
Y_2O_3	—	8.65	0.01	—	—	—	—	—	—	23.19	2.26	2.78	0.13
La_2O_3	—	0.16	1.07	—	—	—	—	—	—	—	5.13	—	0.95
Ce_2O_3	—	1.44	3.42	—	—	—	—	—	—	1.73	5.91	3.94	5.69
Pr_2O_3	—	0.59	0.35	—	—	—	—	—	—	—	—	—	—
Nd_2O_3	—	3.11	1.26	—	—	—	—	—	—	—	—	—	1.26
Sm_2O_3	—	1.05	0.10	—	—	—	—	—	—	—	—	—	—
Eu_2O_3	—	0.05	0.14	—	—	—	—	—	—	—	—	—	—
Gd_2O_3	—	1.16	0.44	—	—	—	—	—	—	—	—	—	—
Sb_2O_3	—	—	—	—	—	—	—	—	—	—	—	—	—
Tb_2O_3	—	0.44	0.01	—	—	—	—	—	—	—	—	—	—
Dy_2O_3	—	2.05	—	—	—	—	—	—	—	—	—	—	—
Ho_2O_3	—	0.48	0.01	—	—	—	—	—	—	—	—	—	—
Er_2O_3	—	1.11	0.12	—	—	—	—	—	—	—	2.26	—	—
Tm_2O_3	—	0.20	—	—	—	—	—	—	—	—	—	—	—
Yb_2O_3	—	0.70	0.02	—	—	—	—	—	—	—	—	—	—
Lu_2O_3	—	0.26	1.09	—	—	—	—	—	—	—	—	—	—
ΣREE	4.49	—	—	2.03	—	—	—	—	29.96	—	—	—	—
HfO_2	—	0.91	—	—	—	—	—	—	—	—	—	—	—
UO_2	—	0.37	—	—	—	—	—	—	—	—	—	2.23	—
U_3O_8	—	—	—	—	—	—	—	—	—	—	—	—	—
ThO_2	2.52	0.68	—	—	—	—	—	—	1.00	1.28	3.92	1.03	0.56
PbO	—	0.54	—	—	—	—	—	—	—	—	0.39	—	—
V_2O_5	—	—	—	—	—	—	—	—	—	—	—	—	—
Nb_2O_5	7.04	1.84	1.40	43.90	2.73	1.24	0.05	—	11.50	7.63	11.90	46.61	54.62
Ta_2O_5	0.41	0.22	—	—	—	—	—	—	0.53	7.63	0.53	9.60	1.33
F	0.60	—	—	—	—	—	—	—	—	—	—	1.62	3.32
H_2O^-	2.65	—	—	—	—	—	—	—	—	0.43	—	3.22	—
H_2O^+	—	—	—	—	2.98	—	—	—	0.46	3.34	0.28	4.70	0.32
Total	99.12	99.67	100.84	98.15	99.73	98.17	100.05	98.20	99.60	100.84	101.39	99.13	104.15

[a] Zirconolite, anal. 1–2 (Wark et al 1973, Vilasov 1966); Perovskite, anal. 3–4 (Boctor & Boyd 1979, Nickel & McAdam 1963); Freudenbergite, anal. 5–7 (Frenzel 1961, McKie & Long 1970, Haggerty et al 1979); Priderite, anal. 8 (Norrish 1951); Loparite, anal. 9 (Vilasov 1966); Kobeite, anal. 10 (Vilasov 1966); Polymignite, anal. 11 (Vilasov 1966); Pyrochlore, anal. 12–20 (Vilasov 1966, Perrault 1967, Petruck & Owens 1975, Van Wambeke 1978, Cerny et al 1979, Crook 1977); Armalcolite-

14	15	16	17	18	19	20	21	22	23	24	25	26	27	28
—	—	—	—	0.13	—	—	—	—	—	—	—	0.20	—	—
5.57	3.57	5.00	4.44	4.83	16.5	0.43	62.99	60.49	58.68	49.81	73.46	58.06	55.44	52.5
0.41	0.32	0.16	0.46	0.48	—	—	1.83	0.10	0.09	0.07	—	4.38	0.2	7.4
—	—	—	—	1.47	0.49	—	1.36	0.02	0.05	0.76	—	0.96	0.18	—
—	—	—	—	—	—	—	10.68	0.15	0.16	2.49	—	5.97	17.59	13.0
—	—	—	—	0.40	—	2.18	—	21.4	16.20	23.62	10.75	20.01	—	14.6
1.38	2.28	2.75	1.45	0.23	0.6	—	13.63	8.18	8.53	5.31	2.00	—	11.62	0.5
—	—	—	—	0.11	—	0.49	2.07	—	0.02	0.18	—	1.38	3.77	3.1
0.24	0.11	0.15	0.13	0.07	0.6	1.73	0.29	2.78	4.08	0.07	3.45	0.15	0.09	—
18.29	14.17	6.33	12.86	0.60	14.5	2.77	2.68	0.06	0.10	0.52	—	2.50	0.47	(3.0)
—	—	—	—	—	—	—	—	—	—	—	—	0.04	—	—
—	—	—	—	—	—	—	—	—	—	—	—	—	—	—
—	—	—	—	—	2.9	3.81	—	—	—	0.05	9.97	—	0.38	—
0.27	0.61	0.21	0.35	2.81	—	—	—	4.17	0.20	0.02	—	0.01	0.10	2.4
3.69	4.68	1.11	5.82	0.43	0.30	—	—	—	—	0.09	—	0.04	0.01	—
—	—	—	—	2.54	—	—	—	—	—	0.04	—	0.04	0.28	(3.0)
—	—	—	—	—	—	—	—	—	—	—	—	—	—	—
—	—	—	—	0.41	—	—	—	—	—	—	—	—	4.33	9.1
—	—	—	—	0.02	—	9.46	0.54	0.21	0.86	1.86	—	0.18	—	—
0.97	1.15	0.77	1.64	0.11	—	0.29	0.32	0.29	0.06	3.35	—	1.23	—	—
6.47	5.51	3.01	7.82	0.24	—	0.43	1.36	0.27	0.07	—	—	1.12	—	—
—	—	—	—	0.04	—	0.13	—	—	—	—	—	—	—	—
0.79	0.84	0.58	1.15	0.04	—	0.67	0.37	0.02	0.01	0.35	—	0.26	—	—
—	—	—	—	—	—	0.75	—	—	—	—	—	—	—	—
—	—	—	—	—	—	—	—	—	—	—	—	—	—	—
—	—	—	—	0.01	—	1.98	—	—	—	—	—	—	—	—
—	—	—	—	—	23.2	—	—	—	—	—	—	—	—	—
—	—	—	—	—	—	0.31	—	—	—	—	—	—	—	—
—	—	—	—	—	—	1.12	—	—	—	—	—	—	—	—
—	—	—	—	—	—	0.45	—	—	—	—	—	—	—	—
—	—	—	—	—	—	0.89	—	—	—	—	—	—	—	—
—	—	—	—	—	—	0.38	—	—	—	—	—	—	—	—
—	—	—	—	—	—	0.92	—	—	—	—	—	—	—	—
—	—	—	—	—	—	0.64	—	—	—	—	—	0.45	—	—
—	—	—	—	—	—	—	0.46	—	—	0.88	—	0.21	4.00	—
—	—	—	—	—	—	—	—	0.23	0.13	0.51	—	0.14	—	—
—	—	—	—	—	—	—	—	—	—	7.89	—	0.27	—	—
—	8.82	24.79	3.58	0.09	—	4.53	—	—	—	—	—	—	—	—
—	—	—	—	0.14	—	0.64	—	0.01	0.12	0.15	—	0.08	—	—
—	—	—	—	0.02	0.13	0.88	—	0.78	9.21	0.74	—	1.30	—	—
—	—	—	—	0.03	—	—	—	0.61	0.70	1.87	—	—	1.27	—
59.02	56.37	51.45	57.82	73.31	21.6	17.10	—	—	—	—	—	—	—	—
0.16	2.23	3.46	3.08	0.08	19.3	44.52	—	—	—	—	—	—	—	—
—	—	—	—	0.23	—	—	—	—	—	—	—	—	—	—
—	—	—	—	0.48	—	—	—	—	—	—	—	—	—	—
—	—	—	—	9.32	0.44	1.64	—	—	—	0.68	—	—	—	—
97.26	100.66	99.78	100.60	98.67	100.56	99.14	98.58	99.77	99.27	100.63	99.63	98.98	99.73	102.60

like phase (FeMg)Ti_2O_5 or a crichtonite-related mineral, anal. 21 (Cameron 1978); Crichtonite, anal. 22 (Grey et al 1976); Senaite, anal. 23 (Grey et al 1976); Davidite, anal. 24 (Smellie et al 1978); Landauite, anal. 25 (Grey & Gatehouse 1978); Loveringite, anal. 26 (Gatehouse et al 1978, Campbell & Kelly 1978); Phases *X* and *Y* are crichtonite-related minerals, anal. 27–28 (Haggerty 1975, Smyth et al 1978).

Table 3 Summary of radwaste fission products and actinides in selected naturally occurring oxides[a]

	Fission (wt%)	Actinides (wt%)
Zirconolite		
Gabbros (Zr,REE)	32–33	1–2.5
Lunar basalts (Zr,REE)	32–39	0.3–2.5
Kimberlites (Zr,REE)	41–71	0
Perovskite		
Kimberlites (REE)	3–13	—
Crichtonite group		
Crichtonite (Sr)	5.0	—
Senaite (Pb)	10.5	—
Davidite (REE)	10.0	4
Loveringite (REE,Zr)	7.5	0.3
Landauite (Pb)	2.5	—
Phase *X* (Ba,REE)	5–10	—
Phase *Y* (K,REE)	5–10	—
Transitional armalcolites		
(REE,Ba,Sr,Pb)	0.5–3.5	—
Pyrochlore group		
(REE,Pb,Sr,Zr)	10–15	20–25

[a] These data are the total concentrations of the fission (shown in parenthesis) and actinide elements that are present in a variety of naturally occurring minerals. Complete analyses are given in Table 2.

is anticipated that within SYNROC Mo will enter $BaAl_2Ti_6O_{16}$ or perovskite by substituting for Ti, but this has yet to be conclusively demonstrated. Phosphorous will possibly combine with Ba and Ca into an orthophosphate, not unlike that of Supercalcine in which P enters an apatite structure. Molybdenum in Supercalcine is partitioned into a scheelite-structured oxide along with Sr and Ba. Here again, it is entirely possible that the extraordinary flexibility of the *B*-cation framework sites in the hollandite structure, or the tetrahedral and octahedral sites in the crichtonite-series minerals, make these sites likely positions for both Mo and P.

As for the remainder of the radwaste elements, there are no outstanding inconsistencies between the molecularly engineered predicted substitutions and those present in natural minerals. The philosophy and strategy of immobilization into crystalline phases are, essentially, endorsed by nature. However, some refinements to the stabilizing crystalline media are clearly required to confidently establish long-term resistance.

RADIATION-INDUCED METAMICTIZATION

Curiously, a concern that applies to the utilization of glass as an immobilizing first barrier for radwaste calcine is the inverse of a significant problem that applies to the storage of radionuclides into crystalline materials (Ewing 1976). Glass is regarded as being unacceptable because of inherent thermodynamic instability, devitrification, and consequent crystallization. However, the breakdown of crystalline structures into X-ray amorphous glasslike constituents may also be induced by α-particle emissions from incorporated and decaying radioisotopes. Two properties accompany decay, and the cumulative effect is regarded as one of interatomic bond severence. This results from the hypervelocity release of α particles and from the consequent recoil of the emitting atom, which has a lower energy but a substantially larger cross-sectional radius. Solid-state radiation damage in synthetic materials (Chadderton & Torrens 1969, Martin 1966) and in minerals (Pabst 1952, Fleischer et al 1975, Mitchell 1973a,b) has received widespread attention, and metamict properties related specifically to radwaste immobilization have been addressed by Hobbs (1979) and by Ewing and co-authors (Ewing 1974, 1975a,b, Ewing & Ehlmann 1975, Ewing & Haaker 1978, Haaker & Ewing 1979a,b).

It is evident from recent comprehensive reviews of the end-product characteristics of radiation damage that there is considerable uncertainty relative to the types of minerals that are predisposed to metamictization, on the critical proportions of radionuclides that are necessary to induce the metamict state, and on the physical properties of metamict matter. The mechanism of α-particle bombardment and the criteria for recognizing the metamict state (Hobbs 1979, Chadderton & Torrens 1969, Martin 1966, Pabst 1952, Fleischer et al 1975, Mitchell 1973a,b, Ewing 1974, 1975a,b, 1976, Ewing & Ehlmann 1975, Ewing & Haaker 1978, Haaker & Ewing 1979a,b, Primak 1954) are reasonably well agreed upon, but opinions differ on the following. The original concept of radiation-damaged metamict minerals is that an amorphous state is induced by the disruption of coherent bonds, which results in a high state of atomic disorder, akin to amorphous solids such as glass. It has been suggested that the transient high temperatures (10^4 K) resulting from rapidly escaping α particles are sufficient to effect melting. The time-spans (10^{-11} s) are short enough, and the volumes of matter are small enough, that rapid quenching would ensure the maintenance of a highly disordered state similar to that of nonbridging ions in solid metastable liquids that constitute glass. Evidence that the metamict state approaches that of glass is based on the apparent lack of a crystalline structure and on the fact that the crystalline state can be restored by annealing at high temperature. Convincing data, however, have also been presented that α-particle bombardment effects one of two probable states. The first state is a disaggregation of the primary phase, dislocating but not totally disrupting the crystal structure of the mineral into units or domains (Graham & Thornber 1974). This model requires that the domains remain coherent but that they are sufficiently small to appear opaque to determinative X-ray techniques. The second state is similarly one of disaggregation into microdomains but is instead accompanied by decomposition of the mineral into a series of polycomponent phases (Hobbs 1979). In this model, a displacement of atoms into new lattice positions results in point-defect condensation of colloidal-sized particles. These particles are anion-cation pairs of interstitial stoichiometric compounds. Neither model has been validated and both are consistent with a return to the crystalline state upon heating. Nucleation and domain-coarsening are accompanied by a release of "stored energy" in the transformation from the metamict to the crystalline state, and the reaction is strongly exothermic (Kurath 1957).

Regardless of whether it is the inducement of an amorphous state or instead a process of particulate reconstitution, metamictization is rightly of concern in the crystalline stabilization of radwaste. Radiation damage causes a decrease in density, an increase in volume, and an increase in surface area (Holland & Gottfried 1955), properties that are likely to compromise the integrity of the system. Although the properties of insoluble gaseous fission products (He, Xe, Kr, and I) are likely to be relatively small in the containment canister, radiation-induced void swelling caused by the accumulation of gasses along grain boundaries, within lattice vacancies, and in microcapillary fractures cannot be ignored. In the most extreme cases of fast neutron bombardment, volume changes of 8% have been observed before interconnecting gas pockets grow sufficiently large to induce gas release (Hobbs 1979).

The choice of crystalline phases must, therefore, also take the effects of metamictization into account, a precaution that is stressed by Ewing & Haaker (1978). They base their selection of alternate minerals to those proposed for either SYNROC or Supercalcine on the absence of a predisposition to attain the metamict state. They also draw attention to a number of intrinsic properties of minerals prone to metamictization, showing, for example, that the mere presence of a radiogenic element is not in itself a necessary prerequisite for crystal radiation damage. In the two mineral polymorphs of $ThSiO_4$, it is tetragonal thorite that is metamict whereas monoclinic huttonite is crystalline (Pabst 1952). Equally, minerals such as xenotime (YPO_4) and monazite ($CePO_4$), which commonly contain substantial concentrations of uranium and have understandably undergone extensive α-particle bombardment, are rarely metamict. Ewing & Haaker conclude that because most of the minerals in the Nb-Ta-Ti (pyrochlore, polymignite) group (Table 1) are prone to metamictization, these phases are unsuitable candidates as crystalline substitutes in the initial stabilizing barrier.

The effects of metamictization will not be serious during the early stages of radwaste storage. As noted above, restoration to a crystalline state from a metamict state is evidently only a matter of scale transformation. *Polymictization* may in fact be a more appropriate concept to describe the effect. Because heating activates recrystallization, the maintenance of elevated temperatures within the canister ought to prevent, or at least diminish, the effect of reduced crystallinity by radiation. Such heat sources need not be very large, and recrystallization temperatures for many metamict minerals, including those listed in Table 1, show that metamict (*polymict*)-crystal transformations can be induced at temperatures as low as 200°C, with optimal structural improvement in the 600–1000°C range (Pyatenko 1970). These temperatures are somewhat higher than those anticipated (~500°C max) from self-induced thermal radiation produced by β- and γ-emitting nuclei within radwaste. Continuous auto-annealing can be expected for several hundreds of years, but as temperatures gradually decrease the longer-lived and α-emitting nuclei will persist. Decay of the actinides will be most pronounced after ~2000 years, when the effects of α-recoil damage are likely to become substantial. Storage wastes will be subject to α-recoil aging, and Dran et al (1980) have demonstrated that the mobility of loosely bonded elements, specifically in glass, will increase as a function of exposure time or dosage. Corrosion is activated at structural or grain-boundary discontinuities, and even the relatively mild effects of exposure to atmospheric moisture (in contrast to dynamic leaching) can be considerable (Hirsch 1980).

While the concerns of interim storage are alleviated, the problems of permanent disposal are perhaps accentuated. Although there is a voluminous body of literature on the effects of radiation damage in solids, the present status of knowledge of these effects in minerals is minimal, a situation clearly requiring remedial attention if crystalline storage repositories are to be seriously considered for radioactive nuclear waste products.

Some progress toward this goal has been achieved by Ringwood et al (1981) for two of the host minerals in SYNROC—perovskite and zirconolite—that contain α-emitting actinide elements. These minerals evidently behave as closed systems for U, Th, and Pb daughter products when exposed to accelerated radiation and/or aging doses. Naturally occurring zirconolites and perovskites have accumulated radiation doses that can be calculated from their ages and from their U and Th contents. These doses may be translated into a SYNROC-equivalent age relationship by specifying the concentration of incorporated actinides. SYNROC-equivalent ages for a variety of perovskites and zirconolites yield between 10^3 and 4×10^8 years, assuming a 10% high-level radwaste product. Volume changes accompany the annealing of these minerals, but the metamict state remains essentially crystalline with only long-range lattice disorder. Based on these data, it is clear that perovskite and zirconolite are prime candidates in the design of a mineral repository for radwaste.

CRYSTALLOCHEMICAL CONSIDERATIONS IN BARRIER PRODUCTION

From the discussion in the preceding sections it is obviously premature to conclude that the problem of isolating radwaste materials into crystalline solids has been solved. The significant outstanding problems are (*a*) to determine the critical α-particle dosage required for radiation damage; (*b*) to establish the least probable and the most likely mineral candidates that exhibit, or lack, the properties of radiation effects; (*c*) to critically evaluate whether it is indeed crystalline reconstitution (polymictization) or instead a highly disordered state (metamictization) that is prevalent in naturally occurring minerals; and (*d*) to ascertain time-dependent and other parameters (e.g. structure, radionuclide concentrations) that will place constraints on the suitability of minerals in the initial isolation barrier.

Among the minerals present in SYNROC and Supercalcine and within the class of alternative minerals suggested in this review, some are known to be more susceptible to alteration than others. However, it is noteworthy that whereas loveringite exhibits radiation damage in rocks dated at 2.1 b.y., some, but not all, pyrochlores dated at 117 m.y. are unaffected. On the other hand, much older lunar zirconolites (3.8–4.3 b.y.), for example, are crystalline. This may be a matter of conducive and nonconducive structures, or of radioisotope concentrations, but it may also be a matter of the environments of formation. There is evidence that radiation-damaged crystals are more hydrous than their unirradiated counterparts (Ewing & Haaker 1978), but there is considerable uncertainty as to whether hydration is the consequence of or the reason for alteration. It is pertinent that experimental evidence shows that point-source radiation damage in quartz (SiO_2, a mineral that is essentially water-free) is preferentially located around minute hydroxal regions.

Hydroradiolysis (i.e. radiation-induced displacement of atoms around hydroxals) may, therefore, be a key issue in understanding the effects of radiation damage. The sterile lunar environment is known to be essentially anhydrous, suggesting that an equally anhydrous set of conditions may be necessary in the production of the first-barrier isolation system if water is shown to be a triggering component in crystalline modification under the influence of radiation.

The geological distribution of minerals that have been discussed show that the major sources of REE, Zr, Ti, and actinide-bearing minerals are from mafic alkalic rocks and from granite pegmatites. In the case of pyrochlore, for example, the fractionation of the lanthanide elements is dependent on host-rock composition and the REE contents vary as a function of the temperature of formation. Herein may lie another critical clue to the temperatures and the oxygen partial pressures that may be necessary in the production of the crystalline stabilization system. Granite pegmatites are hydrous residual melts that form at lower temperatures and at higher oxygen fugacities (partial pressures) than those of mafic and alkalic suites (see Figure 2). The reference curve in this figure is based on the buffered equilibrium transformation of fayalite (Fe_2SiO_4) to magnetite (Fe_3O_4) + quartz (SiO_2) as a function of T°C (temperature) and fO_2 (oxygen fugacity). Hydrous pegmatites exhibit a considerable displacement relative to alkaline suites and to the buffer curve, reflecting their higher oxidation potential. The relative solubilities of the REE and actinide elements are related to their oxidation states, with higher-valence cations being more susceptible to mobilization than lower-valence equivalents. The maintenance of a relatively low oxygen fugacity in stabilizing radwaste should therefore be considered. It may in fact be desirable to produce the monolithic material at oxygen fugacities comparable to those of lunar samples, which crystallized below the iron-wüstite (FeO) buffer curve in Figure 2. This condition would certainly ensure an anhydrous environment at high encapsulation temperatures.

Current isolation proposals for the initial barrier involve combining the radwaste calcine and the inert additives into a crystalline admixture of components. It is furthermore proposed that this material be canisterized into a second metal barrier. If there are any lessons to be learned from geological settings, one is that this relatively concentrated form does not mimic natural occurrences of actinide and related fission-produced products. The Okla natural reactor (Cowan 1976) is of course an exception. Several of the inherent problems that have been identified so far would be alleviated if the crystalline waste form was further diluted. This can be accomplished without modifying the chemistry of the additives. If the radwaste-charged crystalline product is fragmented and incorporated into

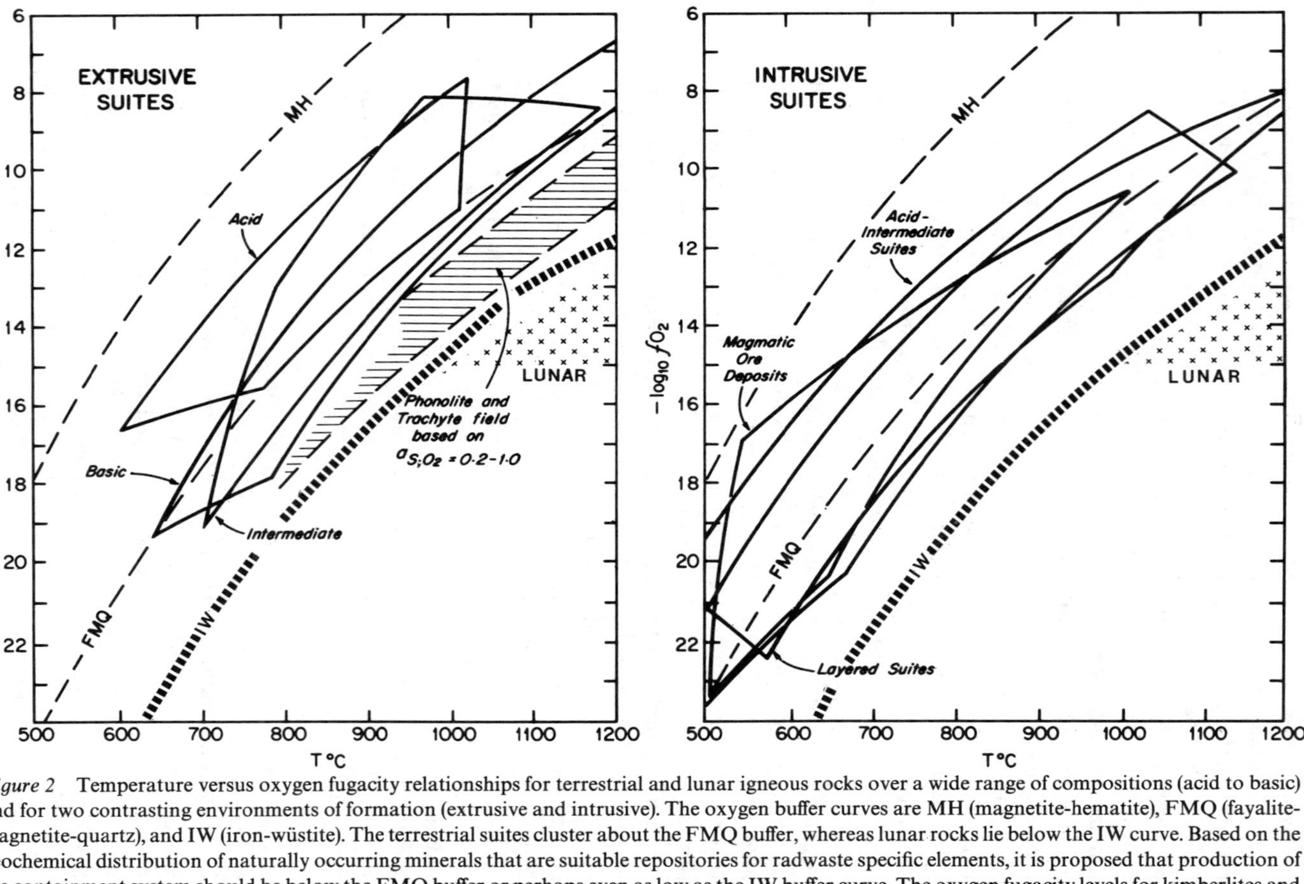

Figure 2 Temperature versus oxygen fugacity relationships for terrestrial and lunar igneous rocks over a wide range of compositions (acid to basic) and for two contrasting environments of formation (extrusive and intrusive). The oxygen buffer curves are MH (magnetite-hematite), FMQ (fayalite-magnetite-quartz), and IW (iron-wüstite). The terrestrial suites cluster about the FMQ buffer, whereas lunar rocks lie below the IW curve. Based on the geochemical distribution of naturally occurring minerals that are suitable repositories for radwaste specific elements, it is proposed that production of the containment system should be below the FMQ buffer or perhaps even as low as the IW buffer curve. The oxygen fugacity levels for kimberlites and carbonatites are between the IW and FMQ curves at lower temperatures than the shaded region shown for phonolites and trachytes. It is within these rock types that the naturally occurring analogs of SYNROC are present.

a second inert matrix, a situation approaching that of ore and gangue (waste rock) will be achieved. This precanistered assemblage will in effect provide an additional barrier, and heat production will certainly be reduced. Although the problems of radiation damage and of solubility will not disappear, they are likely to be substantially reduced. A two-stage process is envisaged. The first stage is crystalline immobilization and the second is suspension of this product into an additional crystalline matrix. The resulting product is then more realistically a rock. And being a rock, crystal-crystal interfaces will facilitate gas release, the system will have radwaste products more evenly dispersed, and a diluted monolithic repository will result, a situation that is more compatible with natural environments. A variant of this matrix dilution mode has been proposed and is referred to as "radiophase encapsulation" (Roy et al 1981).

SUMMARY AND CONCLUSIONS

The safe immobilization of radioactive nuclear waste products is an outstanding problem of major concern to society. It is a responsibility that cannot be ignored, and a consensus on the method and mode of containment must urgently be sought. Although various estimates have been proposed, the durability of the system should be of the order of 100,000 years or more to ensure the containment of the most toxic of the radionucleides that are present in radwaste products.

The SYNROC system of immobilization is attractive from the standpoint of having a simplified mineralogy, a limited number of highly conducive structural types, and sufficient flexibility to accommodate a range of radwaste products. In addition, because two of the three components (perovskite and zirconolite) of SYNROC are present in naturally occurring geological environments, their stability may be assessed on a time-scale far in excess of the idealized 100,000-year containment objective. Several alternate minerals to those suggested for SYNROC are proposed and ought to be considered. Their retention capacities for specific radwaste elements can be ascertained, and are demonstrated to persist, within naturally occurring geological environments.

Among the problems to be addressed for SYNROC or SYNROC-related systems, perhaps the most significant is the containment of Cs. Hervieu & Raveau (1980) recently reported on Cs titanoniobate ($CsTi_2NbO_7$), a synthetic compound not known to occur in nature in which annealing temperatures of 1000–1200°C are required for effective crystallization. This would appear to be a potentially useful phase, because Cs evidently is retained in this Cs-coordinated layered structure at high temperatures. The question of cesium channeling in hollandite-structured $BaAl_2Ti_6O_{16}$, or the presence of a discrete Cs-titanate phase in this mineral, must clearly be resolved. Experimental and geochemical data are also required on the possible existence of Cs-bearing members in the crichtonite mineral group, freudenbergite, priderite, and possibly perovskite-related structures (Benmoussa et al 1982).

A thorough understanding of the cause-and-effect relationship of the processes yielding meta- or polymictization is required to predict the long-term influence that radiation damage will have on crystalline structures. Centerline temperatures within the containment canisters may be employed to maintain the crystalline state and, for minerals susceptible to radiation damage, to inhibit anything more than a transient transformation to the amorphous state.

In summary, consideration of the geochemical distribution of radwaste products in natural settings may provide important suggestions for the mass production and storage of the crystalline containment system. The first suggestion is the use of controlled oxygen fugacity at iron-wüstite; the second is that radwaste active pellets be diluted into an inert media within the canister to more closely approximate a *rock* rather than a concentrated and highly reactive mineral aggregate; third, since freudenbergite ($Na_2FeTi_7O_{16}$) and priderite are two minerals of specific interest to the isolation of alkali-enriched US Defense radwastes, nepheline would be replaced in SYNROC and a system dominated by highly refractory oxides would be maintained; and finally, since the distribution of minerals in SYNROC or alternates are typical of kimberlites intruded into the most stable of the Earth's cratonic regions, the radwaste monoliths should perhaps be returned to the gaping diamond diatremes of the world.

Acknowledgements

Aspects of this study were indirectly supported by NASA (NGR22-010-089) and the National Science Foundation (EAR76-23787, EAR78-02539, and EAR78-02541) and funding is gratefully acknowledged. The comments and active discussions that resulted from presentation of this paper at the Geological Society of America meeting in Atlanta in 1980 aided significantly in formulating the present review. Numerous colleagues and students provided thoughtful insights and helpful comments.

Literature Cited

Allsopp, H. L., Barrett, D. R. 1975. Rb-Sr age determinations on South African kimberlite pipes. *Phys. Chem. Earth* 9:605–19

Angino, E. E. 1977. High-level and long-lived radioactive waste disposal. *Science* 198:885–90

Assistant Secretary for Nuclear Energy. 1980. *Management of Commercially Generated Radioactive Waste: Final Environmental Impact Statement*, Vols. 1, 2, 3. Washington DC: US Dep. Energy. 594 pp., 310 pp., 604 pp.

Bayer, G., Hoffman, W. 1966. Complex alkali titanium oxides $A_x(B_yTi_{8-y})O_{16}$ of the α-MnO_2 structure-type. *Am. Mineral.* 51:511–16

Benmoussa, A., Groult, D., Studer, F., Raveau, B. 1982. Intergrowth of hexagonal tungsten bronze and perovskite-like structures: The oxides $ACu_3M_7O_{21}$ (A = K, Rb, Cs, Ti; M = Nb, Ta). *J. Solid State Chem.* 41:221–26

Boctor, N. Z., Boyd, F. R. 1979. Distribution of rare earth elements in perovskite from kimberlites, *Carnegie Inst. Washington Yearb.* 78:572–74

Borodin, L. S., Nazarenko, I. I., Richter, T. L. 1957. The new mineral zirconolite—a complex oxide of the AB_3O_7 type. *Am. Mineral.* 42:581–82

Borodin, L. S., Bykova, A. V., Kapitonova, T. A., Pyatenko, Y. A. 1961. New data on zirconolite and its niobian variety. *Am. Mineral.* 46:465

Busche, F. D., Prinz, M., Keil, K., Kurat, G. 1972. Lunar zirkelite: A uranium-bearing phase. *Earth Planet. Sci. Lett.* 14:313–21

Butler, J. R., Hall, R. 1960. Chemical characteristics of davidite. *Econ. Geol.* 55:1541–50

Butterman, W. C., Foster, W. R. 1967. Zircon stability and the ZrO_2-SiO_2 phase diagram. *Am. Mineral.* 52:880–85

Cameron, E. N. 1978. An unusual titanium-rich oxide mineral from the Eastern Bushveld Complex. *Am. Mineral.* 63:37–39

Campbell, I. H., Kelly, P. R. 1978. The geochemistry of loveringite, a uranium-rare-earth-bearing accessory phase from the Jimberlana Intrusion of Western Australia. *Mineral. Mag.* 42:187–97

Cerny, P., Hawthorne, G. C., LaFlamme, J. H. G., Hinthorne, J. R. 1979. Stibiobetafite, a new member of the pyrochlore group from Vezná, Czechoslovakia. *Can. Mineral.* 17:583–88

Chadderton, L. T., Torrens, I. Mc. 1969. *Fission Damage in Minerals.* New York: Methuen & Co. 265 pp.

Cowan, C. A. 1976. A natural fission reactor. *Sci. Am.* 234:36–47

Crook, W. W. 1977. New data on yttrocrasite from the Clear Creek pegmatite, Burnet County, Texas. *Am. Mineral.* 62:1009–1011

Davis, G. L. 1978. Zircons from the mantle. *Carnegie Inst. Washington Yearb.* 77:895–97

de Marsily, G., Ledoux, E., Barbreau, A., Margat, J. 1977. Nuclear waste disposal: Can the geologist guarantee isolation? *Science* 197:519–28

Dosch, R. G., Lynch, A. W., Headley, T. J., Hlava, P. F. 1981. Titanate waste forms for high level waste—an evaluation of materials and processes. In *Scientific Basis for Nuclear Waste Management*, ed. J. G. Moore, 3:123–30. New York/London: Plenum. 632 pp.

Dran, J. C., Maurette, M., Petit, J. C. 1980. Radioactive waste storage materials: Their α-recoil aging. *Science* 209:1518–19

El Goresy, A. 1956. Baddeleyite and its significance in impact glasses. *J. Geophys. Res.* 70:3453–56

Ewing, R. C. 1974. Spherulitic recrystallization of metamict polycrase. *Science* 184:561–62

Ewing, R. C. 1975a. The crystal chemistry of complex niobium and tantalum oxides. IV. The metamict state: discussion. *Am. Mineral.* 60:728–33

Ewing, R. C. 1975b. Alteration of metamict, rare-earth, AB_2O_6-type Nb-Ta-Ti oxides. *Geochim. Cosmochim. Acta* 39:521–30

Ewing, R. C. 1976. Metamict mineral alteration: an implication for radioactive waste disposal. *Science* 192:1336–37

Ewing, R. C., Ehlmann, A. J. 1975. Annealing study of metamict, orthorhombic, rare earth, AB_2O_6-type, Nb-Ta-Ti oxides. *Can. Mineral.* 13:1–7

Ewing, R. C., Haaker, R. F. 1978. The metamict state: radiation damage in crystalline phases. In *Proc. Conf. High-Level Radioact. Solid Waste Forms*, ed. L. A. Casey, pp. 651–76

Fleischer, R. L., Price, P. B., Walker, R. M. 1975. *Nuclear Tracks in Solids: Principles and Applications.* Berkeley: Univ. Calif. Press. 605 pp.

Frenzel, G. 1961. Ein neues Mineral: Freudenbergit ($Na_2Fe_2Ti_7O_{18}$). *Neues Jahrb. Mineral. Monatsh.* 1:12–22

Gatehouse, B. M., Grey, I. E., Campbell, I. H., Kelly, P. 1978. The crystal structure of loveringite—a new member of the crichtonite group. *Am. Mineral.* 63:28–36

Gold, D. P. 1966. The minerals of the Oka carbonatite and alkaline complex, Oka, Quebec. *Mineral. Soc. India, Int. Mineral. Assoc. Gen. Meet., 4th*, pp. 109–25

Graham, J., Thornber, M. R. 1974. The crystal chemistry of complex niobium and tantalum oxides. IV. The metamict state. *Am. Mineral.* 59:1047–50

Grey, I. E., Gatehouse, B. M. 1978. The crystal structure of landauite, $Na[MnZn_2(Ti,Fe)_6Ti_{12}]O_{38}$. *Can. Mineral.* 16:63–68

Grey, I. E., Lloyd, D. J. 1976. The crystal structure of senaite. *Acta Crystallogr.* B32:1059–1513

Grey, I. E., Lloyd, D. J., White, J. S. 1976. The structure of crichtonite and its relationship to senaite. *Am. Mineral.* 61:1203–12

Haaker, R. F., Ewing, R. C. 1979a.

Differential thermal analysis of some irradiated materials: discussion. *Am. Mineral.* 64:1131–32

Haaker, R. F., Ewing, R. C. 1979b. The metamict state radiation damage in crystalline materials. In *Ceramics in Nuclear Waste Management, Tech. Inf. Cent., US Dep. Energy Conf. 790420*, pp. 305–9

Haggerty, S. E. 1973. Armalcolite and genetically associated opaque minerals in the lunar samples. *Proc. Lunar Sci. Conf., 4th. Geochim. Cosmochim. Acta* 1 (Suppl. 4): 777–97

Haggerty, S. E. 1975. The chemistry and genesis of opaque minerals in kimberlites. *Phys. Chem. Earth* 9:295–307

Haggerty, S. E. 1976. Opaque mineral oxides in terrestrial rocks. In *Oxide Minerals, Short Course Notes, Mineral. Soc. Am.*, ed. D. Rumble III, 3:101–300. Washington DC: Mineral. Soc. Am.

Haggerty, S. E., Bence, R. J., McMahon, B. M. 1979. Kimberlites in Western Liberia: II. Mineral chemistry. *2nd Kimberlite Symp., Cambridge Univ.*, 2:50–55

Hatch, L. P. 1953. Ultimate disposal of radioactive wastes. *Am. Sci.* 41:410–21

Hervieu, M., Raveau, B. 1980. A layer structure: The titaniobate $Cs_3Ti_2NbO_4$. *J. Solid State Chem.* 32:161–65

Hirsch, E. H. 1980. A new irradiation effect and its implications for the disposal of high-level radioactive waste. *Science* 209:1520–22

Hobbs, L. W. 1979. Application of transmission electron microscopy to radiation damage in ceramics. *J. Am. Ceram. Soc.* 62:267–78

Hogarth, D. D. 1977. Classification and nomenclature of the pyrochlore group. *Am. Mineral.* 62:403–10

Holland, H. D., Gottfried, D. 1955. The effect of nuclear radiation on the structure of zircon. *Acta Crystallogr.* 8:291–300

Jantzen, C. M., Clarke, D. R., Morgan, P. E. D., Harker, A. B. 1982. Leaching of polyphase nuclear waste ceramics: Microstructural and phase characterization. *J. Am. Ceram. Soc.* 65:292–300

Keil, K., Fricker, P. 1974. Baddeleyite in gabbroic rocks from Axel Fleiberg Island, Canadian Arctic Archipelago. *Am. Mineral.* 59:249–53

Kelly, P. R., Campbell, I. H., Grey, I. E., Gatehouse, B. M. 1979. Additional data on loveringite (CA,REE) $(Ti,Fe,Cr)_{21}O_{38}$ and mohsite discredited. *Can. Mineral.* 17:635–38

Kresten, P. 1974. Uranium in kimberlites and associated rocks, with special reference to Lesotho occurrences. *Lithos* 3:171–80

Krugmann, H., von Hippel, F. 1977. Radioactive wastes: A comparison of U.S. military and civilian inventories. *Science* 197:883–85

Kurath, S. F. 1957. Storage of energy in metamict minerals. *Am. Mineral.* 42:91–99

Lee, K. N. 1980. A Federalist strategy for nuclear waste management. *Science* 208:679–84

Martin, D. G. 1966. Radiation damage effects in crystals. *Sci. Prog. (Oxford)* 54:209–25

McCarthy, G. J. 1976. High-level waste ceramics. *Trans. Am. Nucl. Soc.* 23:168–69

McCarthy, G. J., Davidson, M. T. 1975. Ceramic nuclear waste forms: I. Crystal chemistry and phase formation. *Am. Ceram. Soc. Bull.* 54:782–86

McCarthy, G. J., Davidson, M. T. 1976. Ceramic nuclear waste forms: II. A ceramic-waste composite prepared by hot pressing. *Am. Ceram. Soc. Bull.* 55:190–94

McCarthy, G. J., White, W. B., Roy, R., Scheetz, B. E., Komarneni, S., Smith, D. K., Roy, D. M. 1978. Interactions between nuclear waste and surrounding rock. *Nature* 273:216–17

McKie, D., Long, J. V. P. 1970. The unit-cell contents of freudenbergite. *Z. Kristallogr. Kristallgeom. Kristallphys. Kristallchem.* 132:157–60

McMahon, B. M., Haggerty, S. E. 1979. In *Kimberlites, Diatremes, and Diamonds: Their Geology, Petrology, and Geochemistry*, ed. F. R. Boyd, H. O. A. Meyer, 1:382–92. Washington DC: Am. Geophys. Union

Merritt, W. F. 1976. High-level waste glass: Field leach test. *Trans. Am. Nucl. Soc.* 23:167–68

Meyer, H. O. A., Boctor, N. Z. 1974. Opaque mineraology: Apollo 17, rock 75035. *Proc. Lunar Sci. Conf., 5th. Geochim. Cosmochim. Acta* 1 (Suppl. 5): 707–16

Mitchell, R. S. 1973a. Metamict minerals: a review. *Mineral. Rec.* 4:177–82

Mitchell, R. S. 1973b. Metamict minerals: a review. *Mineral. Rec.* 4:214–23

Morgan, P. E. D., Cirlin, E. H. 1982. The magnetoplumbite crystal structure as a radwaste host. *J. Am. Ceram. Soc.* 65:114–15

Morgan, P. E. D., Clarke, D. R., Jantzen, C. M., Harker, A. B. 1981. High-alumina tailored nuclear waste ceramics. *J. Am. Ceram. Soc.* 64:249–57

Morris, J. B., Boult, K. A., Dalton, J. T., Delve, M. H., Gayler, R., Herring, L., Hough, A., Marples, J. A. C. 1978. Durability of vitrified highly active waste from nuclear reprocessing. *Nature* 273:215–16

National Academy of Sciences. 1975. *Interim Storage of Solidified High-Level Radioactive Wastes.* Washington DC: Natl. Acad. Sci. 82 pp.

Newnham, R. E. 1967. Crystal structure and optical properties of pollucite. *Am. Mineral.* 52:1515–18

Nickel, E. H., McAdam, R. C. 1963. Niobian perovskite from Oka, Quebec: A new classification for minerals of the perovskite group. *Can. Mineral.* 7:683–97

Norrish, K. 1951. Priderite, a new mineral from the leucite-lamproites of the west Kimberley area, Western Australia. *Mineral. Mag.* 29:406–501

Pabst, A. 1952. The metamict state. *Am. Mineral.* 37:137–57

Perrault, G. 1967. La composition chimique et la structure cristalline du pyrochlore d'Oka, P. Q. *Can. Mineral.* 9:383–402

Petruck, W., Owens, D. R. 1975. Electron microprobe analyses for pyrochlores from Oka, Quebec. *Can. Mineral.* 13:282–85

Primak, W. 1954. The metamict state. *Phys. Rev.* 95:837

Pyatenko, Y. A. 1970. Behavior of metamict minerals on heating and the general problems of metamictization. *Geochem. Int.* 7:758–63

Raber, E., Haggerty, S. E. 1979. Zircon-oxide reactions in diamond bearing kimberlites. In *Kimberlites, Diatremes and Diamonds: Their Geology, Petrology, and Geochemistry*, ed. F. R. Boyd, H. O. A. Meyer, 1:229–40. Washington DC: Am. Geophys. Union

Ramdohr, P. 1973. *The Opaque Minerals in Stony Meteorites.* Amsterdam/New York: Elsevier. 245 pp.

Ringwood, A. E. 1978. *Safe Disposal of High-Level Nuclear Reactor Wastes: A New Strategy.* Canberra: Aust. Natl. Univ. Press. 64 pp.

Ringwood, A. E., Kesson, S. E., Ware, N. G., Hibberson, W. O., Major, A. 1979a. The SYNROC process: A geochemical approach to nuclear waste immobilization. *Geochem. J.* 13:141–65

Ringwood, A. E., Kesson, S. E., Ware, N. G., Hibberson, W., Major, A. 1979b. Immobilization of high level nuclear reactor wastes in SYNROC. *Nature* 278:219–23

Ringwood, A. E., Oversby, V. M., Kesson, S. E., Sinclair, W., Ware, N., Hibberson, W., Major, A. 1981. Immobilization of high-level nuclear reactor wastes in SYNROC: A current appraisal. In *Nuclear and Chemical Waste Management.* Canberra: Res. Sch. Earth Sci., Aust. Natl. Univ. 1475:52 pp. In press

Rochlin, G. I. 1977. Nuclear waste disposal: Two social criteria. *Science* 195:23–31

Roy, R. 1977. Rational molecular engineering of ceramic materials. *J. Am. Ceram. Soc.* 60:350–63

Roy, R. 1980. Nuclear waste—The Penn State connection. *Earth Mineral. Sci. Bull. Penn. State Univ.* 50(2):14–24

Roy, R., Vance, E. R., McCarthy, G. J., White, W. B. 1981. Matrix-encapsulated waste forms: Application to idealized systems, commercial and SRP/INEL wastes, hydrated radio phases and encapsulant phases. In *Scientific Basis for Nuclear Waste Management*, ed. J. G. Moore. 3:155–63. New York/London: Plenum. 632 pp.

Scientific Basis for Nuclear Waste Management. 1979–1981. Vol. 1, ed. G. J. McCarthy, 563 pp.; Vol. 2, ed. C. Northrup, 432 pp.; Vol. 3, ed. J. G. Moore, 632 pp. New York/London: Plenum

Shafigullah, M., Tupper, W. M., Cole, T. J. S. 1969. K-Ar age of the carbonatite complex, Oka, Quebec. *Can. Mineral.* 10:541–52

Smellie, J. A. T., Cogger, N., Herrington, J. 1978. Standards for quantitative microprobe determination of uranium and thorium with additional information on the chemical formulae of davidite and euxenite-polycrase. *Chem. Geol.* 22:1–10

Smyth, J. R., Erlank, A. J., Rickard, R. S. 1978. A new Ba-Sr-Cr-Fe titanate mineral from a kimberlite nodule. *EOS, Trans. Am. Geophys. Union* 59:394 (Abstr.)

Van Der Veen, A. H. 1963. A study of pyrochlore. In *Verhandelingen van het Koninklijk Nederlands Geologisch Mijnbouwkundig Genootschap. Geol. Ser. 22.* Leiden: J. J. Groen & Zn. N.V. 188 pp.

Van Wambeke, L. 1971. Pandaite, baddeleyite and associated minerals from the Bingo niobium deposit, Kivu, Democratic Republic of Congo. *Miner. Deposita* 6:153–55

Van Wambeke, L. 1978. Kalipyrochlore, a new mineral of the pyrochlore group. *Am. Mineral.* 63:528–30

Verhoogen, J. 1962. Distribution of titanium between silicates and oxides in igneous rocks. *Am. J. Sci.* 260:211–20

Vilasov, K. A. 1966. *Geochemistry and Mineralogy of Rare Elements and Genetic Types of Deposits.* Vols. 1, 2, 3. Jerusalem: Israel Prog. Sci. Transl. 688 pp., 945 pp., 916 pp.

Wark, D. A., Reid, A. F., Lovering, J. F., El Gorsey, A. 1973. Zirconolite (*versus* Zirkelite) in lunar rocks. *Proc. Lunar Sci. Conf., 4th*, pp. 764–66 (Abstr.)

Westerman, R. E. 1981. Development of structural engineered barriers for the long-term containment of nuclear waste. In *Scientific Basis for Nuclear Waste Management*, ed. J. G. Moore, 3:515–22. London/New York: Plenum. 632 pp.

Williams, C. T. 1978. Uranium-enriched minerals in mesostasis areas of the Rhum layered pluton. *Contrib. Mineral. Petrol.* 66:29–39

Ann. Rev. Earth Planet. Sci. 1983. 11: 165–93

HOTSPOT SWELLS

S. Thomas Crough†
Department of Geosciences, Purdue University, West Lafayette, Indiana 47907

INTRODUCTION

Our understanding of large-scale vertical motions, particularly those occurring beneath the oceans, has increased considerably in the aftermath of the plate-tectonics revolution. The breakthrough came with the recognition that ocean depth is largely dependent on crustal age and explainable by boundary-layer cooling of the upper mantle (Menard 1969, Sclater et al 1971). Density is apparently so sensitive to temperature that an average change of 600°C in a 100-km-thick lithosphere can change the surface elevation by 3 km, thus causing the observed seafloor subsidence from ridgecrest to old ocean basin. Although some still maintain that ridges are dynamically supported by mantle convection currents, this view has lost favor as the boundary-layer models have explained the spatial patterns of elevation, free-air gravity (Cochran & Talwani 1977), heat flow (Lister 1977), and deep seismic structure (Leeds 1975). Thermal models of oceanic lithosphere have been so successful as to spawn a number of attempts to apply the same physical principles to similar problems. Thermal perturbations are now seen as possible explanations for continental margins, continental basins, and the broad volcano-capped uplifts that are the subject of this review—hotspot swells.

Most of the volcanic regions known as hotspots occur in the oceans and most of what has been deduced about swells has come from oceanographic work, so this paper has a distinct marine emphasis. No slight is intended to those who have worked so long on continental swells; it is only that seafloor processes seem simpler to analyze and understand than similar processes on the continents. Part of this simplicity may be only apparent and due to 5 km of seawater that prevents detailed observation, but most is real and reflects the youthfulness of the seafloor. Less has happened to complicate the structure because there has been less time for processes to work.

† Professor S. Thomas Crough died on December 3, 1982, shortly after the manuscript for this review went to press.

0084–6597/83/0515–0165$02.00

DEPTH ANOMALIES AND HOTSPOTS

Although cooling of the oceanic upper mantle is the first-order control of oceanic bathymetry, many large bathymetric features are not related to standard cooling and subsidence. To highlight these regions, it is useful to subtract the expected depth based on crustal age from the observed depth, thus creating a new variable called the *depth anomaly* (Menard 1973), which is plotted in Figure 1. The results for the North Atlantic and Indian oceans are taken directly from the pioneering global survey of Cochran & Talwani (1977) with some additional data near the Cape Verde Islands (Crough 1982). The central Pacific values are taken from Crough & Jarrard (1981) and Crough (1978). The remainder of the data in the Atlantic and Pacific are as yet unpublished (Crough, in preparation, 1982), but the North Pacific anomalies are quite similar to those derived previously by Menard (1973) and Watts (1976) for more restricted areas. Depth anomalies are calculated relative to the age-depth curve estimated by Parsons & Sclater (1977), which includes a departure from square root of age subsidence on old seafloor. Individual depth-anomaly values are accurate to roughly ± 200 m, so the map is contoured at the $+250$-m and -250-m values and not at 0 m in order to suppress small variations due to errors in depth, age, or sediment thickness.

The map's main characteristic is a prevalence of high, wide, anomalously shallow regions that are often elongate in form. A few places are markedly deep, notably the seafloor between Australia and Antarctica and the Argentine Basin of the South Atlantic, but most of the negative areas are less than 400 m below expected depth and they comprise a relatively small fraction of the seafloor area. The shallow areas, by contrast, often exceed 1200 m in height and occupy large areas—almost the entire North Atlantic and most of the western Pacific that has been mapped.

The most striking feature of the shallow areas is their spatial coincidence with the major volcanic edifices as indicated in Figure 1. Almost every volcanic island, seamount, or seamount chain surmounts a topographic swell of broad extent, and vice versa. To evaluate whether this good visual correlation is statistically significant, consider whether it might arise by chance if swells and volcanoes were not geometrically related. Because depth anomalies higher than 250 m occupy 34% of the mapped Pacific area and 55% of the mapped Atlantic area, by random chance alone these areas would be expected to encompass these same percentages of the total seamount population in each ocean. Instead, by rough estimate the Pacific swells include 77% by area of that region's seamounts and the Atlantic swells incorporate 94%. The binomial distribution gives the probability of

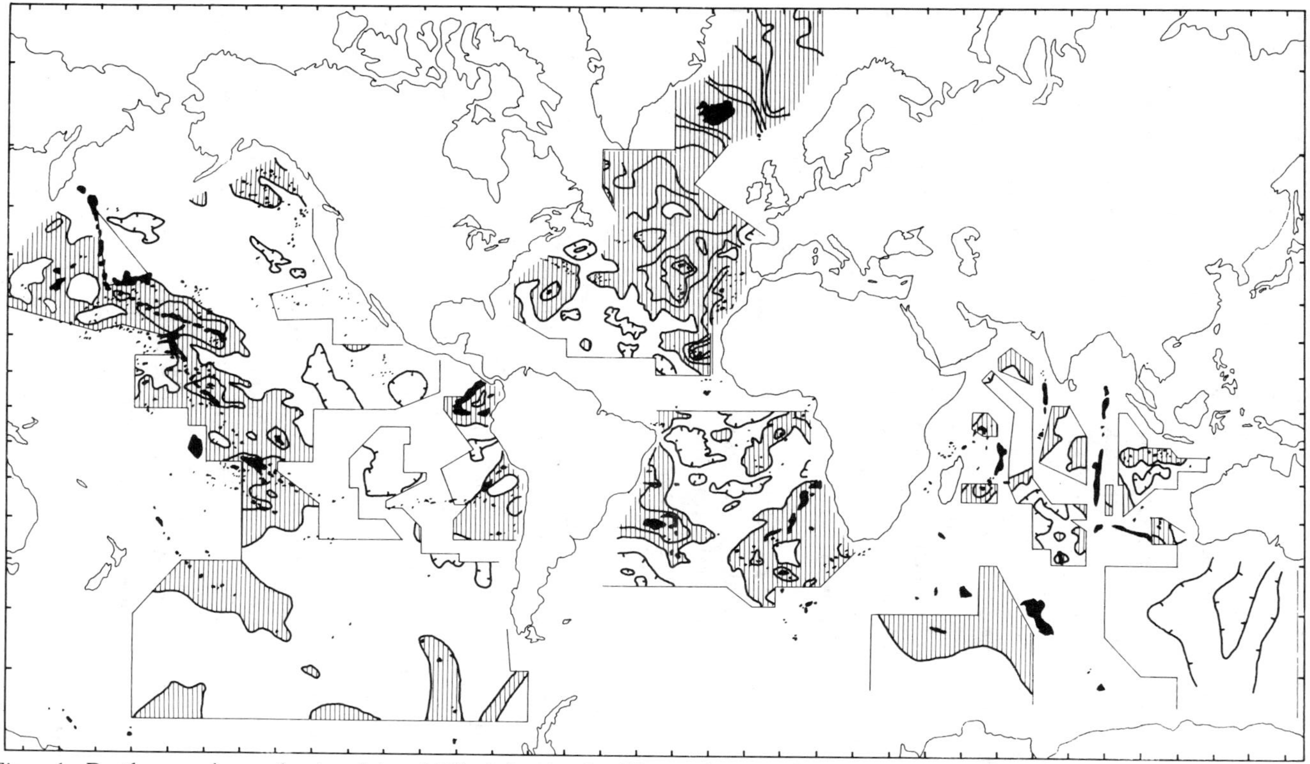

Figure 1 Depth-anomaly map (contour interval 500 m) showing the difference between observed depth and that predicted by standard subsidence. Shaded areas are anomalously shallow by 250 m or more; hachured contours enclose areas anomalously deep by 250 m or greater. Light outlines show regions where anomalies have been calculated and approximately delineate the areas of reliable crustal ages. Volcanic edifices, taken from Chase (1975), are shown in black.

observing this high degree of correlation if it is only a random occurrence. Given the large total number of seamounts observed, there is essentially no chance that this concentration could be accidental—the probability is 10^{-37} for the Pacific region and 10^{-13} for the Atlantic.

Further support for a genetic relation comes from the spatial relation of volcanoes and depth anomalies on individual swells. The volcanic islands, lineaments, and platforms are labeled and interpreted as hotspot traces (Wilson 1963, Morgan 1971) in Figure 2. Comparing this figure with Figure 1 shows that each elongate swell has the same orientation as its overlying volcanic trace, which is the local azimuth of plate-hotspot motion. On profiles taken perpendicular to this azimuth, the volcanic trace occupies the maximum elevation and the center of the bathymetric high. On profiles taken parallel to this azimuth, the swell extends only slightly in front of the most recent volcano. For example, there are no prominent bathymetric highs on the seafloor that in the next ten million years will pass over the Hawaii, Cape Verde, or Tristan da Cunha hotspots. Mammerickx (1981), however, has noted the presence of a small rise, about 300 m high and 300 km wide, to the southeast of the Hawaiian Swell and has suggested that it is a precursor to the main swell. This rise is apparent in Figure 1 as a prong of shaded area extending in front of Hawaii, but the other hotspots do not show this feature so its significance is not clear. The large swells occur directly beneath the volcanic centers and extend away from them in the downstream direction of plate-hotspot motion. The simplest explanation of this pattern is that swells and volcanoes form simultaneously when the plate moves over the hotspot location in the underlying mantle.

Geologic evidence for swell age, although limited to just three swells, also suggests that swells and their associated volcanics form contemporaneously. Cretaceous seamount summits on the Hawaiian Swell are anomalously shallow for their age and expected subsidence, but are at the proper depth if the seafloor on the swell was elevated within the past 25 Ma (Crough 1983). Eocene fossils recovered at sea level on Makatea Island in the Pacific are consistent with island motion over the Society hotspot causing 1100 m of uplift in the past 5 m.y. (Crough 1983). The change from turbidite to pelagic sedimentation in the past 10 m.y. on the Cape Verde Rise is interpreted as a response to regional uplift coeval with island volcanism (Lancelot et al 1977).

Although the correlation between swells and volcanism is too significant to be fortuitous, a few volcanic features exist without associated swells and some positive depth anomalies have no seamounts. Anomalously shallow regions about 400 km wide border the Kuril Trench in the western Pacific, the Aleutian Trench in the North Pacific, and the Chile Trench in the

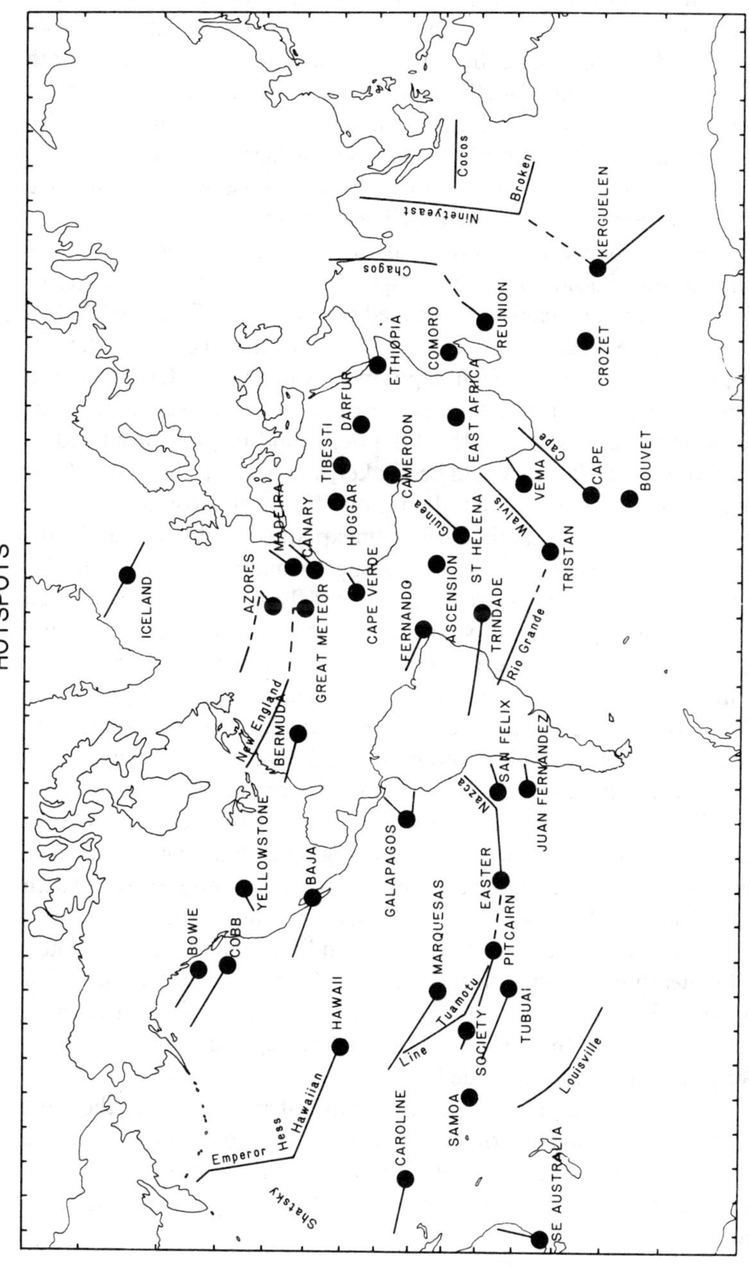

Figure 2 Global distribution of hotspots (*dots*) and hotspot traces (*lines*). Sources include Morgan (1982), Burke & Wilson (1976), and Jarrard & Clague (1977). Some traces are well documented and are labeled with the name of the volcanic feature; others, such as Bermuda's, are hypothetical, showing the predicted path if the hotspot had remained active for a long period. Dashed lines connect trace segments in gaps where no trace should be observed, usually because of ridge movement over hotspots.

eastern Pacific, and are best explained as flexural uplifts of the oceanic lithosphere prior to subduction (Walcott 1970a, Watts & Talwani 1974). Small positive depth anomalies lie in the northwestern Indian Ocean south of Pakistan, in the western North Atlantic near the Caribbean, and in the southernmost South Pacific, and their origin is presently unknown. Major volcano lineaments without swells include the northern end of the Emperor Seamounts, the Cobb Seamount group off the west coast of North America, and the Easter Island trace on the East Pacific Rise.

The correlation between hotspots and swells depends somewhat on the chosen definition of a hotspot. If a hotspot is any amount of midplate or anomalous ridgecrest volcanism, as defined by Burke & Wilson (1976), then every seamount in Figure 1 is a hotspot trace and the correlation is excellent, as has been shown. If a hotspot must also be persistent in time so as to generate a linear trace, then some of the isolated seamounts, for instance Bermuda and Vema in the Atlantic, would not be considered as hotspots and the correlation would be weaker, although still too good to be fortuitous. This more restricted definition, however, would leave the Bermuda and Vema rises as the largest unexplained highs on the depth-anomaly map. Because these rises have approximately the same height, width, and shape as the swells beneath more widely accepted hotspots such as Hawaii, it is likely that they have the same origin. Therefore, my personal definition is that a hotspot is a region of midplate or anomalous ridgecrest volcanism that is either persistent or accompanied by a broad topographic swell. Either one of these attributes suggests that the region is something more than a random lithospheric crack through which magma rises.

It should be noted that this relationship between volcanism and swells is not a new discovery. In an examination of an early bathymetric map of the Pacific, Betz & Hess (1942) noticed the regional shoaling around the Hawaiian Islands and concluded that the swell formed simultaneously with the islands. Dietz & Menard (1953) later observed similar swells beneath the Gulf of Alaska and Mid-Pacific seamounts, and inferred that swells would be found beneath most seamount groups. Morgan (1971) noted that most hotspots are characterized by regionally shallow seafloor suggestive of an active mantle process. In one sense, a decade of depth-anomaly mapping has merely confirmed and extended these earlier observations.

What is new, however, is the revelation that hotspot swells are the most significant and widespread form of uplift on the seafloor. Aside from the volcanic swells, there is very little relief to the map in Figure 1. The obvious inference is that any other uplift process must have only a minor effect on topography.

SUPPORT OF SWELLS

One of the fundamental questions about swells is how they are supported at depth. Vertical crustal movements, in general, have long been a puzzle to earth scientists, and swells, in particular, have stimulated much speculation as to their structure and probable origin. Different proposals place the isostatic support at depths anywhere from the sea surface to the asthenosphere and can be conveniently grouped into the five categories depicted in Figure 3.

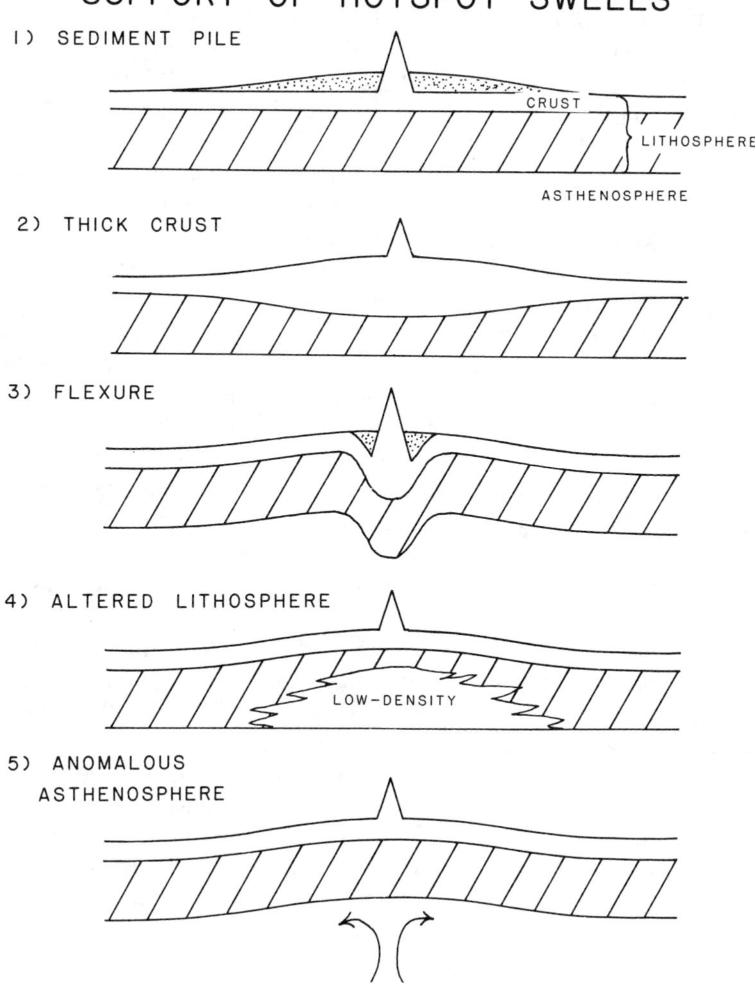

Figure 3 Cross sections of models proposed to explain the isostatic support of hotspot swells.

The simplest origin is the one sometimes proposed for the Cape Verde Rise—a thick accumulation of sediments (Lancelot et al 1977). Sedimentary deposits thicken toward many volcanoes forming archipelagic aprons (Menard 1964), but this effect is insufficient to explain the swells. Sediment thickness can be measured accurately by seismic-reflection profiling, so depth-anomaly maps such as Figure 1 are corrected to give the seafloor depth if there were no isostatic loading by sediments. If the swells were sediment piles and the sediment correction were done correctly, then the swells would not appear in Figure 1. There is a chance that errors in the sediment correction could obscure basement topography or create false depth anomalies, but this should only be a significant problem where sediment thicknesses exceed 1 km and that is primarily at continental margins (Ewing et al 1973). Drilling and multichannel seismic-reflection work show that the Cape Verde Rise is a true basement uplift and that its thick sediment cover is typical of its position near the African coast (Lancelot et al 1977). If any feature in Figure 1 is probably a sedimentary artifact, it is the previously mentioned high in the northwestern Indian Ocean, coincident with the thick Indus Fan.

Another shallow support mechanism was originally suggested by Betz & Hess (1942) to explain the Hawaiian Swell—a broad thickening of the oceanic crust by lava flows emanating from the islands. There is no doubt that this mechanism can create topographic highs, but the best available evidence indicates that it does not cause the swells in Figure 1. According to seismic-refraction surveys, the crust on the Hawaiian Swell has uniform thickness, except in the immediate presence of the islands (Shor & Pollard 1964). In fact, the peak of the Hawaiian Swell near Oahu was selected as the site for Project Mohole partly because the Moho is elevated in parallel with the seafloor and therefore lies at relatively shallow depth. The other swells are not known in as much detail, but it is presumed that the Hawaiian results are typical.

Many bathymetric features are supported by thick crust, but these were omitted from the depth-anomaly map. The islands and seamounts themselves are piles of basalt and could be considered depth anomalies because they are shallower than standard seafloor. However, because seamount support is well understood and the purpose of Figure 1 is to highlight more subtle regional trends, no depth-anomaly values were calculated over seamounts, and contours were interpolated through those regions of no coverage. Data over the high, steep-sided plateaus of the western Pacific such as the Shatsky Rise (Figure 2) were also omitted, as seismic results over such features also indicate anomalously thick crust (Hussong et al 1979). In addition, no depth anomalies were calculated for the Phoenix lineations in the western Pacific because drilling has revealed thick sills intruded into the crust (Larson & Schlanger 1981). Given accurate seismic results and complete coverage, one could correct observed basement depths for the isostatic effect of crustal-thickness variations as was done for gravity measurements by Yoshii (1973). However, in most cases this process adds large errors and it is presently more accurate to avoid regions of known crustal complexity.

The third shallow support mechanism was developed by Walcott (1970b) to explain the Hawaiian Swell—flexural uplift caused by volcanic loading. According to beam theory, the flexural depression beneath and adjacent to the Hawaiian Islands (Figure 4) should be accompanied by upwarping at greater distances from the islands. This hypothesis can explain the shape and width of the swell but not the observed height and free-air gravity anomaly. The most straightforward model, that of an unbroken elastic plate, produces an uplift of only 100 m, and a broken plate, which produces the maximum uplift, still yields only 600 m, whereas the swell is at least 1200 m high. Because a flexural uplift has no low-density isostatic root beneath it, it generates a free-air gravity high with an amplitude of approximately $2\pi G\rho h$, where G is the gravitational constant, ρ is the density contrast between the uplift and its surroundings, and h is the height of the uplift. For flexure beneath sea level, the predicted anomaly is about 90 mGal per km of

HAWAIIAN SWELL

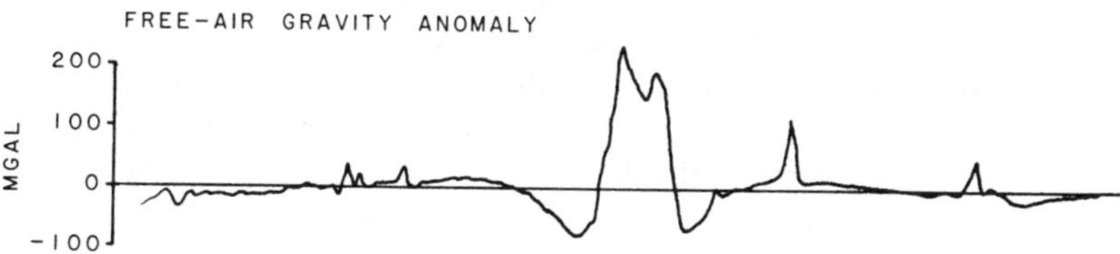

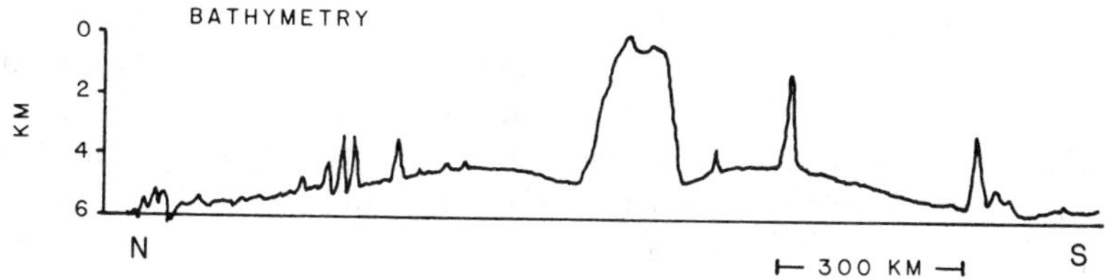

Figure 4 Profiles of free-air gravity anomaly and bathymetry across the Hawaiian Swell (from Watts 1976), which is the regional rise in seafloor depth from 5.8 km to 4.2 km over a total width of 1200 km. The down-bowing of the plate beneath Hawaii is clearly seen, but the small-amplitude gravity anomaly over most of the swell is inconsistent with flexural uplift.

uplift, which is the ratio observed over the flexural rises seaward of subduction zones (Watts & Talwani 1974). However, the gravity/height ratio over the Hawaiian Swell (Figure 4) is 20 to 25 mGal km^{-1} (Watts 1976, Detrick & Crough 1978), which is much too small for this model. Some flexural uplift must occur, but it is probably less than 200 m.

Because the shallow-compensation models are all unsatisfactory, recent work has focused on deeper support mechanisms, which for convenience can be grouped into two categories—support within the lithosphere and support beneath the lithosphere. Models of internal lithospheric support include underplating basalt to the base of the lithosphere (Burke & Whiteman 1973), underplating depleted peridotite (Jordan 1979), altering the lithosphere's composition (Haxby & Turcotte 1978), and reheating the lithosphere (Detrick & Crough 1978). Models of deeper support usually involve dynamic support by upwelling mantle convection (Morgan 1971, Menard 1973, Anderson et al 1973), but anomalous composition or temperature in the asthenosphere is also possible (Cochran & Talwani 1978). Because there are few measurements capable of directly yielding the temperature, structure, or composition of the upper mantle, testing these proposed models requires a wide variety of observations and a great deal of inference.

Observational Constraints

MORPHOLOGY Swell planform provides the simplest discriminant between the lithospheric and sublithospheric support hypotheses. As Menard (1973) noted, if the support is fixed in the asthenosphere then lateral movement will take the lithosphere off the support and into deeper water. If, on the other hand, the support is within the plate then no amount of lateral motion can displace a seamount from a swell. Because the lithosphere moves relative to hotspots, the test is to see whether swells persist along hotspot traces or are confined to the loci of active volcanism.

As mentioned earlier and as is evident in Figures 1 and 2, the swells have the same spatial arrangement as the volcanic chains, suggesting that they are fixed relative to the lithosphere once they are formed. As examples, the Hawaiian Swell extends as far as the Hawaiian-Emperor bend, the Gulf of Alaska Swells parallel the local seamount chains, and, as first remarked by Cochran & Talwani (1977), swells beneath the Walvis Ridge and Rio Grande Rise follow these two traces of the Tristan da Cunha hotspot. In addition to the other correlations which can be observed in the figures, Hyndman (1973) has noted that the Labrador Sea, which is not mapped in Figure 1, is anomalously shallow along the probable path of the Iceland hotspot.

Although it has been suggested that this pattern is consistent with dynamic uplift by elongate convective roll-cells (Marsh & Marsh 1976, Watts 1976), this hypothesis raises at least three major difficulties. Firstly, if upwelling convection occurs beneath the entire Hawaiian island chain it is problematical why volcanism is restricted to the southeast extremity of the swell. Secondly, there is no explanation for the termination of swells. At the Hawaiian-Emperor bend, depth anomalies decrease by 750 m from the island chain westward over a distance of several hundred kilometers before beginning to increase again toward the Shatsky Rise. The westward decrease has the same magnitude as the decrease from the bend southward, perpendicular to the island trend, and most probably indicates the western edge of the swell. If swells are caused by the contemporary pattern of mantle convection, it must be a coincidence that the Hawaiian Swell ends exactly at the western end of the island chain rather than anywhere else—at Midway Island or Japan, for example. Thirdly, there is no explanation for swells that do not parallel the recent plate-hotspot motions. Laboratory experiments show that convection cells in the presence of a superposed shear flow may line up in the direction of shearing (Richter & Parsons 1975). Swells beneath the Nazca Ridge and the New England Seamounts (Figures 1 and 2), however, display the azimuth of plate-hotspot motion existing at the time the seamounts were formed, indicating again that swells are probably locked into the lithosphere.

The heights of swells also provide evidence for their origin. If swells are formed by reheating, their crests cannot be shallower than midocean ridges, which are at the depth expected if the lithosphere were totally converted to asthenosphere. All midplate swells are smaller than the maximum allowed and thus consistent with the thinning mechanism. However, the reheating model also predicts a uniform depth at ridgecrests where there is little or no lithosphere to reheat. Therefore, swells such as those surrounding Iceland and the Azores are impossible to explain by this model.

The height of midplate swells varies according to the age of the seafloor they occur on (Figure 5). In general, the smaller swells are located on younger seafloor and the larger ones are on older crust, but with recent depth-anomaly mapping of additional swells there is now greater scatter in this relationship than was seen earlier. Formerly, it appeared that all midplate swells reach a uniform depth below sea level, approximately 4250 m (Crough 1978), but the data are as well fitted by Menard & McNutt's (1982) proposal that swell height is proportional to the square root of crustal age (Figure 5). Either relation suggests some control of swell height by the lithosphere and favors a support mechanism based on lithosphere-hotspot interaction. The simplest explanation may be some type of reheating mechanism; the older the crust, the thicker the lithosphere susceptible to reheating, and the higher the possible swell.

Swells on ridgecrests have heights that are inversely proportional to the local spreading rate, with Iceland, the largest swell, on the slowest opening ridge and Easter, the smallest, on the fastest (Vogt 1976). As Vogt (1976) has shown, this is consistent with a special type of dynamic support wherein relatively shallow flow along ridge axes away from hotspots controls elevation. The width of the subridge asthenospheric channel available for this flow is proportional to the local spreading rate, so flow along a fast-spreading ridge encounters less viscous resistance than flow along a slower ridge. The pressure gradient necessary to drive the lateral flow manifests itself as a topographic gradient, and because slow-spreading ridges require the largest pressure gradients they support the highest swells. The major problem with this support mechanism is, of course, that it only functions near a ridgecrest and cannot explain why some swells that originally formed at ridgecrests still persist after tens of millions of years.

An alternative explanation for the relation between swell height and spreading rate is a constant flux of anomalously low-density material from

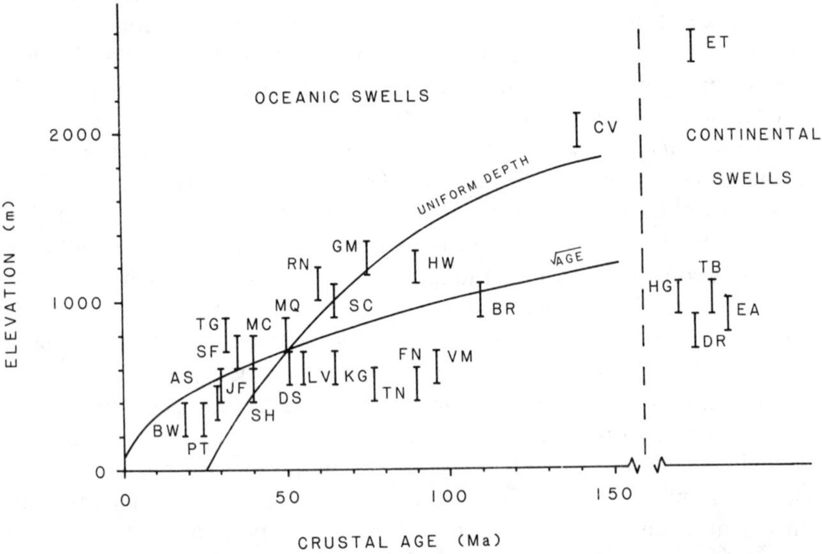

Figure 5 Heights of midplate hotspot swells, from Figure 1, versus local lithospheric age. Ascencion (AS), Bermuda (BR), Bowie (BW), Cape Verde (CV), Darfur (DR), Discovery (DS), East Africa (EA), Ethiopia (ET), Fernando de Noranha (FN), Great Meteor (GM), Hawaii (HW), Hoggar (HG), Juan Fernandez (JF), Kerguelen (KG), Louisville (LV), MacDonald (MC), Marquesas (MQ), Pitcairn (PT), Reunion (RN), St. Helena (SH), San Felix (SF), Society (SC), Tibesti (TB), Trindade (TN), Tristan da Cunha-Gough (TG), Vema (VM). Curves give expected relations if all swells reach a uniform depth (4250 m) or have heights proportional to the square root of age. Continental lithosphere is not well dated, but is generally older than oceanic lithosphere.

hotspots. Spreading ridges are generally viewed as passive phenomena, drawing up asthenosphere as needed to replace the rock removed by newly created lithosphere. Mantle hotspots may be active sources, bringing rock of a slightly different temperature and/or composition upward into the ridge system, and each hotspot may have roughly the same strength or supply rate. On slow-spreading ridges the hotspot flux may supply new material as fast or faster than the laterally moving plates can accrete it, whereas on fast-spreading ridges the same hotspot flux will obviously supply a smaller percentage of plate need. Material must replace the moving lithosphere, so on faster-spreading ridges the anomalous hotspot component will be diluted with passively upwelling normal asthenosphere. The faster the plate movement, the greater the dilution and the smaller the swell. Because this anomalous asthenosphere becomes part of the lithosphere as the plate cools and thickens, a swell formed in this manner will remain fixed beneath its associated hotspot trace.

The cross-sectional shape of swells may also furnish useful information. Vogt (1976) is perhaps the only worker who has addressed this problem, showing that flow along ridges can explain the profiles of several ridgecrest swells. Most midplate swells look like normal distributions with broad, relatively flat summits and gently sloping sides. Any successful model should be able to explain this simple appearance.

GRAVITY AND GEOID HEIGHT Because many of the proposed compensation models predict different gravity and geoid height anomalies over swells, these measurements furnish another useful test. Although gravity measurements cannot be inverted to yield a unique density structure and thus prove that any particular proposal is correct, they can show that some models are inconsistent and, therefore, incorrect.

Lithospheric support models are the easiest to evaluate for they predict local isostatic balance with a compensating mass deficiency or root at standard lithospheric depths. At sea level the elevated swell and its root produce a small positive free-air gravity anomaly and geoid height anomaly resulting from the root's greater depth. Geoid height anomalies are particularly simple to interpret, for their amplitude over long-wavelength, isostatically balanced uplifts should be $2\pi G\rho hD/g$ (Haxby & Turcotte 1978), where D is the compensating root's mean depth below seafloor, g is the acceleration of gravity at the Earth's surface, and the other symbols are as used earlier.

Measured geoid height anomalies over midplate hotspot swells are consistent with any mechanism that creates a root in the middle of the lithosphere. A profile over the Cape Verde Rise (Figure 6) is quite typical. A local geoid height anomaly that is spatially coincident with the swell can be isolated from the regional field. This geoid height anomaly has the same shape as the swell and a maximum amplitude of 8 m. For lithospheric uplifts underwater, the isostatic relation predicts 0.1 m of geoid height anomaly per km of depth anomaly per km of mean root depth, so the Cape Verde measurements are explainable by a root only 40 km below the seafloor. The geoid height anomalies over the Bermuda Rise (Haxby & Turcotte 1978), Marquesas-Line Swell (Crough & Jarrard 1981), and Hawaiian Swell (Crough 1978, Sandwell & Poehls 1980) are consistent with roots 40–60 km deep. For each of these regions, the lithosphere should be at least 100 km thick as calculated from standard thermal models and as inferred from geoid height anomalies over ridgecrests (Haxby & Turcotte 1978). Therefore the observed anomalies are inconsistent with the underplating models.

Gravity anomalies should and do give the same results, although they are more difficult to analyze because their amplitude depends on wavelength as well as root depth. Profiles over the Hawaiian, Bermuda, Cape Verde, and Cook-Austral rises are consistent with root depths of 40–70 km (Crough 1978). One study concluded that the compensation beneath the Hawaiian Swell was 120 km deep on average (Watts 1976), but this finding is inconsistent with the geoid measurements and apparently was caused by overestimating the swell's wavelength (Detrick & Crough 1978).

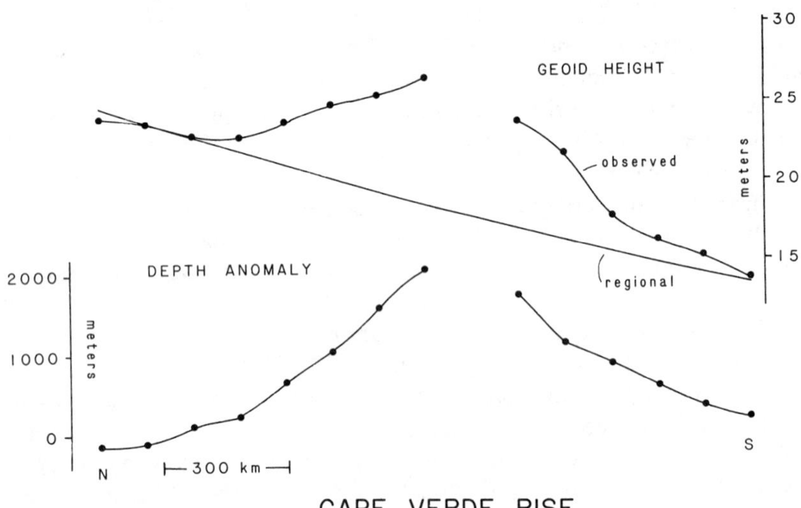

Figure 6 Profiles of depth anomaly and geoid height anomaly from north to south across the Cape Verde Rise at longitude 25°W (Crough 1982b). Curves are interpolations between calculated 1° × 1° averages of both variables; data are omitted over the seamounts at the swell's crest. Removing a regional field leaves a geoid height anomaly with the same shape and width as the topographic swell and whose amplitude is 4 m per km of swell height.

It is presently uncertain whether these gravity and geoid height observations might also be consistent with dynamic support. Some simple numerical simulations (McKenzie et al 1974) predicted that upwelling convection currents would create positive gravity anomalies with an amplitude of about 30 mGal per km of uplift. The relation between depth anomaly and gravity along ridgecrests (Anderson et al 1973) and in the North Atlantic (Sclater et al 1975) has approximately this value and was considered to confirm the model. However, more recent simulations have shown that the amplitude and even the sign of the predicted gravity anomaly depend critically on mantle properties such as the temperature dependence of viscosity (McKenzie 1977), making it necessary to recalculate gravity anomalies using present best-estimates of mantle rheology and boundary conditions. As a further complication, Cochran & Talwani (1978) noted that some variation of gravity along ridgecrests is due to standard plate cooling combined with variable spreading rate; when the plate effect is removed, the 90-mGal high predicted by the numerical flow experiments is not seen over the Iceland Swell, which is the largest depth anomaly in the oceans. Marsh & Marsh (1976) noted the elongate shape of many free-air gravity anomalies, especially when filtered to remove the longest wavelengths, and concluded that this was caused by elongate upwellings in the mantle. However, there is a global correlation between depth anomaly and gravity (Cochran & Talwani 1977), so the planform of the gravity anomalies primarily reflects the elongate shape of the depth anomalies (Figure 1) and is not a significant additional constraint on their origin.

SUBSIDENCE AND HEAT FLOW A permanent compositional change and a temporary reheating within the lithosphere are distinguishable by observing the subsidence and surface heat flow of swells. If swells are permanently supported, then they and the seamounts atop them will subside only by continued standard cooling of the lithosphere beneath. However, if they are supported by a temperature increase, then they will subside anomalously fast as this extra heat diffuses through the seafloor.

On the Hawaiian Swell, the only rise where heat-flow observations have been made in detail, the results are consistent with the reheating hypothesis (Figure 7). Heat flow is approximately normal for lithospheric age on the swell crest near Hawaii, but increases slightly toward Midway rather than decreasing as the seafloor becomes older (Detrick et al 1981). The swell's depth below sea level reaches a minimum of about 4.2 km near Oahu and then increases steadily along the seamount chain, becoming over a kilometer greater at the Hawaiian-Emperor bend. Seamount summits near the bend are approximately 1100 m below sea level, suggesting that the westward decrease in swell height is caused by subsidence (Detrick & Crough 1978, Crough 1978). The subsidence along the chain is larger by a factor of three than predicted by standard cooling and the heat flux on the older part of the swell is anomalously high by 25%. Both observations are consistent with a sudden temperature increase in the lower two thirds of the lithosphere immediately over the mantle hotspot, followed by cooling. The normal surface heat flux near Hawaii may indicate that the reheating process begins at the base of the lithosphere and moves upward, leaving the near-surface temperatures unaffected immediately. Then as the heat input diffuses away, it gradually raises shallow temperatures, creating an increased surface heat flux as the swell cools and subsides.

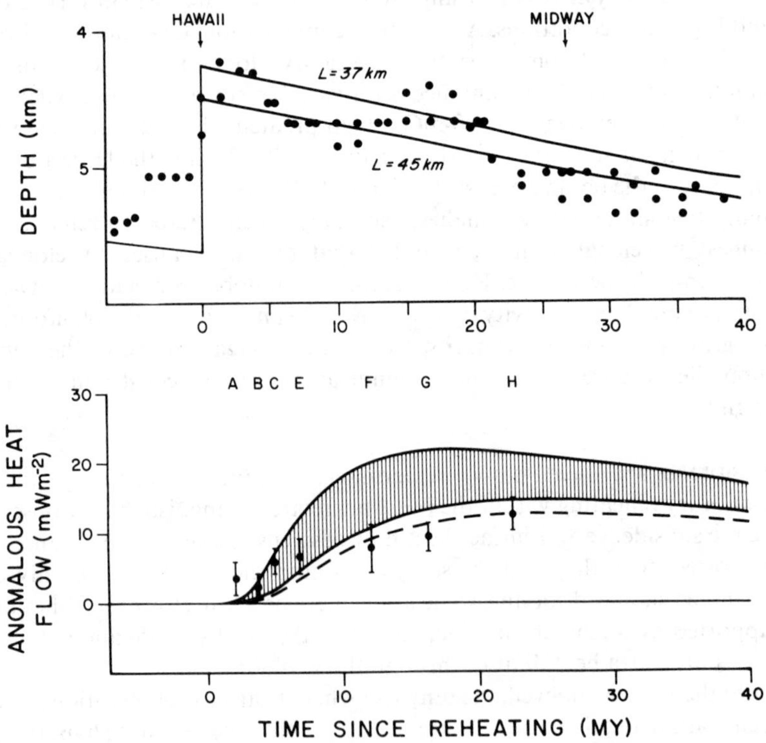

Figure 7 Minimum depth, 1° × 1° averages, and anomalous surface heat flow along the Hawaiian Swell from southeast of Hawaii northwestward to the Hawaiian-Emperor bend (Detrick et al 1981). Ages are times since the local seafloor passed over the hotspot. Solid lines are the predictions for instantaneous reheating of the lower part of 90 m.y. old lithosphere at Hawaii, followed by standard cooling. Upper curve in both plots is for a new lithospheric thickness (*L*) of 37 km after reheating; lower is for 45 km. The lower curve is a better fit to both data sets. The dashed curve considers the possible effect of an age offset across the Molokai fracture zone near Hawaii and is for reheated 80 m.y. old lithosphere.

These heat-flow results are important, for they are inconsistent with every proposed mechanism except reheating. Material underplated to the lithosphere would be at approximately the same temperature as the asthenosphere and would not create a heat-flow anomaly. Even if the added material were hotter, it would take ten times longer to alter the surface gradient than the thinning model in Figure 7 predicts. Similarly, dynamic uplift should not affect the surface heat in the short time scale observed. A compositional change within the lithosphere could cause a heat flow anomaly, but only if the change is associated with temperature increases and, in that case, the change is also a reheating event. Given that heat flow is such a useful constraint, more measurements on swells are a high-priority need. For example, the anomalies in Figure 7 are calculated relative to the expected heat flow for crustal age (Lister 1977), so a profile of heat-flow measurements across the Hawaiian Swell perpendicular to the island chain near Midway would better establish the size of the anomalous flux and test whether it correlates with swell height.

The subsidence of the Hawaiian Swell is somewhat less of a constraint because it can be explained—although not easily—by other processes. The underplated root beneath the lithosphere might flow laterally causing the swell to subside. The strength of the dynamic uplift might be greatest at Hawaii and gradually decrease along the chain. The amount of low-density material added to the lithosphere might have increased in time, so that the older parts of the Hawaiian Swell were never as high as the younger parts. In that case, the swell does not subside anomalously fast and the seamount depths, rather than indicating seafloor subsidence, would have to be reinterpreted as signifying sinking of the volcanoes relative to the seafloor, perhaps by viscous creep.

The subsidence becomes more of a constraint when it is seen that other midplate swells and seamounts behave similarly to Hawaii. Plots of crustal swell depth for the Cook-Austral (Crough 1978) and Marquesas-Line (Crough & Jarrard 1981) rises show the same pattern of an abrupt swell uplift at the active hotspot and then a gradual increase of depth along the presumed hotspot trace, with depth versus age mostly following the same curve that can be drawn through the Hawaiian Swell depths. Complications, however, exist on both swells, with the depths near the Line Islands being somewhat shallower and those westward of the Cook Islands being somewhat deeper than expected. Volcanic basement in the Eniwetok and Bikini atolls is too deep to be explained by normal subsidence, but is at the predicted depth if the seamounts formed on a reheated swell. The subsidence history of the atolls, which is known from their drilled carbonate caps, matches that of the Hawaiian Swell almost exactly (Crough 1978). Guyots near Wake Island are at depths explained by subsidence of reheated lithosphere (Crough 1978) and the regional shoaling of much of the western Pacific also is consistent with this concept (Larson & Schlanger 1981). In fact, most of the available seamount-subsidence data support the reheating hypothesis (Menard & McNutt 1982). The only two known exceptions are Midway Island, which has moved downward less than expected by reheating but consistently with standard cooling, and Bermuda, which has not even subsided as much as standard cooling predicts (Crough 1978).

Swells that originally formed at spreading centers show no evidence of subsidence in excess of that predicted by standard cooling and, therefore, are consistent with permanent support. No one has yet examined the depths of these swells, but Detrick et al (1977) compiled the summit depths of the volcanic edifices occurring on their crests. To a good first approximation, the depths follow the standard subsidence curve as a function of age, implying that the swells presently beneath these volcanic features must have been there from the time of eruption and must have maintained their present height above the seafloor. For if the swells appeared later, the seamounts should presently be shallower than otherwise predicted and if the swells changed their height, this would also cause a misfit. On average, the seamounts are slightly shallower than predicted, but it is a small difference compared to the elevation of the swells and can be attributed to the time necessary to erode a volcano down to sea level (Detrick et al 1977).

CONTINENTAL SWELLS Although only oceanic data have been discussed to this point, there are several volcanic swells on the continents that also have been studied and have contributed toward an understanding of deep structure and support. The most familiar examples are the great highlands of Africa, including the Ethiopian and East African plateaus and the Hoggar and Tibesti massifs, but other likely hotspot swells include the high plains and northern Rocky Mountains surrounding Yellowstone in the western United States and the Brazilian highlands with their associated Cretaceous intrusions. Many of the theories advanced to explain the oceanic swells were proposed at even earlier times by those workers concerned with continental uplifts. Faure (1971) suggested that the African domes were dynamically maintained by bumps on the top of the asthenosphere, while Sowerbutts (1969) suggested reheating and thinning of the lithosphere.

In shape and size, these continental swells are quite similar to their oceanic counterparts. The Hoggar Massif, for example, is a gentle volcano-capped rise of Precambrian basement over 1000 km across and standing 1 km higher than the surrounding lowlands. The Ethiopian and East African plateaus have similar widths, but are transected by rifts that obscure their domal shape. However, with the exception of the Yellowstone uplift, none of these continental domes are noticeably elongate in form—perhaps they are one-pulse uplifts like the Bermuda Rise or perhaps erosion rapidly removes their traces. It is, of course, the absence of erosion below sea level that makes bathymetric maps so useful in delineating oceanic hotspot traces and swells.

The gravity signature of these swells is similar to that observed in the oceans. Large-amplitude Bouguer gravity lows over the African uplifts imply isostatic compensation at depth (Sowerbutts 1969, Brown & Girdler 1980) and the small free-air gravity highs over the Kenya Dome (Banks & Swain 1978), Hoggar Massif (Crough 1981a), and Darfur Swell (Crough 1981b) are consistent with mean root depths of 40–100 km.

The subsidence and heat-flow data near Yellowstone are consistent with lithospheric reheating. Both Suppe et al (1975) and Brott et al (1978) have noted that the elevation of the Snake River Plain, which is the trace of the Yellowstone hotspot, (Morgan 1972) decreases westward from Yellowstone and can be explained by cooling. However, unlike Hawaii, the crust subsides even faster than a standard midocean ridge and, therefore, the reheating must occur in the uppermost part of the lithosphere rather than the lower part. As predicted by this reheating model, the heat flux is anomalously high near Yellowstone and decreases westward (Brott et al 1978, 1981). The remaining question is whether this thermal anomaly is confined to the immediate vicinity of the Snake River Plain, which is only 100 km wide, or whether it extends over a distance comparable to the width of oceanic swells. If confined to the plain alone, this reheating might be due to local rifting and volcanic intrusions, and might not be the process creating regional uplift.

Because of their greater accessibility, it has been possible to collect seismic information concerning the mantle structure of continental swells, whereas this information is still lacking for oceanic swells. Fairhead & Reeves (1977) examined teleseismic P-wave residuals as a function of station elevation in Africa. Relative to lower regions, the plateau uplifts have later arrivals which the authors interpreted as indicating thin lithosphere. Banks & Swain (1978) used arrivals at an East African array to infer a broad thinning of the lithosphere beneath the Kenya Dome. More recently, seismic instruments were deployed across the Darfur Swell and the preliminary results also indicate delays associated with the uplift (Bermingham et al 1983). As with gravity, however, these interpretations are not unique and it is worth examining whether seismic delays also might be consistent with some compositional change in the lithosphere or some anomalous property in the asthenosphere.

UPLIFT RATE One of the most remarkable aspects of hotspot swells is the speed at which they form. Although it is maddeningly difficult to measure accurate uplift rates for most areas, on hotspot swells the rates of elevation are given directly by the depth anomalies and the speed of plate movement. For example, from the closest seafloor at standard depth in front of the Hawaiian Swell to the swell crest near Oahu is a distance of 600 km. With a Pacific plate velocity of 9 cm yr^{-1} relative to the Hawaiian hotspot (Jarrard & Clague 1977), this distance is traveled in 6.6 m.y. The seafloor rises 1200 m in this time interval, for a mean uplift rate of 0.2 mm yr^{-1}. The rates are different for different swells and this Hawaiian value is one of the largest, but 1 km in 5 m.y. is a geologically rapid rate of uplift that any support mechanism must be able to explain.

Most hypotheses are consistent with fast uplift, but the lithospheric reheating model has major problems. Any model relying directly on some material attribute of the hotspot, such as dynamic topography or low-density asthenosphere, can explain almost any rate of uplift observed. The swell support either always exists or is continually produced at the hotspot location and the plate is elevated immediately as it passes onto this support. However, any mechanism requiring some interaction between the hotspot and the lithosphere may have the uplift velocity limited by the rate at which this interaction can occur. In particular, rocks are poor thermal conductors, so that reheating can be very slow.

A conductive reheating model that treats the hotspot as a temperature or heat-flow increase at the base of the lithosphere is not satisfactory. Gass et al (1978) and Mareschal (1981) treated the problem with a constant-thickness plate and found that it takes approximately 100 m.y. to create an uplift the size of the Hawaiian Swell if the heat flow or temperature at the base of the lithosphere is increased by 50%. Crough & Thompson (1976) let the base migrate upward as the thinning proceeds, but this increases the reheating rate by only a small factor. Their result for an initial 100-km-thick plate is that uplift rate equals 0.02 $(W-1)$ mm yr^{-1}, where W is the ratio of the hotspot heat flux to the background heat flux. Uplift as rapid as that observed near Hawaii is possible only if the hotspot heat flux is ten times normal. This flux seems excessively high because it would thin the lithosphere to a steady-state thickness of about 10 km and the resultant swell would stand as high as a typical ridgecrest. Recently Spohn & Schubert (1982) have suggested that such high heat fluxes are typical of hotspots, but that the Pacific plate is moving so fast that it never remains over the hotspot long enough to attain steady-state thinning and elevation. A comparison of midplate swell height versus plate velocity, however, does not show the inverse relation predicted by this idea (Figure 8). There are two large swells on the slow-moving African plate, but, aside from these, there is little or no correlation between size and speed. Either the hotspots beneath fast-moving lithosphere happen to be several times as hot as those beneath slower plates or the conductive-thinning model fails.

Conduction is, of course, the slowest possible way to transfer heat in rocks; other, faster processes may do the reheating. Heestand & Morgan (1981) and Emerman & Turcotte (1982) considered whether lateral flow at the top of a plume might entrain the lower lithosphere and sweep it away, but found that the lithosphere is too viscous for this to be effective. Bird (1979) considered the possible detachment of the mantle lithosphere from the base of the crust, and Withjack (1979) considered the diapiric uprise of magma; both concluded that these mechanisms could function in reasonably short times. Volcanoes provide evidence that penetrative magmatism occurs at a swell's center, and magma can transfer heat almost instantaneously. However, if swell-wide intrusion is the mechanism for reheating, one would expect to see surface volcanics over most of the swell, not just the middle. Some must get through the lithosphere to provide evidence of the remainder that does not. The Ethiopian Plateau is covered with basaltic shield volcanoes and the East African Plateau is riddled with Tertiary kimberlites, but the oceanic swells are known in less detail. Miocene plugs and sills have been drilled on the Cape Verde Rise and a Pleistocene volcano was sampled and dated on the flanks of the Hawaiian Swell (Dymond & Windom 1968), but these may not be typical. A dredging program aimed at understanding the many small seamounts on swells might help resolve this matter. The volume of magma necessary to furnish

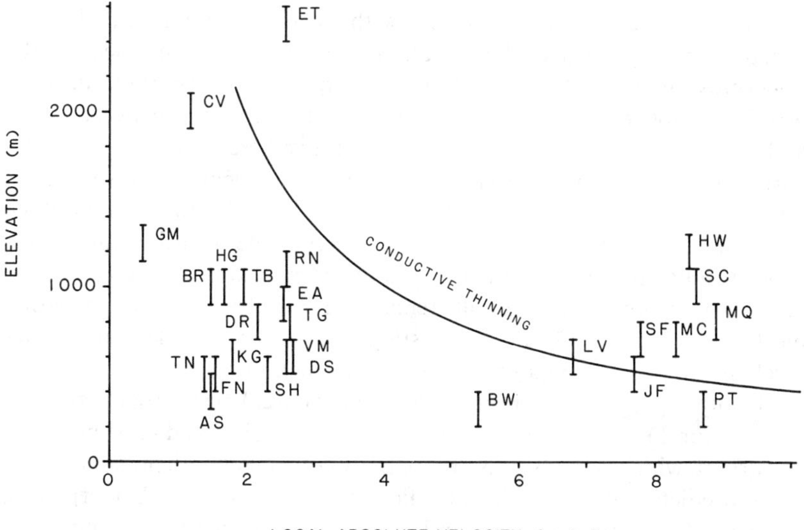

Figure 8 Height of midplate swells versus local plate velocity relative to fixed hotspots (Morgan 1972). Curve gives shape of inverse relation predicted by conductive-reheating model of swell formation.

all the heat required to raise a swell is quite large—about half the volume of the swell itself. However, the magma need only supply enough heat to disaggregate the lithosphere so the asthenosphere can replace it.

Model Appraisal

Because no single hypothesis satisfies all the constraints, my preferred model is a hybrid, with a pillow of anomalously low-density asthenosphere supporting ridgecrest swells and thinned lithosphere supporting midplate swells. At first thought, the idea of hotspots forming one type of swell when beneath plates and a different type when beneath spreading centers seems too complicated to be right. However, the zone of anomalous asthenosphere could exist beneath midplate swells but be dominated by the thermal uplift generated by hotspot-plate interaction. It is well known that plate thickness has an effect on hotspot volcanism, with mantle hotspots at ridgecrests erupting magma at a near-constant rate, thus forming continuous ridges, and with midplate hotspots erupting magma intermittently, resulting in separate volcanoes. This difference in edifice structure does not necessarily imply that ridgecrest hotspots are fundamentally different than midplate ones, but only that magma has difficulty penetrating the lithosphere. It is possible that swells show a similar effect and that the surficial differences simply reflect varying responses to the same underlying process.

This combination of low-density asthenosphere and reheated lithosphere seems capable of explaining all the crucial observations. Because the uppermost asthenosphere at ridgecrests is quickly incorporated into the growing lithosphere, the exact and persistant correlation between swells and volcanoes is guaranteed. Reheated lithosphere produces the same correlation for midplate swells. The gravity and geoid observations in midplate regions are satisfied, but further study is needed to determine if the anomalies over ridgecrest swells are consistent. The heat-flow and subsidence results are also explained, as are the correlations between plate age and swell height. Rapid uplift remains a problem for the reheating mechanism, but the difficulty is in testing the various possible heat-transfer mechanisms, not in imagining geologically quick processes.

The other proposed support models all have weaknesses that I consider to be major. Dynamic uplift fails to satisfy the simplest and most powerful constraint, which is that swells are spatially coincident with hotspot traces, and it predicts neither the high heat flow nor the subsidence. Underplating is inconsistent with the gravity and heat-flow determinations on midplate swells.

IMPLICATIONS

Hotspot swells are clearly a major morphological feature of the Earth's surface. The Atlantic and Pacific swells in Figure 1, as defined by the 250-m contour of depth anomaly, comprise 11% of the planet's surface area. The percentage covered by all swells is unknown, but the swells in Figure 1 constitute 42% of the region mapped for depth anomaly. This number may serve as a rough estimate, even though the true percentage should be somewhat lower, given that the oceans contain a greater proportion of hotspots than the continents. Any process that is this common must have significant effects.

Continental Epeirogeny

Continental movement over mantle hotspots may have produced many of the uplift and subsidence events seen in the geological record. If it is assumed that hotspots are randomly located or that continents move randomly over them, then for any single continental region the probable duration between episodes of hotspot uplift is a decreasing exponential function of time (Crough 1979). The mean interval between hotspot encounters is A/NDV, where A is the Earth's surface area, N is the number of hotspots, D is the diameter of swells, and V is the continental velocity relative to the hotspots. Using $N = 40$, $D = 1200$ km, and $V = 2$ cm yr^{-1} yields an interval of 530 m.y., which is short enough to imply that most continental regions have passed over several hotspots during their geological evolution. Judging from the lack of sedimentary cover atop the present African swells, erosion induced by hotspot uplift may explain the general lack of preserved Precambrian platform deposits in continental interiors. Detailed examination of Mesozoic and Tertiary continental hotspot traces, as attempted by Crough (1981c) for New England, will test whether hotspots do indeed cause epeirogenic motion.

Hotspot swells may also affect continents by initiating rifting, breakup, and dispersal. Rift valleys on the East African and Ethiopian swells suggest a causal relation between uplift and normal faulting, with the presence of unrifted domes, such as the Hoggar, giving evidence that swells come first and generate the rifts (Burke & Whiteman 1973). Plate reconstructions show that hotspots are often present when continents split to form a new ocean (Morgan 1982), and the immediate cause of failure is probably the deviatoric horizontal tension induced by the swells (Bott & Kusznir 1979). The challenge at present is to quantify these hotspot-generated stresses and to explain why so few hotspot tracks generate new rifts.

Heat Flow and Mantle Convection

If midplate swells are formed by reheating, then hotspots put a large amount of energy into the lithosphere; in fact, they may supply all the heat inferred to flow from the deeper mantle to the base of the plates. The same model used to calculate the recurrence of continental uplift also gives the mean surface heat flux expected from repeated reheating events. If newly thinned continental lithosphere has the same thermal structure as 50 m.y. old oceanic lithosphere, then the average continental flux due to this process alone is 27 $mW\ m^{-2}$ (Crough 1979). This value, which is probably accurate to within 50% given the uncertainties in the parameter values used, matches the reduced heat flow observed in most shield areas, thus providing a simple explanation of the Earth's background heat flux.

Although the origin of hotspots remains unknown, the heat-flow results support Morgan's (1971) hypothesis of mantle plumes. At the high Rayleigh number inferred for the mantle, the dominant mode of heat transfer should be convection in the form of irregularly spaced and temporally varying plumes and thermals (Elder 1976). The swells suggest that hotspots are a major heat-transport mechanism, and the hotspot population (Figure 2) seems to have the chaotic distribution expected of convection.

Note that this interpretation of the origin of hotspots depends critically on the interpretation of swell support. If swells are maintained by dynamic uplift, then upwelling convection occurs beneath every positive depth anomaly, whereas if swells are the result of lithosphere-hotspot interaction, then upwelling occurs only beneath the active volcanic areas. Thus those who favor dynamic uplift see Figure 1 as evidence for shallow roll-cell convection (McKenzie et al 1980), and those who favor interaction find evidence for plume flow. As always, inferences about deeper processes require knowledge of shallower structure.

Oceanic Age-Depth Relation

Given that almost half the seafloor shown in Figure 1 is part of a swell, one may wonder how a standard age-depth curve could be defined or calculated. Almost every bit of ocean floor will pass near a hotspot and become anomalously shallow at sometime during its residence at the Earth's surface, and consequently there is no single subsidence curve that all seafloor, or even most seafloor, follows. There are two ways of handling this additional complexity: (*a*) One can calculate an average-subsidence curve, which includes all seafloor and therefore incorporates the effects of hotspot uplift, or (*b*) one can specifically exclude hotspot swells and examine the subsidence that occurs when seafloor does not override a hotspot. Unfortunately, most existing age-depth curves, including the one used to compute depth anomalies in Figure 1, are neither of the two preferred types. Most were calculated before the influence of hotspots was fully realized and are based on samples of data that are too areally restricted to reflect average behavior, yet too near hotspots to be unbiased. For example, the Parsons & Sclater (1977) curve includes points from the Bermuda and Hawaiian rises.

As a first step toward establishing more rigorously defined subsidence curves, Heestand & Crough (1981) examined the North Atlantic ocean floor. The mean age-depth relation is a simple square root of age function, and the depths at all ages are shallower than previously calculated in the Pacific, a difference that is probably attributable to the greater spatial density of hotspots in the Atlantic. The subsidence without hotspot influence was estimated by plotting depth versus age for data at fixed distance ranges from the nearest hotspot. It is impossible to find any seafloor older than 100 m.y. that has not passed within 800 km of a hotspot, but all the data ranges agree with square root of age subsidence. The farthest distances, which should approach unaltered seafloor, have a depth of about 2900 m + 300 m $t^{1/2}$ (m.y.), a relation that also seems to apply at large distances from hotspots in the Pacific (Crough, in preparation, 1982).

If this new age-depth relation is correct, it will change the depth anomalies on younger seafloor in Figure 1. The reference curve used in Figure 1 for seafloor less than 70 m.y. old is 2500 m + 350 m $t^{1/2}$ (m.y.), which is consistently shallower than the curve derived for nonhotspot regions. To adjust to the new standard, all ridgecrest depth anomalies in Figure 1 would have 400 m added to them and older anomalies would have progressively smaller corrections. The effect would be to eliminate most of the negative anomalies in the figure and to enlarge the percentage of shaded area. One significant implication of this change is that the lows in the present figure may not be basins, namely areas of downwarping relative to some standard state, but merely regions that have never been upwarped by hotspots.

Other Planets

Hotspot swells occur on Mars and Venus, reinforcing the inference that swells are caused by some basic planetary process like thermal convection. Mars has one enormous rise, the Tharsis region, which is similar in morphology, but not in scale, to the swells on Earth. Tharsis is approximately 3000 km wide and 7 km high, surmounted by three volcanoes (Carr 1974). Venus has some 34 swells, by my count, with characteristic widths and heights of 1500–3000 km and 2 km respectively (Masursky et al 1980). At the moment the altimetric resolution precludes knowing whether each of these rises is associated with volcanoes, but some certainly seem to be. Why the number and size of swells vary from planet to planet is unknown, but a correlation with planet size seems obvious. Mars is the smallest planet and has the fewest and largest swells. Venus is slightly smaller than Earth and has about the same number of swells, but they are slightly larger.

It would be surprising if these swells were formed differently than those on Earth, but at present there are too few data constraints to be certain of their structure. Gravity anomalies over the swells are almost the only observational evidence pertinent to support. They are consistent with lithospheric reheating (Sleep & Phillips 1979, Morgan & Phillips 1982), but only if there is crustal variation as well and if the standard lithosphere on each planet is several hundred kilometers thick. Such thicknesses are plausible, especially if hotspots supply heat to the lithosphere and these lithospheres are static or slow-moving; however, direct observational evidence is unavailable.

SUMMARY

Most hotspot volcanic areas surmount gentle topographic swells with typical widths of 1000–1500 km and characteristic heights of 500–1200 m. Hotspot swells are both ubiquitous, located on crust of all geological ages in both continents and oceans, and widespread, occupying an estimated 11–42% of the Earth's surface area. Aside from standard plate cooling, they are the major form of vertical motion in the ocean basins, and they may also be an important agent of continental uplift.

According to a variety of geologic and geophysical evidence, ridgecrest swells are isostatically supported by low-density asthenosphere and midplate swells by reheated lithosphere. The key observation is that swells persist beneath hotspot traces of all ages and orientations, implying that rises become fixed to the lithosphere and move with it. Gravity and geoid height anomalies over swells are consistent with isostatic balance by a root at midlithospheric depths. The anomalously fast subsidence and high heat flow of midplate swells are the predicted expressions of decay of a lithospheric thermal pulse. Swell height, width, shape, and rapid uplift rate, however, are still not satisfactorily explained and offer continuing tests for origin hypotheses. More direct observational evidence of deep structure, through seismic or electromagnetic methods, would also help in understanding swell formation.

Swells are probably the indirect surface manifestations of ascending mantle plumes. If the reheating hypothesis is correct, then the creation of swells may be the main heat-transfer process between the Earth's lithosphere and deeper mantle. Similarly shaped swells on Mars and Venus, which can also be attributed to reheating events, suggest that this heat-transport mechanism is not restricted to our globe, but is a common planetary process.

ACKNOWLEDGMENTS

I thank Fred Vine for first awakening my interest in large-scale earth structures, Henry Pollack and George Thompson for helping me focus that interest, and my wife, Faith, for continued forebearance while I pursued it. This work was supported by the National Science Foundation through grants EAR-8021126 and OCE-8109458.

Literature Cited

Anderson, R. N., McKenzie, D., Sclater, J. G. 1973. Gravity, bathymetry, and convection in the earth. *Earth Planet. Sci. Lett.* 18:391–407

Banks, R. J., Swain, C. J. 1978. The isostatic compensation of East Africa. *Philos. Trans. R. Soc. London Ser. A* 364:331–52

Bermingham, P. M., Fairhead, J. D., Stuart, G. W. 1983. The relevance of intraplate volcanic centres to rifting: an example from Jebel Marra, western Sudan. *Tectonophysics.* In press

Betz, F., Hess, H. H. 1942. The floor of the North Pacific Ocean. *Geogr. Rev.* 32:99–116

Bird, P. 1979. Continental delamination and the Colorado Plateau. *J. Geophys. Res.* 84:7561–71

Bott, M. H. P., Kusznir, N. J. 1979. Stress distributions associated with compensated plateau uplift structures with application to the continental splitting mechanism. *Geophys. J. R. Astron. Soc.* 56:451–59

Brott, C. A., Blackwell, D. D., Mitchell, J. C. 1978. Tectonic implications of the heat flow of the western Snake River Plain, Idaho. *Geol. Soc. Am. Bull.* 89:1697–707

Brott, C. A., Blackwell, D. D., Ziagos, J. P. 1981. Thermal and tectonic implications of heat flow in the eastern Snake River Plain, Idaho. *J. Geophys. Res.* 86:11709–34

Brown, C., Girdler, R. W. 1980. Interpretation of African gravity and its implication for the breakup of the continents. *J. Geophys. Res.* 85:6443–55

Burke, K., Whiteman, A. J. 1973. Uplift, rifting and the break-up of Africa. In *Implications of Continental Drift to the Earth Sciences*, ed. D. H. Tarling, S. K. Runcorn, 2:735–55. New York: Academic

Burke, K. C., Wilson, J. T. 1976. Hot spots on the earth's surface. *Sci. Am.* 235:46–57

Carr, H. J. 1974. Tectonism and volcanism of the Tharsis region of Mars. *J. Geophys. Res.* 79:3943–49

Chase, T. E. 1975. Topography of the oceans. *IMR Tech. Rep. TR57*, Scripps Inst. Oceanogr., La Jolla, Calif.

Cochran, J. R., Talwani, M. 1977. Free-air gravity anomalies in the world's oceans and their relationship to residual elevation. *Geophys. J. R. Astron. Soc.* 50:495–552

Cochran, J. R., Talwani, M. 1978. Gravity anomalies, regional elevation, and the deep structure of the North Atlantic. *J. Geophys. Res.* 83:4907–24

Crough, S. T. 1978. Thermal origin of mid-plate hot-spot swells. *Geophys. J. R. Astron. Soc.* 55:451–69

Crough, S. T. 1979. Hotspot epeirogeny. *Tectonophysics* 61:321–33

Crough, S. T. 1981a. Free-air gravity over the Hoggar Massif, northwest Africa: evidence for alteration of the lithosphere. *Tectonophysics* 77:189–202

Crough, S. T. 1981b. The Darfur Swell, Africa: gravity constraints on its isostatic compensation. *Geophys. Res. Lett.* 8:877–79

Crough, S. T. 1981c. Mesozoic hotspot epeirogeny in eastern North America. *Geology* 9:2–6

Crough, S. T. 1982. Geoid height anomalies over the Cape Verde Rise. *Mar. Geophys. Res.* 5:263–71

Crough, S. T. 1983. Seamounts as recorders of hotspot epeirogeny. *Geol. Soc. Am. Bull.* In press

Crough, S. T., Jarrard, R. D. 1981. The Marquesas-Line Swell. *J. Geophys. Res.* 86:11763–71

Crough, S. T., Thompson, G. A. 1976. Numerical and approximate solutions for lithospheric thickening and thinning. *Earth Planet. Sci. Lett.* 31:397–402

Detrick, R. S., Crough, S. T. 1978. Island subsidence, hot spots, and lithospheric thinning. *J. Geophys. Res.* 83:1236–44

Detrick, R. S., Sclater, J. G., Thiede, J. 1977.

The subsidence of aseismic ridges. *Earth Planet. Sci. Lett.* 34:185–96

Detrick, R. S., von Herzen, R. P., Crough, S. T., Epp, D., Fehn, U. 1981. Heat flow on the Hawaiian Swell and lithospheric reheating. *Nature* 292:142–43

Dietz, R. S., Menard, H. W. 1953. Hawaiian Swell, Deep, and Arch and subsidence of the Hawaiian Islands. *J. Geol.* 61:99–113

Dymond, J., Windom, H. L. 1968. Cretaceous K-Ar ages from Pacific Ocean seamounts. *Earth Planet. Sci. Lett.* 4:47–52

Elder, J. 1976. *The Bowels of the Earth.* New York: Oxford Univ. Press

Emerman, S. H., Turcotte, D. L. 1982. Stagnation flow with a temperature-dependent viscosity. *J. Geophys. Res.* In press

Ewing, M., Carpenter, C., Windisch, C., Ewing, J. 1973. Sediment distribution in the oceans: the Atlantic. *Geol. Soc. Am. Bull.* 84:71–88

Fairhead, J. D., Reeves, C. V. 1977. Teleseismic delay times, Bouguer anomalies and inferred thickness of the African lithosphere. *Earth Planet. Sci. Lett.* 36:63–76

Faure, H. 1971. Relations dynamiques entre la croute et le manteau d'apres l'etude de l'evolution paleogeographique des bassins sedimentaires. *C. R. Acad. Sci. Paris* 272:3239–42

Gass, I. G., Chapman, D. S., Pollack, H. N., Thorpe, R. S. 1978. Geological and geophysical parameters of mid-plate volcanism. *Philos. Trans. R. Soc. London Ser. A* 288:581–97

Haxby, W. F., Turcotte, D. L. 1978. On isostatic geoid anomalies. *J. Geophys. Res.* 83:5473–78

Heestand, R. L., Crough, S. T. 1981. The effect of hot spots on the oceanic age-depth relation. *J. Geophys. Res.* 86:6107–14

Heestand, R. L., Morgan, W. J. 1981. Mechanical and thermal thinning of the lithosphere by a hotspot. *EOS, Trans. Am. Geophys. Union* 62:1028 (Abstr.)

Hussong, D. M., Wipperman, L. K., Kroenke, L. W. 1979. The crustal structure of the Ontong Java and Manikiki oceanic plateaus. *J. Geophys. Res.* 84:6003–10

Hyndman, R. D. 1973. Evolution of the Labrador Sea. *Can. J. Earth Sci.* 10:637–44

Jarrard, R. D., Clague, D. A. 1977. Implications of Pacific island and seamount ages for the origin of volcanic chains. *Rev. Geophys. Space Phys.* 15:57–76

Jordan, T. H. 1979. Mineralogies, densities and seismic velocities of garnet lhergolites and their geophysical implications. In *Proc. 2nd Int. Kimberlite Conf.*, Vol. 2, ed. F. R. Boyd, H. O. A. Meyer. Washington DC: Am. Geophys. Union

Lancelot, Y., Seibold, E., et al. 1977. *Initial Reports of the Deep Sea Drilling Project*, Vol. 41. Washington DC: GPO

Larson, R. L., Schlanger, S. O. 1981. Geological evolution of the Nauru Basin, and regional implications. In *Initial Reports of the Deep Sea Drilling Project*, ed. R. L. Larson et al, 61:841–62. Washington DC: GPO

Leeds, A. R. 1975. Lithospheric thickness in the western Pacific. *Phys. Earth Planet. Inter.* 11:61–64

Lister, C. R. B. 1977. Estimators for heat flow and deep rock properties based on boundary layer theory. *Tectonophysics* 41:157–71

Mammerickx, J. 1981. Depth anomalies in the Pacific: active, fossil and precursor. *Earth Planet. Sci. Lett.* 53:147–57

Mareschal, J. C. 1981. Uplift by thermal expansion of the lithosphere. *Geophys. J. R. Astron. Soc.* 66:535–52

Marsh, B. D., Marsh, J. G. 1976. On global gravity anomalies and two-scale mantle convection. *J. Geophys. Res.* 81:5267–80

Masursky, H., Eliason, E., Ford, P. G., McGill, G. E., Pettengill, G. H., Schaber, G. G., Schubert, G. 1980. Pioneer Venus radar results: geology from images and altimetry. *J. Geophys. Res.* 85:8232–60

McKenzie, D. P. 1977. Surface deformation, gravity anomalies and convection. *Geophys. J. R. Astron. Soc.* 48:211–38

McKenzie, D. P., Roberts, J. M., Weiss, N. O. 1974. Convection in the earth's mantle: towards a numerical simulation. *J. Fluid Mech.* 62:465–538

McKenzie, D. P., Watts, A., Parsons, B., Roufosse, M. 1980. Planform of mantle convection beneath the Pacific Ocean. *Nature* 288:442–46

Menard, H. W. 1964. *Marine Geology of the Pacific.* New York: McGraw-Hill

Menard, H. W. 1969. Elevation and subsidence of oceanic crust. *Earth Planet. Sci. Lett.* 6:275–84

Menard, H. W. 1973. Depth anomalies and the bobbing motion of drifting islands. *J. Geophys. Res.* 78:5128–37

Menard, H. W., McNutt, M. 1982. Evidence for and consequences of thermal rejuvenation. *J. Geophys. Res.* In press

Morgan, P., Phillips, R. J. 1982. Hot spot heat transfer: its application to Venus and implications to Venus and the Earth. *J. Geophys. Res.* In press

Morgan, W. J. 1971. Convection plumes in the lower mantle. *Nature* 230:42–43

Morgan, W. J. 1972. Plate motions and deep

mantle convection. *Geol. Soc. Am. Mem.* 132:7–22

Morgan, W. J. 1982. Hotspot tracks and the opening of the Atlantic and Indian Oceans. In *The Sea*, ed. C. Emiliani, Vol. 10. New York: Wiley

Parsons, B., Sclater, J. G. 1977. An analysis of the variation of ocean floor bathymetry and heat flow with age. *J. Geophys. Res.* 82:803–27

Richter, F. M., Parsons, B. 1975. On the interaction of two scales of convection in the mantle. *J. Geophys. Res.* 80:2529–41

Sandwell, D. T., Poehls, K. A. 1980. A compensation mechanism for the central Pacific. *J. Geophys. Res.* 85:3751–58

Sclater, J. G., Anderson, R. N., Bell, L. 1971. Elevation of ridges and evolution of the central eastern Pacific. *J. Geophys. Res.* 76:7888–7915

Sclater, J. G., Lawver, L. A., Parsons, B. 1975. Comparison of long-wavelength residual elevation and free air gravity anomalies in the North Atlantic and possible implications for the thickness of the lithosphere plate. *J. Geophys. Res.* 80:1031–52

Shor, G. G., Pollard, D. D. 1964. Mohole site selection studies north of Maui. *J. Geophys. Res.* 69:1627–37

Sleep, N. H., Phillips, R. J. 1979. An isostatic model for the Tharsis province. *Geophys. Res. Lett.* 6:803–6

Sowerbutts, W. T. C. 1969. Crustal structure of the East African Plateau and rift valleys from gravity measurements. *Nature* 223:143–46

Spohn, T., Schubert, G. 1982. Convective thinning of the lithosphere: a mechanism for the initiation of continental rifting. *J. Geophys. Res.* 87:4669–81

Suppe, J., Powell, C., Berry, R. 1975. Regional topography, seismicity, Quaternary volcanism, and the present-day tectonics of the western United States. *Am. J. Sci.* 275A:397–436

Vogt, P. R. 1976. Plumes, subaxial pipe flow, and topography along the mid-oceanic ridge. *Earth Planet Sci. Lett.* 29:309–25

Walcott, R. I. 1970a. Flexural rigidity, thickness, and viscosity of the lithosphere. *J. Geophys. Res.* 75:3941–54

Walcott, R. I. 1970b. Flexure of the lithosphere at Hawaii. *Tectonophysics* 9:435–46

Watts, A. B. 1976. Gravity and bathymetry in the central Pacific Ocean. *J. Geophys. Res.* 81:1533–53

Watts, A. B., Talwani, M. 1974. Gravity anomalies seaward of deep-sea trenches and their tectonic implications. *Geophys. J. R. Astron. Soc.* 36:57–90

Wilson, J. T. 1963. A possible origin of the Hawaiian Islands. *Can. J. Phys.* 41:863–70

Withjack, M. 1979. A convective heat transfer model for lithospheric thinning and crustal uplift. *J. Geophys. Res.* 84:3008–22

Yoshii, T. 1973. Normal ocean, marginal seas and hot spots. *Nature Phys. Sci.* 244:92–93

Ann. Rev. Earth Planet. Sci. 1983. 11: 195–214

OCEANIC INTRAPLATE SEISMICITY

Emile A. Okal
Department of Geology and Geophysics, Yale University, Box 6666, New Haven, Connecticut 06511

Introduction

The occurrence of seismicity at the surface of the globe largely along preferential lines, now recognized as mid-oceanic ridges and subduction zones, was key evidence for the development of the theory of plate tectonics in the 1960s. Indeed, many plate boundaries, especially some of the southern, less accessible ridges, were drawn, at least initially, on the basis of this evidence alone. The mere concept of intraplate seismicity may then appear as somewhat of a paradox, if not as a failure of the whole theory. However, and as shown later in this paper, the relatively low level of this seismicity warrants its consideration as a perturbation in the general framework of nearly rigid plates. Additionally, intraplate earthquakes are of great value since, in the oceanic environment, they are our only clue to the state of stress of the lithosphere, thereby providing us with critical insight into the forces responsible for its motion. Understanding parameters that control these earthquakes is made easier (with respect to a continental situation) by the generally younger age of the ocean floor and the much simplified tectonic history of any given oceanic province. On the other hand, seismic detection capabilities are greatly reduced at sea, because of both logistics and generally higher seismic attenuation; thus, our knowledge of many aspects of oceanic intraplate seismicity is still rudimentary. It is in this general framework that the present paper reviews the following points:

1. Definition of intraplate seismicity (What?)
2. The level of seismicity in the oceans (How much?)
3. Location, including depth (Where and when?)
4. Mechanism of intraplate seismicity. Inferences about stresses (Why?)
5. Preferential siting of the stress release on the plate (How?)

0084–6597/83/0515–0195$02.00

Definition of Intraplate Earthquakes (What?)

In characterizing intraplate earthquakes, we must recognize at once several types of such events. The broadest definition of an intraplate earthquake would be an event not involving displacement between two plates. As such, all stress-controlled events, even those located in the immediate vicinity of a plate boundary, would qualify as intraplate. So would all decoupling events (of the type of the 1977 Indonesian or 1933 Sanriku earthquakes), involving rupture of the lithosphere seaward of the trench, in a zone where it undergoes bending just prior to subduction. However, following Sykes & Sbar (1974), we call this type of event "boundary-related," and do not include it in our definition. This category would also include smaller events controlled by the flexure of the plate seaward of a trench, such as those described by Chen & Forsyth (1978).

Additionally, some earthquakes occurring inside oceanic plates, such as the 1975 Kalapana event on Hawaii, are clearly associated with volcanism. Such events are not representative of the usual processes associated with the evolution and cooling of the lithospheric plate, and they are excluded from the definition of intraplate earthquakes used in this paper. As discussed later, this exclusion is easily made for well-documented seismicity in the vicinity of well-known hotspots, but would be difficult to extend to low magnitudes in poorly surveyed areas.

Finally, some areas of oceanic basins not involving major plate boundaries have been found to exhibit substantial seismicity, at a level clearly incompatible with the assumption of rigidity of the lithospheric plate. Examples are the Ninetyeast Ridge area in the Indo-Australian plate, whose seismicity, compiled by Stein & Okal (1978), included several magnitude 7 earthquakes since the 1910s, and the Caroline wedge of the Pacific plate between New Guinea and the Mariana Trench (Weissel & Anderson 1978). The amount of deformation documented in these areas suggests that they involve genuine plate boundaries, leading to the concept of two independent plates (Indian and Australian) in the case of the Ninetyeast Ridge, and of a Caroline miniplate north of New Guinea. Similarly, a 1964 event located east of the Caribbean arc (Liu & Kanamori 1980), characterized by NNW-SSE compression and a relatively deep focus (23 km), may be representative of the convergence between North and South America predicted by kinematic models such as Minster & Jordan's (1978). It is clear that these situations are peculiar, and that they define a separate type of seismicity.

Thus, we adopt a very restrictive view, and define intraplate oceanic earthquakes as events not controlled in their location and mechanism by present plate boundaries or by phenomena of a clearly extraordinary nature in the morphology of the oceanic plate, such as hotspot volcanism or large-scale deformation.

Level of Intraplate Seismicity in the Oceans (How much?)

The earliest attempt at recognizing the seismicity of the so-called stable blocks, including the Pacific Ocean Basin, is found in Gutenberg & Richter's *Seismicity of the Earth* (1941). Since the recent progress in detection brought about by the World-Wide Standardized Seismic Network [WWSSN] in the 1960s, we now have documented evidence for magnitude 5 or greater earthquakes in all of the world's oceans.

The systematic study of intraplate events, both continental and oceanic, was initiated by Sykes & Sbar (1973), after focal mechanisms of interplate earthquakes provided a spectacular qualitative confirmation of plate kinematics. More recently, Bergman & Solomon (1980) compiled a catalog of 159 oceanic intraplate events covering the period 1939–79. This catalog can be used to estimate the fraction of the world's seismicity located in the interior of oceanic plates, although such an estimate will be high since these authors use a definition of intraplate events less stringent than ours. Using Kanamori's (1977) moment-magnitude relations, one finds a total seismic moment release of 1.3×10^{27} dyn-cm (or slightly less than 10^{26} dyn-cm/yr) for the portion of the Bergman & Solomon catalog corresponding to the years since 1964, when the worldwide detection capabilities were upgraded substantially. This figure is to be compared with 6×10^{30} dyn-cm (or approximately 8×10^{28} dyn-cm/yr) for earthquakes of all types (but mostly interplate) compiled from Kanamori's (1977) list for 1904–76. It is immediately apparent that oceanic intraplate seismicity is only a minor contribution to worldwide seismicity, representing small-scale deformation in otherwise rigid plates. Similarly, Okal (1981) estimated that the deformation taken up through intraplate seismicity in the northern part of the Antarctica plate was only 2% of the rate of accretion of the plate. These figures warrant treating oceanic intraplate seismicity as a small perturbation of the rigid plate framework, and actually save the plate tectonics concept.

Oceanic intraplate seismicity is a universal feature of the world's oceans, occurring in all major plates, both wholly oceanic and continent-bearing. Figure 1 is adapted from Bergman & Solomon's paper, and was obtained by removing from their catalog events clearly associated with either volcanism or large-scale intraplate deformation. The epicenters removed were mostly located in the Ninetyeast, Hawaii, and Caroline areas. Table 1 is a list of some of the most significant oceanic intraplate earthquakes. It is not intended to substitute for a complete catalog of intraplate earthquakes [for this, the reader is referred to Richardson et al (1979) or Bergman &

Solomon (1980), and reminded once again of their less stringent definition of intraplate seismicity], but merely presents data on the largest events known in each oceanic area, and on a few earthquakes of particular importance that are used in this review. These epicenters are identified in Figure 2, together with the plate boundary system. Of particular interest in the Pacific plate are the seismic clusters at the so-called Regions A and C, studied in detail by Okal et al (1980). Region A, east of the Line Islands, was the site of 86 earthquakes during 1968–76; Region C, 500 km northeast of Pitcairn, underwent a swarm of 98 events during 1976–79, and has been quiescent at the magnitude 3 level ever since. Except for event 2 in the

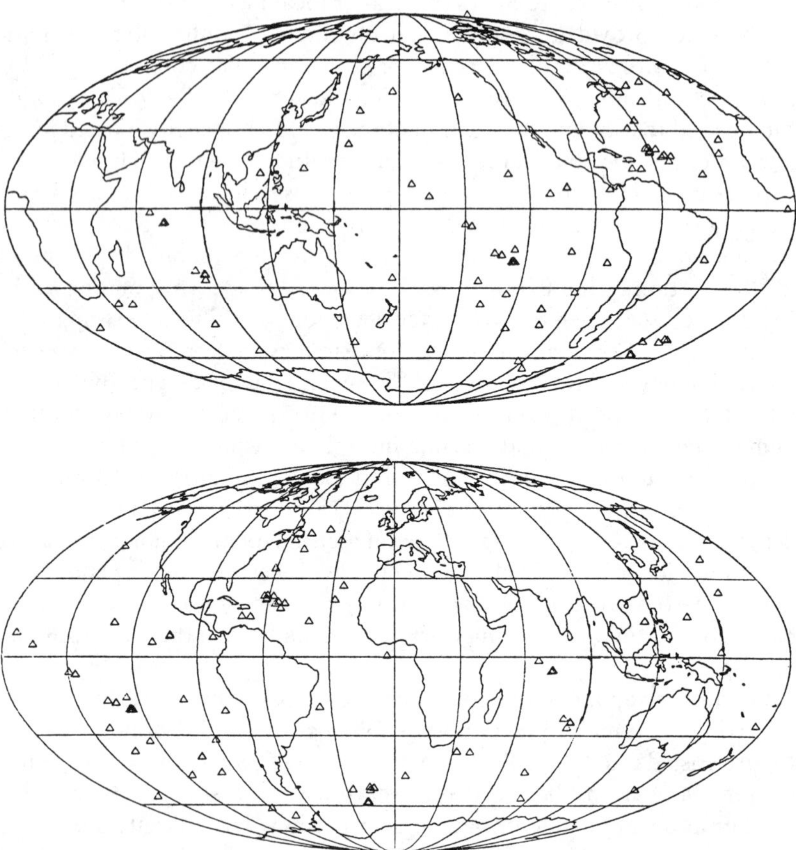

Figure 1 Oceanic intraplate seismicity of the Earth displayed on a Mollweide equal-area projection. This map is adapted from Bergman & Solomon (1980) by excluding from their catalog events associated with hotspot volcanism and large-scale internal deformation of the plates.

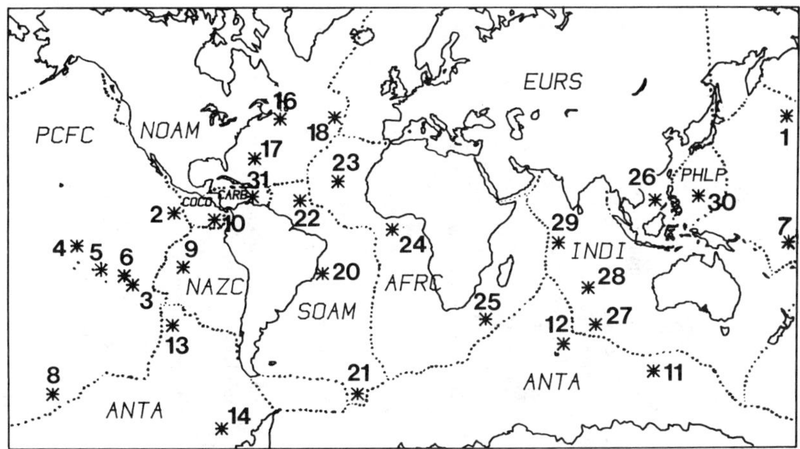

Figure 2 Mercator projection map of significant oceanic intraplate earthquakes used in the present review. Oceanic plate boundaries shown as dotted lines. Numbers refer to Table 1. Events 15 and 19, in northern part of North American plate, are not shown.

easternmost part of the plate, these two regions have contributed more than 90% of the seismic energy release in the Pacific plate during the period studied by Okal et al.

Magnitudes

The only magnitude 7 earthquakes confirmed inside oceanic plates are events 15 and 16 at the continental margins of North America, event 11 in the Indian Ocean (very little is known about this earthquake; see Okal 1981), and event 21 in the South Atlantic, a slow, complex earthquake (Creaven et al 1979), whose mechanism, not fully understood, may involve an on-going ridge jump (Farmer et al 1982). On the other hand, Table 1 shows that most plates exhibit some magnitude 6 seismicity. Of particular interest, however, is the case of the Pacific plate, by far the largest oceanic one, where the only magnitude 6 or greater seismicity is concentrated in the extreme southern tip of the plate (event 8), in the immediate vicinity of the East Pacific Rise (events 2 and 3), and at the southeastern tip of the Gilbert Island chain (event 7; Okal & Lay 1982). No earthquake with a body-wave magnitude m_b greater than 5.5 is known elsewhere in the plate.

A frequency-magnitude (b-value) investigation by Bergman & Solomon (1980) has revealed no significant departure from averages computed using both intraplate and interplate earthquakes ($b = 0.9$), except in the Pacific Ocean ($b = 1.3$). Higher b-values mean that stress release takes place through smaller, but more numerous events (a property often interpreted as weakening of the rock) that are known to be characteristic of volcanic

Table 1 Significant oceanic intraplate earthquakes

Number	Date Y M D	Epicenter °N	Epicenter °E	Magnitude	Focal type[a]	Age of plate (m.y.)	Ref.[b]	Remarks
				Pacific plate				
1	1968 04 28	44.8	174.5	$5.5m_b$	t	80–100	1	Emperor Trough
2	1970 01 21	7.0	−104.2	$6.8M_s$	t	3	2	Close to East Pacific Rise
3	1955 11 22	−24.5	−123.0	$6.5M_s$	n	10	3	Normal faulting
4	1969 08 06	−7.6	−148.3	$5.0m_b$	ss	73	4	Region A, 86 events 1965–1979
5	1965 03 06	−18.4	−132.9	$5.5m_b$	ss	34	4	Region B, 13 events 1965–1981
6	1978 07 25	−20.7	−126.8	$5.4m_b$	ss	20	4	Region C, 98 events 1976–1979
7	1982 05 23	−03.4	177.4	$6.0m_b$	t	110–120	5	Gilbert Islands
8	1947 12 15	−59.5	−159.5	$6.9M_s$	t	8	3	Largest event known in plate
				Nazca plate				
9	1965 11 25	−17.1	−100.2	$5.8m_b$	t	?	6	
				Cocos plate				
10	1976 03 29	4.0	−85.9	$6.5M_s$	ss	9	7	Slow event
				Antarctic plate				
11	1947 12 24	−54	114	$7.0M_s$		15	8	Largest event known in plate
12	1973 05 03	−46.1	73.2	$5.5m_b$	n	25	8	Off Kerguelen plateau
13	1971 05 09	−39.8	−104.9	$6.0M_s$	t	13	9	Similar event in 1925
14	1977 02 05	−66.5	−82.5	$6.4M_s$	t	56	10	Bellingshausen Sea
				North American plate				
15	1933 11 20	73.3	−70.7	$7.3M_s$	t		11	Passive margin, Baffin Bay
16	1929 11 18	44	−56	$7.2M_s$			12	Passive margin, Grand Banks
17	1978 03 24	29.9	−67.4	$6.1M_s$	t	117	13	Bermuda event
18	1964 09 17	44.5	−31.3	$5.6m_b$	t	10	2	
19	1978 01 04	85.7	−23.8	$5.0m_b$	t	10	2	Arctic Ocean

				South American plate				
20	1955 03 01	−19.9	−36.7	$6.5M_s$	t		14	Passive margin, Brazil
21	1977 08 26	−59.5	−20.6	$7.1M_s$	ss	3–4	15	Slow event
22	1968 02 20	12.4	−46.9	$5.6m_b$	n(?)	7	16	
				African plate				
23	1972 10 20	20.6	−29.7	$5.8M_s$	ss	100	17	
24	1971 09 30	−0.5	−4.8	$6.0m_b$	t		18	*P*-axis perpendicular to Guinea F.Z.
25	1968 09 03	−37.8	38.0	$5.0m_b$		80	2	
				Eurasian plate				
26	1965 10 07	12.5	114.5	$5.8m_b$	t	30	19	South China Sea
				Indian plate				
27	1968 10 08	−39.8	87.7	$6.0m_b$	n	6	16	
28	1974 06 25	−26.0	84.3	$6.2M_s$	t	35	20	
29	1965 09 12	−6.5	70.8	$6.0M_s$	n	35	21	Chagos-Laccadive Rise, swarm
				Philippine plate				
30	1974 04 12	14.3	134.4	$5.5m_b$	ss	37	2	
				Caribbean plate				
31	1972 08 14	14.2	−68.5	$4.7m_b$	ss	UK[c]	22	

[a] Focal mechanisms:
t: thrust fault
n: normal fault
ss: strike-slip

[b] References:
1. Stein (1979)
2. Bergman & Solomon (1980)
3. Okal & Greenberg, in preparation
4. Okal et al (1980)
5. Okal & Lay (1982)
6. Mendiguren (1971)
7. Okal & Stewart (1982)
8. Okal (1981)
9. Forsyth (1973)
10. Okal (1980)
11. Stein et al (1979)
12. Gutenberg & Richter (1941)
13. Stewart & Helmberger (1981)
14. Mendiguren & Richter (1978)
15. Creaven et al (1979)
16. Sykes & Sbar (1974)
17. Richardson & Solomon (1977)
18. Liu & Kanamori (1980)
19. Wang et al (1979)
20. Stein & Okal (1978)
21. Stein (1978)
22. Kafka & Weidner (1979)

[c] Age:
UK: Upper Cretaceous

seismicity (Mogi 1963). Because of the scarcity of magnitude 6 events in the Pacific plate (Bergman & Solomon did not include events 3 and 8 in their catalog), and of the extremely small range of magnitudes covered, their conclusions must be taken with caution. Okal et al (1980) conducted local b-value studies, using swarms of earthquakes at the most active locations in the southcentral Pacific, including events down to local magnitudes $M_L = 3.0$. Their results show b-values that are, if anything, lower than world averages, suggesting that Bergman & Solomon's high Pacific b-values are biased by their sampling. Okal et al then used their results ($b \leq 0.86$) to argue against a volcanic origin for the seismicity at Regions A, B (event 5 in Figure 2), and C.

Detection and Location of Oceanic Intraplate Seismicity (Where and when?)

One of the major problems encountered in assessing the seismicity of the interior of oceanic plates is the poor detection capability provided by present station coverage. At least three factors contribute significantly to this situation. First, the great majority of permanent seismic stations are still based on islands, and despite technological advances, this will probably remain a fact of life for some time: ocean-bottom seismometers (OBSs) still have recording capabilities strongly limited in time, and as such, have been used primarily in aftershock campaigns (for instance, following the 1978 Bermuda earthquake), during submersible dives usually at mid-oceanic ridges, and in the immediate vicinity of coastlines. Second, because of the swell-generated noise in the 1–5 second band, most standard oceanic stations operate at short-period gains considerably less than continental ones. Finally, higher attenuation along oceanic paths further reduces detection capabilities. A notable exception to this pattern is provided by the French Polynesian seismic network of 15 short-period stations. This network, which was described in detail by Talandier & Kuster (1976) and Okal et al (1980), uses narrow-band rejection filters to operate at gains of up to 150,000 at 1 Hz, comparable to those achieved on continents, thus allowing detection of $M_L = 4$ seismicity over an area of 1.5×10^7 km^2, and of $M_L = 3$ over 3×10^6 km^2. It has provided considerable insight into the intraplate seismicity of the southcentral Pacific. By contrast, in other oceanic areas, the absence of adequate recording facilities prevents the detection of any seismicity below the worldwide magnitude threshold of about $m_b = 4.7$.

Through the use of high-gain teleseismic records, epicentral locations for intraplate oceanic events can be achieved with reasonably good precision, usually at the level of ± 15 km for $m_b = 5.0$ earthquakes (Jordan & Sverdrup 1981). Although this precision may not be sufficient to interpret the seismicity in the context of available bathymetry, it must be considered good in view of the problems associated with hypocentral depth determination.

Because of the general paucity of arrival-time data at short distances, the trade-off between depth and origin time leads to singularity in hypocentral inversions, and very little depth information can be obtained from arrival times alone. Also, the rare arrival times at regional distances are of little if any help, since crustal structure is often poorly known in remote parts of the ocean. As a result, very little was known until recently about the depth of oceanic intraplate earthquakes. As late as 1965, event 9 in the Nazca plate was given a focal depth of 143 km by the US Geological Survey. Using surface wave spectral characteristics, Mendiguren (1971) was able to relocate this event no deeper than 13 km. A variety of techniques, including comparative surface wave excitation (Okal 1981), body-wave modeling (Wang et al 1979), identification of water-reflected phases (Okal et al 1980), and excitation of high-frequency interface waves controlled by sedimentary layering (Okal & Talandier 1981) have been used to constrain intraplate earthquake depths. All suggest that intraplate oceanic foci are located in the shallowest portion of the plate.

Using long-period body-wave modeling to constrain the depths of 18 teleseismically recorded oceanic earthquakes in the magnitude 6 range, Wiens & Stein (1983) have shown that the seismically active layer of the lithosphere thickens with age to a maximum of about 40 km. Thus, it coincides with the elastic layer, as defined by the lithosphere's response to loading over long periods of time (Watts et al 1978, McNutt & Menard 1978), and is much thinner than the 100 km or so associated with the response to faster loading, in the frequency range of seismic waves. Wiens & Stein have shown further that they can interpret this downward boundary of the occurrence of seismicity as the 600°C isotherm, corresponding to the 100 MPa olivine failure limit. Studies of smaller events, in the magnitude 5 range, have indicated even shallower foci (Harkrider & Okal 1982), located in the first 2–3 km of hard rock below sediments. It should be mentioned that some of the deeper earthquakes sampled by Wiens & Stein belong to deformed areas, such as the Caroline miniplate; this could act to bias their study slightly toward a thicker seismically active layer.

Exceptions to the pattern of very shallow oceanic intraplate seismicity are found mostly in cases involving continental margins (event 15 in Baffin Bay) or hotspot structures (a priori eliminated from this study).

Since only the uppermost part of the oceanic lithosphere is capable of rupture, and the fault width is thus strongly reduced, earthquake scaling laws (Geller 1976) are violated by oceanic intraplate earthquakes. Indeed, recent studies (Liu & Kanamori 1980) have suggested that stresses released in intraplate earthquakes are larger than their interplate counterparts; as a result, oceanic earthquakes exhibit a definite $m_b : M_s$ anomaly, their 20-s surface wave magnitude M_s being deficient by as much as 0.5 unit with respect to their 1-s body-wave magnitude m_b (Wiens & Stein 1983).

Mechanisms of Oceanic Intraplate Seismicity and Interpretation (Why?)

Sykes & Sbar (1974) were the first to gather focal mechanism data for oceanic intraplate earthquakes. Their studies were entirely based on first-motion *P*-wave data. In the oceanic environment, one generally faces a lack of stations at short distances; additionally, at regional distances, oceanic P_n is usually emergent and of very little use. This results in the concentration of most of the data at the center of the focal hemisphere. Under these conditions, Sykes & Sbar usually obtained only the general character of the event (normal, thrust, or strike-slip), but in many cases could not constrain the orientation of the focal planes. Their results indicated that while normal fault earthquakes are found in the youngest part of the oceans, most oceanic intraplate seismicity is of the thrust fault type once the plate has reached the age of approximately 15 m.y.

Later studies, involving the polarization of *S* waves (Forsyth 1973), modeling of surface waves (Mendiguren 1971) or body waves (Stewart & Helmberger 1981), or a combination of these techniques along the lines of Stein & Okal (1978), have not only confirmed the predominance of thrust fault solutions, but have allowed for the determination of the full earthquake mechanisms and of their principal stress axes. Because of greater station distances, it should be noted that the absolute threshold for reliable focal mechanisms remains around $m_b = 4.8$ for solutions relying purely on body waves; Stein (1979) has extended surface wave radiation pattern methods down to $M_s = 5.2$, below which results can be expected to be strongly affected by local (and often unknown) crustal structure. In the well documented Caribbean area, Kafka & Weidner (1979) have used event pair inversion techniques for events down to $M_s = 4.7$. Significant focal mechanisms are indicated (by type of faulting) in Table 1 and sketched in Figure 3.

A number of studies, in particular Okal et al (1980), have shown that while focal mechanisms can vary at a given seismic locality, they usually share a common direction of principal compressional stress over large areas, indicating that these earthquakes involve the release of horizontal tectonic compressional stress accumulated in the plate as a result of large-scale plate dynamics. A correlation between the azimuth of the axis of compressional stress and the direction of motion of the plate away from the ridge has been reported in the following areas: Antarctic plate panhandle

(Forsyth 1973), Bellingshausen Sea (Antarctic plate; Okal 1980), south-central Pacific (Okal et al 1980), Nazca plate (Mendiguren 1971), Canary Basin (African plate; Richardson & Solomon 1977), Brazil Basin (South American plate; Mendiguren & Richter 1978). Other events, including event 2, close to the East Pacific Rise, and event 19, in the Arctic Ocean (Bergman & Solomon 1980), have thrust fault mechanisms which, although not fully constrained, are compatible with this pattern. Many small events have thrust fault mechanisms, whose focal planes cannot be constrained (Okal et al 1980).

On continents, it has been possible to use independent in situ stress measurements to confirm that stresses derived from earthquake focal solutions are indeed representative of the ambient tectonic stress accumulated in the plate as the result of its motion over the asthenosphere (Sbar & Sykes 1973, McGarr & Gay 1978). Similar measurements cannot be taken in the oceanic environment; however, the consistency of the compressional stresses released by intraplate oceanic earthquakes over large distances is taken as a strong argument for postulating that they are indeed representative of the tectonic state of stress of the lithosphere. This is also supported by the generally good agreement found in continent-bearing plates between

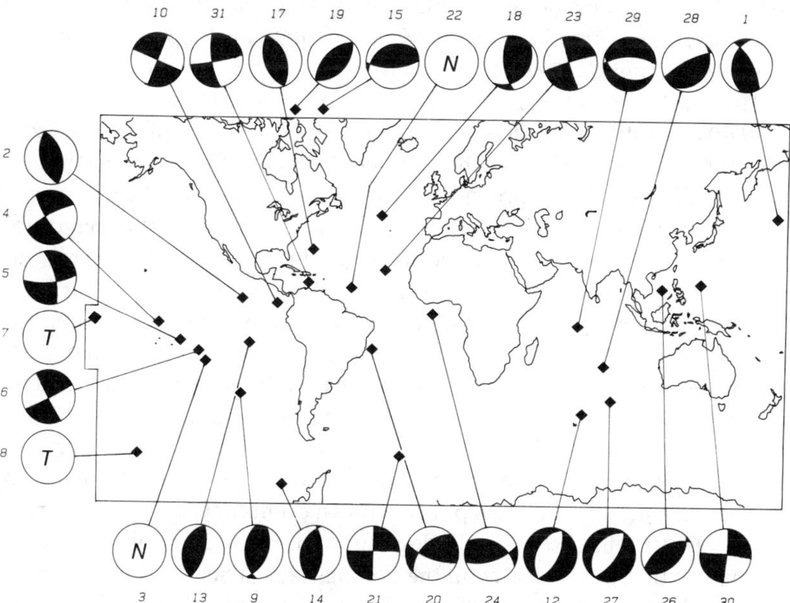

Figure 3 Focal mechanisms of oceanic intraplate earthquakes, shown by stereographic projections of lower-hemisphere focal quadrants (compressional areas in black; dilatational ones in white). N: unconstrained normal fault mechanism; T: unconstrained thrust fault mechanism. Numbers refer to Table 1 and Figure 2.

the direction of compressional stresses released during oceanic intraplate earthquakes, and of the tectonic stresses in their continental parts, as determined by either seismology or in situ measurements. Typical examples are North America (Sbar & Sykes 1977) and South America (Mendiguren & Richter 1978).

Stresses obtained from oceanic intraplate focal mechanisms may then be used to infer the relative importance of the forces involved in driving the plates. In a recent study, Richardson et al (1979) used finite element techniques to obtain estimates of the various forces acting on the plates. They were able to confirm earlier results by Forsyth & Uyeda (1975) suggesting that forces resistive to slab penetration nearly balance the strong gravitational pull of the slab at subduction zones; this leaves the state of stress in the plate controlled largely by ridge-push (a simple name for the complex process of gravitational sliding of the lithosphere as it ages away from the ridge; see Frank 1972) and by whatever drag forces exist on the lithosphere-asthenosphere boundary. Data from the Bellingshausen Sea earthquake in the Antarctic plate (which has no subduction zones and is practically fixed in the hotspot frame of reference) suggest that ridge-push is the dominant contributor (Okal 1980), a theory also upheld by the variation of the stress inside continental North America (Sbar & Sykes 1977).

However, as pointed out by McKenzie (1969), in the presence of fossil structures acting as preferential fault planes, there can be a significant deviation of the direction of released stress from the ambient one. Raleigh et al (1972) quantify this effect to a maximum $\pm 25°$. An example is believed to be event 1 (Stein 1979); another one could be the Bermuda earthquake (event 17), for which compressional stress is 23° away from the direction of ridge-push (Stewart & Helmberger 1981). Thus, the fact that a favorable morphology of the ocean floor may affect the direction of stress release must be kept in mind when interpreting earthquake mechanisms.

Another potential source of compressional stress in the plate could be asthenospheric drag in the direction of the plate's absolute motion over the mantle. Richardson et al (1979) have shown that this force is passive and has little effect on oceanic lithosphere. However, Wesnousky & Scholz (1980) have proposed that in the vicinity of a continent the craton's deeper roots distort the stress field in the oceanic lithosphere. Such a mechanism may be involved in a number of large, poorly known events at continental margins (Sykes 1978). In the case of motionless Antarctica, however, Okal (1980) has shown that the stress released in the Bellingshausen Sea earthquake, in the vicinity of the continental shelf, could be interpreted only by ridge-push.

Stresses in plates with little or no ridge boundaries must be of a different origin: these include the Caribbean and Philippine plates and the South China Sea portion of the Eurasian plate. In the Caribbean plate, event 31, studied by Kafka & Weidner (1979), shows release of a compressional stress oriented WNW-ESE, whose origin is probably related to subduction at the Lesser Antilles and convergence with South America along the Venezuelan coast. Similarly, Bergman & Solomon's (1980) solution for event 30 inside the Philippine plate involves compression along the NW-SE axis, representative of the convergence involved at the Mariana and Philippine trenches, after allowing for possible reorientation of the released stress by the Palau-Kyushu Ridge. Wang et al's (1979) focal solution in the South China Sea, an area where spreading stopped 17 m.y. ago, was also interpreted in terms of convergent tectonics.

Most earthquakes whose mechanism is not explained by ridge-push or convergence have been interpreted as phenomena associated with continental margins (e.g. Baffin Bay seismicity; Stein et al 1979), large intraplate asthenospheric flow (near Kerguelen Island; Okal 1981), or continued deformation along a major bathymetric feature (Chagos Bank; Stein 1978). This is probably also the case of the recent Gilbert Islands earthquake swarm: event 7 is located at the southeastern tip of the island chain, and its tentative focal solution is a thrust fault involving NNE-SSW compressional stress (Okal & Lay 1982).

However, some problems remain unsolved. They are basically of two types: tensional events involving normal faulting, and thrust mechanisms whose compressional axes do not fit known stress fields, even when adjusted 25° or less for preferential release along existing faults. Tensional events are usually found close to the ridges (e.g. the 1955 earthquake near Easter Island). This prompted Sykes & Sbar (1973) to propose that the plate remains under tensional stress for about 10 to 20 m.y., until it cools down sufficiently and goes to a compressional state of stress. However, as shown by the data in Table 1, thrust events are found in extremely young lithosphere (event 2 is only 200 km, or 3 m.y., west of the East Pacific Rise; the large 1947 earthquake, event 8, is 300 km, or 8 m.y., from the South Pacific Ridge; and event 18 is 100 km, or 10 m.y., from the Mid-Atlantic Ridge); in contrast, normal events are found as far as 10 m.y. away from ridges (event 3). Thus the cooling process invoked by Sykes & Sbar (1974) cannot be uniquely dependent on age, as opposed to many other cooling-controlled phenomena (Tréhu et al 1976). One cannot discard the alternative possibility that normal fault events reflect local factors, such as increased asthenospheric flow, or volcanism, potentially of "plume" origin, which may also be responsible for features such as gravity and bathymetry highs, especially in the vicinity of volcanic islands, such as Easter. As discussed by Okal & Bergeal (1983), event 3 is indeed located in the vicinity of seamounts. It is also unlikely that such tensional stresses are due to loading effects (in the absence of major island chains) or to membrane tectonics (Turcotte & Oxburgh 1973), since the lithosphere involved in the case of event 3, for example, has basically remained at the same latitude since it was created.

Some thrust events also exhibit compressional stress directions that are incompatible with the direction of motion of the plate away from its ridge: The 1971 event on the Guinea Fracture Zone, in the African plate, exhibits a compressional axis more or less perpendicular to the fracture zone (Liu & Kanamori 1980), whose origin must lie in locally created conditions, poorly understood at present.

Additionally, a few strike-slip events, whose mechanisms are not readily interpreted in the context of known tectonic stresses, add to the complexity of the picture: Event 10 in the Cocos plate north of the Galapagos Ridge (in an area about 9 m.y. old) has a compressional axis trending approximately N70°E (Okal & Stewart 1982), about 60° away from the direction of ridge-push; event 21, northeast of the South Sandwich Islands, in the South American plate, has a vertical strike-slip mechanism with compressional axis N135°E (Creaven et al 1979), 45° away from ridge-push. Both of these events exhibit "slow" mechanisms, characterized by an increase of magnitude with period and a complex source rupture process, which have led Okal & Stewart (1982) to propose that large-scale intraplate deformation, possibly related to hotspot or other magmatic activity, may be taking place in these areas.

In conclusion, we now understand most of the stresses responsible for oceanic intraplate seismicity. Ridge-push plays an apparently predominant role; convergent tectonics at the plate boundaries also contribute to horizontal compressional stresses inside the plates. The nature of occasionally observed tensional stresses, as well as the origins of some isolated cases of strike-slip faulting of an apparent local nature, remain unclear.

Finally, we should repeat that the above discussion is limited to earthquakes of magnitude $m_b \geq 4.8$. In continental areas, focal mechanism solutions have been found consistent across the magnitude scale (Sbar & Sykes 1977); however, an interpolation of the above results to lower magnitudes may not be warranted in the oceanic environment. In particular, volcanic seismicity, at a level escaping teleseismic detection except in the form of T waves, has been documented, notably at Macdonald Volcano (Johnson 1970). Despite some present insight into the characteristics of volcanic seismicity (e.g. Klein 1982), the identification of the origin of a seismic source as tectonic or volcanic remains a difficult problem in the case of low-level seismicity in unsurveyed areas. Such identification involves a study of the bathymetric characteristics of the epicentral area, and is best discussed in the framework of the next section.

Siting of Seismicity in Correlation with Bathymetry (How?)

Having described what is now considered a reasonably well understood picture of the stresses involved in major oceanic intraplate earthquakes, we must now address the question of the factors governing the siting of seismicity on the plate, in other words "How is the stress released?"

In a monumental review paper, Sykes (1978) examined the question of the preferential location of intraplate seismicity, and other forms of tectonism, onto otherwise "stable" blocks, and concluded that there exist zones of weakness, such as sutured fracture zones, representative of previous cycles of active tectonism, along which a number of phenomena, such as intraplate seismicity and kimberlites, occur preferentially. His study was, however, mainly concerned with continents. In the oceanic environment, it has also been suggested that zones of weakness, such as fracture zones outside of their active transform segments, may be areas of preferential intraplate seismicity. The Ninetyeast Ridge, one of the longest known suture zones, is indeed the preferred site of large-scale deformation induced in the Indo-Australian plate by the Himalayan collision (Stein & Okal 1978). Other areas of preferential weakness could include former plate boundaries and the traces of former hotspots, evidenced in present-day bathymetry as seamount chains. This has prompted Bergman & Solomon (1980) to investigate systematically the correlation between seismicity and bathymetry for a data set of 83 intraplate oceanic earthquakes. Their results suggest that the larger earthquakes are often associated with old fracture zones, but that the correlation with other large bathymetric features is poor; their most compelling results, however, are taken from the Ninetyeast area of the Indian Ocean, which is not fully representative of genuine intraplate oceanic seismicity. A good correlation of intraplate oceanic seismicity with major fracture zones has also been reported for events 11, 14, 17, 21, and 24, all of them major shocks. Smaller events were found by Bergman & Solomon to correlate less significantly with known bathymetry.

Okal (1981) has shown that a number of earthquakes in the northern panhandle of the Antarctic plate are aligned in the vicinity of the line of maximum age of the plate, which is the wake of the Easter Island triple junction, and as such a zone of suture and of potential weakness. More recently, Okal & Bergeal (1983) have shown that at least 6 foci in the southcentral Pacific (including Region C) are located on the boundary line of lithosphere generated at the old Farallon Ridge, before it jumped and reoriented itself along the present East Pacific Ridge sometime in the early Miocene (Herron 1972). This boundary is a line of age discontinuity on the present Pacific plate, and probably represents a zone of weakness. Epicenter 9 and two other seismic sites belong to a similar boundary in the Nazca plate. Stein (1979) has also discussed preferential occurrence of seismicity along fossil features such as the Emperor Trough, but indicated that the stress field released in their vicinity may be considerably distorted with respect to the tectonic stress in the plate. Under some circumstances, it may become difficult to distinguish between mechanisms believed to be distorted by the bathymetry and mechanisms involving local stress regimes, such as in the Chagos area.

However, old fracture zones and other weak lines are far from being the only repositories of oceanic intraplate seismicity. Most of the seismic localities identified by Okal et al (1980) in the southcentral Pacific lack a definite correlation with bathymetric features; one of the most active sites, Region A, east of the Line Islands, was the target of an on-ship survey in 1979, which failed to reveal any substantial bathymetric feature, let alone a major fracture zone. This region (the second most active in the southcentral Pacific) lies about 130 km south of the large Galápagos Fracture Zone, and some 60 km north of a much smaller one, and is not apparently associated with these features (Sverdrup 1981). Similarly, the large-scale bathymetry in Region C was explored by Sailor & Okal (1983) at long wavelengths, using satellite radar altimetry; they discovered only one fracture zone, located about 70 km to the south, too far away to permit an association with the epicenter. As discussed by Sverdrup (1981), the geomorphological features present in Region A are oriented parallel to the local regime of fracture zones, derived from the old Farallon Ridge, at approximately 45° from the tectonic stress created by the present East Pacific Ridge. This favorable situation may explain the relatively low magnitude level of the seismicity at Region A and the strike-slip focal solutions.

In particular, it is interesting to note that this discrepancy between existing large-scale tectonic directions and the orientation of present stress can exist only in a plate having undergone a change in its accretion pattern. In practice, this applies only to the bulk of the Pacific plate, which was generated at the old Farallon Ridge. Significantly, as mentioned earlier, the only magnitude 6 events known in the Pacific plate are located in areas that did not go through the reorientation process: the young, easternmost fringe of the plate, generated since the ridge jump (events 2 and 3); the region south of the Louisville Ridge (event 8); and the Gilbert Islands area, generated at the Pacific-Phoenix boundary (Hilde et al 1977). This could provide an explanation for the absence of magnitude 6 events in the bulk of the Pacific plate, and for the low *b*-values obtained by Bergman & Solomon (1980) in this plate.

Correlation with Volcanism?

The association of seismicity with small-scale bathymetric features such as seamounts is even more difficult to investigate because the mapping of seamounts in remote areas of the oceans is still very incomplete: Seismic swarms were responsible for the discovery of Macdonald Volcano in 1967 (Johnson 1970) and of the suspected volcanoes Rocard and Moua Pihaa in the Tahiti-Mehetia area (Talandier & Kuster 1976). Recent seismic swarms in the Society Islands also suggest volcanic activity on the flanks of Mehetia (Talandier 1981) and at Teahitia, only 65 km east of Tahiti (Talandier & Okal 1982). It is then possible, at least in principle, to suspect a similar phenomenon in the case of a sustained swarm of seismicity in an uncharted oceanic area, such as Regions A or C. Geophysicists can make use of two major tools to assert the volcanic vs tectonic origin of seismicity: knowledge of the geomorphology and bathymetry in the epicentral area, or analysis of the patterns of recorded seismicity, followed by comparison with well-documented volcanism, such as that of Kilauea and Loihi in Hawaii (Klein 1982). The former kind of information is nonexistent for most oceanic seismic foci, but large-scale bathymetry could be obtained at little cost from satellite data (Sailor & Okal 1983); the latter will require precise knowledge of the seismicity at low magnitudes, and in particular of the hypocentral depths.

The complexity of this issue can best be judged on the example of Region C: Sailor & Okal (1983) have rejected the possibility of a major volcanic edifice, a result also confirmed by an on-site survey (Francheteau, personal communication, 1981), which has, on the other hand, indicated topography on the order of 500 m and an unusually high level of low-magnitude, largely unexplained, seismic activity (Pascal, personal communication, 1981). It is clear that the concentration of seismic activity (which at the $m_b \geq 5.2$ level is compatible with release of tectonic, ridge-push type stresses) at Region C is presently not understood.

Conclusion: Our State of Ignorance, and Recommendations for Future Research Orientation

Despite the fundamental progress made over the past 12 years in our knowledge and understanding of oceanic intraplate seismicity, including the origin of the stresses released in the major events, at least one major problem remains unsolved: *What is the level of low-magnitude seismicity in the ocean basins, far away from plate boundaries, hotspot edifices, and other islands?* High-gain local networks have not been deployed permanently in the oceanic environment, with the exception of the French Polynesian array, whose existence has allowed the recognition of some 30 unsuspected epicenters. Consequently, we have no estimate of the seismicity below magnitude 4.7 for most parts of the ocean floor. For example, no seismicity other than event 11 is known east of Kerguelen in the Antarctic plate, all the way to Balleny Islands, and the question of whether this pattern of a "quiet" zone extends to lower magnitudes remains open. Instrumentation of more oceanic islands with high-gain permanent stations is thus a primordial step in furthering our knowledge of oceanic intraplate seismicity. Had not comparable networks existed on continents, the seismicity of such areas as New England would be practically undocumented.

OBSs have until now been deployed mostly in areas of major geophysical interest, such as mid-oceanic ridges, continental margins, fracture zones, or epicentral areas of major earthquakes, and always for limited periods of time. The relatively low level of intraplate seismicity is such that only a permanently deployed network can retrieve significant information about it. It would be extremely useful to develop an OBS system capable of permanent recording, and to run a prolonged OBS campaign in "average" areas on the floor of an oceanic basin, in order to get a clear picture of the level of the low-magnitude ($m_b = 2$–4) seismicity of the interior of the plate, independently of any disturbing tectonic influence. This potential seismicity presently escapes detection.

Assuming that we can improve our knowledge of oceanic intraplate seismicity down to this level of magnitudes, its interpretation will be possible only in the context of other aspects of marine geophysics, in particular of the small-scale bathymetry of the ocean floor. This is a second, and formidable, unknown; but we have seen that the large-scale bathymetry, available now at little cost from satellite data, does not always provide a clue as to the siting of the seismicity in the plate. Only detailed knowledge of the geomorphology and of the tectonic history of such areas as Region C will allow a better understanding of the relationships between oceanic seismicity, volcanism, and possibly other forms of active tectonism of a lesser intensity, which may be involved along weak lines, as suggested by the microseismicity recorded at Region C.

ACKNOWLEDGMENTS

Many of the ideas expressed in this paper evolved from several years of stimulating collaboration with Jacques Talandier and Tom Jordan. Discussions and preprint exchanges with Seth Stein are also gratefully acknowledged [even though our studies in Stein & Okal (1978) no longer qualify as intraplate in the strict sense!]. I am grateful to the staff of Hawaii Volcano Observatory for their hospitality and many discussions on volcanic seismicity, and to Sean Solomon for permission to adapt Figure 1 from Bergman & Solomon (1980). I thank Jean Francheteau and Georges Pascal for access to preliminary data obtained at Region C. This research was supported by the Office of Naval Research, under Contract N00014-79-C-0292, and also in part by the National Science Foundation, under Grant EAR-81-06106.

Literature Cited

Bergman, E. A., Solomon, S. C. 1980. Oceanic intraplate earthquakes: implications for local and regional intraplate stress. *J. Geophys. Res.* 85:5389–5410

Chen, T., Forsyth, D. W. 1978. A detailed study of two earthquakes seaward of the Tonga Trench: Implications for mechanical behavior of the lithosphere. *J. Geophys. Res.* 83:4995–5003

Creaven, C. T., Kanamori, H., Fujita, K. 1979. A large intraplate event near the Scotia arc. *Eos, Trans. Am. Geophys. Union* 60:894 (Abstr.)

Forsyth, D. W. 1973. Compressive stress between two mid-oceanic ridges. *Nature* 243:78–79

Forsyth, D. W., Uyeda, S. 1975. On the relative importance of the driving forces of plate motion. *Geophys. J. R. Astron. Soc.* 43:163–200

Farmer, R. A., Fujita, K., Stein, S. 1982. Seismicity and tectonics of the Scotia arc area. *Eos, Trans. Am. Geophys. Union* 63:440 (Abstr.)

Frank, F. C. 1972. Plate tectonics, the analogy with glacier flow and isostasy. In *Flow and Fracture of Rocks*, ed. H. C. Heard. *Geophys. Monogr. Ser.* 16:285–92

Geller, R. J. 1976. Scaling relations for earthquake source parameters and magnitudes. *Bull. Seismol. Soc. Am.* 66:1501–23

Gutenberg, B., Richter, C. F. 1941. Seismicity of the Earth. *Geol. Soc. Am. Spec. Pap. 34*

Harkrider, D. G., Okal, E. A. 1982. Propagation and generation of high-frequency sediment-controlled Rayleigh modes following shallow earthquakes in the southcentral Pacific. *Eos, Trans. Am. Geophys. Union* 63:1025 (Abstr.)

Herron, E. M. 1972. Sea-floor spreading and the Cenozoic history of the eastcentral Pacific. *Geol. Soc. Am. Bull.* 83:1671–92

Hilde, T. W. C., Uyeda, S., Kroenke, L. 1977. Evolution of the western Pacific and its margin. *Tectonophysics* 38:145–65

Johnson, R. H. 1970. Active submarine volcanism in the Austral Islands. *Science* 167:977–79

Jordan, T. H., Sverdrup, K. A. 1981. Teleseismic location techniques and their application to earthquake clusters in the southcentral Pacific. *Bull. Seismol. Soc. Am.* 71:1105–30

Kafka, A. L., Weidner, D. J. 1979. The focal mechanisms and depths of small earthquakes as determined from Rayleigh-wave radiation patterns. *Bull. Seismol. Soc. Am.* 69:1379–90

Kanamori, H. 1977. The energy release in great earthquakes. *J. Geophys. Res.* 82:2981–87

Klein, F. W. 1982. Earthquakes at Loihi submarine volcano and the Hawaiian hotspot. *J. Geophys. Res.* 87:7719–26

Liu, H.-L., Kanamori, H. 1980. Determination of source parameters of mid-plate earthquakes from the waveforms of body waves. *Bull. Seismol. Soc. Am.* 70:1989–2004

McGarr, A., Gay, N. C. 1978. State of stress in the Earth's crust. *Ann. Rev. Earth Planet. Sci.* 6:405–36

McKenzie, D. P. 1969. The relationship between fault plane solutions for earthquakes and the direction of the principal stresses. *Bull. Seismol. Soc. Am.* 59:591–601

McNutt, M. K., Menard, H. W. 1978. Lithospheric flexure and uplifted atolls. *J. Geophys. Res.* 83:1206–12

Mendiguren, J. A. 1971. Focal mechanism of a shock in the middle of the Nazca plate. *J. Geophys. Res.* 76:3861–79

Mendiguren, J. A., Richter, F. M. 1978. On the origin of compressional intraplate stress in South America. *Phys. Earth Planet. Inter.* 16:318–26

Minster, J. B., Jordan, T. H. 1978. Present-day plate motions. *J. Geophys. Res.* 83:5331–54

Mogi, K. 1963. Some discussions on aftershocks, foreshocks, and earthquake swarms, the fracture of a semi-infinite body caused by an inner stress origin and its relation to the earthquake phenomena, 3. *Bull. Earthquake Res. Inst. Tokyo Univ.* 41:615–58

Okal, E. A. 1980. The Bellingshausen Sea earthquake of February 5, 1977: evidence for ridge-generated compression in the

Antarctic plate. *Earth Planet. Sci. Lett.* 46:306–10

Okal, E. A. 1981. Intraplate seismicity of Antarctica and tectonic implications. *Earth Planet. Sci. Lett.* 54:397–406

Okal, E. A., Bergeal, J.-M. 1983. Mapping the Miocene Farallon ridge jump on the Pacific plate. *Earth Planet. Sci. Lett.* In press

Okal, E. A., Lay, T. 1982. The Gilbert Islands earthquake swarm, 1981–1982. *Eos, Trans. Am. Geophys. Union* 63:1138–39 (Abstr.)

Okal, E. A., Stewart, L. M. 1982. Slow earthquakes along oceanic fracture zones: evidence for asthenospheric flow away from hotspots? *Earth Planet. Sci. Lett.* 57:75–87

Okal, E. A., Talandier, J. 1981. Dispersion of one-second Rayleigh modes through oceanic sediments following shallow earthquakes in the southcentral Pacific. In *Bottom-Interacting Ocean Acoustics*, ed. W. A. Kuperman, F. B. Jensen, *NATO Ser. 4*, 5:345–58. New York: Plenum

Okal, E. A., Talandier, J., Sverdrup, K. A., Jordan, T. H. 1980. Seismicity and tectonic stress in the southcentral Pacific. *J. Geophys. Res.* 85:6479–95

Raleigh, C. B., Healy, H. J., Bredehoeft, D. J. 1972. Faulting and crustal stress at Rangely, Colorado. In *Flow and Fracture of Rocks*, ed. H. C. Heard. *Geophys. Monogr. Ser.* 16:275–84

Richardson, R. M., Solomon, S. C. 1977. Apparent stress and stress drop for intraplate earthquakes and tectonic stress in the plates. *Pure Appl. Geophys.* 115:317–31

Richardson, R. M., Solomon, S. C., Sleep, N. H. 1979. Tectonic stress in the plates. *Rev. Geophys. Space Phys.* 17:981–1008

Sailor, R. V., Okal, E. A. 1983. Applications of SEASAT altimeter data in seismotectonic studies of the southcentral Pacific. *J. Geophys. Res.* 88: In press

Sbar, M. L., Sykes, L. R. 1973. Contemporary compressive stress and seismicity in eastern North America: an example of intraplate tectonics. *Geol. Soc. Am. Bull.* 84:1861–82

Sbar, M. L., Sykes, L. R. 1977. Seismicity and lithospheric stress in New York and adjacent areas. *J. Geophys. Res.* 82:5771–86

Stein, S. 1978. An earthquake swarm on the Chagos-Laccadive Ridge and its tectonic implications. *Geophys. J. R. Astron. Soc.* 55:577–88

Stein, S. 1979. Intraplate seismicity on bathymetric features: The 1968 Emperor Trough earthquake. *J. Geophys. Res.* 84:4763–68

Stein, S., Okal, E. A. 1978. Seismicity and tectonics of the Ninetyeast Ridge area: evidence for internal deformation of the Indian plate. *J. Geophys. Res.* 83:2233–45

Stein, S., Sleep, N. H., Geller, R. J., Wang, S.-C., Kroeger, G. C. 1979. Earthquakes along the passive margin of eastern Canada. *Geophys. Res. Lett.* 6:537–40

Stewart, G. S., Helmberger, D. V. 1981. The Bermuda earthquake of March 24, 1978: A significant oceanic intraplate event. *J. Geophys. Res.* 86:7027–36

Sverdrup, K. A. 1981. *Seismotectonic studies in the Pacific Ocean Basin.* PhD thesis. Univ. Calif., San Diego. 436 pp.

Sykes, L. R. 1978. Intraplate seismicity, reactivation of preexisting zones of weakness, alkaline magmatism and other tectonism postdating continental fragmentation. *Rev. Geophys. Space Phys.* 16:621–88

Sykes, L. R., Sbar, M. L. 1973. Intraplate earthquakes, lithospheric stresses and the driving mechanism of plate tectonics. *Nature* 245:298–302

Sykes, L. R., Sbar, M. L. 1974. Focal mechanism solutions of intraplate earthquakes and stresses in the lithosphere. In *Geodynamics of Iceland and the North Atlantic Area*, ed. Kristjansson, pp. 207–24. Dordrecht: Reidel

Talandier, J. 1981. 1981 Seismic crisis under Mehetia Island, French Polynesia. *Eos, Trans. Am. Geophys. Union* 62:949 (Abstr.)

Talandier, J., Kuster, G. T. 1976. Seismicity and submarine volcanic activity in French Polynesia. *J. Geophys. Res.* 81:936–48

Talandier, J., Okal, E. A. 1982. 1982 Seismic crisis at Teahitia Seamount, French Polynesia. *Eos, Trans. Am. Geophys. Union* 63:1092 (Abstr.)

Tréhu, A. M., Sclater, J. G., Nábělek, J. 1976. The depth and thickness of the ocean crust and its dependence upon age. *Bull. Soc. Géol. Fr. Sér.* 7 18:917–30

Turcotte, D. L., Oxburgh, E. R. 1973. Mid-plate tectonics. *Nature* 244:337–39

Wang, S.-C., Geller, R. J., Stein, S., Taylor, B. 1979. An intraplate thrust earthquake in the South China Sea. *J. Geophys. Res.* 84:5627–32

Watts, A. B., Bodine, J. H., Steckler, M. S. 1978. Observations of flexure and the state of stress in the oceanic lithosphere. *J. Geophys. Res.* 85:6369–76

Weissel, J. K., Anderson, R. N. 1978. Is there a Caroline plate? *Earth Planet. Sci. Lett.* 41:143–58

Wesnousky, S. G., Scholz, C. H. 1980. The craton: its effect on the distribution of seismicity and stress in North America. *Earth Planet. Sci. Lett.* 48:348–55

Wiens, D. A., Stein, S. 1983. Age dependence of oceanic intraplate seismicity and implications for lithospheric evolution. *J. Geophys. Res.* 88: In press

Ann. Rev. Earth Planet. Sci. 1983. 11: 215–240

CREEP DEFORMATION OF ICE

Johannes Weertman
Department of Materials Science and Engineering and
Department of Geological Sciences, Northwestern University,
Evanston, Illinois 60201

INTRODUCTION

Knowledge of the creep properties of ice often is needed in field and theoretical studies of the flow and deformation of glaciers, ice shelves, and ice sheets. Recently many of the icy moons in the solar system have become much better known as a result of the successful deep-space probe missions. An understanding of the physical processes that take place within them requires information about the plastic deformation behavior of ice. Unlike ice within glaciers and ice sheets, the crystal structure of ice deep within an icy satellite may be one of the high-pressure polymorphs rather than that of the more familiar hexagonal ice Ih. At the present time, there exists a great deal of creep data on ordinary ice Ih but very little on other forms of ice.

The aim of this article is to present our basic knowledge on the creep properties of ice. This includes information derived from laboratory experiments as well as gleaned from field measurements made on glaciers, ice sheets, and ice shelves. No attempt is made here to review in a complete manner all the literature on this subject. However, new results from recent papers on ice creep that are not covered in earlier and more complete reviews are included. The reader is referred to the reviews of Hooke (1981; called review H hereafter), Paterson (1977; called review P), Glen (1974, 1975; called review G), and Weertman (1973; called review W) for a more complete coverage of the earlier literature. Review H compares experimental results on ice creep with field measurements on glaciers and other large bodies of ice. Review P summarizes closure data on boreholes in the Antarctic and Greenland ice sheets and Canadian ice caps. Review G covers most aspects of the mechanics and physics of ice. A tutorial-type review on polycrystalline ice has been given by Mellor (1980; called review

0084–6597/83/0515–0215$02.00

M). Hooke et al (1980) have reviewed the research that still needs to be carried out on the mechanical properties of ice, rather than what has been accomplished up to now. The book by Hobbs (1974) on the physics of ice also covers ice creep. Ice flow law determinations made from measurements of the increase with time of the tilt of vertical boreholes in glaciers are summarized by Raymond (1980) and by Paterson (1981). Goodman et al (1981) have reviewed creep data and creep mechanisms of ice for the purpose of making creep diagrams (deformation maps).

Theoretical explanations of the creep process are given in this article, but in only limited detail. Obviously, the degree of risk involved in extrapolating creep results outside the stress-strain-temperature-time region in which measurements actually were made is a function of how well the creep processes are understood.

Unless otherwise stated, the word "ice" in this review means ordinary ice Ih that exists under ambient pressures and whose hexagonal crystalline structure gives rise to the classic form of snowflakes. The hexagonal crystal form is also the origin of the large anisotropy of the creep properties of ice. Slip occurs readily across the basal planes (normal to the crystal *c*-axis) of an ice crystal. If stress of magnitude sufficient to produce plastic deformation is applied to an ice single crystal, the ice crystal is "softer" if the applied stress has a shear component resolved across the basal planes that is relatively large. If the resolved basal shear stress is relatively small, the ice crystal is "harder" and the amount of plastic deformation in a given time period is smaller. In the former situation the ice crystal is in an "easy" glide orientation in which lattice dislocations are moved easily across the basal slip planes to produce slip. In the latter situation the ice crystal is in a "hard" glide orientation because deformation requires slip on nonbasal planes. The deformation of the grains of polycrystalline ice generally requires a hard-glide component and thus polycrystalline ice usually is "hard" too.

Our review is restricted to the creep properties of ice of grain size sufficiently large (>0.5 mm) that deformation occurs primarily by dislocation motion. Only the creep observed at moderate to large strains (greater than, say, 0.1%) is considered. Excluded are small-strain anelastic deformation and creep produced by the mechanisms of the creep of very fine grain material.

TIME DEPENDENCE OF CREEP STRAIN FOR CONSTANT STRESS TESTS

Two types of curves of creep strain ε versus time t are found for ice specimens made to creep by application of a constant stress. These are shown in Figure 1. Both polycrystalline ice and ice single crystals in a hard orientation usually exhibit a creep rate $\dot{\varepsilon} \equiv d\varepsilon/dt$, which decelerates as the creep strain increases (see reviews G and W). At larger creep strains the creep rate approaches a constant, or quasi-constant, steady-state value. The creep rate of ice single crystals that are in an easy-glide orientation usually accelerates toward the steady-state creep rate that exists at the larger plastic strains (see reviews G and W).

The two types of creep curves shown in Figure 1 also describe the creep behavior of other crystalline material. Germanium and silicon (Alexander & Haasen 1968), as well as some alloys that are called class I alloys by Sherby & Burke (1967) and class A (Alloy type) by Yavari et al (1981), have creep curves that accelerate into the steady-state region. (Class I/A alloys have a power-law creep exponent n with a value $n \simeq 3$.) The creep curves of pure metals and alloys called class II by Sherby & Burke [pure metal and class II alloys are called class M (Metal type) by Yavari et al] always follow a decelerating approach of the creep rate into steady state. But some class I/A alloys also show this latter behavior. (Class II/M metals and alloys have a power-law exponent of $n \simeq 4$ to 5.) In the case of rocks and rock minerals, our knowledge of the behavior in the transient-creep region is somewhat uncertain because of the experimental difficulties in applying pure hydrostatic pressure in a test run to prevent microcracking of a specimen. Data reviewed by Carter & Kirby (1978) showed decelerating transient creep.

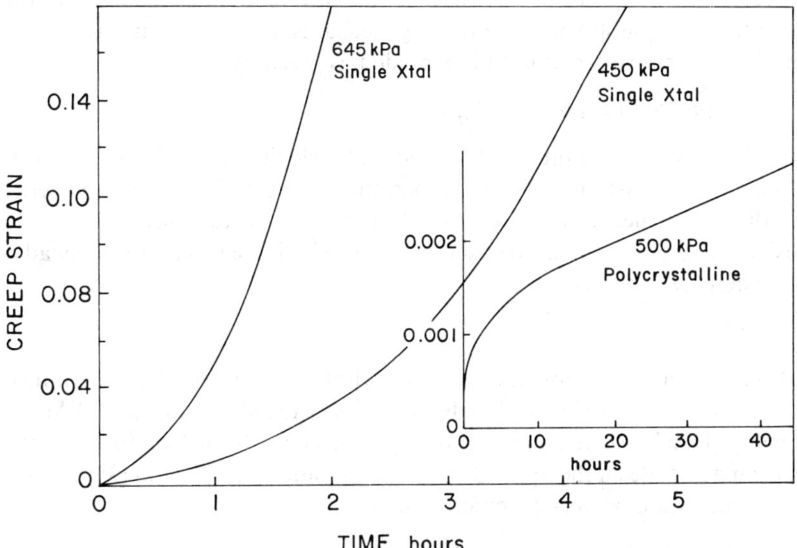

Figure 1 Creep strain versus time under constant stress for polycrystalline ice at −7°C (after Duval & Le Gac 1980) and easy-glide single crystals at ≃ −10.5°C (after Ramseier 1972).

However, studies on olivine (Durham et al 1977, 1979), which has $n \simeq 3$, indicate that the transient creep could be of the accelerating type.

Ice is somewhat unusual in that both accelerating and decelerating transient creep are found for it. But metal alloys that are on a borderline between class I/A and class II/M behavior also exhibit both types of transient-creep behavior (Oikawa et al 1977). [In the alloys examined by Oikawa et al, as well as those looked at by Yavari et al (1981), a change from $n \simeq 3$ to 5 could be accomplished by altering the magnitude of the applied stress. Vagarali & Langdon (1982 and private communication) have shown even more clearly for a hcp Mg-Al alloy the switch from accelerating to decelerating transient creep, the increase of n from 3 to 4, and the existence (or nonexistence) of the upper yield point with increase of stress.]

There is a simple qualitative explanation for the creep curves of Figure 1. Material in which creep starts out with an accelerating rate presumably are those in which the initial dislocation density is small and in which dislocation glide motion is hindered by any number of drag mechanisms. Dislocation multiplication, which requires dislocation motion, is also hindered. The creep rate is given by the equation

$$\dot{\varepsilon} = \alpha \rho b v, \tag{1}$$

where v is the average dislocation velocity, b is the length of the burgers vector, ρ is the dislocation density, and α is a dimensionless geometric factor whose exact value, usually of order of magnitude one, depends upon the orientation of the slip planes. For drag mechanisms the dislocation velocity, until some critical stress level is exceeded, is given by

$$v = v_0(\sigma b^3/kT) \exp(-Q/kT), \tag{2}$$

where v_0 is a constant of dimension of velocity, σ is the stress, k is Boltzmann's constant, T is the temperature, and Q is the activation energy of the drag mechanism. As the dislocations move, more and more dislocations are created. The dislocation density increases up to the steady-state density given by

$$\rho = \beta\sigma^2/b^2\mu^2, \tag{3}$$

where β is a dimensionless constant whose experimentally determined value (Weertman 1975) is of order 1, and μ is the shear modulus. [At the density given by (3), the internal stress produced by the dislocations is of the same magnitude as the applied stress.] Combining (1), (2), and (3) gives for the steady-state power-law creep rate

$$\dot{\varepsilon} = \dot{\varepsilon}_0(\sigma/\mu)^n(\mu b^3/kT) \exp(-Q/kT), \tag{4}$$

where $n = 3$ and $\dot{\varepsilon}_0 = \alpha\beta v_0/b$. This equation is that of a class I/A alloy.

An explanation of the decelerating creep rate seen in class II/M material is that the dislocations experience a smaller resistive drag to their glide motion compared with that of their climb motion. Thus when a stress is applied, dislocations multiply almost instantaneously to give a density of the order of (3). A large amount of glide motion occurs and produces an instantaneous plastic strain. Because the dislocation morphology that exists after the "instantaneous" plastic deformation presumably is not the most stable dislocation arrangement possible, the dislocations start to rearrange themselves into a more stable pattern. The creep rate slows down until steady-state creep is attained. The creep rate presumably is controlled by dislocation climb, which in turn is controlled by self-diffusion. At high stresses and/or lower temperatures and moderate stresses, it is likely to be controlled by dislocation pipe self-diffusion.

One difficulty remains to be resolved for this simple picture of climb-controlled creep for class II/M material: If no ad hoc assumptions are made in the dislocation models used to calculate the creep rate, the power-law exponent is equal to $n = 3$ (Weertman 1975) when climb is controlled by bulk diffusion, rather than the experimentally observed $n = 4$ to 5. [When pipe diffusion, i.e. diffusion down dislocation cores, controls the climb, the value of n is increased by 2 to $n = 5$ (Robinson & Sherby 1969, Evans & Knowles 1977, Langdon 1978, Langdon & Mohamed 1978, Sherby & Weertman 1979, Spingarn et al 1979). However, experimentally $n \simeq 6$ to 7 in this situation.] An experimental determination that $n = 3$ is, therefore, no guarantee that a dislocation drag mechanism controls the creep rate. But a value of $n = 3$, coupled with the observation that the creep rate accelerates during the transient period, should give strong support that such a mechanism does control the creep rate. Because the steady-state creep of ice single crystals in an easy-glide orientation has an exponent $\simeq 2$–3.9 (see Table 1), and because of results found on the motion of individual dislocations that are considered later, it appears likely that a dislocation drag process does control the creep of these crystals.

The implication of this explanation of accelerating creep is that if the strength of the drag process could be reduced sufficiently the transient behavior of easy-glide crystals should change from accelerating to decelerating behavior. Doping ice crystals does increase the dislocation velocity and decrease the drag force (Mai et al 1978). Riley et al (1978) have made the interesting observation that NaCl-doped easy-glide single crystals have a decelerating transient creep. (They did not determine the value of n because the steady-state creep region was not reached in their experiments.)

The value of n for polycrystalline ice and ice single crystals of hard orientation also is approximately equal to three (see Table 1). The rate-

Table 1 Power law exponent *n* and creep activation energy *Q*

n	*Q* (kJ $mole^{-1}$)	Comment	Reference
Polycrystalline Ice below $-10°C$			
3.1	59.9	—	Ramseier (1972)
3	59.9	Frazil ice	Ramseier (1972)
2.5	59.9	Columnar ice	Ramseier (1972)
3.1	77.9	—	Barnes et al (1971)
3.1–3.6	83.8	—	Steinemann (1958) with reanalysis by Barnes et al (1971)
—	50.3	Friction exp.	Bowden & Tabor (1964)
3.5	41.9	—	Bender et al (1961)
3.2	41.9	D_2O ice	Bender et al (1961)
3.5	44.8	—	Mellor & Smith (1967)
—	68.7	—	Mellor & Testa (1969a)
—	64.9	Columnar ice	Gold (1973)
2.9–3.05	67	Columnar ice	Sinha (1978, 1982)
3	—	Creep rate of floating ice shelves	Thomas (1971)
3	54	Secondary creep rate determined from closure rates in boreholes	Paterson (1977)
3	$\simeq 74$	Our estimate of *Q* from their data	Russell-Head & Budd (1979)
—	78	Antarctic ice	Duval & Le Gac (1982)
Polycrystalline Ice above $-10°C$			
3.2	134	—	Glen (1955)
2.8–3.2	134	—	Steinemann (1958) with reanalysis by Barnes et al (1971)
3.2	122	—	Barnes et al (1971)
—	$\simeq 168$	—	Mellor & Testa (1969a) with review W analysis
3	—	Glacier tunnel closing	Nye (1953)
3	$\simeq 200$	Our estimate of *Q* from their data	Russell-Head & Budd (1979)
3	—	Triaxial stress tests	Duval (1976)
Single crystals in easy glide			
2.5	—	—	Butkovitch & Landauer (1958, 1959)
4	—	—	Glen (1952)
1.5–3.9	—	—	Steinemann (1954)
1.6	66.2	Bend test	Higashi et al (1965)
—	65.4	$-50°C$ to $-10°C$	Jones & Glen (1968)
—	39.8	$-90°C$ to $-50°C$	Jones & Glen (1968)
4	—	$-50°C$	Glen & Jones (1967)

Table 1 *(continued)*

n	Q (kJ mole^{-1})	Comment	Reference
$\simeq 2$	59.9	—	Ramseier (1972)
—	—	−20°C to −0.2°C No activation energy change, value not stated	Gold (1977)
1.9	78	−30°C to −4°C	Homer & Glen (1978)
2.9	75	Bicrystals −30°C to −4°C	Homer & Glen (1978)
2	70	−20°C to −0.2°C	Jones & Brunet (1978)
1.8–2.6	81	−5°C to −7°C	Nakamura (1978)
Single crystals in hard glide			
3	—	—	Butkovitch & Landauer (1958, 1959)
—	57.0	—	Ramseier (1967a,b)
—	69.1	—	Mellor & Testa (1969a)

controlling process for this material is not certain. The fact that $n \simeq 3$ lends support to a drag mechanism but, as already mentioned, this value does not necessarily rule out a dislocation climb control mechanism.

CONSTANT STRAIN RATE TESTS

If ice is deformed at a constant strain rate (rather than under a constant stress) a curve of stress versus strain (rather than strain versus time) is obtained. Figure 2 shows the type of stress-strain curve that is found for single crystals of ice in an easy-glide orientation and in some tests on polycrystalline ice (reviews G and W). As the strain is increased, the stress reaches a maximum, "upper yield point" value, falls off, and finally reaches a steady-state value. In the case of ice crystals of hard orientation and in other polycrystalline ice tests, the stress-strain curve obtained in a constant strain rate test is different (reviews G and M, Higashi 1967, Hawkes & Mellor 1972, Shoji 1978). The curve is essentially the same as a metal tested at constant strain rate at a high temperature. The stress increases with strain and reaches a quasi-saturation, quasi-steady-state level. (A slight decrease in stress in high strain rate tests, which seems to be produced by cracking, may occur at large strains.) It should be noted that once the stress reaches a steady-state value the constant strain rate test is the equivalent of a constant stress test when the creep rate has attained a steady-state value (Mellor & Cole 1982).

A stress-strain curve of the type shown in Figure 2 is readily explained (see reviews G and W and the literature quoted there) if dislocation motion, and thus creep, is controlled by a drag mechanism. In the initial stage of a test there are only a few dislocations present. The strain is almost all elastic. The stress must increase as the elastic strain is increased when the total strain rate is held constant. As the stress increases, the dislocations that are present move faster, multiply more rapidly, and add a greater plastic component to the total strain. Eventually enough dislocations exist that all of the strain rate can be produced by their movement at a smaller average velocity and at a lower stress. Hence the stress falls off to a much lower level.

The shapes of the curves of easy glide for the ice single crystals of Figures 1 and 2 thus can both be explained qualitatively if drag controls the glide motion of dislocations and the creep rate. Similarly, for the case of hard-glide crystals and some polycrystalline ice the fact that the creep curves show deceleration and the stress-strain curve shows no marked upper yield point might imply that dislocation climb, rather than drag, controls the creep rate. If so, it is puzzling why it is so much more difficult to make ice slip on nonbasal planes. We leave this as an exercise for the reader to solve; a second part of the exercise is given when borehole closure results are considered. A partial answer is given later.

Stress-strain curves have also been obtained on columnar ice under application of a stress that increases at a constant stress rate (Sinha 1982). No upper yield point is observed. The stress increases monotonically with

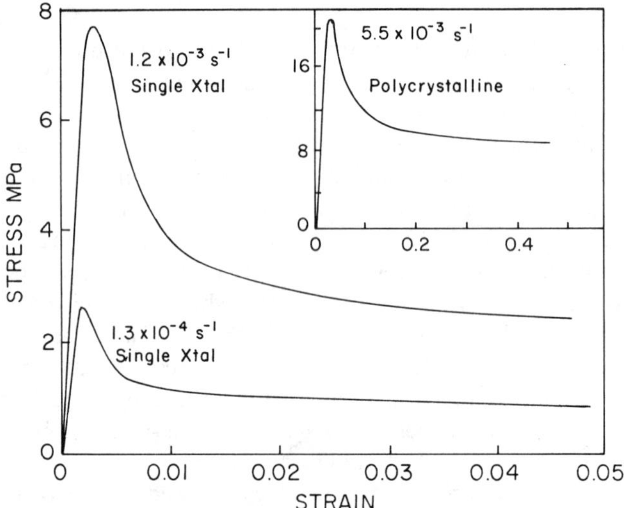

Figure 2 Stress versus strain at constant strain rate for polycrystalline ice at −11°C (after Jones 1982) and easy-glide ice single crystals at −10°C (after Ramseier 1972).

increasing strain but with a decreasing work-hardening coefficient (defined to be the slope $d\sigma/d\varepsilon$ of the stress-strain curve). Cracking within the ice makes the results at later stages ambiguous.

STRESS DEPENDENCE OF STEADY-STATE CREEP RATE

At moderate stresses, as the reader must have deduced from the discussion of a previous section, the steady-state creep rate of ice is given by the power-law creep equation

$$\dot{\varepsilon} = B\sigma^n \exp(-Q/kT). \tag{5}$$

Here B is a constant. In the glaciological literature this equation is known as Glen's creep law for ice because he showed first that it describes the creep of ice (Glen 1955).

In Table 1 are listed measured values of the power exponent n and the creep activation energy Q obtained from both laboratory experiments and field measurements. Figure 3 is a log-log plot of stress and creep rate for some of these data.

It should be noted in Table 1 and Figure 3 that the power exponent determined in most tests is approximately equal to 3. Also note that the activation energy for creep of single crystals up to the melting temperature (and above $-50°C$) is approximately 60 kJ mole^{-1}. This value is, within experimental error, the same as that found for the self-diffusion of hydrogen and of oxygen in ice (see reviews G and W). Hydrogen and oxygen diffuse at identical rates in ice. The equality of the creep and diffusion activation energies cannot be used as proof that the rate-controlling process is dislocation climb (which in turn is controlled by self-diffusion) because drag mechanisms can also involve the motion of hydrogen in the lattice.

Polycrystalline ice has the same activation energy as single crystal ice at temperatures below $-10°C$, but it has a value somewhat greater than twice this value at temperatures above $-10°C$. The larger activation energy of polycrystalline ice at the warmer temperatures generally is attributed to the softening that is produced by the significant amounts of recrystallization and grain growth that occur simultaneously with the creep deformation. The experiment of Jones & Brunet (1978) on single crystals was done for the express purpose of ascertaining whether there was a true change in activation energy in the dislocation deformation process. The activation energy was observed not to change. The ratio $\dot{\varepsilon}_{-1°C}/\dot{\varepsilon}_{-10°C}$ for polycrystalline ice is 7.6. For single crystal ice the same ratio is 2.9. Thus although the change in activation energy of polycrystalline ice is quite large, the maximum increase in the creep rate it produces is relatively small when it is considered that Q enters the creep rate in an exponential term.

It should also be noted in Figure 3 that the creep rate at a given stress is much larger for easy-glide crystals than it is for hard-glide crystals and for polycrystalline ice. The difference is a factor of about 500.

At large stress levels the data of Barnes et al (1971) show that the power

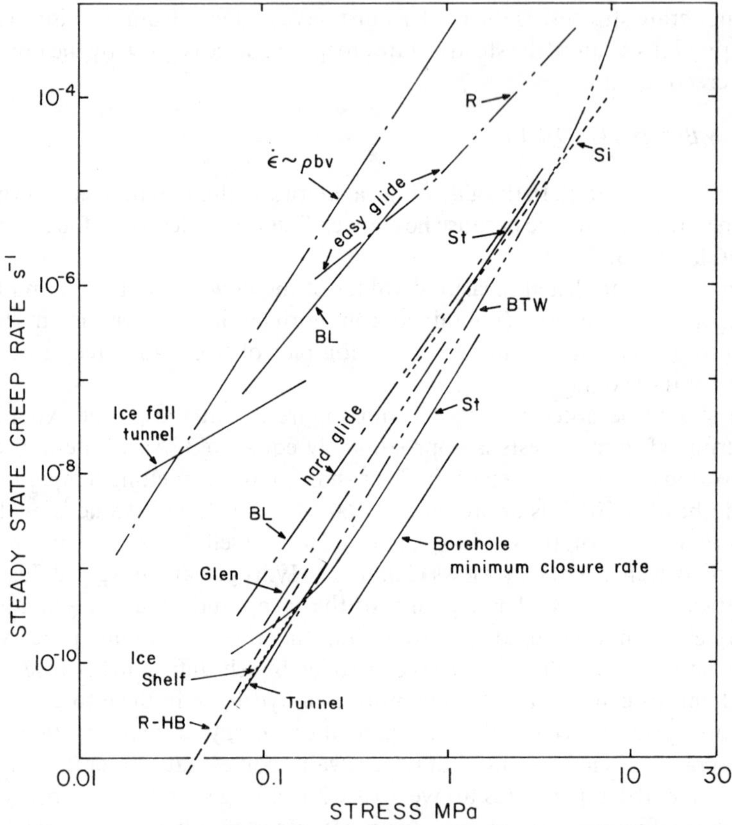

Figure 3 Log-log plot of steady-state or the minimum creep rate versus stress. Creep rate is normalized to $-10°C$, with $Q = 134$ kJ mole^{-1} for polycrystalline ice above $-10°C$ and 60 kJ mole^{-1} below this temperature or the activation energy reported by experimenter. Creep rate and stress are for, or have been changed to, the equivalent of a uniaxial tension or compression test. Data from the following sources. BL: Butkovitch & Landauer (1958, 1959); BTW: Barnes et al (1971); Glen: Glen (1955); St: Steinemann (1954); R: Ramseier (1972); Si: Sinha (1982); R-HB: Russell-Head & Budd (1979); ice shelf: Thomas (1971); glacier tunnel closing (Veslskautbre in Norway and Z'Mutt in Switzerland): Nye (1953); ice fall tunnel closing (Austerdalsbre in Norway): Glen (1956); borehole closing (Byrd in Antarctica, Site 2 in Greenland, and Meighen and Devon in Canada): Paterson (1977).

law breaks down, in agreement with observations on other crystalline material. Over a stress range that includes large stresses, creep data, including those of ice, follow the more general power sinh equation of Garofalo: $\dot{\varepsilon} \simeq [\sinh(\sigma/\sigma^*)]^n$, where σ^* is a constant. The sinh function reduces to its argument when its argument is small, and thus when $\sigma < \sigma^*$ the Garofalo equation reduces to the power-law equation.

Creep data on ice obtained at very low stresses and creep rates, which are not plotted in Figure 3 or tabulated in Table 1, generally indicate that the creep rate is proportional to stress (Mellor & Testa 1969b). The creep rates at a given stress level show scatter between different investigators. Because the creep strains are so small, these data should not be compared with those of Figure 3, which are obtained at much larger strains. In fact, the data of Russell-Head & Budd (1979) shown in Figure 3 obey a third-power relationship, whereas small-strain experiments in the lower part of the same stress region give a power closer to one. The Russell-Head & Budd work is remarkable in that they carried out experimental runs up to two years in time in order to obtain larger strains. [A linear relationship at low stresses and small strain can be accounted for with a grain boundary sliding mechanism, the Nabarro-Herring or Coble mechanism of mass diffusion of point defects between grain boundaries, and a dislocation mechanism using (1) and (2) if there is no significant change in the dislocation density upon and after application of the stress.]

Field Data

In Figure 3 are plotted data obtained from field measurements of the deformation of floating ice shelves, and from the closure of tunnels excavated in glaciers and boreholes drilled in ice sheets and ice caps. An unconfined floating ice shelf deforms in a very simple manner. It thins under its weight in a manner that is exactly equivalent to that of a laboratory sheet specimen that is pulled in two orthogonal directions in biaxial tension. [If the ice shelf is unconstrained in both horizontal directions, the two stresses of the biaxial test are equal to each other. If the ice shelf can spread in only one direction, the two biaxial stresses are not equal to each other. One has a deviatoric value (see a later section) equal to zero.]

The theory for the spreading of ice shelves is particularly simple. Thomas (1971) has used this theory, part of which he has helped develop, to analyze spreading-rate measurements reported from different ice shelves. In Figure 3 is plotted the creep rate stress relationship for ice, which is deduced from his calculations. It agrees reasonably well with the slower laboratory creep data.

Tunnel-closing data have been analyzed by Nye (1953), who developed the theory for tunnel closing. These data also are plotted in Figure 3, and agree with Thomas' results and the slower laboratory data. The Nye theory for tunnel closing applies equally well to the closure rate of boreholes drilled into ice. In Figure 3, borehole data from different sources, analyzed by Paterson (1977) using Nye's theory, lie considerably below all the other data. The only difference between the tunnels and the boreholes is that the tunnels were in ice near the melting point and the boreholes were in cold ice. A man or woman could enter the tunnels but a borehole could only accommodate one of their arms or legs. However, the closure curves of the boreholes exhibit a curious behavior. The closure, converted into a true creep rate, generally took place at an increasing rate, although a minimum rate occurred shortly after the start of closure. In other words, the polycrystalline ice around the borehole has a creep curve somewhat like that of Figure 1 for easy-glide ice single crystals. The creep rate plotted in Figure 3 for the borehole data is the minimum creep rate. The creep rate increased by factors of 3 to 10 during the course of closure. Hence the steady-state creep rate for the ice around boreholes should lie much closer to the tunnel-closing data and the ice shelf data. Why this polycrystalline ice from the Arctic and the Antarctic shows this acceleration is the second part of the exercise mentioned earlier.

Field data on the change with time of the vertical tilt of boreholes in glaciers, ice sheets, and ice caps also give information from which the creep properties of ice can be deduced. The creep within a glacier is primarily a shear deformation that increases from the upper surface to the lower surface. Thus a vertical borehole in time is tilted. The amount of tilt increases with depth. Extracting the ice creep properties from these data is made difficult partly because of multi-stress component effects on the creep rate. (These effects are discussed in a later section. The ice fall tunnel line in Figure 3 is an example of such an effect.) The uncertainty in the estimates of the shear-stress magnitude at different depths below the surface is also a problem.

Creep rate data determined from boreholes in glaciers [summarized by Raymond (1980) and by Paterson (1981)] have not been included in Figure 3. However, Table 2 lists values of the stress exponent n and values of the creep rate at 0.1 MPa (creep rate and stress appropriate for uniaxial tension or compression tests) for data reanalyzed by Raymond (1980). The creep rate is normalized again to $-10°C$ with use of the activation energy $Q = 134$ kJ mole^{-1}. The ice of the glaciers is close to the melting point. Also listed in Table 2, for comparison purposes, is the average creep rate at 0.1 MPa of the ice shelf, glacier tunnel closing, Russell-Head & Budd, and Barnes et al data of Figure 3. It can be seen that the exponent value given by the tilt results is approximately three. The absolute values of the creep rates are approximately the same as the smaller creep rate lines of

Table 2 Creep data for borehole tilt measurements after reanalysis by Raymond (1980)[a]

Glacier	n	$\dot{\varepsilon}$ ($s^{-1} \times 10^{10}$)	Source
Salmon (British Columbia, Canada)	2.8	1.3	Mathews (1959)
Athabasca (Alberta, Canada)	4	0.9	Paterson & Savage (1963)
Athabasca	3.6	0.9	Raymond (1973)
Blue (Washington, USA)	3.3	0.9	Shreve & Sharp (1970)
Figure 3 data (slower creep rate data)	3	1	—

[a] Creep rate normalized to $-10°C$, with $Q = 134$ kJ mole^{-1}. Creep rate is for the equivalent of a longitudinal strain rate under a uniaxial compressive or tensile stress of 0.1 MPa.

Figure 3. These lower creep rates probably are the best ones to use in theoretical calculations of the flow of glaciers and ice sheets. Goodman et al (1981) have used a creep diagram and the Byrd borehole data to successfully account for the horizontal ice movement measured at Byrd station, Antarctica.

Complications

The steady-state creep of ice actually is a bit more complicated and not as neat and tidy as has been indicated so far. Curves of creep rate versus creep strain at constant stress in the so-called steady-state region generally show a minimum in the creep rate, which is followed by somewhat higher strain rates at larger strains (review M, Mellor & Cole 1982). Budd (unpublished data) and Mellor (review M) have pointed out that the minimum creep strain under a fixed stress is reached at longer times, but at the same strain, as the test temperature is lowered. They also point out that some apparent conflicts in reported experimental and field results may simply be a consequence of not taking the position of the minimum creep rate into account. That is, one experiment may have been carried out on one side of the minimum and another on the other side. Steady-state creep rates should be regarded as quasi-steady-state creep rates. Figure 4 illustrates one rather extreme example of a large increase in creep rate of polycrystalline ice that follows the minimum creep rate. This increase is apparently produced by the development of a fabric, with the ice crystals becoming more favorably oriented for creep. The lower curve exhibits irregularities in the strain rate of another sample that is produced by simultaneous recrystallization. The increase in the activation energy of polycrystalline ice, which is produced presumably by recrystallization, has already been mentioned. Detailed studies of the development under stress of crystal orientation fabrics have been carried out by Kamb (1972), Budd (1972), and Wilson & Russell-Head (1982).

Because ice crystal orientation fabrics develop within ice sheets at different depths, the creep properties of ice in large ice masses can be a strong function of the creep flow itself. Russell-Head & Budd (1979) have analyzed the shear in creep from borehole tilt measurements in the Law Dome, Antarctica. They find that the shear-strain rates are anomalously large because of fabric development. At intermediate depths, a higher fraction of the ice has its c-axis oriented near a vertical direction. Analysis of the ice fabrics of the Camp Century and Dye 3, Greenland, borehole ice (Herron 1982, Herron et al 1982) and the Byrd Station, Antarctica, borehole ice by Gow & Williamson (1976) reveals a strong increase with depth of the fraction of ice with a near-vertical c-axis.

Creep experiments carried out by Shoji & Langway (1982) on the strongly oriented polycrystalline ice (Herron et al 1982) obtained from the Dye 3 borehole close to the bottom of the 2000-m core, as well as ice from

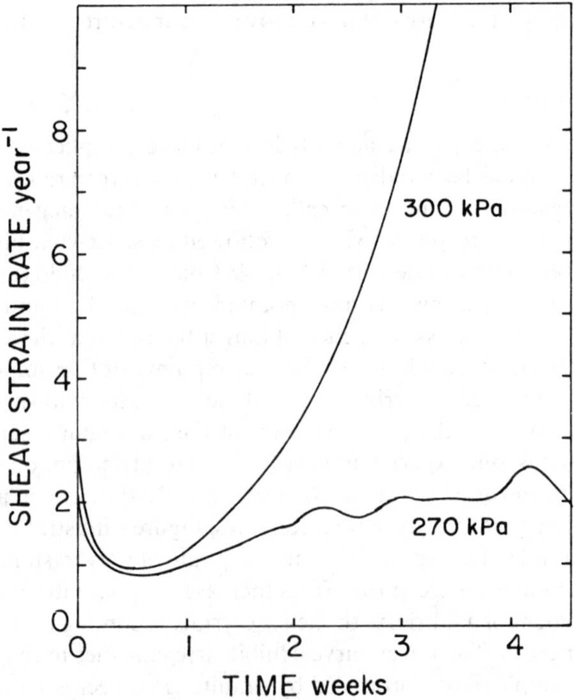

Figure 4 Strain rate versus time for polycrystalline ice at $-1°C$ under constant stress that is undergoing recrystallization and fabric development (after Duval 1981).

the Barnes Ice Cap margin (Baker 1981), gave creep rates that are a factor of four larger than those of the ice shelf, glacier tunnel, R-HB, and BTW lines in Figure 3 of randomly oriented polycrystalline ice. The experiments were carried out in a field laboratory immediately after the ice was brought up from the borehole. The same factor of four enhancement was found by Russell-Head & Budd (1979) in the creep rate of the strongly oriented ice that occurs at intermediate depths in the Law Dome boreholes. Russell-Head & Budd determined the creep rates from borehole tilt measurements as well as from laboratory experiments. It appears reasonable from these results to use an enhancement factor of four in ice-modeling work wherever strongly oriented ice is expected or is known to occur.

An implicit assumption in the plots of Figure 3 for polycrystalline ice is that there is no significant grain size effect on the creep rate. For large grain material a grain size effect is not expected. However, the experimental

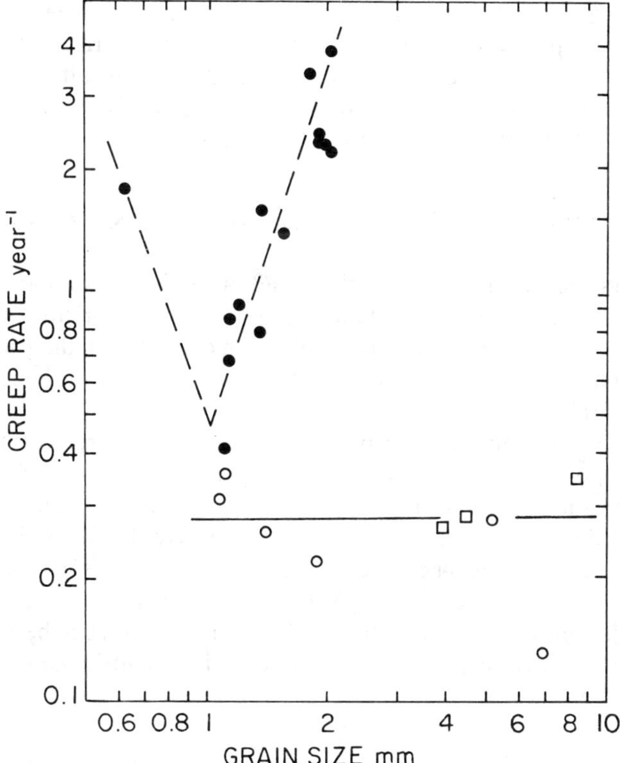

Figure 5 Creep rate versus grain size. ● : data of Baker (1978) for −7.1°C to −7.2°C at stress of 560 kPa; ○, □ : data of Duval & Le Gac (1980) for −7.0°C to −7.2°C at stress of about 500 kPa. The open squares are for Antarctic ice.

evidence on grain size effect is sparse and somewhat in conflict. Baker (1978, 1981) reported that there is a strong grain size dependence for grain sizes between about 0.6 mm to 6 mm. His data, some of which are plotted in Figure 5, show that the creep rate goes through a minimum at a grain size near 1 mm. However, Duval & Le Gac (1980) failed to find a grain size dependence for steady-state creep in their tests on specimens of grain sizes between about 1 mm and 10 mm, although a weak dependence was found for transient creep.

Anomalously large creep rates (up to almost an order of magnitude larger) have been measured in experiments carried out on polycrystalline ice whose temperature is extremely close to the melting temperature (within 0.01°C) (Barnes et al 1971, review H, Duval 1977, Budd, unpublished data). Duval (1977) looked at this problem by measuring the creep rate of ice that is close to its melting temperature as a function of the water content of a sample. He found that the creep rate tripled when the water content increased from 0.01% to 0.8%. The variation of creep rate with water content was approximately linear. Duval attributed his results to the fact that samples with different water contents had different salt contents and different melting temperatures. Impurity-doped ice is known to creep appreciably faster than pure ice (reviews G and W).

CREEP UNDER MULTI-COMPONENT STRESS

In glaciers, ice sheets, and ice shelves, and around boreholes and tunnels excavated in them, the stress state is almost always a multi-component stress state. Thus in any complete formulation of ice flow in these ice bodies a more general form of the creep equation than the one-component stress equation (4) is needed.

Suppose acting on isotropic, polycrystalline ice there are more than one stress components σ_{ij} that have nonnull values. More than one of the tensor strain rate components $\dot{\varepsilon}_{ij}$ in steady-state creep will have nonnull values too, except for the case of pure hydrostatic pressure. Nye (1953) obtained for an incompressible solid a generalized creep equation for multi-stress by using the analogue of plasticity theory's Lévy-von Mises relations. Since the material is considered to be isotropic, the creep equation can be a function of the three stress invariants. The first stress invariant is the hydrostatic pressure P given by

$$P = -(\sigma_{11}+\sigma_{22}+\sigma_{33})/3 = -\sigma_{ii}/3. \tag{6}$$

The effect of P is taken into account in (5) by rewriting this equation as

$$\dot{\varepsilon} = B\sigma^n \exp(-Q/kT) \exp(-VP/kT), \tag{7}$$

where V is the activation volume for creep, a quantity that can be determined experimentally.

Since creep does not take place under pure hydrostatic pressure in fully dense material, it is convenient in the above equations to use the deviatoric stress components σ'_{ij} defined by

$$\sigma'_{ij} = \sigma_{ij} + \delta_{ij}P, \tag{8}$$

where $\delta_{ij} = 0$ for $i \neq j$ and $\delta_{ij} = 1$ for $i = j$. (Note that $\sigma'_{ij} = 0$ for pure hydrostatic pressure.)

The second stress invariant τ can be defined by

$$2\tau^2 = \sigma'_{ij}\sigma'_{ij}, \tag{9}$$

where the right-hand side is summed over all the deviatoric stress components. (When $i \neq j$, σ'_{ij} and σ'_{ji} are each counted.) The analog of the Lévy-von Mises relations requires for any two stress components and the corresponding strain rate components that

$$\sigma'_{ij}/\sigma'_{mn} = \dot{\varepsilon}_{ij}/\dot{\varepsilon}_{mn}, \tag{10}$$

where $\dot{\varepsilon}_{ij}$ are the tensor strain rate components. [The effective strain rate $\dot{\varepsilon}_{\mathrm{eff}} = (\dot{\varepsilon}_{ij}\dot{\varepsilon}_{ij}/2)^{1/2}$ is also invariant. For an incompressible solid in plastic deformation, $\dot{\varepsilon}_{11} + \dot{\varepsilon}_{22} + \dot{\varepsilon}_{33} = 0$.]

Suppose in a uniaxial tension or compression test the relationship between longitudinal creep rate $\dot{\varepsilon}$ and the tensile or compressive stress is given by a general equation

$$\dot{\varepsilon} = f(\sigma), \tag{11}$$

where $f(\sigma)$ is an arbitrary function. Power-law creep is one special case. [If pressure $P \neq 0$, its effect is taken into account through the expression $\exp(-PV/kT)$ in the function f.] Equations (10) and (11) are satisfied if

$$\dot{\varepsilon}_{ij} = f(\sqrt{3}\tau)(\sqrt{3}/2\tau)\sigma'_{ij}. \tag{12}$$

(Note that $\sigma = \sqrt{3}\tau$ for the uniaxial stress situation. If σ_{11} is the tensile stress σ, then $\sigma'_{11} = 2\sigma_{11}/3$, $\sigma'_{22} = \sigma'_{33} = -\sigma_{11}/3$. Also, $\dot{\varepsilon} = \dot{\varepsilon}_{11} = -2\dot{\varepsilon}_{22} = -2\dot{\varepsilon}_{33}$.) Equation (12) reduces to (11) for uniaxial tensile creep, and the transverse strain rate components have their correct values, too. Thus (12) is a reasonable generalization for creep under multi-component stress for isotropic material, and has been used extensively in analyses of glacier flow problems. For power-law creep the function f is

$$f(\sigma) = B\exp(-Q/kT)\exp(-PV/kT)\sigma^n. \tag{13}$$

The third stress invariant, which is $\sum\sigma'_{ij}\sigma'_{jk}\sigma'_{ki}/3$, has been considered for equations of multi-component creep (Glen 1958). It has not been clearly demonstrated that it is necessary to include this invariant in the creep equation for isotropic polycrystalline ice (see review G). Creep experiments of Duval (1976) for polycrystalline ice under triaxial stress conditions strongly support Nye's use of the second stress invariant. However, the third stress invariant appears to be needed to account for the creep properties of anisotropic polycrystalline ice (Budd, private conversation, Lile 1978).

Nye (1953) used the above equations to calculate the expected closure rate s (s = radial wall velocity divided by wall radius) of round tunnels and boreholes in ice bodies. For the case of power-law creep with $n = 3$, Nye's results reduce to

$$s = B \exp(-Q/kT) \exp(-PV/kT)P^3/6, \tag{14}$$

where P is the overburden pressure. The value of τ given by (9) is $\tau = -P/3$. Since $\sigma = \sqrt{3}\tau$ in a uniaxial stress, it is necessary in comparing uniaxial creep date with closure data to set $\sigma = -P/\sqrt{3}$ and $\dot{\varepsilon} = -2s/\sqrt{3}$. This is done in Figure 3.

A very striking example of the effect that one stress component has on the strain rate of a different component is the closure rate, measured by Glen (1956), of a tunnel created in an ice fall in one of the Norwegian glaciers. When a tunnel is dug into an ordinary glacier to depths in which the dominant stress is primarily the hydrostatic pressure P (equal to the ice overburden), the closure rate according to the theory of Nye (1953) is equal to CP^n, where n is the power-law exponent and C is a constant. However, large stress components other than hydrostatic pressure exist in an ice fall. Glen estimated that a compressive stress $\bar{\sigma}$ of the order of 300 kPa existed in the vicinity of his tunnel. This stress is larger in magnitude than hydrostatic pressure. In such a situation it is reasonable to expect, from the theory of creep under multi-stress components of this section, that the closure rate will be of the order of $C\bar{\sigma}^{(n-1)}P$. Thus the closure rate is linearized with respect to P and the closure rate is increased. In Figure 3 are shown Glen's data, which are treated as if $\bar{\sigma}$ were equal to zero. It can be seen that the closure rate is indeed approximately proportional to P, and is considerably larger than the usual closure rate that is proportional to P^3.

In glacier and ice sheet flow problems similar multi-stress component effects occur. Moreover, different ice fabrics develop under different stress components. In a glacier there exist at least two stress components: a shear stress that acts across any plane that is parallel to the bed of the glacier, and a longitudinal compressive or tensile stress component that acts along the length of the glacier. They both influence the flow that causes a borehole to tilt, making it more difficult to interpret tilt measurements.

Effect of Hydrostatic Pressure

The effect that hydrostatic pressure has on the creep rate is not well established at the present time. The melting temperature of ice changes in an anomalous manner with pressure. It decreases, rather than increases, with pressure at a rate of -7.4×10^{-8}°C Pa^{-1}. Thus it is reasonable to expect that increasing the pressure at a constant test temperature will increase the creep rate because the melting temperature is brought closer to the test temperature. The first determinations of the activation volume gave $V = -2 \times 10^{-5}$ m^3 mole^{-1} $= -3.3 \times 10^{-29}$ m^3 (review W). These two experiments are preliminary (see review W), but they do show that the creep rate is increased by an amount expected from the decrease in the melting temperature. However, Higashi & Shoji (1979) find in a constant strain rate experiment on Antarctic ice that the maximum stress of the monotonically increasing stress-strain curve increases with the hydrostatic pressure. This result implies that hydrostatic pressure decreases the creep rate. Higashi & Shoji attribute the hardening increase with pressure to the fact that air bubbles in the ice are reduced in volume and cleavage cracks are closed by the higher pressure.

Very recent experiments by Jones & Chew (1982) on polycrystalline ice tested at -9.6°C under a compressive stress of 0.525 MPa gave the surprising result that the creep rate *decreased* from 8.3×10^{-9} s^{-1} to 7×10^{-9} s^{-1} when the superimposed hydrostatic pressure was increased from 0.1 to 15 MPa. The creep rate went through a minimum at pressures between 15 and 30 MPa. Above 30 MPa the creep rate *increased* with increase in pressure, and reached a value of 17×10^{-9} s^{-1} at a pressure of 60 MPa. The activation volume changed from $+3.2 \times 10^{-5}$ m^3 mole^{-1} $=$ 5.3×10^{-29} m^3 to -5.5×10^{-5} m^3 mole^{-1} $= -9.2 \times 10^{-29}$ m^3 as the pressure was increased. Jones points out that these results imply that several processes are controlling the creep rate. (We have already seen that different processes probably control the rate of easy glide and hard glide. Perhaps Jones' results involve both these mechanisms.) Recent results of Durham et al (1982) indicate that the activation volume is negative.

The effect of hydrostatic pressure on grain growth has been studied by Azuma & Higashi (1982). The rate of growth increases with pressure as well as temperature. The activation energy is 77 kJ mole^{-1} and the activation volume is -6.68×10^{-5} m^3 mole^{-1} $= -11 \times 10^{-29}$ m^3. Paterson (1981, p. 18) has determined an activation energy of grain growth in polar firn of only 42.3 kJ mole^{-1}.

VELOCITY OF INDIVIDUAL DISLOCATIONS

The velocity of individual dislocations in ice can be measured directly. In Figure 6 are plotted measured velocities of dislocations on the basal plane of pure ice versus applied basal shear stress. The velocity is proportional to the stress. The activation energy for motion determined by varying the temperature is approximately the same as the creep activation energy. [Mai et al (1978) have found that dislocations move about 50% faster in ice that is doped with HF.]

A partly phenomenological, partly theoretical creep rate can be found by using the Figure 6 data in creep equation (1) and by using (3) to estimate the steady-state dislocation density. In (1) and (3) use $\alpha = \beta = 1$, $b = 4.5 \times 10^{-10}$ m, $\mu = 3 \times 10^9$ Pa, and assume that the Figure 6 data are reasonably accounted for if $v = 6 \times 10^{-7}$ m s^{-1} at a shear stress of 0.1 MPa. The creep rate versus stress plot calculated this way is shown in Figure 3 (where shear creep rate and shear stress have been converted to the longitudinal creep rate and stress appropriate for uniaxial

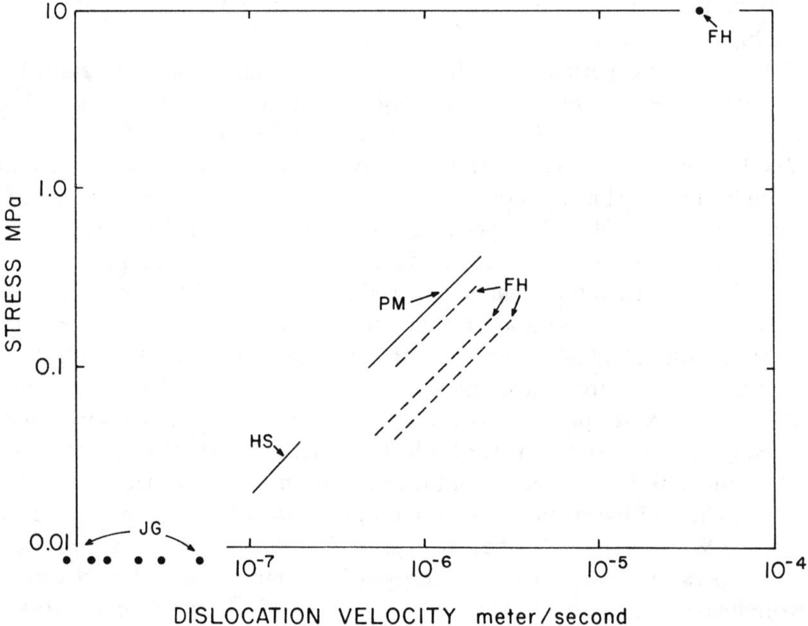

Figure 6 Dislocation velocity (normalized to $-10°C$ using $Q = 60$ kJ/mole) versus stress. Data from the following sources. FH: Fukuda & Higashi (1973); JG: Jones & Gilra (1973); HS: Higashi & Sake (given in FH); PM: Perez et al (1978) and Mai et al (1978).

tension and compression). This calculated creep rate can only be accurate to about a factor of about 5 because of the rough nature of the analysis. It appears, therefore, that easy glide can be accounted for since the easy-glide curve and the calculated curve lie within a factor of 2 to 3 of each other in Figure 3.

A complete theory also would account for the velocity-stress relationship of Figure 6. Mechanisms that give a linear velocity-stress relationship involving stress-induced order of protons in the stress field of a dislocation and bond-breaking proton rearrangements and/or defect reorientations in the core regions of dislocations have been proposed. The stress-induced order mechanism predicts a dislocation velocity of 5×10^{-7} m s^{-1} at $-10°C$ and at a stress of 0.1 MPa (see review W). This value agrees quite well with the data of Figure 6. (However, in review W it is pointed out that this mechanism probably can be ruled out if the creep rate indeed increases with pressure.)

The core region mechanism, which has been analyzed repeatedly (Glen 1968, Perez et al 1975, Whitworth et al 1976, Frost et al 1976, Forouhi & Bloomer 1978, Whitworth 1980, 1982, Goodman et al 1981) initially appeared to give a basal dislocation glide velocity much smaller than observed. However, more recent developments of the theory make it appear likely that dislocations controlled by this mechanism do move with about the right velocity.

Polycrystalline ice and ice in a hard orientation creep more slowly by a factor of about 500 than ice of a soft orientation. If the second glide-controlled mechanism does apply to glide on nonbasal planes with predicted dislocation velocities as low as found in the original theories, it might account for the lower creep rates in the harder ice.

Dislocation climb, however, cannot be ruled out as the rate-controlling process of the harder ice. The climb velocity of an isolated edge dislocation is approximately equal to (Weertman 1975)

$$v \simeq (D/2b)(\sigma\Omega/kT), \tag{15}$$

where $\Omega = 3.2 \times 10^{-29}$ m^3 is the molecular volume for ice, $k = 1.38 \times 10^{-23}$ J K^{-1} = Boltzmann's constant, and $D = D_0 \exp(-Q/kT)$ is the self-diffusion coefficient for ice. Here $D_0 \simeq 1.5 \times 10^{-3}$ m^2 s^{-1} and $Q =$ 60 kJ $mole^{-1}$ for both hydrogen and oxygen diffusion (see review W). At $T = 263$ K ($-10°C$) and $\sigma = 0.1$ MPa the climb velocity v is equal to $v = 2.0 \times 10^{-9}$ m s^{-1}. This velocity is a factor of 300 smaller than the experimentally determined glide velocity of 6×10^{-7} m s^{-1} given above, a factor of similar magnitude as that between the creep rate of the soft ice and the hard ice. Therefore the climb mechanism can account quantitatively for the creep of the harder ice. The deceleration in the creep rate of the creep curve and, in the constant strain rate test, the monotonic increase of the stress-strain curve of the hard ice are explained if the glide motion is faster at the same stress level than the climb motion. Thus the exercise given earlier is partly solved. But now another problem remains to be solved: If the climb velocity is so small compared with the glide velocities of Figure 6, why doesn't climb control the creep rate of easy-glide single crystals and cause the creep rate of such crystals to be about the same as that of polycrystalline ice? [Duval & Ashby (1982) have concluded that glide-controlled creep is not important for polycrystalline ice and that recovery dislocation processes, which I assume involve dislocation climb, may be.]

Discussions of other creep mechanisms proposed for ice are given by Landgon (1973), Goodman et al (1977, 1981), Gilra (1974), Perez et al (1978), and in reviews G and W.

CREEP OF ICE OTHER THAN ICE Ih

Only very limited creep results on ice in a crystalline form other than ice Ih have been obtained up to now. Poirier et al (1981) measured an effective viscosity of ice VI at room temperature at pressures in the range of 1.1–1.2 GPa in a sapphire anvil press. They found a viscosity of 10^{14} poise (10^{13} Pa s). The shear stress where this effective viscosity is determined is of the order of 90 MPa and the creep rate in shear of the order of $3 \times 10^{-6}\ s^{-1}$. Thus this ice is considerably harder than ice Ih. If a third power law is assumed to extrapolate the creep rate to stresses of the order of 1 to 10 MPa (10 to 100 bars), the effective viscosity is increased to 10^{17} to 10^{20} poise.

Poirier (1982) has briefly described unpublished preliminary results of Echelmeyer & Kamb on the effective viscosity of ice II and ice III. Ice II is somewhat more viscous than ice Ih at a given temperature (but different pressure), but ice III is three orders of magnitude less viscous than ordinary ice. Durham et al (1982, private conversation) have studied the ductile and the brittle deformation of ice at pressures up to 350 MPa and temperatures between 77 and 195 K. The brittle strength of a higher density ice phase was found to be 156 MPa at 158 K and 350 MPa. Lack of X-ray data prevented identification of the phase, which probably is ice IX but could be ice II.

CONCLUSION

The quasi-steady-state creep rate of coarse grain and single crystal ice at moderate stress levels and at relatively large strains is best described by a power-law creep equation with a power exponent equal to three. The activation energy of creep, 60 kJ/mole, is the same as that of self-diffusion of hydrogen and oxygen. The creep rate of single crystals oriented for easy-glide, basal slip can be accounted for with experimentally measured values of the glide velocity of individual basal dislocations. The creep mechanism of polycrystalline ice and ice single crystals in a hard orientation in which basal slip is not possible is not certain. Although a dislocation climb mechanism can account for the observed creep rates for this harder ice, it is not possible to conclude that this mechanism actually is rate controlling.

For purposes of ice modeling the best constants to use in the equivalent of a uniaxial tension or compression test appear to be a creep rate of 10^{-10} s^{-1} under a stress of 0.1 MPa and a temperature of $-10°C$. To obtain creep rates at other stress levels, a power exponent of $n = 3$ can be used. To obtain the creep rates at other temperatures, the activation energy to use is 60 kJ $mole^{-1}$ for single crystal ice and polycrystalline ice below $-10°C$. For polycrystalline ice above $-10°C$ the activation energy to use is 134 kJ $mole^{-1}$. These values of creep rate, power exponent, and activation energy are almost identical to those recommended by Paterson (1981, p. 39, and private communication) when his constants are converted into those appropriate to the uniaxial stress condition. In polycrystalline ice with the *c*-axis strongly aligned to the vertical, the creep rate in shear parallel to the glacier bed ought to be increased by a factor of four. The effect of hydrostatic pressure in any ice sheet or glacier on the creep rate should be rather small, and except for the thickest ice sheets can ordinarily be ignored.

Acknowledgment

This research was supported by the National Science Foundation under NSF Grant DMR79-12136. I wish to thank P. Duval, A. Higashi, Y. Hiki, K. Itagaki, S. J. Jones and R. W. Whitworth for making available their abstracts or preprint papers for the 6th International Symposium on the Physics and Chemistry of Ice, Rolla, Missouri, 2–6 August 1982. In addition, I appreciate the many helpful suggestions made in conversations and correspondence by W. F. Budd, H. C. Heard, R. LeB. Hooke, T. G. Langdon, M. Mellor, W. S. B. Paterson, and J. P. Poirier.

Literature Cited

Alexander, H., Haasen, P. 1968. Dislocations and plastic flow in the diamond structure. *Solid State Phys.* 22:27–148

Azuma, N., Higashi, A. 1982. Effect of the hydrostatic pressure on the rate of grain growth in Antarctic polycrystalline ice. *6th Int. Symp. Phys. Chem. Ice, Rolla, Mo., Aug.* (Abstr.)

Baker, R. W. 1978. The influence of ice-crystal size on creep. *J. Glaciol.* 21:485–500

Baker, R. W. 1981. Textural crystal-fabric anisotropies and the flow of ice masses. *Science* 211:1043–44

Barnes, P., Tabor, D., Walker, J. C. F. 1971. The friction and creep of polycrystalline ice. *Proc. R. Soc. London Ser. A* 324:127–55

Bender, J., Walker, G., Weertman, J. 1961. Unpublished data quoted in Weertman 1973

Bowden, F. P., Tabor, D. 1964. *The Friction*

and Lubrication of Solids, Part II, p. 128. Oxford: Clarendon

Budd, W. F. 1972. The development of crystal orientation fabrics in moving ice. *Z. Gletscherkd. Glazialgeol.* 8:65–105

Butkovitch, T. R., Landauer, J. K. 1958. The flow law for ice. *Symp. Chamonix, Int. Assoc. Sci. Hydrol./Int. Union Geod. Geophys., IASH Publ.* 47:318–25

Butkovitch, T. R., Landauer, J. K. 1959. The flow law for ice. *Ice Permafrost Res. Establ. Res. Rep. No. 56*

Carter, N. L., Kirby, S. H. 1978. Transient creep and semibrittle behavior of crystalline rocks. *Pure Appl. Geophys.* 116:807–39

Durham, W. B., Goetze, C., Blake, B. 1977. Plastic flow of oriented single olivine 2. Observations and interpretations of the dislocation structures. *J. Geophys. Res.* 82:5755–70

Durham, W. B., Froidevaux, C., Jaoul, O. 1979. Transient and steady-state creep of pure forsterite at low stress. *Phys. Earth Planet. Inter.* 19:263–74

Durham, W. B., Heard, H. C., Kirby, S. H. 1982. Deformation of ice at pressures to 350 MPa at 77 to 195 K. *EOS, Trans. Am. Geophys. Union* 63:1094 (Abstr.)

Duval, P. 1976. Lois du fluage transitoire ou permanent de la glace polycristalline pour divers états contrainte. *Ann. Géophys.* 32:335–50

Duval, P. 1977. The role of the water content on the creep rate of polycrystalline ice. *Isotopes and Impurities in Snow and Ice, Int. Assoc. Sci. Hydrol./Int. Union Geod. Geophys., IASH Publ.* 118:29–33

Duval, P. 1981. Creep and fabrics of polycrystalline ice under shear and compression. *J. Glaciol.* 27:129–40

Duval, P., Ashby, M. F. 1982. Rate-controlling processes during creep of polycrystalline ice. *6th Int. Symp. Phys. Chem. Ice, Rolla, Mo., August* (Abstr.)

Duval, P., Le Gac, H. 1980. Does the permanent creep-rate of polycrystalline ice increase with crystal size? *J. Glaciol.* 25:151–57

Duval, P., Le Gac, H. 1982. Mechanical behavior of Antarctic ice. *Ann. Glaciol.* 3:92–95

Evans, H. E., Knowles, G. 1977. A model of creep in pure materials. *Acta Metall.* 25:963–75

Forouhi, A. R., Bloomer, I. 1978. A quantum mechanical approach to the velocity of dislocations in ice. *Phys. Status Solidi B* 89:309–12

Frost, H. J., Goodman, D. J., Ashby, M. F. 1976. Kink velocities on dislocation in ice. A comment on the Whitworth, Paren and Glen model. *Philos. Mag.* 33:941–61

Fukuda, A., Higashi, A. 1973. Dynamical behavior of dislocations in ice crystals. *Cryst. Lattice Defects* 4:203–10

Gilra, N. K. 1974. Non-basal glide in ice. *Phys. Status Solidi A* 21:323–27

Glen, J. W. 1952. Experiments on the deformation of ice. *J. Glaciol.* 2:111–14

Glen, J. W. 1955. The creep of polycrystalline ice. *Proc. R. Soc. London Ser. A* 228:519–38

Glen, J. W. 1956. Measurements of the deformation of ice in a tunnel at the foot of an ice fall. *J. Glaciol.* 2:735–46

Glen, J. W. 1958. The flow law of ice: A discussion of the assumptions made in glacier theory, their experimental foundations and consequences. *Symp. Chamonix, Int. Assoc. Sci. Hydrol./Int. Union Geod. Geophys., IASH Publ.* 47:171–83

Glen, J. W. 1968. The effect of hydrogen disorder on dislocation movement and plastic deformation in ice. *Phys. Kondens. Mater.* 7:43–51

Glen, J. W. 1974. The physics of ice. *Cold Reg. Res. Eng. Lab. Monogr. II-C2a*, Hanover, N.H.

Glen, J. W. 1975. The mechanics of ice. *Cold Reg. Res. Eng. Lab. Monogr. II-C2b*, Hanover, N.H.

Glen, J. W., Jones, S. 1967. The deformation of ice single crystals at low temperatures. In *Physics of Snow and Ice*, ed. H. Ôura, 1(1):267–75. Sapporo: Inst. Low Temp. Sci., Hokkaido Univ.

Gold, L. W. 1973. Activation energy for creep of columnar-grained ice. In *Physics and Chemistry of Ice*, ed. E. Whalley, S. J. Jones, L. W. Gold, pp. 362–64. Ottawa: R. Soc. Can.

Gold, L. W. 1977. Deformation of ice single crystals close to the melting point. *EOS, Trans. Am. Geophys. Union* 58:901 (Abstr.)

Goodman, D. J., Frost, H. J., Ashby, M. F. 1977. The effect of impurities on the creep of ice Ih and its illustration by the construction of deformation maps. *Isotopes and Impurities in Snow and Ice, Int. Assoc. Sci. Hydrol./Int. Union Geod. Geophys., IASH Publ.* 118:29–33

Goodman, D. J., Frost, H. J., Ashby, M. F. 1981. The plasticity of polycrystalline ice. *Philos. Mag.* 43A:665–95

Gow, A. J., Williamson, T. 1976. Rheological implications of the internal structure and crystal fabrics of the West Antarctic ice sheet as revealed by deep core drilling at Byrd Station. *Geol. Soc. Am. Bull.* 87:1665–77

Hawkes, I., Mellor, M. 1972. Deformation and fracture of ice under uniaxial stress. *J. Glaciol.* 11:103–31

Herron, S. L. 1982. *Physical properties of the*

deep ice core from Camp Century, Greenland. PhD thesis. State Univ. N.Y., Buffalo

Herron, S. L., Langway, C. C. Jr., Brugger, K. A. 1982. Ultrasonic velocities and crystalline anisotropy in the ice core from Dye 3, Greenland. *EOS, Trans. Am. Geophys. Union* 63:297 (Abstr.)

Higashi, A. 1967. Mechanisms of plastic deformation in ice single crystals. In *Physics of Snow and Ice*, ed. H. Ôura, 1(1):277–89. Sapporo: Inst. Low Temp. Sci., Hokkaido Univ.

Higashi, A., Shoji, H. 1979. Mechanical tests of Antarctic deep core ice under hydrostatic pressure. *Ôhyobutsuri* (*Appl. Phys.*) 48:41–47

Higashi, A., Konimua, S., Mae, S. 1965. Plastic yielding in ice single crystals. *Jpn. J. Appl. Phys.* 4:575–82

Hobbs, P. V. 1974. *Ice Physics.* Oxford: Clarendon

Homer, D. R., Glen, J. W. 1978. The creep activation energies of ice. *J. Glaciol.* 21:429–44

Hooke, R. LeB. 1981. Flow law for polycrystalline ice in glaciers: Comparison of theoretical predictions, laboratory data, and field measurements. *Rev. Geophys. Space Phys.* 19:664–72

Hooke, R. LeB., Mellor, M., Budd, W. F., Glen, J. W., Higashi, A., Jacka, T. H., Jones, S. J., Lile, R. C., Martin, R. T., Meier, M. F., Russell-Head, D. S., Weertman, J. 1980. Mechanical properties of polycrystalline ice: An assessment of current knowledge and priorities for research. Report prepared for the International Commission on Snow and Ice, with support from the U.S. National Science Foundation. *Cold Reg. Sci. Technol.* 3:263–75

Jones, S. J. 1982. The confined compressive strength of polycrystalline ice. *J. Glaciol.* 28:171–77

Jones, S. J., Brunet, J.-G. 1978. Deformation of ice single crystals close to the melting point. *J. Glaciol.* 21:445–55

Jones, S. J., Chew, H. A. M. 1982. Creep of ice under hydrostatic pressure. *6th Int. Symp. Phys. Chem. Ice, Rolla, Mo., Aug.* (Abstr.)

Jones, S. J., Gilra, N. K. 1973. Dislocations in ice observed by X-ray topography. In *Physics and Chemistry of Ice*, ed. E. Whalley, S. J. Jones, L. W. Gold, pp. 344–49. Ottawa: R. Soc. Can.

Jones, S. J., Glen, J. W. 1968. The mechnical properties of single crystals of ice at low temperatures. *General Assembly of Bern, Commission of Snow and Ice, Int. Union Geod. Geophys., IASH Publ.* 79:326–40

Kamb, B. 1972. Experimental recrystallization of ice under stress. In *Flow and Fracture of Rocks*, ed. H. C. Heard, I. Y. Borg, N. L. Carter, C. B. Raleigh, pp. 211–41. *Geophys. Monogr. No. 16.* Washington DC: Am. Geophys. Union

Langdon, T. G. 1973. Creep mechanisms in ice. In *Physics and Chemistry of Ice*, ed. E. Whalley, S. J. Jones, L. W. Gold, pp. 362–64. Ottawa: R. Soc. Can.

Langdon, T. G. 1978. Recent developments in deformation mechanism maps. *Met. Forum* 1:59–70

Langdon, T. G., Mohamed, F. A. 1978. A new type of deformation mechanism map for high-temperature creep. *Mater. Sci. Eng.* 32:103–12

Lile, R. C. 1978. The effect of anisotropy on the creep of polycrystalline ice. *J. Glaciol.* 21:475–83

Mai, C., Perez, J., Tatibouët, J., Vassoille, R. 1978. Vitesses des dislocations dans la glace dopée avec HF. *J. Phys. Lett.* 39:L307–9

Mathews, W. H. 1959. Vertical distribution of velocity in Salmon Glacier, British Columbia. *J. Glaciol.* 3:448–54

Mellor, M. 1980. Mechanical properties of polycrystalline ice. In *Physics and Mechanics of Ice*, ed. P. Tryde, pp. 217–45. Berlin: Springer-Verlag

Mellor, M., Cole, D. M. 1982. Deformation and failure of ice under constant stress or constant strain-rate. *Cold Reg. Sci. Technol.* 5:201–19

Mellor, M., Smith, J. H. 1967. Creep of snow and ice. In *Physics of Snow and Ice*, ed. H. Ôura, 1(2):843–55. Sapporo: Inst. Low Temp. Sci., Hokkaido Univ.

Mellor, M., Testa, R. 1969a. Effect of temperature on the creep of ice. *J. Glaciol.* 8:131–45

Mellor, M., Testa, R. 1969b. Creep of ice under low stress. *J. Glaciol.* 8:147–52

Nakamura, T. 1978. Mechanical properties of impure ice single crystals at high temperature. *Proc. IAHR Symp. Ice Probl., Int. Assoc. Hydraul. Res.*, pp. 273–91

Nye, J. F. 1953. The flow law of ice from measurements in glacier tunnels, laboratory experiments and the Jungfraufirn borehole experiment. *Proc. R. Soc. London Ser. A* 219:477–89

Oikawa, H., Kuriyama, N., Mizukoshi, D., Karashima, S. 1977. Effect of testing modes on deformation behavior at stages prior to the steady states at high temperatures in Class I alloys. *Mater. Sci. Eng.* 29:131–35

Paterson, W. S. B. 1977. Secondary and tertiary creep of glacier ice as measured by borehole closure rates. *Rev. Geophys. Space Phys.* 15:47–55

Paterson, W. S. B. 1981. *The Physics of Glaciers.* Oxford: Pergamon. 2nd ed.

Paterson, W. S. B., Savage, J. C. 1963.

Measurements on Athabasca Glacier relating to the flow law of ice. *J. Geophys. Res.* 68:4537–43

Perez, J., Tatibouët, J., Vassoille, R., Gobin, P.-F. 1975. Comportement dynamique des dislocations dans la glace. *Philos. Mag.* 31:985–99

Perez, J., Mai, C., Vassoille, R. 1978. Cooperative movements of H_2O molecules and dynamic behavior of dislocation in ice Ih. *J. Glaciol.* 21:361–74

Poirier, J. P. 1982. The rheology of ices: A key to the tectonics of the ice moons of Jupiter and Saturn. *Nature* 299:683–87

Poirier, J. P., Sotin, C., Peyronneau, J. 1981. Viscosity of high pressure ice VI and evolution and dynamics of Ganymede. *Nature* 292:225–27

Ramseier, R. O. 1967a. Self-diffusion in ice monocrystals. *Cold Reg. Res. Eng. Lab. Res. Rep. No. 232*

Ramseier, R. O. 1967b. Self-diffusion of tritium in natural and synthetic ice monocrystals. *J. Appl. Phys.* 38:2553–56

Ramseier, R. O. 1972. *Growth and mechanical properties of river and lake ice.* PhD thesis. Laval Univ., Quebec

Raymond, C. F. 1973. Inversion of flow measurements for stress and rheological parameters in a valley glacier. *J. Glaciol.* 12:19–44

Raymond, C. F. 1980. Temperate valley glaciers. In *Dynamics of Snow and Ice Masses*, ed. S. C. Colbeck, pp. 79–139. New York: Academic

Riley, N. W., Noll, G., Glen, J. W. 1978. The creep of NaCl doped ice. *J. Glaciol.* 21:501–7

Robinson, S. L., Sherby, O. D. 1969. Mechanical behavior of polycrystalline tungsten at elevated temperature. *Acta Metall.* 17:109–25

Russell-Head, D. S., Budd, W. F. 1979. Ice-sheet flow properties derived from borehole shear measurements combined with ice-core studies. *J. Glaciol.* 24:117–30

Sherby, O. D., Burke, P. M. 1967. Mechanical behavior of crystalline solids at elevated temperature. *Progr. Mater. Sci.* 13:324–90

Sherby, O. D., Weertman, J. 1979. Diffusion controlled creep: A defense. *Acta Metall.* 27:387–400

Shoji, H. 1978. Stress strain tests of ice core drilled at Mizuho Station, East Antarctica. *Mem. Natl. Inst. Polar Res.*, Spec. Issue 10:95–101

Shoji, H., Langway, C. C. Jr. 1982. Mechanical property of fresh ice from Dye 3, Greenland. *EOS, Trans. Am. Geophys. Union* 63:397 (Abstr.)

Shreve, R. L., Sharp, R. P. 1970. Internal deformation and thermal anomolies in lower Blue Glacier, Mount Olympus, Washington, U.S.A. *J. Glaciol.* 9:65–86

Sinha, N. K. 1978. Rheology of columnar-grained ice. *Exp. Mech.* 18:464–70

Sinha, N. K. 1982. Constant strain- and stress-rate compressive strength of columnar-grained ice. *J. Mater. Sci.* 17:785–802

Spingarn, J. R., Barnett, D. M., Nix, W. D. 1979. Theoretical description of climb controlled steady state creep at high and intermediate temperatures. *Acta Metall.* 27:1549–61

Steinemann, S. 1954. Results of preliminary experiments on the plasticity of ice crystals. *J. Glaciol.* 2:404–12

Steinemann, S. 1958. Experimentelle Untersuchungen zur Plastizität von Eis. *Beitr. Geol. Schweiz, Geotech. Ser. Hydrol. No. 10.* 72 pp.

Thomas, R. H. 1971. Flow law for Antarctic ice shelves. *Nature Phys. Sci.* 232:85–87

Vagarali, S. S., Langdon, T. G. 1982. Deformation mechanisms in h.c.p. metals at elevated temperature. II. Creep behavior of a Mg-0.8% Al solid solution alloy. *Acta Metall.* 30:1157–70

Weertman, J. 1973. Creep of ice. In *Physics and Chemistry of Ice*, ed. E. Whalley, S. J. Jones, L. W. Gold, pp. 320–37. Ottawa: R. Soc. Can.

Weertman, J. 1975. High temperature creep produced by dislocation motion. In *Rate Processes in Plastic Deformation of Materials*, ed. J. C. M. Li, A. K. Mukherjee, pp. 315–36. Metals Park, Ohio: Am. Soc. Met.

Whitworth, R. W. 1980. The influence of the choice of glide plane on the theory of the velocity of dislocations in ice. *Philos. Mag.* 41:521–28

Whitworth, R. W. 1982. The velocity of dislocations in ice on various glide planes. *6th Int. Symp. Phys. Chem. Ice, Rolla, Mo., Aug.* (Abstr.)

Whitworth, R. W., Paren, J. G., Glen, J. W. 1976. The velocity of dislocations in ice—a theory based on proton disorder. *Philos. Mag.* 33:409–26

Wilson, C. J. L., Russell-Head, D. S. 1982. Steady-state preferred orientation of ice deformed in plane strain at −1°C. *J. Glaciol.* 28:145–60

Yavari, P., Mohamed, F. A., Langdon, T. G. 1981. Creep and substructure formation in an Al-5% Mg solid solution alloy. *Acta Metall.* 29:1495–507

Ann. Rev. Earth Planet. Sci. 1983. 11: 241–68

RECENT DEVELOPMENTS IN THE DYNAMO THEORY OF PLANETARY MAGNETISM

F. H. Busse
Department of Earth and Space Sciences and Institute of Geophysics and Planetary Physics, University of California, Los Angeles, California 90024

1. INTRODUCTION

During the past decade the dynamo hypothesis of the origin of the Earth's magnetic field has been developed from an abstract concept to a theory that can be related to specific observational evidence. The relationship between theoretical models and geomagnetic data is still rather tenuous. Important parameters, such as the strength of the toroidal component of the magnetic field inside the Earth's core, are not known even within an order of magnitude. But there is a growing appreciation of the various ways in which paleomagnetic data or observations of secular variations can be used to test theoretical ideas about dynamical processes in the core.

Attempts at improving the contact between theory and observations have focused on two problems. First, the idea of interpreting secular variations of the geomagnetic field in terms of fluid motions in the outermost core has been revived and new methods of data analysis have been applied to this problem. Second, physically more realistic models of fluid motions in the liquid core have been introduced into the dynamo problem and attempts are being made to model particular episodes, such as reversals in the evolution in time of the geodynamo.

Another factor that has stimulated interest in dynamo theory is the discovery and measurement of the magnetic fields of other planets. Jupiter, Saturn, and Mercury exhibit magnetic fields that appear to be caused by a dynamo process in the electrically conducting liquid cores of those planets. Because of the close similarity of the relevant physical conditions in the four planetary cores believed to possess an active dynamo, any theoretical

0084–6597/83/0515–0241$02.00

model should be applicable in all four cases with only minor modifications. In particular, the dependence of the field strength on other measured or inferred planetary parameters will provide useful constraints on dynamo theories of planetary magnetism.

In some respects this article represents a sequel to an earlier review by the author (Busse 1978). For this reason theoretical concepts of dynamo theory are outlined only briefly in Section 2 and recent developments are emphasized. In the following section the subject of symmetry properties of planetary magnetic fields is considered. That planetary fields are dipolar in character is usually taken for granted, but theoretical analysis shows that dipolar and quadrupolar classes of fields are basically equivalent and that only secondary effects cause distinctions in the generation mechanism. Observational data and various methods for using them in inferring properties of the geodynamo are discussed in Section 4. A detailed description of recent approaches toward mathematical models of the geodynamo would go much beyond the scope of this paper. Instead, attention is focused in Section 5 on some typical problems in the modern theory of planetary dynamos. The effect of an azimuthal magnetic field on buoyancy-driven motions in a rotating sphere is a special aspect of the hydromagnetic dynamo problem that can be understood without recourse to numerical computations. In Section 6 it is shown how the annulus model exhibits some typical features of this problem. The article closes with an outlook on future developments in planetary dynamo theory.

2. OUTLINE OF DYNAMO THEORY

The dynamo process can be regarded as an instability. Hydrodynamic instabilities often provide access to new degrees of freedom of motion, which permit a more effective dissipation of available energy. Among instabilities of fluid flow, the dynamo process is distinguished by the property that the state bifurcating from the unstable primary state of motion is characterized not only by a new degree of freedom, but by a new physical quantity—the magnetic field. In this respect the generation of magnetic flux by the growing instability is analogous to the generation of vorticity by the Rayleigh-Bénard instability of a static fluid layer heated from below. In both cases new physical processes are introduced for the dissipation of energy. The fact that the dynamo process represents a special category among fluid-dynamical instabilities is reflected in the mathematical methods used in dynamo theory.

The mathematical description of the generation of magnetic fields by fluid motion is based on the equation of induction,

$$\frac{\partial}{\partial t}\mathbf{B}-\nabla\times(\lambda\nabla\times\mathbf{B})=\nabla\times(\mathbf{v}\times\mathbf{B}), \tag{1}$$

which can be derived from Ohm's law in a moving medium and from Maxwell's equations after the displacement current has been neglected. The magnetic diffusivity λ is defined by $\lambda \equiv (\mu\sigma)^{-1}$, where σ is the electrical conductivity and μ is the magnetic permeability of the fluid. The difference between μ and the permeability μ_0 of vacuum can usually be neglected in electrically conducting fluids such as liquid metals. The magnetic diffusivity in liquid metals, as well as in the Earth's core, is of the order 1 $m^2 s^{-1}$.

In the kinematic problem the velocity field **v** is arbitrarily prescribed and a dynamo is found when a growing solution for **B** of Equation (1) is obtained for appropriate boundary conditions. In the physical dynamo problem **v** is given as the result of physically realistic forces, while in the hydromagnetic dynamo problem the Lorentz forces modifying the velocity field are taken into account. Because the velocity vector depends on the magnetic flux density **B** in the latter case, Equation (1) no longer describes a linear homogeneous problem. An analysis of the hydromagnetic dynamo problem is required in order to derive expressions for the equilibrium amplitude of the magnetic field.

As in the theory of hydrodynamic instabilities, there exist only necessary but no general sufficient conditions for the dynamo process. The following statements represent the most important general results.

1. A necessary condition for a dynamo in a sphere of radius r_0 with constant diffusivity λ is

$$\mathrm{Re_m} \equiv V_0 r_0/\lambda > \pi \tag{2}$$

(Backus 1958, Childress 1969), where V_0 is the maximum velocity inside the fluid sphere. Proctor (1977b) has improved the bound (2), but his results do not change the fact that the estimate (2) for the critical magnetic Reynolds number $\mathrm{Re_m}$ is far below those required for realistic dynamos. A physical interpretation of condition (2) is that the time required by a fluid parcel to traverse the distance r_0/π must be shorter than the magnetic decay time $r_0^2/\lambda\pi^2$. This time is about 25×10^3 yr for the Earth's core.

2. Cowling's (1934) theorem states that growing axisymmetric or two-dimensional solutions of Equation (1) do not exist. Cowling's proof of this theorem has been extended by many authors; for a discussion, see James et al (1980). The interesting generalization that growing axisymmetric poloidal fields do not exist even if time-dependent and compressible fluid flows with arbitrary spatial distributions of the diffusivity λ are admitted has recently been proven by Lortz & Meyer-Spasche (1982a) and by Hide & Palmer (1982) using different mathematical methods. These proofs put to rest the speculation that compressibility may be important in the generation of Saturn's magnetic field, which exhibits a high degree of axisym-

metry, at least as seen from the outside (Todoeschuck & Rochester 1980). More recently, Lortz & Meyer-Spasche (1982b) have extended their analysis and obtained a proof that the toroidal component of an axisymmetric or two-dimensional field decays as well. Cowling's theorem thus appears to hold under the most general conditions.

Cowling's theorem has exerted a strong influence on the development of dynamo theory, but its physical implications tend to be exaggerated. From a mathematical point of view the property of axisymmetry is singular; dynamos with arbitrarily small deviations from axisymmetry can be constructed. Braginsky's (1975, 1978) model for the geodynamo is actually based on an analysis of small perturbations of a basic axisymmetric field.

3. The toroidal theorem first noticed by Elsasser (1946) and proven rigorously by Bullard & Gellman (1954) states that velocity fields of the form $\mathbf{u} = \nabla \times \mathbf{r}\psi$ cannot generate a magnetic field if λ is a function of $|\mathbf{r}|$ only, where $\mathbf{r}$ is the position vector. The important geophysical implication of this theorem is that a radial component of the velocity must be maintained over periods of the order of magnetic decay time. A lower bound on the radial velocity in terms of the ratio between poloidal and toroidal field strengths has been derived by Busse (1975a).

There is no need to discuss dynamo theory in more detail, since monographs on the subject by Moffatt (1978) and Parker (1979) have appeared in recent years. Among the various methods for solving Equation (1), we mention only the most widely used one based on the concept of mean-field electrohydrodynamics (MFE), which was developed by Steenbeck et al (1966; see also Krause & Rädler 1980). In the MFE theory, it is assumed that the velocity field can be separated into two parts $\mathbf{v} = \bar{\mathbf{v}} + \check{\mathbf{v}}$, which are well separated by either their spatial or their time scale of variation. By taking the average of Equation (1) over the small spatial scale or the short time scale and assuming that $\check{\mathbf{v}}$ obeys the statistical properties of isotropic but not mirror-symmetric turbulence, an equation of the form

$$\frac{\partial}{\partial t}\bar{\mathbf{B}} - \nabla \times \lambda_\ell(\nabla \times \bar{\mathbf{B}}) = \nabla \times (\bar{\mathbf{v}} \times \bar{\mathbf{B}}) + \nabla \times (\alpha\bar{\mathbf{B}}) \tag{3a}$$

can be obtained. In the simplest case of MFE theory, the assumption of a turbulent velocity field with vanishing large scale or time-average component $\check{\mathbf{v}}$ gives rise to an equation of the form

$$\frac{\partial}{\partial t}\bar{\mathbf{B}} - \nabla \times \lambda_\ell(\nabla \times \bar{\mathbf{B}}) = \nabla \times \alpha\bar{\mathbf{B}}, \tag{3b}$$

where $\bar{\mathbf{B}}$ denotes an averaged magnetic field. The effects of the homogeneous isotropic turbulence are represented by the eddy diffusivity λ_ℓ and the scalar function α, which is proportional to the helicity of the turbulent velocity field. A finite value of α thus requires that the statistical properties of the turbulent velocity field are not mirror symmetric.

A large number of solutions have been obtained for spherical dynamos, based on Equation (3b) or equations in which the effects of an axisymmetric velocity field have been added. While these solutions elucidate properties of the kinematic dynamo problem, they are of little help for the understanding of planetary dynamos. The assumption of isotropic turbulence that is not mirror symmetric owing to the effects of rotation is in contradiction to the strong anisotropy caused by the Coriolis force. Anisotropic tensors α_{ik}, which replace α in the general case, change the dynamo process significantly (Busse & Miin 1979). When physically reasonable solutions for the velocity field are obtained, the spatial dependence of the α_{ik} tensor can be determined; however, little progress in this direction has been made. The first-order smoothing assumption (Roberts & Soward 1975), which is used in the derivation of Equations (3a) and (3b), is also not well justified in planetary cores where axisymmetric and nonaxisymmetric components of the magnetic fields are of comparable orders of magnitude. But this assumption probably does less violence to the physics of the problem than arbitrary assumptions about the α_{ik} tensor.

3. DYNAMOS IN PLANETARY CORES

Because so little is known about the magnetic fields of other planets and about their time dependence in particular, we focus this discussion on the Earth's dynamo, keeping in mind that most of our considerations apply to other planetary dynamos as well. For a review of known properties of planetary magnetic fields, see Russell (1980). Any theory of the geodynamo must attempt to model the main features of the Earth's magnetic field, which are listed in Table 1. Primary and secondary properties have been distinguished in Table 1, but this does not imply that secondary features are less important. In fact, they are probably more important in discriminating between competing models of the geodynamo once models with sufficient details have been developed. At the present stage of analysis, models are too crude to exhibit much more than the primary features. But even this reduced problem is not trivial, as is evident in the following.

In discussing magnetic-field generation in spherical cores it is convenient to adopt the general representation of the solenoidal vector $\mathbf{B}$ in terms of poloidal and toroidal parts:

$$\mathbf{B} = \nabla \times (\nabla \times \mathbf{r}h) + \nabla \times \mathbf{r}g,$$

where $\mathbf{r}$ is the position vector with respect to the center of the sphere and h and g are scalar functions. When the condition is imposed that the average

Table 1 Properties of the geomagnetic field

Primary features	Secondary features
Dipole component dominates	Higher-order multipole components are nearly comparable to dipole component at the core surface.
Alignment of dipole axis and rotation axis.	Persistent angle of the order of 10° between the two axes.
Magnetic moment is stationary on the time scale τ_m of magnetic diffusion.	Time scales of geomagnetism: Secular variation of nondipole field $\approx 2 \times 10^3$ yr; dipole oscillations $\approx 10^4$–10^5 yr; reversals $\approx 5 \times 10^5$ yr.
Magnetic moment representing poloidal part is 8×10^{15} T m^3.	Ohmic dissipation is dominated by higher harmonics of the field. The toroidal field inside the core is likely to be larger than the poloidal field.

of h and g vanishes over spherical surfaces, h and g are uniquely determined by the components of the magnetic field and vice versa (Backus 1958).

By multiplying Equation (1) and its curl by $\mathbf{r}$, two equations for h and g are obtained:

$$\left(\frac{\partial}{\partial t} - \lambda\nabla^2\right)L_2 h = \mathbf{r}\cdot\nabla \times (\mathbf{v} \times \mathbf{B}), \tag{4a}$$

$$\left(\frac{\partial}{\partial t} - \lambda\nabla^2\right)L_2 g = \mathbf{r}\cdot\nabla \times (\nabla \times (\mathbf{v} \times \mathbf{B}), \tag{4b}$$

where L_2 is the negative Laplacian on the unit sphere, i.e. in spherical coordinates (r, θ, ϕ)

$$L_2 = -\sin^{-1}\theta\frac{\partial}{\partial\theta}\sin\theta\frac{\partial}{\partial\theta} - \sin^{-2}\theta\frac{\partial^2}{\partial\phi^2}.$$

For simplicity a constant diffusivity λ has been assumed. It is generally believed that dynamos with a purely poloidal field h or a purely toroidal field g do not exist, although mathematical proofs of these hypotheses are not yet available. A dynamo in a spherical geometry thus depends on the coupling of Equations (4a) and (4b) provided by the right-hand sides. This coupling can be discussed in more detail when the axisymmetric components $\bar{h},\bar{g}$ of the fields h,g are considered. By taking the average over the ϕ-coordinate and indicating it by a bar, the following equations for $\bar{h}$ and $\bar{g}$

are obtained:

$$\left(\frac{\partial}{\partial t} - \lambda\nabla^2\right)L_2\bar{h} = \mathbf{r}\cdot\nabla\times(\bar{\mathbf{v}}\times\bar{\mathbf{B}}) + \mathbf{r}\cdot\nabla\times\overline{(\check{\mathbf{v}}\times\check{\mathbf{B}})}, \tag{5a}$$

$$\left(\frac{\partial}{\partial t} - \lambda\nabla^2\right)L_2\bar{g} = \mathbf{r}\cdot\nabla\times(\nabla\times(\bar{\mathbf{v}}\times\bar{\mathbf{B}})) + \mathbf{r}\cdot\nabla\times(\nabla\times\overline{(\check{\mathbf{v}}\times\check{\mathbf{B}})}), \tag{5b}$$

where the definitions

$$\bar{\mathbf{B}} = \nabla\times(\nabla\times\mathbf{r}\bar{h}) + \mathbf{k}\times\mathbf{r}\sin^{-1}\theta r^{-1}\frac{\partial}{\partial\theta}\bar{g}$$

$$\check{\mathbf{B}} = \mathbf{B} - \bar{\mathbf{B}}, \qquad \bar{\check{\mathbf{B}}} \equiv 0,$$

$$\check{\mathbf{v}} = \mathbf{v} - \bar{\mathbf{v}}, \qquad \bar{\check{\mathbf{v}}} \equiv 0,$$

have been used and $\mathbf{k}$ denotes the unit vector in the direction of the polar axis. Only the ϕ-component of $\bar{\mathbf{v}}\times\bar{\mathbf{B}}$ contributes in Equation (5a), and thus $\bar{g}$ does not enter this equation. The generation of $\bar{h}$, therefore, depends on the term $\overline{\check{\mathbf{v}}\times\check{\mathbf{B}}}$. Using the MFE notation, this term can be written in the case of isotropic but not mirror-symmetric turbulence in the form of $\alpha\bar{\mathbf{B}}$. On the other hand, both $\check{h}$ and $\bar{h}$ enter Equation (5b) and two possibilities for the generation of $\bar{g}$ exist. When the term $\overline{\check{\mathbf{v}}\times\check{\mathbf{B}}}$ dominates, one speaks of an α^2 dynamo since this term is used twice, in Equation (5b) as well in Equation (5a), for the dynamo process. When the term $\bar{\mathbf{v}}\times\bar{\mathbf{B}}$ dominates, an $\alpha\omega$ dynamo is obtained because the azimuthal part of $\bar{\mathbf{v}}$ given by the differential rotation $\omega\mathbf{k}\times\mathbf{r}$ is the only one entering $\bar{\mathbf{v}}\times\bar{\mathbf{B}}$.

In solving the dynamo problem for a spherical symmetry, one naturally attempts to use the available symmetries. But the possibilities are limited. Axisymmetric solutions are eliminated by Cowling's theorem. Axisymmetric velocity fields are possible and have been used (Gubbins 1973, Bullard & Gubbins 1977), but they necessarily yield magnetic fields with $\bar{\mathbf{B}} \equiv 0$, which are not of geophysical interest. Moreover, it is well known that velocity fields generated in rapidly rotating systems by buoyancy forces are nonaxisymmetric (Roberts 1968, Busse, 1970). The physically preferred form of convection flow exhibits, however, a mirror symmetry about the equatorial plane

$$\{v_r(r,\theta,\phi), v_\theta(r,\theta,\phi), v_\phi(r,\theta,\phi)\}$$
$$= \{v_r(r,\pi-\theta,\phi), -v_\theta(r,\pi-\theta,\phi), v_\phi(r,\pi-\theta,\phi)\}, \tag{6}$$

at least outside the cylindrical surface touching the inner solid boundary at its equator (Busse & Cuong 1977). This symmetry is a consequence of the Proudman-Taylor theorem, which states that steady, small-amplitude flows in an inviscid rotating fluid must not vary in the direction of the axis of rotation. This theorem cannot be strictly satisfied by any convection flow in a sphere, but motions of the form (6) exhibit a minimal dependence on the coordinate in the direction of the axis of rotation (Busse 1970). Even in a turbulent state of motion at high Rayleigh numbers (representing the strength of buoyancy forces), the alignment of convection motion with the axis of rotation and thus the symmetry property (6) appears to be approximately preserved, as indicated by laboratory experiments (Busse & Carrigan 1976).

The symmetry property (6) permits the separation of two classes of solutions of Equation (1), provided the boundary conditions for the magnetic field and the distribution λ are symmetric with respect to the equatorial plane. The first class of solutions $\mathbf{B}$ exhibits the same symmetry as the velocity field $\mathbf{v}$ and is called the *quadrupolar class*. The second is the *dipolar class* of solutions, which exhibits the opposite symmetry corresponding to a negative sign in front of the right-hand side in relationship (6). Using h and g, the symmetry properties can be expressed more simply:

quadrupolar class: h is a symmetric, g an antisymmetric function of $\mathbf{k} \cdot \mathbf{r}$, (7a)

dipolar class: h is an antisymmetric, g a symmetric function of $\mathbf{k} \cdot \mathbf{r}$. (7b)

Although the term "quadrupolar" may suggest a higher mode relative to a dipolar solution, the two classes of solution are indeed quite equivalent. When the poloidal component of the magnetic field described by h exhibits a quadrupolar character, the associated axisymmetric azimuthal component of the magnetic field $\bar{B}_\phi$ is symmetric with respect to the equatorial plane. On the other hand, a dipolar poloidal magnetic field is associated with an antisymmetric $\bar{B}_\phi$. This dual nature of the two classes of solutions is reflected in the closeness of the critical values of the magnetic Reynolds numbers for their generation. In fact, it can be shown in simple cases involving unphysical boundary conditions that the dynamo problems of the two classes of solutions are adjoint to each other (Proctor 1977a).

4. OBSERVATIONAL EVIDENCE

Spherical Harmonic Analysis

The form of the geomagnetic field and its changes in time are best described by an expansion in terms of spherical harmonics. Because the field $\mathbf{B}$ can be written as the gradient of a potential outside the conducting core, the following representation is obtained.

$$\mathbf{B} = -\nabla V, \tag{8}$$

$$V = a \sum_{\ell=1}^{\infty} \sum_{m=0}^{\ell} \left(\frac{a}{r}\right)^{\ell+1} [g_\ell^m \cos m\phi + h_\ell^m \sin m\phi] P_\ell^m (\cos\theta), \tag{9}$$

where a is the radius of the Earth and P_ℓ^m denotes Schmidt quasi-normalized spherical harmonics. Contributions from currents external to the Earth's surface can be eliminated by a proper analysis of the data and are not represented in expression (9). Since these contributions are small, they are neglected in paleomagnetic work.

If the geomagnetic field were a pure dipolar field in the sense of definition (7b), the Gauss coefficients g_ℓ^m, h_ℓ^m would vanish whenever $m+\ell$ is an even integer. Conversely, a pure quadrupolar field would exhibit only coefficients with even $m+\ell$. Obviously the geomagnetic field is predominantly dipolar, and it has long been known that the quadrupolar coefficients with $\ell = 2$ are relatively small. But coefficients of higher order indicate little preference for one symmetry or the other. The axisymmetric part of the radial component of the present magnetic field at the core surface (shown in Figure 1) shows typical deviations from equatorial symmetry. This asymmetry is strongly evident in the polar regions.

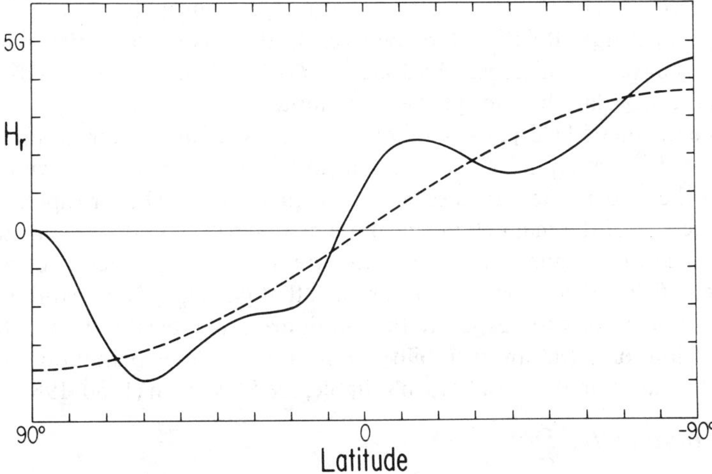

Figure 1 The radial component of the axisymmetric part of the geomagnetic field at the core surface. The zonal coefficients g_ℓ^0 ($\ell = 1, \ldots, 8$) of the International Geomagnetic Reference Field 1980 have been used in the extrapolation to the core-mantle boundary. The dashed line indicates the dipole field given by g_1^0 alone for comparison.

Because of statistical fluctuations, planetary magnetic fields cannot be expected to exhibit simple symmetries. But persistent deviations from a particular symmetry may indicate physical processes that are not taken into account in the usual dynamo theory. The understanding of the origin of asymmetric features is thus not only revealing about the nature of the dynamo process itself, but may also provide information about physical properties of planetary cores that is not available by other means.

There are a number of possible causes of asymmetries aside from statistical fluctuations. The symmetry (6) of the velocity field about the equatorial plane cannot be expected to hold inside the cylindrical surface touching the equator of an inner spherical boundary. Buoyancy-driven flows do not even approximately obey the Proudman-Taylor condition in this region (Busse & Cuong 1977) and an interaction of dynamo processes with different symmetries may result. All planets that appear to have dynamos are also believed to possess impenetrable inner cores. In the case of the Earth, the 1215-km radius of the inner core is a significant fraction of the core radius of 3485 km. The assumption of equatorial symmetry of motions can further be violated by asymmetric heat-transport conditions at the core-mantle boundary. Because of mantle convection, the geothermal heat flux shows strong lateral variations. These variations have little effect on the temperature of the core-mantle boundary because of the high conductivity of the liquid iron core. Yet thermal convection will be strongly increased at places where steep temperature gradients in the mantle require a high heat flux. Relationships between heat transport and the dynamo process have been discussed by Jones (1977), but a quantitative analysis of the pertinent data has not yet been attempted.

The magnetic field per se cannot be the cause of asymmetries, since the Lorentz force is a quadratic expression and does not destroy the symmetry (6) of the velocity field in cases of either quadrupolar (7a) or dipolar (7b) symmetries of the magnetic field. But a symmetric state may be unstable, and once the magnetic field loses its symmetry the symmetry (6) of the velocity field will also be affected. The fact that planetary fields exhibit a tilt of the dipole axis with respect to the rotation axis of typically 10° may be an indication of such an instability. For a possible explanation of the exceptionally small tilt of Saturn's dipole, see Stevenson (1980, 1982).

Paleomagnetic Data

Paleomagnetic measurements of the geomagnetic field are of special interest because they permit us to take averages over long times, which tend to eliminate the effects of temporal fluctuations. Since the nonaxisymmetric components of the time-averaged field are very small, paleomagnetic data are expressed in terms of the axisymmetric coefficients g_ℓ^0 in the representation (9). The scatter of the data and the difficulty of approaching global coverage cause large uncertainties in the results, even for low-order harmonics. The ratios g_3^0/g_1^0 have been determined by a number of authors from measurements of inclinations. Merrill & McElhinny (1977) and Coupland & Van der Voo (1980) have reviewed the observational evidence of the last 10^8 yr and show that g_2^0 is relatively small for the last 50 m.y.—where the data are best—and that g_3^0 always appears to have the same sign as g_1^0. This agrees with a dynamo mechanism generating a field of predominantly dipolar symmetry, except for the slightly disconcerting property that the present value of g_3^0 has the opposite sign of g_1^0.

Since dynamo theory admits a solution $-\mathbf{B}$ for any solution $\mathbf{B}$ of the basic equations, the symmetry between normal and reversed states is the most stringent test for the dynamo origin of the Earth's magnetic field. The increasing evidence (Wilson 1970, Merrill & McElhinny 1977, Phillips 1977) for asymmetries between normal and reversed states cannot be reconciled within the domain of dynamo theory. Merrill et al (1979) have given a thorough discussion of the problem and dismiss the effects that might be caused by thermoelectric currents and remanent fields in the crust. But their alternative suggestion that two different states of the velocity field exist and that transitions between these states occur only at times of reversals appears to be farfetched. While a correlation between polarity changes and changes in the velocity field over the time span of a few reversals cannot be excluded, the persistent influence of an initial condition over the lifetime of the geodynamo is extremely unlikely. The problem of differences between normal and reversed states must thus be regarded as a puzzling and unresolved problem requiring further analysis of paleomagnetic data as well as detailed theoretical studies. Perhaps even the small influence of a crustal field can be crucial in a critical stage of the reversal.

Geomagnetic Reversals

In the past decade it has become possible to follow the process of the reversal through paleomagnetic measurements at different sites. A review of the observational evidence and the various possibilities for theoretical interpretations has been given by Fuller et al (1979). Additional data from the Southern Hemisphere appear to be crucial for determining the dominant higher harmonic at the time when the dipole component changes sign. The reversal process offers the potential opportunity for a distinction between various dynamo theories, but since hydromagnetic dynamo models are still in their infancy the complex process of a reversal has not yet been described by a deductive theory.

It has long been realized that statistical aspects of reversals can be simulated by two coupled disk dynamos (Rikitake 1958, Allan 1962, Cook

& Roberts 1970). Malkus (1972) pointed out that adding a shunt to the circuit of a single disk dynamo also leads to reversals of the magnetic fields. In fact, the equations of the latter system are formally identical to the Lorenz system, which was the first example of three coupled ordinary differential equations that exhibit chaotic solutions (Lorenz 1963). Today it is generally recognized that chaotic behavior of solutions for systems of three or more coupled nonlinear differential equations is the rule rather than the exception, and there is a burgeoning mathematical literature on strange attractors, as the complex attracting manifolds in the solution space are called (Ruelle & Takens 1971). These recent developments have demonstrated that aperiodic reversals are not a specific property of the dynamo process but occur in a wide variety of nonlinear physical systems. The problem becomes complicated by the fact that properties, such as aperiodic solutions, exhibited by systems with few degrees of freedom often disappear when more degrees of freedom are added to the system (Marcus 1981). Nevertheless, the description of reversal sequences by solutions of simple systems of ordinary differential equations remains an interesting possibility (Robbins 1977), especially if features such as the reversal frequency can be correlated with physical parameters of the system.

The dramatic changes in the frequency of reversals over periods of the order of 5×10^7 yr (Cox 1975) represent one of the most fascinating aspects of paleomagnetic data. Tectonic events (Vogt 1975) and changes in the geothermal heat flux (Jacobs 1981) have been associated with the changes in the reversal frequency. Since the strength of convective motions in the core and in the mantle must be correlated because of the continuity of the heat flux originating in the core, it is not surprising that a relationship may exist between the otherwise quite distinct phenomena of plate tectonics and reversal frequency. The origin of the long-period time dependence must arise from the mechanism of mantle convection, which is not yet well understood.

Secular Variation

By far the largest amount of data on the Earth's magnetic field has been collected by geomagnetic observatories and by satellite measurements. While the time span for which data are available is too short to obtain meaningful results on the time dependence of the geodynamo, the observed secular variation can provide some information about the velocity at the core surface. Backus (1968) pointed out the difficulties of this approach and demonstrated that earlier attempts by Kahle et al (1967) failed because of problems of nonuniqueness. But recently work on this subject has flourished again.

The analysis of the problem is usually based on the radial component of

Equation (1) without Ohmic dissipation:

$$0 = \frac{\partial}{\partial t} B_r + \nabla \cdot (\mathbf{v} B_r) - \nabla \cdot (\mathbf{B} v_r). \tag{10}$$

The neglection of flux diffusion can be justified for time scales short compared with the free ohmic decay time of the smallest scale features considered (Roberts & Scott 1965). In order to extrapolate the magnetic field and its variation in time as observed at the Earth's surface down to the core-mantle boundary, an insulating mantle must be assumed. The finite conductivity of the mantle filters out magnetic variations on a time scale of a few years and precludes any information about short-period fluctuations that may be important for the understanding of the dynamics of the core. At the core-mantle boundary, v_r vanishes and (10) can be replaced by the simpler equation

$$\frac{\partial}{\partial t} B_r = -\nabla_s \cdot (\mathbf{v}_s B_r), \tag{11}$$

where the subscript s refers to the horizontal component of the respective vector quantity. Of particular interest are the contours at the core-mantle boundary on which B_r vanishes (which are also called *null flux curves*). The flux F_n enclosed by these contours obeys the equation (Backus 1968)

$$\frac{\mathrm{d}}{\mathrm{d}t} F_n \equiv \frac{\mathrm{d}}{\mathrm{d}t} \iint_{C_n} B_r \,\mathrm{d}S = \iint_{C_n} \frac{\mathrm{d}}{\mathrm{d}t} B_r \,\mathrm{d}S = 0, \tag{12}$$

where the integral is extended over the part of the spherical surface enclosed by the contour C_n. Since any solenoidal vector field intersecting a spherical surface has at least one contour of vanishing radial component, Equation (12) provides a powerful constraint on the secular variation under the assumption of negligible dissipation. Unfortunately, it is difficult to test Equation (12) because secular variation data depend mainly on measurements at observatories and thus suffer from poor global coverage. The problem is exacerbated by the fact that higher harmonics exhibit much stronger variation in time than lower ones, such that the time derivatives of the coefficients g_n^m, h_n^m increase with degree n when extrapolated to the core surface (Booker 1969). Alternate methods of data analysis avoiding the use of spherical harmonics are presently being investigated (Whaler & Gubbins 1981, Shure et al 1982).

A different test of the assumption that the magnetic field is frozen in the core fluid on the time scale of decades has been proposed by Hide (1978). Since Equation (12) is valid at the core surface but not, in general, at any other radius, Hide suggested that the radius of the highly conducting core of a planet can be determined as the point at which (12) becomes satisfied for the first time when the magnetic field is extrapolated in small steps downward from the surface. Hide's suggestion was motivated by the earlier attempt of Elphic & Russell (1978) to extract information about the core radius from a measured planetary magnetic field by correlating it with the surface at which the energy spectrum of the extrapolated magnetic field becomes white. Although not based on a well-founded physical principle, this latter method appears to work quite well and does not require the knowledge of time derivatives, as Hide's potentially much more accurate method does.

Hide did not use Equation (12), but instead used the pole strength

$$P \equiv \iint_S |\mathbf{B} \cdot \mathbf{n}| \, \mathrm{d}S. \tag{13}$$

Bondi & Gold (1950) showed that P is a constant in time if the integral is extended over a closed surface separating a fluid of infinite conductivity from the insulating exterior. Obviously P is equivalent to $\sum_n |F_n|$. The summation involved in P reduces the scatter of the data, and Hide & Malin (1981) found that the radius of the Earth's core could be determined magnetically within 10% of the much more accurately determined seismic value. Recent improvements of the analysis based on the MAGSAT satellite observations have yielded values within 2% of the correct radius (Voorhies & Benton 1982). These results demonstrate that most of the geomagnetic secular variation can be understood in terms of a rearrangement of the magnetic flux intersecting the core-mantle boundary. The alternative mechanism of the ejection of toroidal flux by diffusion above upwelling regions in the core (Allan & Bullard 1966) appears to be insignificant, at least on the decade time scale, either because the toroidal field in the core is not as large (300 G) as Allan & Bullard assumed or because the magnitude of the radial velocity is much less than the typical velocity of 0.3 cm s^{-1} inferred from secular variation.

This latter possibility is supported by an independent analysis of secular variation by Whaler (1980). Adapting a suggestion of Roberts & Scott (1965), Whaler plotted lines of vanishing $\partial B_r/\partial t$ at the core surface and found that these lines pass close by nearly all points where B_r assumes an extremum. According to Equation (11), this implies that $\nabla \cdot \mathbf{v}_s$ nearly vanishes at these points.

Since there are 17 extremal points fairly evenly distributed over the core surface, Whaler's results suggest that there is little upwelling or downwelling near the core-mantle boundary. Results obtained earlier by Benton

& Muth (1979) from the consideration of null flux contours agree with this conclusion. It is too early to deduce a stably stratified layer in the outermost core from these results, since motions dominated by the Coriolis force also tend to have small radial components (as Whaler points out).

The information about $\nabla_s \cdot \mathbf{v}_s$ is but one example of the variety of constraints on motions and magnetic fields near the core-mantle boundary that can be derived from sufficiently accurate secular variation data. For more detailed discussions, we refer the reader to the papers by Benton (1979, 1981) and Gubbins (1982).

Note added in proof In a recent paper Backus (1982) has considered the problem of inferring the electric field $\mathbf{E}^{(g)}$ produced by the geodynamo in the Earth's mantle. This problem is complicated by the unknown distribution of the conductivity σ in the lower mantle. In the geophysically interesting case when σ is a function of the radius r only and satisfies $-r\partial \ln \sigma/\partial r \gg 1$, Backus has been able to show the $\mathbf{r} \times \mathbf{E}^{(g)}$ is a function of θ and ϕ, but not of r, throughout the mantle. If data for $\mathbf{E}^{(g)}$ could be obtained by measuring the electric field near the Earth's surface (preferably at the ocean bottoms as Backus suggests) and by subtracting externally induced fields, then the tangential velocity $\mathbf{v}_s$ near the core-mantle boundary could be determined according to the relationship $\mathbf{v}_s = \mathbf{E}^{(g)} \times \mathbf{r}/r\mathbf{B}$. (The author is indebted to Paul H. Roberts for referring him to the work by Backus.)

Core-Mantle Coupling

Changes in the length of the day (l.o.d.) of the order of up to 4×10^{-3} s over the period of several years have long puzzled geophysicists. Because of the decade time scale and the considerable amplitude of the l.o.d. fluctuations, an exchange of angular momentum between the mantle and core of the Earth is their only feasible cause. A recent analysis of the l.o.d. data by Morrison (1979) has shown that torques of the order 10^{18} Nm exerted by the core on the mantle are sufficient to account for the observations. These torques can be explained by electromagnetic core-mantle coupling if a lower-mantle conductivity σ_{m} of the order of 10^2 mho m^{-1} is assumed (Rochester 1960, Roberts 1972b). Because the electromagnetic torque exerted on the mantle,

$$L = \iiint_{V_{\mathrm{m}}} \mathbf{r} \times (\mathbf{j} \times \mathbf{B})\, \mathrm{d}^3 V, \tag{14}$$

where V_{m} denotes the volume of the mantle, depends on the leakage currents $\mathbf{j}$ from the core, σ_{m} is a critical parameter of the theory.

Earlier analysis of l.o.d. data had suggested torques up to 10^{19} Nm, which would have required unreasonably high values of σ_{m}. This difficulty motivated Hide (1969) to suggest topographic core-mantle coupling as an alternative mechanism for angular-momentum exchange. Because of the dominating influence of rotation, forces exerted by bumps on the core-mantle boundary are primarily directed at a right angle to the relative motion. The computation of the drag caused by topographic coupling thus requires a detailed analysis of secondary effects. The Lorentz force generally has the effect of releasing the constraint of rotation, and Anufriyev & Braginsky (1977) have come to the conclusion that because of the influence of the magnetic field, topographic core-mantle coupling is relatively unimportant.

While electromagnetic core-mantle coupling appears to be capable, in principle, of accounting for the observed l.o.d. changes, little progress has been made until recently in the establishment of quantitative relationships between the variations of the magnetic field and the rotation of the Earth. The theory of electromagnetic coupling is complicated because of the different ways in which fluctuating and axisymmetric components of the magnetic field can contribute (Rochester 1960, Roberts 1972b, Acheson 1975). Correlations between various observational data have led to ambiguous conclusions, mainly because of the uncertainties in the geomagnetic secular variation data. But the strong correlation found between l.o.d. changes and accelerations in the secular variation of the east component of the magnetic field (Le Mouël et al 1981) has given rise to optimism. In an accompanying theoretical paper (Le Mouël & Courtillot 1982), it is shown that the observations can be interpreted in terms of an extension of Bullard's model. Bullard (1949) regards the westward drift of the geomagnetic field as representing the material velocity of an outer layer of the core while an inner part of the liquid core is more rapidly rotating. Le Mouël & Courtillot show that a torque between inner and outer parts of the liquid core produces an electromagnetic interaction between the three participating bodies that can explain the observed correlation between variations of the westward drift and of the l.o.d., including the time lag of about a decade between the two phenomena. The modeling of the core in terms of two rigidly rotating spherical shells is too crude to permit more detailed information on core-mantle coupling. But it indicates the way in which electromagnetic mantle coupling can provide constraints on more complete models of the geodynamo.

A model of the core consisting of an outer, stratified fluid shell and an inner, unstably stratified part has been proposed independently by Yukutake (1968, 1981). His finding that the drift rate is the same for different sectorial harmonics supports the idea that the westward drift reflects the relative rotation of the outer core, rather than a wave phenomenon as suggested by Hide (1966).

5. MODELS OF THE GEODYNAMO

In the past two decades dynamo theory has developed into a thriving subject of theoretical fluid dynamics and applied mathematics, but few attempts have been made to derive models for the dynamo in the Earth's core on the basis of the fundamental equations. While the numerical difficulties of solving the equation of induction (1) in a fluid sphere have largely been overcome, the hydromagnetic dynamo problem of the combined equations of motion and of induction coupled by the Lorentz force still poses a formidable challenge.

The difficulties start with the choice of forces driving motions in the core. Buoyancy forces of chemical or thermal origin are the most likely contenders. The growing solid inner core acts both as a source of heat and of low-density fluid, which are liberated as iron freezes onto its surface. The presence of light elements in the liquid core is well established, but the exact nature of these elements is a topic of much speculation (Stevenson 1981). Since the thermodynamics and energetics of the Earth's core have been discussed in recent reviews (Gubbins & Masters 1979, Verhoogen 1980, Loper & Roberts 1982), there is no need to discuss the subject here in detail. For the dynamics of the core the exact nature of the buoyancy force is not likely to be important since the Coriolis force and the Lorentz force dominate the energy-producing force. Because thermal buoyancy is more familiar to most workers in the field, it is the favored choice.

The main difficulty of the geodynamo problem is the multiplicity of solutions of the basic equations describing stationary equilibria and the likelihood of their instability. It is obvious that strictly steady solutions cannot be expected, since the problem of convection in a rotating sphere without magnetic field already leads to time-dependent solutions (Roberts 1968, Busse 1970). But numerous solutions appear to exist in which at least the axisymmetric components of velocity and magnetic field are steady. Some of these solutions may exhibit poloidal fields resembling the geomagnetic field, but differ with respect to the toroidal field inside the core. The strong variations of the Earth's magnetic field on the magnetic diffusion time scale τ_m (McElhinny & Senanayake 1982) suggest that all stationary states are unstable and that the dynamo process is intrinsically time dependent. Still, the exploration of stationary states is an important goal because the geodynamo may be characterized either by fluctuations about a stationary state or by nonlinear oscillations between different stationary states.

Aside from numerical integrations of the basic equations forward in time, which are barely feasible on today's computers and which require artificially high diffusivities for numerical stability, two methods appear to be suitable for determining stationary equilibrium states of the geodynamo. The first approach, which has been employed quite successfully by Braginsky (1965, 1976, 1978), is based on dominant mean azimuthal components of velocity and magnetic fields. Braginsky's analysis is based on an expansion of the variables in powers of $\mathrm{Re}_m^{-1/2}$, where Re_m is the magnetic Reynolds number formed with the azimuthal velocity. This is an attractive method because the magnetic Reynolds number in planetary cores is likely to be high and because a large part of the solution can be obtained in terms of analytical expressions. The analysis has been improved and elucidated by Soward (1972); for a detailed account, see Moffatt (1978). Since the buoyancy-driven nonaxisymmetric motions enter the analysis in relative high order, it is difficult to obtain a "closure" of the problem. It has not yet been demonstrated that the fluctuating motions in the α-effect for the generation of the poloidal field from the toroidal field are identical to those obtained as solutions of the equations of motion. Another difficulty mentioned by Braginsky (1978) is the displacement of the toroidal flux from the generation region. This effect causes $\alpha\omega$ dynamos such as Braginsky's to be oscillatory in general (Roberts 1972a); special conditions are required for steady solution. Braginsky overcomes this difficulty by concentrating the α-effect in the equatorial region.

An alternative approach to the nonlinear hydromagnetic dynamo problem is to follow the bifurcating solutions. Starting with the static solution for an internally heated self-gravitating fluid sphere, the first bifurcation leads to the onset of convection without magnetic field. This problem is well understood, and asymptotic analytical (Roberts 1968, Busse 1970, Soward 1977) as well as numerical (Gilman 1975, Cuong 1979) treatments of the problem exist. Convection in a rotating sphere is highly nonaxisymmetric and thus well suited to generate a magnetic field with a strong axisymmetric component. The onset of dynamo action occurs when the magnetic Reynolds number exceeds a critical value Re_{mc}. An analytical treatment of this problem is possible only when the spherical geometry is deformed into an annular configuration (Busse 1975b). But numerical computations of the spherical dynamo are in qualitative agreement with the analytical model (Cuong & Busse 1981). As the strength of the magnetic field increases, stable stationary states are possible as long as the magnetic energy density M is small. Since only effects of first order in M have been taken into account so far, little can be inferred from the analysis about the case of a strong magnetic field with a Lorentz force comparable to the Coriolis force. It is likely that new bifurcations occur as the solutions are extended to higher values of M. While the second approach permits a systematic exploration of equilibrium states of planetary dynamos, it has not yet been carried far enough to encounter two conditions that constrain planetary dynamos in the limit of low diffusivities. These conditions are the Taylor constraint and the balance of Coriolis and Lorentz forces governing the minimum Rayleigh number for convection.

The Taylor Constraint

For slow, steady motions of a rotating, electrically conducting, inviscid fluid within an axisymmetric container, Taylor (1963) derived the condition

$$\iint_C [(\nabla \times \mathbf{B}) \times \mathbf{B}]_\phi \, dS \approx 0, \tag{15}$$

where the integral is extended over any coaxial cylindrical surface and the subscript ϕ refers to the azimuthal component. Condition (14) reflects the fact that only the Coriolis force associated with radial motions is available for balancing the azimuthally averaged ϕ-component of the Lorentz force. But the average of the azimuthal Coriolis force over the coaxial cylindrical surface must vanish because inflow and outflow through the surface are in balance. This property leads to condition (15).

Mean zonal flows are generated if condition (15) is not satisfied. These flows inhibit the dynamo process and prevent the amplitude of the magnetic field from growing (Ierley 1982). The field strength is essentially limited by a balance between the Lorentz force and viscous friction. Only when the magnetic Reynolds number based on a meridional component of motion exceeds a second critical value can growing magnetic fields be expected of the class restricted by condition (15).

In planetary cores, condition (15) does not have to be satisfied exactly, because motions have finite amplitude and are not steady. Braginsky (1976) uses Ekman-layer suction to balance the geostrophic component of the Lorentz force, as the expression on the left-hand side of (15) may be called. The component is small because the component B_s of $\mathbf{B}$, which is perpendicular to the axis of rotation, nearly vanishes in Braginsky's *model Z*. The name "model Z" reflects this property: because $B_s \ll B_z$ the meridional field lines inside the core are parallel to the z- (or rotation) axis. Other possible balances involving the geostrophic component of the Lorentz force have been discussed by Childress (1982) and Ierley (1982).

Minimum Rayleigh-Number Convection

The fact that a magnetic field can facilitate convective flow in a rotating system is generally regarded as the physical reason for the occurrence of dynamos in planetary cores. Chandrasekhar (1961) found that the critical Rayleigh number for onset of convection in a horizontal layer rotating with angular velocity Ω about a vertical axis reaches a minimum as a function of the flux density B_0 of an imposed vertical magnetic field when the balance

$$\Lambda \approx 1 \tag{16}$$

is attained, where Λ is defined by

$$\Lambda \equiv \frac{B_0^2}{2\Omega\rho_0\mu\lambda} \equiv \frac{B_0^2\sigma}{2\Omega\rho_0}. \tag{17}$$

The balance between Coriolis and Lorentz forces, described by condition (16), is not a complete one. But the Rayleigh number is strongly reduced to a value of the order $\Omega d^2/\nu$ rather than $(\Omega d^2/\nu)^{4/3}$, where ν is the kinematic viscosity and d is the thickness of the layer. Eltayeb & Roberts (1970) suggested that condition (16) determines the field strength in planets and in stars. In the Earth's core, condition (16) corresponds to a modest field strength of 15 G of the toroidal field if the recent estimate $\sigma = 6 \times 10^5$ mho m^{-1} for the conductivity of the core is used (Stevenson 1981). In Jupiter's core, σ is about the same but ρ_0 is less than in the Earth's core, such that Λ exceeds the balance (16) already when the extrapolated strength of the poloidal field is used for B_0.

The difficulty of interpreting measured planetary fields in terms of the balance (16) has led to a search for alternative balances. Busse (1976) proposed as the characteristic planetary flux density the maximum field strength for which thermal Rossby waves can exist with sufficiently high wave number to accomplish dynamo action. Since the latter criterion involves the radius of the planetary core as a length scale, it appears to fit the observations much better than condition (16). But both criteria fail to take into account some nonlinear processes that may be decisive in determining the equilibrium strength of the magnetic field. The high magnetic Reynolds numbers realized in the cores of the major planets are likely to lead to flux expulsion from the convective eddies, for example, and a condition of the form (16) may indicate the equilibrium strength more closely if the reduced dynamic influence of the expelled magnetic field is taken into account.

Because these nonlinear effects are difficult to incorporate, most of the recent work on convection in the presence of rotation and magnetic fields has focused on the influence of the spherical geometry on the critical Rayleigh number for onset of convection (Eltayeb & Kumar 1977, Fearn 1979a,b). This aspect of the problem is well understood and can be presented in terms of simple analytical models. Because of the insights that can be gained from explicit expressions for wave number, frequency, and buoyancy parameters, we next discuss such a model in some detail.

6. A SIMPLE MODEL OF CONVECTION IN A ROTATING MAGNETIC SPHERE

Typical properties of convection in rotating spherical shells can be modeled by the simpler problem of convection in a rotating annulus with gravity perpendicular to the axis of rotation (Busse 1970). Here we use the annulus model to exhibit the effects of an azimuthal magnetic field on convection. As shown in Figure 2, the annulus is formed by two cylindrical surfaces with radii $r_0/4$ and $3r_0/4$, which are coaxial to the axis of rotation. The conical end surfaces are chosen to be tangential to the sphere at a distance $r_0/2$ from the axis. The temperatures T_2 and T_1 ($T_1 < T_2$) are prescribed on the inner and outer cylinders.

The problem of onset of convection in the annulus is characterized by five dimensionless parameters:

Buoyancy parameter $$B \equiv \frac{\beta 2(T_2 - T_1)g}{\Omega^2 r_0}, \tag{18a}$$

Ekman number $$E \equiv \nu/\Omega r_0^2, \tag{18b}$$

Prandtl number $$\mathrm{Pr} = \nu/\kappa, \tag{18c}$$

magnetic Ekman number $$\tau = \lambda/\Omega r_0^2, \tag{18d}$$

dimensionless magnetic energy $$\mathrm{C} = H_A^2/\rho_0 \mu \Omega^2 r_0^2, \tag{18e}$$

where β, g, and κ denote the coefficient of thermal expansion, the acceleration of gravity, and the thermal diffusivity, respectively. Ω is the angular velocity of rotation, and H_A is the flux density of the azimuthal field, which is assumed to be constant for simplicity. The equations of

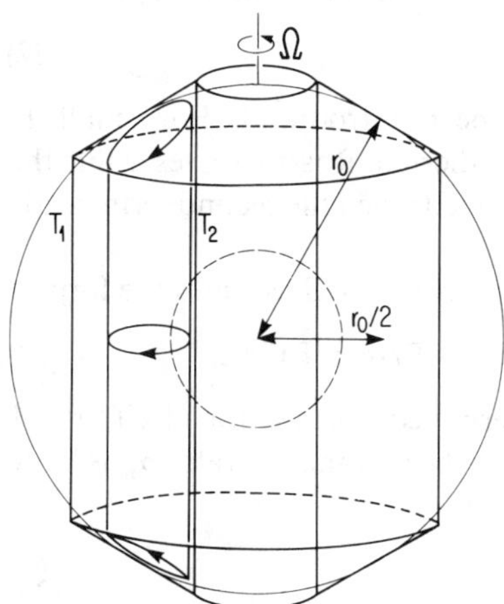

Figure 2 The geometrical configuration of the annulus used as a model for the dynamics in a thick rotating spherical shell.

motion and of induction can easily be solved in their linearized form after an exponential dependence on time and on azimuthal angle ϕ has been assumed:

$$\exp\,[i\omega\Omega t + i\alpha\phi/2].$$

The following expressions for ω and B are obtained on the basis of the small-gap approximation (Busse 1976; below referred to by B76):

$$\omega(1+\mathrm{Pr}) + 2\eta\alpha a^{-2} + \alpha^2 C\omega(\lambda/\kappa - 1)(\omega^2 + \tau^2 a^4)^{-1} = 0, \tag{19}$$

$$B = \mathrm{Pr}^{-1}E^2 a^6/\alpha^2 + \omega^2 a^2 \mathrm{Pr}/\alpha^2 + Ca^2\lambda/\kappa(\omega^2 + (E/\mathrm{Pr})^2 a^4)(\omega^2 + \tau^2 a^4)^{-1}. \tag{20}$$

While the small-gap approximation is not quite appropriate in the present case, only minor changes result if the correct solution is used.

Relationships (19) and (20) are identical to those given in B76, except that the length scale r_0 (instead of ℓ) has been used in the present case. As a result η corresponds to tan χ/cos χ, where $\chi = 30°$ is the angle inclination of the conical end surfaces of the annulus. Since the lowest radial wave number is $\gamma_1 = 2\pi$, a is given by

$$a^2 = 4\pi^2 + \alpha^2.$$

According to Equation (20), the buoyancy force is balanced by viscous friction, by inertial forces caused by the effects of rotation, and by the Lorentz force. The dispersion relation (19) can be simplified if the property

$$|\omega| \ll \tau a^2 \tag{21}$$

is used, which holds for the convection modes minimizing B in the geophysically interesting limit $E \to 0$, $\lambda/\kappa \to 0$. Accordingly, we obtain

$$\omega = -2\eta\alpha[a^2(1+\mathrm{Pr}) + \Gamma\alpha^2/a^2]^{-1}, \tag{22}$$

where the definition $\Gamma = C\lambda/\kappa\tau^2$ has been introduced. For small Γ, expression (22) describes the dispersion of thermal Rossby waves, while the frequency for slow magnetic waves is obtained when the second term inside the brackets dominates the first one.

After introducing relation (22), the expression for B assumes the form

$$B = (\mathrm{Pr}a^4 + \Gamma\alpha^2)[(Ea/\alpha\mathrm{Pr})^2 + 4\eta^2 a^2(a^4(1+\mathrm{Pr}) + \alpha^2\Gamma)^{-2}], \tag{23}$$

the minimum of which can readily be determined as a function of α if $\mathrm{Pr} = 1$ is assumed. For sufficiently low values of Γ the minimizing value α_m is large compared to unity and obeys the equation

$$(2\alpha_m^2 + \Gamma)^4 - \frac{4\eta^2}{E^2}(2\alpha_m^2 + 3\Gamma) = 0, \tag{24}$$

which yields

$$\alpha_m \approx \left(\frac{\eta}{\sqrt{2E}}\right)^{1/3} \tag{25}$$

for $\Gamma \ll E^{-2/3}$. Expression (25) governs the thermal Rossby wave regime, where α_m decreases only slowly with Γ and B stays about constant, as shown in Figure 3. As Γ approaches the order of magnitude $E^{-2/3}$, however, α_m^2 begins to change rapidly according to Equation (24). As α_m^2 becomes small compared to Γ, the assumption $a^2 = \alpha^2$ is no longer appropriate in determining the minimum, even though $\alpha \gg 1$ is still valid. After dropping sharply, but continuously, α_m begins to follow the relationship

$$\alpha_m^2 = 4\pi\eta/E\Gamma, \tag{26a}$$

and B obeys the simple law

$$B \approx 4\eta^2/\Gamma. \tag{26b}$$

Expressions (26a) and (26b) are valid within the magnetic Rossby wave regime

$$E^{-2/3} \lesssim \Gamma \lesssim E^{-1}. \tag{26c}$$

As α_m approaches values of order unity, it must be taken into account that α possesses a lower bound $\alpha_\ell = 2$ for which a single wave length fits into the annular region. Without considering the property that α assumes only

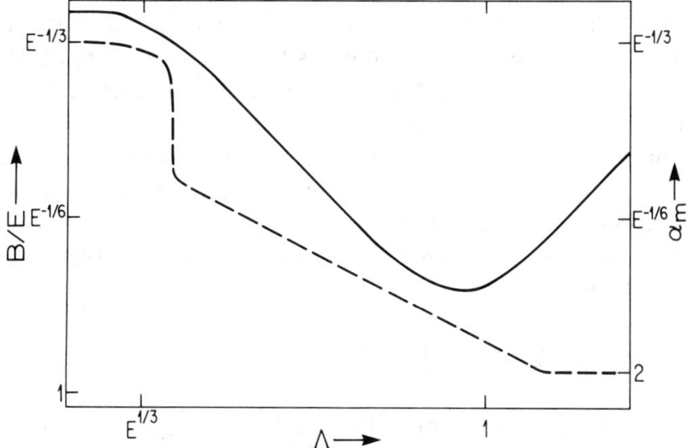

Figure 3 Buoyancy parameter B (solid line) and azimuthal wavenumber α of columnar convection modes as a function of the parameter Λ. This parameter is related to Γ by the relationship $\Lambda = \Gamma E/2$. The graph has been drawn for $E = 10^{-15}$ using the asymptotic relationships discussed in the text. The scales are logarithmic.

integer values, we note that B reaches its minimum

$$B_{\min} = 16\pi E\eta \tag{27}$$

for $\Gamma = \eta/\pi E$, $\alpha \approx 2\pi$. As Γ increases further, B increases according to the relationship

$$B = 4(\pi^2+1)(\Gamma E^2+\eta^2\Gamma^{-1}). \tag{28}$$

Since $\Gamma E = 2\Lambda$, the position of the minimum is in approximate agreement with the findings mentioned in the preceding section. The other regimes also agree well with the numerical results obtained by Fearn (1979a,b) for a spherical geometry, except for the magnetically induced instability caused by the inhomogenous magnetic field in Fearn's problem. The frequency ω decreases monotonically with Γ. Although it reaches very low values it never changes sign, as it does in Fearn's spherical case for $\Lambda \approx 10$. This latter property is probably connected with the fact that the minimizing convection mode is no longer a columnar mode aligned with the axis of rotation, as has been assumed in the annulus model. As Λ exceeds 4 a mode inclined with respect to the axis of rotation becomes preferred (Eltayeb 1972).

The low frequencies and the associated low growth rates of the slow magnetic modes indicate that periods much in excess of the magnetic decay time are required to establish the convection mode minimizing B. Because of the higher growth rate of the magnetic field, the changeover from thermal Rossby modes to magnetic modes may occur only gradually and field strength in excess of the optimum strength given by $\Lambda \approx 1$ may be attained. As the slow magnetic modes finally become established a new problem occurs, in that these modes are not likely to be generators of the magnetic field because of their low azimuthal wave number (Cuong & Busse 1981). Thus the magnetic field will start to decay and thermal Rossby waves will become important again. Whether cycles of this nature occur locally or globally in planetary cores is a question that cannot be answered at this time. Since the slow magnetic modes tend to be particularly sensitive to finite amplitude effects, a more thorough understanding of the problem will require a nonlinear theory.

7. CONCLUDING REMARKS

In this review only a brief introduction has been given to some of the research areas that are presently studied with the general goal to improve the understanding of the process by which planetary magnetic fields are generated. The high level of activity, as shown by the increasing rate of publications, indicates that much more will be learned in the coming years. In the past decade considerable progress has been made in the theory of hydromagnetic dynamos and in the understanding of convective motion subject to both Coriolis and Lorentz forces. But important nonlinear properties of planetary dynamos still seem to be beyond the grasp of deductive analysis. In returning to Table 1, we note that most of the main properties of geomagnetism have found explanations. The alignment of convective eddies with the axis of rotation produces the alignment of the dipolar axis. Dipolar fields seem to be preferred in comparison to quadrupolar fields because of the action of differential rotation in planetary cores (Cuong & Busse 1981). The approximately stationary state of the geodynamo could be the result of a dominant α^2-dynamo process or it could be achieved by an $\alpha\omega$-dynamo process working within certain constraints, as in Braginsky's (1978) model. The theoretical prediction of the strength of planetary magnetism has not been possible so far, mainly because of the unknown relationship between observed poloidal and unobservable toroidal fields in planetary cores. The variability of the geomagnetic dipole over the magnetic decay time scale is of special interest in this respect, since observational evidence for oscillations could provide a crucial piece of information.

Models of convection in the Earth's core predict that the strongest motion occurs outside the cylindrical surface touching the solid inner core at its equator (Busse & Cuong 1977, Cuong & Busse 1981). This property leads to equatorial field strengths in excess of that of a pure dipole, in agreement with the presently observed field (Figure 1). But paleomagnetic measurements tend to indicate an enhanced strength at the poles. Enhanced convection at high latitudes could be a result of inhomogeneous conditions at the core-mantle boundary introduced by mantle convection. The interaction of mantle convection and core dynamics is a most fascinating problem, but little is known about it yet.

While individual problems in the area of geomagnetism are often not well constrained and do not offer unique interpretations of the observational evidence, there are good reasons to believe that the theoretical constraints derived for various properties of the geodynamo will eventually fit the observations within a fairly unique framework. The problem of geomagnetism has often been compared to the problem of the general circulation of the atmosphere, not only because maps of secular variation look like weather maps but, more importantly, because the Earth's rotation plays a dominant role in both problems. The problem of geomagnetism is made difficult by the lack of important information; yet because of the greater homogeneity of conditions in the core, it is likely to have a simpler solution than the problem of atmospheric motions. The problem of turbulence, which complicates the understanding of the dynamics in both the atmosphere and the core, is much less severe in the latter case because of the dominant effect of rotation and the moderate value of the magnetic Reynolds number. It thus appears that a coherent theory of the diverse geophysical observations related to core processes is within reach of present scientific capabilities.

Literature Cited

Acheson, D. 1975. On the theory of the geomagnetic westward drift and rapid fluctuations in the earth's rotation rate. *Pageoph* 113:611–24

Allan, D. W. 1962. On the behaviour of systems of coupled dynamos. *Proc. Cambridge Philos. Soc.* 58:671–93

Allan, D. W., Bullard, E. C. 1966. The secular variation of the earth's magnetic field. *Proc. Cambridge Philos. Soc.* 62:783–809

Anufriyev, A. P., Braginsky, S. I. 1977. Effects of irregularities of the boundary of the earth's core on the speed of the fluid and on the magnetic field. III. *Geomagn. Aeron.* 17:492–97

Backus, G. E. 1958. A class of self sustaining dissipative spherical dynamos. *Ann. Phys.* 4:372–447

Backus, G. E. 1968. Kinematics of geomagnetic secular variation in a perfectly conducting core. *Philos. Trans. R. Soc. London Ser. A* 263:239–66

Backus, G. E. 1982. The electric field produced in the mantle by the dynamo in the core. *Phys. Earth Planet. Inter.* 28:191–214

Benton, E. R. 1979. Magnetic probing of planetary interiors. *Phys. Earth Planet. Inter.* 20:111–18

Benton, E. R. 1981. Inviscid, frozen-flux velocity components at the top of earth's core from magnetic observations at earth's surface: Part 1, a new methodology. *Geophys. Astrophys. Fluid Dyn.* 18:157–74

Benton, E. R., Muth, L. A. 1979. On the strength of electric currents and zonal magnetic fields at the top of the earth's core: Methodology and preliminary estimates. *Phys. Earth. Planet. Inter.* 20:127–33

Bondi, J. R., Gold, T. 1950. On the generation of magnetism by fluid motion. *MNRAS* 110:607–11

Booker, J. R. 1969. Geomagnetic data and core motions. *Proc. R. Soc. London* 309:27–40

Braginsky, S. I. 1965. Theory of the hydromagnetic dynamo. *Soviet Phys. JETP* 20:1462–71

Braginsky, S. I. 1975. An almost axially symmetrical model of the hydromagnetic dynamo of the earth I. *Geomagn. Aeron.* 15:122–27

Braginsky, S. I. 1976. On the nearly axially-symmetrical model of the hydromagnetic dynamo of the earth. *Phys. Earth Planet. Inter.* 11:191–99

Braginsky, S. I. 1978. Nearly axially symmetric model of the hydromagnetic dynamo of the earth. *Geomagn. Aeron.* 18:225–31

Bullard, E. C. 1949. The magnetic field within the earth. *Proc. R. Soc. London Ser. A* 197:433–53

Bullard, E. C., Gellman, H. 1954. Homogeneous dynamos and terrestrial magnetism. *Philos. Trans. R. Soc. London Ser. A* 247:213–78

Bullard, E. C., Gubbins, D. 1977. Generation of magnetic fields by fluid motions of global scale. *Geophys. Astrophys. Fluid Dyn.* 8:43–56

Busse, F. H. 1970. Thermal instabilities in rapidly rotating systems. *J. Fluid Mech.* 44:441–60

Busse, F. H. 1975a. A necessary condition for the geodynamo. *J. Geophys. Res.* 80:278–80

Busse, F. H. 1975b. A model of the geodynamo. *Geophys. J. R. Astron. Soc.* 42:437–59

Busse, F. H. 1976. Generation of planetary magnetism by convection. *Phys. Earth Planet. Inter.* 12:350–58

Busse, F. H. 1978. Magnetohydrodynamics of the Earth's dynamo. *Ann. Rev. Fluid Mech.* 10:435–62

Busse, F. H., Carrigan, C. R. 1976. Laboratory simulation of thermal convection in rotating planets and stars. *Science* 191:81–83

Busse, F. H., Cuong, P. G. 1977. Convection in rapidly rotating spherical fluid shells. *Geophys. Astrophys. Fluid Dyn.* 8:17–44

Busse, F. H., Miin, S. W. 1979. Spherical dynamos with anisotropic α-effect. *Geophys. Astrophys. Fluid Dyn.* 14:167–81

Chandrasekhar, S. 1961. *Hydrodynamic and Hydromagnetic Stability*. Oxford: Clarendon. 652 pp.

Childress, S. 1969. Théorie magnétohydrodynamique de l'effet dynamo. *Dép. Méch. Fac. Sci., Paris, Rep.*

Childress, S. 1982. The macrodynamics of spherical dynamos. In *Stellar and Planetary Magnetism*, ed. A. M. Soward. New York: Gordon & Breach

Cook, A. E., Roberts, P. H. 1970. The Rikitake two-disk dynamo system. *Proc. Cambridge Philos. Soc.* 68:547–69

Coupland, D. H., Van der Voo, R. 1980. Long-term nondipole components in the geomagnetic field during the last 130 m.y. *J. Geophys. Res.* 85:3529–48

Cowling, T. G. 1934. The magnetic field of sunspots. *MNRAS* 94:39–48

Cox, A. 1975. The frequency of geomagnetic reversals and the symmetry of the non-dipole field. *Rev. Geophys. Space Phys.* 13:35–51

Cuong, P. G. 1979. *Thermal convection and magnetic field generation in rotating spherical shells.* Dissertation. Univ. Calif., Los Angeles

Cuong, P. G., Busse, F. H. 1981. Generation of magnetic fields by convection in a rotating sphere I. *Phys. Earth Planet. Inter.* 24:272–83

Elphic, R. C., Russell, C. T. 1978. On the apparent source depth of planetary magnetic fields. *Geophys. Res. Lett.* 5:211–14

Elsasser, W. M. 1946. Induction effects in terrestrial magnetism. *Phys. Rev.* 69:106–16

Eltayeb, I. A. 1972. Hydromagnetic convection in a rapidly rotating fluid layer. *Proc. R. Soc. London Ser. A* 326:229–54

Eltayeb, I. A., Kumar, S. 1977. Hydromagnetic convective instability in a rotating self-gravitating fluid sphere containing a uniform distribution of heat sources. *Proc. R. Soc. London Ser. A* 353:145–62

Eltayeb, I. A., Roberts, P. H. 1970. On the hydromagnetics of rotating fluids. *Astrophys. J.* 162:699–701

Fearn, D. R. 1979a. Thermal and magnetic instabilities in a rapidly rotating fluid sphere. *Geophys. Astrophys. Fluid Dyn.* 14:103–26

Fearn, D. R. 1979b. Thermally driven hydromagnetic convection in a rapidly rotating sphere. *Proc. R. Soc. London Ser. A* 369:227–42

Fuller, M., Williams, I., Hoffman, K. A. 1979. Paleomagnetic records of geomagnetic field reversals and morphology of the transitional field. *Rev. Geophys. Space Phys.* 17:179–203

Gilman, P. A. 1975. Linear simulations of Boussinesq convection in a deep rotating spherical shell. *J. Atmos. Sci.* 32:1331–52

Gubbins, D. 1973. On large magnetic Reynolds number dynamos. *Geophys. J. R. Astron. Soc.* 33:57–64

Gubbins, D. 1982. Finding core motions from magnetic observations. *Philos. Trans. R. Soc. London Ser. A* 306:247–54

Gubbins, D., Masters, T. G. 1979. Driving mechanisms for the Earth's dynamo. *Adv. Geophys.* 21:1–50

Hide, R. 1966. Free hydromagnetic oscillations of the earth's core and the theory of geomagnetic secular variation. *Philos. Trans. R. Soc. London Ser. A* 259:615–50

Hide, R. 1969. Interaction between the Earth's liquid core and solid mantle. *Nature* 222:1055

Hide, R. 1978. How to locate the electrically conducting fluid core of a planet from external magnetic observations. *Nature* 271:640–41

Hide, R., Malin, S. R. C. 1981. On the determination of the size of the earth's core from observations of the geomagnetic secular variation. *Proc. R. Soc. London Ser. A* 374:15–33

Hide, R., Palmer, T. N. 1982. Generalization of Cowling's Theorem. *Geophys. Astrophys. Fluid Dyn.* 19:301–9

Ierley, G. R. 1982. Macrodynamics of Alpha-Dynamos. *Bull. Am. Phys. Soc.* 27:549 (Abstr.)

Jacobs, J. A. 1981. Heat flow and reversals of the earth's magnetic field. *J. Geomagn. Geoelectr.* 33:527–29

James, R. W., Roberts, P. H., Winch, D. E. 1980. The Cowling anti-dynamo theorem. *Geophys. Astrophys. Fluid Dyn.* 15:149–60

Jones, G. M. 1977. Thermal interaction of the core and the mantle and long-term behavior of the geomagnetic field. *J. Geophys. Res.* 82:1703–9

Kahle, A. B., Ball, R. H., Vestine, E. H. 1967. Comparison of estimates of surface fluid motions of the earth's core for various epochs. *J. Geophys. Res.* 72:4917–25

Krause, F., Rädler, K.-H. 1980. *Mean-Field Magnetohydrodynamics and Dynamo Theory.* Oxford: Pergamon. 270 pp.

Le Mouël, J.-L., Courtillot, V. 1982. On the outer layers of the core and geomagnetic secular variation. *J. Geophys. Res.* 87:4103–8

Le Mouël, J.-L., Madden, T. R., Ducruix, J., Courtillot, V. 1981. Decade fluctuations in geomagnetic westward drift and Earth rotation. *Nature* 290:763–65

Loper, D. E., Roberts, P. H. 1982. Compositional convection and gravitationally powered dynamo. In *Stellar and Planetary Magnetism*, ed. A. M. Soward. New York: Gordon & Breach

Lorenz, E. N. 1963. Deterministic nonperiodic flow. *J. Atmos. Sci.* 20:130–41

Lortz, D., Meyer-Spasche, R. 1982a. On the decay of symmetric dynamo fields. *Math. Meth. Appl. Sci.* 4:91–97

Lortz, D., Meyer-Spasche, R. 1982b. On the decay of symmetric toroidal dynamo fields. *Z. Naturforsch. Teil A* 37:736–40

Marcus, P. S. 1981. Effects of truncation in modal representations of thermal convection. *J. Fluid Mech.* 103:241–55

Malkus, W. V. R. 1972. Reversing Bullard's dynamo. *EOS, Trans. Am. Geophys. Union* 53:617 (Abstr.)

McElhinny, M. W., Senanayake, W. E. 1982. Variations in the geomagnetic dipole I:

The past 50,000 years. *J. Geomagn. Geoelectr.* 34:39–51

Merrill, R. T., McElhinny, M. W. 1977. Anomalies in the time-arranged paleomagnetic field and their implications for the lower mantle. *Rev. Geophys. Space Phys.* 15:309–23

Merrill, R. T., McElhinny, M., Stevenson, D. J. 1979. Evidence for long-term asymmetries in the Earth's magnetic field and possible implications for dynamo theories. *Phys. Earth Planet. Inter.* 20:75–82

Moffatt, H. K. 1978. *Magnetic Field Generation in Electrically Conducting Fluids.* Cambridge Univ. Press. 343 pp.

Morrison, L. V. 1979. Re-determination of the decade fluctuations in the rotation of the Earth in the period 1861–1978. *Geophys. J. R. Astron. Soc.* 48:349–60

Parker, E. N. 1979. *Cosmical Magnetic Fields.* Oxford: Clarendon. 860 pp.

Phillips, J. D. 1977. Time variations and asymmetry in the statistics of geomagnetic reversal sequences. *J. Geophys. Res.* 82:835–43

Proctor, M. R. E. 1977a. The role of mean circulation in parity selection by planetary magnetic fields. *Geophys. Astrophys. Fluid Dyn.* 8:311–24

Proctor, M. R. E. 1977b. On Backus' necessary condition for dynamo action in a conducting sphere. *Geophys. Astrophys. Fluid Dyn.* 9:89–93

Rikitake, T. 1958. Oscillations of a system of disk dynamos. *Proc. Cambridge Philos. Soc.* 54:89–105

Robbins, K. A. 1977. A new approach to subcritical instability and turbulent transitions in a simple dynamo. *Math. Proc. Cambridge Philos. Soc.* 82:309–25

Roberts, P. H. 1968. On the thermal instability of a rotating-fluid sphere containing heat sources. *Philos. Trans. R. Soc. London Ser. A* 263:93–117

Roberts, P. H. 1972a. Kinematic dynamo models. *Philos. Trans. R. Soc. London Ser. A* 272: 663–98

Roberts, P. H. 1972b. Electromagnetic core-mantle coupling. *J. Geomagn. Geoelectr.* 24:231–59

Roberts, P. H., Scott, S. 1965. On analysis of the secular variation. 1. A hydromagnetic constraint: theory. *J. Geomagn. Geoelectr.* 17:137–51

Roberts, P. H., Soward, A. M. 1975. A unified approach to mean field electrodynamics. *Astron. Nachr.* 296:49–64

Rochester, M. G. 1960. Geomagnetic westward drift and irregularities in the earth's rotation. *Philos. Trans. R. Soc. London Ser. A* 252:531–55

Ruelle, D., Takens, F. 1971. On the nature of turbulence. *Commun. Math. Phys.* 20:167–92

Russell, C. T. 1980. Planetary magnetism. *Rev. Geophys. Space Phys.* 18:77–106

Shure, L., Parker, R. L., Backus, G. E. 1982. Harmonic splines for geomagnetic modeling. *Phys. Earth Planet. Inter.* 28:215–29

Soward, A. M. 1972. A kinematic theory of large magnetic Reynolds number dynamos. *Philos. Trans. R. Soc. London Ser. A* 272:431–62

Soward, A. M. 1977. On the finite amplitude thermal instability in a rapidly rotating fluid sphere. *Geophys. Astrophys. Fluid Dyn.* 9:19–74

Steenbeck, M., Krause, F., Rädler, K.-H. 1966. A calculation of the mean emf in an electrically conducting fluid in turbulent motion, under influence of Coriolis forces. *Z. Naturforsch. Teil A* 21:369–76

Stevenson, D. J. 1980. Saturn's luminosity and magnetism. *Science* 208:746–48

Stevenson, D. J. 1981. Models of the earth's core. *Science* 214:611–19

Stevenson, D. J. 1982. Reducing the non-axisymmetry of a planetary dynamo and an application to Saturn. *Geophys. Astrophys. Fluid Dyn.* 21:113–27

Taylor, J. B. 1963. The magneto-hydrodynamics of a rotating fluid and the earth's dynamo problem. *Proc. R. Soc. London Ser. A* 274:274–83

Todoeschuck, J. P., Rochester, M. G. 1980. The effect of compressible flow on anti-dynamo theorems. *Nature* 284:250–51

Verhoogen, J. 1980. *Energetics of the Earth.* Washington DC: Natl. Acad. Sci. 139 pp.

Vogt, P. R. 1975. Changes in geomagnetic reversal frequency at times of tectonic changes: evidence for coupling between core and upper mantle processes. *Earth Planet. Sci. Lett.* 25:313–21

Voorhies, C. V., Benton, E. R. 1982. Pole-strength of the earth from MAGSAT and magnetic determination of the core radius. *Geophys. Res. Lett.* 9:258–61

Whaler, K. A. 1980. Does the whole of the earth's core convect? New evidence from geomagnetism. *Nature* 287:528–30

Whaler, K. A., Gubbins, D. 1981. Spherical harmonic analysis of the geomagnetic field: an example of a linear inverse problem. *Geophys. J. R. Astron. Soc.* 65:645–93

Wilson, R. L. 1970. Permanent aspects of the earth's non-dipole magnetic field over Upper Tertiary times. *Geophys. J. R. Astron. Soc.* 19:417–39

Yukutake, T. 1968. The drift velocity of the geomagnetic secular variation. *J. Geomagn. Geoelectr.* 20:403–14

Yukutake, T. 1981. A stratified core motion inferred from geomagnetic secular variation. *Phys. Earth Planet. Inter.* 24:253–58

Ann. Rev. Earth Planet. Sci. 1983. 11 : 269–98

RATES OF BIOGEOCHEMICAL PROCESSES IN ANOXIC SEDIMENTS

William S. Reeburgh
Institute of Marine Science, University of Alaska, Fairbanks, Alaska 99701

INTRODUCTION

A new type of study has appeared in the interstitial water literature during the last five years. Guided by microbial ecologists and aided by newly developed analytical techniques from organic geochemistry, direct measurements of a number of remineralization rates have been made in sediments using incubation and stable and radioisotope tracer techniques. These measurements have enormous potential to both confirm and extend the diagenetic models that have been used so successfully during the past decade. These techniques will probably emerge as one of our most effective tools in elucidating the controls on early diagenetic reactions. Since these reaction rate measurements are just beginning, and since the potential is high for great leaps in understanding (as well as misunderstanding), it is important that they be consolidated and presented as an easily compared unit.

This review has several goals. First, since most of these direct rate measurements are scattered throughout the literature of microbial ecology and oceanography, it summarizes and consolidates the measurements so that comparisons between environments and studies can be made conveniently. Second, it compares the results of these direct rate measurements with those predicted by models. Third, it attempts to show the importance of various reactions to total sediment metabolism. Fourth, it presents some perspectives and insights on how future rate measurements should be conducted to insure ready comparison between studies and environments.

The results of the studies summarized here are presented in tables; I have attempted to include the most recent work on sediments. The reported rates

0084–6597/83/0515–0269$02.00

or rate constants have not been extended beyond conversion to uniform units. Since the water contents of most of the sediments considered here are high, no distinctions have been made in results reported in interstitial water and sediment volume units. Studies of the turnover of specific compounds are only included in the tables when they are accompanied by pool size or concentration measurements. Turnover measurements are invaluable in determining reaction pathways and products, but without knowledge of the ambient concentration of the compound studied they give no information on the importance of a transformation to the sediment system. This review does not emphasize a number of important and related subjects, namely, inorganic reactions such as those reviewed recently by Gieskes (1975, 1981) and Manheim & Sayles (1974) for Deep Sea Drilling Project sediments, chronologies (Goldberg & Bruland 1974), sampling methods (Kriukov & Manheim 1982), bioturbation (Aller 1978, Guinasso & Schink 1975), irrigation (Grundmanis & Murray 1977), and studies of bacterial identities, numbers, and physiology (Jørgensen 1978c, Karl 1982).

Studies on the composition of interstitial waters provide a key to many problems in early diagenesis. Interstitial waters are particularly attractive for study because conditions in the interstitial environment, such as limited mixing and circulation, a high surface: volume ratio, and an abundant supply of organic detritus, support large microbial populations, which in turn lead to large chemical composition changes. Studies on interstitial waters from a variety of lacustrine, estuarine, and deep-sea environments are reported in an extensive literature that has been periodically reviewed in chapters and articles (Glasby 1973, Manheim 1976, Gieskes 1981) and in a number of books (Berner 1971, 1980, Kaplan 1974, McCave 1976, Fanning & Manheim 1982). Fenchel & Jørgensen's (1977) review of the role of bacteria in detritus food chains, Fenchel & Blackburn's (1979) book, and Karl's (1982) article are particularly valuable in the context of this review.

ENVIRONMENTS

Since many of the studies summarized here have been conducted in a limited number of coastal environments, some of the important characteristics of the most familiar and frequently cited are summarized in Table 1. It should be noted that the most important parameter in driving the remineralization processes summarized in this paper—the flux of organic carbon—is rarely measured directly; it is usually estimated from budgets and sedimentation rates. This, plus a lack of information on the composition and properties of the organic matter that actually reaches the sediments, is one of our major knowledge gaps.

These environments are not important from a global mass balance

Table 1 Characteristics of frequently studied anoxic sediment environments

Location	Depth (m)	Sedimentation rate (cm yr^{-1})	Carbon content (% dry wt)	Temperature (°C)	Water column O_2	Bioturbation	$SO_4^=$ reducing zone thickness (cm)	Carbon flux to sediments (mmole cm^{-2} yr^{-1})
Chesapeake Bay (Reeburgh 1969)	30.4 (858-C) 15.2 (858-D)	0.1–1.0	3	4–25	oxic	+	30	—
Santa Barbara Basin (Sholkovitz 1973)	590 (450–475 m sill)	0.4	2–3	6.2–6.4	0.05–0.1 ml L^{-1}	–	200–300	—
Limfjorden (Jørgensen 1977)	4–12	0.2	1–13	0–20	summer anoxia	+	>140	1.42
Cariaco Trench (Reeburgh 1976)	1300–1400 (150-m sill)	0.05	4	16.7	permanent anoxia	–	45–50	0.1
Long Island Sound (Goldhaber et al 1977, Martens & Berner 1977, Rosenfeld 1981)	9 (FOAM)	0.3	1.5–2	4–28	oxic	+	20	—
Cape Lookout Bight (Martens & Klump 1980, Chanton 1979, Klump 1980	10	8.4–11.6	3.5–4	5–28	oxic	+ winter – summer	10–20	11.1
Saanich Inlet (Murray et al 1978, Anderson & Devol 1973)	225 (70-m sill)	1.3	3–4	9	anoxic ~8 mo yr^{-1}	–	20	0.54
Skan Bay (Reeburgh 1980)	65 (10-m sill)	0.7–0.8	2–3	4	intermittent anoxia (late summer)	–	20	0.36

standpoint, but they are important because they represent a range of end-member environments where sulfate reduction and methanogenesis are occurring. Further, because they are convenient to study and since so much is already known about them, they seem logical environments for future studies of the remineralization of organic matter. Even though several environments appear to be nearly isothermal (Santa Barbara Basin, Saanich Inlet), the fact that their sediments are varved indicates that inputs are not uniform in time and that seasonal variations in the organic and inorganic inputs are a characteristic of all of the environments listed. Only a few seasonal studies on sediments have been performed.

REACTION SEQUENCE AND CAPACITY

It is generally recognized that organic matter degradation in sediments proceeds using the available oxidant producing the greatest free energy. Several authors have summarized energy yields and reaction sequences by considering oxidation of glucose (Claypool & Kaplan 1974) or hypothetical compounds such as CH_2O (Berner 1980) and the Redfield molecule, $(CH_2O)_{106}(NH_3)_{16}H_3PO_4$ (Froelich et al 1979, Emerson et al 1980). Organic matter degradation has been observed to follow this sequence of reactions, with each successive reaction starting when the previous oxidant is either exhausted or chemical and biological conditions allow a particular community of organisms to become active.

Portraying the oxidizing capacity of a partially open dynamic system like sediments is difficult, but it can be approached by considering a closed system. The distribution and concentration of species under a range of redox conditions may be shown with log concentration vs $p(e)$ diagrams (Stumm & Morgan 1981, Breck 1974) or with log concentration vs pH $+p(e)$ diagrams (Lindsay 1979), but these tend to become complicated when the commonly encountered oxidants and their products are plotted together. The above approaches imply that we know more about reactions in sediments than we actually do.

Table 2 illustrates the organic matter oxidizing sequence and capacity of marine sediments by summarizing $p(e)$, Eh, and energy yield values from previous references and by also considering a hypothetical sediment that presents dissolved and solid-phase oxidants in comparable concentration units. This sediment has an oxygen-saturated seawater content of 85% and a density of 1.1 g cm^{-3}. Whole sediment concentrations for the hypothetical sediment were obtained using appropriate proportions of the typical dissolved and solid-phase constituents. The oxidation capacity values were derived using Berner's (1980) stoichiometry for CH_2O oxidation. This example is also unrealistic because it considers a closed, homogeneous

Table 2 Sequence and capacity of organic matter oxidation processes important in marine sediments

Reaction	$p(e)$	Eh (mv)	Energy yield (KJ $mole^{-1}$ CH_2O[a])	Typical concentration (mM)	Concentrations in hypothetical sediment (mmole L^{-1} sed.)	CH_2O oxidizing capacity (mmole L^{-1} sed.)	Depth scales in typical environments: Low CH_2O flux	Depth scales in typical environments: High CH_2O flux
O_2 reduction	12.1–12.5	720–740	−475	0–0.09	0–0.85	0–0.85	0–1 cm	
Denitrification	~12	710	−448	0–0.04	0–0.037	0–0.03	15 cm	
Mn(IV) → Mn(II)	8.0	470	−349	ΣMn:				
				≥1.5% in deep sea (0.27 mmole g^{-1})	43.2	21.6 (?)	15–30 cm	
				0.15% in Skan Bay (0.027 mmole g^{-1})	4.3	2.2 (?)	≥1 m	~1–5 cm
Fe(III) → Fe(II)	1.0	60	−114	ΣFe:				
				≥4.0% in deep sea (0.7 mmole g^{-1})	112	28 (?)	20+ cm	
				2.0% in Skan Bay (0.35 mmole g^{-1})	56	14 (?)		
Sulfate reduction	−3.8	−200	−77	0–30	0–28.2	56.4		10+ cm
Methanogenesis (fermentation)	−4.2	−250	−58			(?)		25+ cm
			Dissolved organic carbon	1–7	0.95–6.5			
			Particulate organic carbon	4% (3.3 mmole g^{-1})	528			

[a] 1 KJ = 0.239 Kcal.

system, neglects diffusive supply and recycling, and overestimates the concentration of dissolved oxygen and nitrate. Table 2 shows clearly how minor additions of organic matter can overwhelm the oxygen reduction and denitrification capacity. It also shows that sulfate reduction, which is the principal difference between freshwater and marine sediments, is the dominant process because of its large capacity. The comparison of Sørensen et al (1979) experimentally demonstrates the dominant oxidizing capacity of sulfate reduction in marine sediments. The table also points out our poor understanding of the role of Mn and Fe oxides in these sediments. We lack knowledge of how much of the total amount reported is available for redox reactions and whether microbial mediation of Mn and Fe is important in oxidation of organic matter.

The right-hand column of Table 2 shows typical depth scales for this sequence of generalized reactions in two end-member environments; the low organic flux example is the pelagic sediment suboxic diagenesis study of Froelich et al (1979), and the high organic flux example is typical of the environments summarized in Table 1. The depth interval over which sulfate reduction takes place is shown in more detail in Table 1. Bioturbation and irrigation have the effect of increasing rates (Aller 1978, Aller & Yingst 1980) and physically thickening the zones of all reactions preceding sulfate reduction in high organic flux coastal sediments.

MODELS

Two broad classes of models are used in the interpretation of diagenetic processes in interstitial waters: one-dimensional diffusion-advection-reaction models, and analog models that consider decomposition processes in sediments using the rumen (Hungate 1975, Wolin 1979) and sewage digesters (Hobson et al 1974) as model systems. The diagenetic models describe depth distributions of reactants or products without regard to specific reactions or mechanisms, while the analog models draw on a wealth of experience to predict intermediates and products.

The diagenetic models have been developed and extended in a series of papers and books by Berner (1964, 1971, 1974, 1976a, 1980, 1981). These models consider diffusion, advection due to sedimentation, and reaction at steady state in equations similar to the following:

$$\frac{dC}{dt} = \frac{D\partial^2 C}{\partial x^2} - \frac{\omega \partial C}{\partial x} - kC = 0,$$

where C is concentration, t is time, x is distance (positive downward), D is diffusivity, ω is sedimentation rate, and k is a first-order rate constant. Corrections for porosity and sorption are usually applied to the diffusivity.

Values for the diffusivity have been measured for common ionic species (Li & Gregory 1974, Goldhaber et al 1977, Krom & Berner 1980a, Hesslein 1980) and hydrocarbons (Sahores & Witherspoon 1970). Adsorption is incorporated in these models (Berner 1974, 1976b) using a term of the form $-(1+K)\frac{\omega\partial C}{\partial x}$, where K is an adsorption coefficient. Adsorption coefficients have been measured experimentally by Rosenfeld (1979), Klump (1980), and Krom & Berner (1980b) and modeled by Murray et al (1978). Sedimentation rates using excess lead-210 (Goldberg & Bruland 1974) have been determined as part of many of the investigations reviewed here. Depth distributions of concentrations are sampled with a variety of squeezing, centrifugation, or equilibration techniques. The ability to measure or readily estimate most of the terms in the diagenetic equation has led to wide use of these mathematical diagenetic models to estimate reaction rates. Variations with depth in diffusivities and reaction rate constants are recognized to be important (Jørgensen 1978b), but are usually not determined.

Since so much is known of the microbial ecology of ruminant digestion and, to a lesser extent, sludge digestion, these systems are convenient analogs in studies of the remineralization of complex organic matter. The range of organic molecules utilized by bacteria that mediate processes like sulfate reduction and methane production is usually limited (Fenchel & Blackburn 1979), so these organisms are dependent on a complex community of fermentative bacteria to supply the necessary substrates. The complex biopolymers present in organic detritus reaching the sediments are initially hydrolyzed to amino acids, simple sugars, and long chain fatty acids. These are in turn converted to volatile fatty acids and eventually to carbon dioxide and methane. The rumen functions with a variety of rations and organisms to maximize microbial biomass and production of volatile fatty acids, which are absorbed by the animal. Digesters are operated to maximize gas production. Microbial communities in sediments appear to operate within narrow environmental and substrate limits, and are probably maximizing their numbers or biomass. Analogies between the rumen, sludge digesters, and sediments break down in several ways, namely (*a*) our lack of knowledge of the amount and composition of carbon entering the sediments, as well as the extent of reaction, (*b*) the presence in marine sediments of large quantities of sulfate, which is a relatively minor rumen and sludge component, (*c*) the high ($\sim 39°C$) and uniform temperature of the rumen, and (*d*) the vast differences in substrate concentrations and supply rates, residence times, and mixing rates. Unmetabolized and recalcitrant organic components remain in the sediments, and molecular weight depth distributions (Krom & Sholkovitz 1977, Krom & Westrich

1981) suggest they repolymerize below the sulfate reducing zone to form high-molecular-weight dissolved organic matter and eventually humic and fulvic materials.

One very important aspect of the above microbiological studies is the availability of specific inhibitors for groups of organisms and specific transformations. The ability to experimentally inhibit and manipulate processes is rare in the Earth Sciences, where the scales of processes force us to be passive observers. The use of specific inhibitors combined with models and careful observations provides a powerful approach to understanding the biogeochemistry of sediments.

RATE MEASUREMENTS

The rate measurements summarized in this review represent an attempt to determine the importance of various transformations in total sediment metabolism, and thus have a system rather than an organism or mechanism focus. They deal with complex mixed bacterial populations and multiple substrates, and while they fail to do justice to previous careful work on enzyme and microbial kinetics (Lehninger 1975), they do give a good picture of chemical dynamics in natural systems (Fenchel & Blackburn 1979). The rate measurements reported here fall into two broad categories: time-series incubations in isolated sediments, and turnover rate measurements involving additions of labeled tracer compounds.

The time-series incubations or jar experiments have been used to determine rates of sulfate reduction (Martens & Berner 1974, Goldhaber et al 1977), ammonia production (Rosenfeld 1981), and methane production (Crill 1981). Homogenized sediment from desired depths is sealed in jars and analyzed sequentially for consumption of oxidant or appearance of end products. These experiments are typically conducted over a period of weeks or months.

Experiments involving addition of stable or radioisotope tracers have much greater sensitivity and thus can be conducted with incubation times of minutes to hours. One of the most compelling reasons for using tracer methods is their ability to measure transformation rates of intermediates, which undergo further reactions and do not accumulate in sediments.

All of these rate determinations using tracers fall under the term *turnover rate*; the guidelines of Zilversmit (1955) are used in this review to eliminate confusion in terminology. The turnover rate is the amount of material transformed per unit time and is equivalent to the amount of material entering or leaving a pool. The turnover rate is the product of the in situ pool size or concentration and a fractional turnover rate or first-order rate constant. For a labeled pool, the fractional turnover rate equals $a/A\Delta t$, where A is the initial pool activity, a is the turned-over activity or the activity of a reaction product, and Δt is the incubation time. Fractional turnover rates determined in tracer experiments are generally referred to as turnover rate constants. Turnover rates are also calculated from experimentally determined first-order rate constants, the slope of a ln tracer activity vs time plot. Fractional turnover rates give no information on the order of a reaction, but are generally equivalent to first-order rate constants for slow reactions or large pools. Since attention is experimentally restricted to one particular molecule, first-order rate constants in complex systems are very likely pseudo-first-order. The turnover time is the reciprocal of the fractional turnover rate. Since there are large variations with depth for most of the experimentally determined rates, they are often integrated with depth and expressed in flux units, permitting comparisons between environments and reactions.

Since there is ample time for competing reactions to occur, results from jar experiments such as ammonia production and methane production probably yield net rates. Rates determined with isotope tracers are probably nearer to gross rates. Sorption of added tracer has been shown to be a serious complication in turnover rate determinations of volatile fatty acids (Christensen & Blackburn 1982) and amino acids (Christensen & Blackburn 1980). Rate constants for sorption were determined in both of these studies using short-term experiments with high specific activity tracers. The overall turnover rates were corrected for sorption. Many of the turnover rate measurements are correctly identified as "potential" (Sørensen 1978a,b) or "apparent" (Sansone & Martens 1981a,b) and should be regarded as such until we can demonstrate with models and more experiments that the measurements themselves do not perturb the sediment system. Karl (1982) has emphasized the importance of determining exactly what these tracer experiments are measuring.

Jørgensen's (1978a,b) papers on sulfate reduction rate determinations in sediments provide some of the clearest descriptions of how these measurements should be conducted and what precautions should be taken. The technique involves injection of a sediment core with microliter quantities of a $^{35}SO_4^{=}$ tracer solution through silicone rubber septa located along a plastic core tube. The core is incubated, killed by freezing, and cut into segments for analysis. The radioactivity of sulfate and sulfide as well as the concentration of sulfate are determined in each segment. Slight modifications involving use of segmented core liners (Reeburgh 1980) or incubation in syringe subcores (Devol & Ahmed 1981) have been reported.

Ideally, the sediment cores should be disturbed as little as possible to preserve zonation and to avoid perturbing the obligate anaerobes present. It is not necessary to homogenize a high specific activity tracer in the sediment so long as the tracer added and the product formed are retained in the core segment and the form and fate of the tracer are known. Jørgensen (1978a) performed parallel sulfate reduction rate measurements using homogenized and diluted sediment (Sorokin 1962) and reported rates 2 to 30 times lower than those obtained with the core injection technique. Ansbaek & Blackburn (1980) reported a decrease in the acetate turnover rate of 50–75% when the sediment was homogenized. Christensen & Blackburn (1980) obtained similar results for core injection and homogenized tracer experiments with alanine. The effects of concentration increases resulting from use of lower specific activity tracer has been investigated by Christensen & Blackburn (1980) and Ansbaek & Blackburn (1980). Increasing tracer concentrations decreased the acetate and alanine rate constants in both of these studies. The rate measurements summarized here are reported in the same order as their occurrence, proceeding downward in sediments. Since the type of experiment is important in interpreting these sediment rate measurements, as much experimental detail as possible is included in the tables. The rates are tabulated as turnover rates (mmole liter^{-1} yr^{-1}; mM yr^{-1}) or as depth integrated rates, which are useful in comparing the magnitudes of different processes and have the same units (μmole cm^{-2} yr^{-1}) as fluxes. Results from the amino acid and volatile fatty acid turnover rate measurements are reported in μM hr^{-1}.

Oxygen Reduction

Dissolved oxygen is supplied to sediments by mixing or diffusion from overlying waters. Oxygen consumption rates have been measured using cores (Pamatmat 1971), or a variety of diver-operated or free-vehicle benthic respirometers (Smith 1978, Hinga et al 1979). Oxygen consumption rates are often partitioned into community and chemical rates by poisoning with formalin. These rates range over three orders of magnitude with depth (Hinga et al 1979, figure 5), from less than 1 to about 1200 μmoles cm^{-2} yr^{-1}. Oxygen reduction is probably confined to the uppermost centimeter of anoxic sediments, and oxygen supplied by the overlying water produces these high integrated rates.

Depth distributions of oxygen concentration have been reported in equatorial red clay and calcareous oozes by Murray & Grundmanis (1980). Oxygen was present in concentrations that were never less than 50 μM in the upper 50 cm of these low organic carbon flux deep-sea sediments. Detailed depth distributions of dissolved oxygen in the surface portions of high organic content sediments have been measured with microelectrodes (Revsbech et al 1980a,b). These electrodes are so small that they require no stirring; depth distributions are obtained by advancing them into sediments with micromanipulators. These electrodes may be used to measure detailed oxygen gradients over distances of less than a centimeter.

Denitrification

Table 3 summarizes recent denitrification rate measurements in marine sediments, covering environments ranging from deep-sea to coastal sediments. Denitrification was reviewed recently by Knowles (1982); it is of limited significance in terms of oxidizing capacity (Table 2) because of low nitrate concentrations in sediments. The perspective in most of the studies reported here is of denitrification as a process important in completing the nitrogen cycle, rather than as a source of oxidizing capacity.

Denitrification rates are measured by a wide variety of methods, including direct observation of increases in N_2 (Wilson 1978, Kaplan et al 1979), labeling with $^{15}NO_3^-$, and experiments involving addition of the inhibitor acetylene, which blocks reduction beyond N_2O (Sørensen 1978a). Excluding the N_2 production and model determinations, all methods involve homogenizing depth intervals of the sediment to distribute either inhibitors or tracers. The NO_3^- pool is usually increased to concentrations well above ambient in experiments involving $^{15}NO_3^-$. Oren & Blackburn (1979) determined Michaelis-Menten kinetic parameters on dilutions of the sediment and corrected the "potential" rate measurements obtained at nitrate saturation to in situ levels.

Denitrification is one of a number of nitrogen transformations taking place near the sediment surface. Billen (1978) modeled ammonification, nitrification, and denitrification rates in North Sea sediments and obtained reasonable agreement with observed rates.

Metal Oxide Reduction

Iron and manganese are added to sediments predominantly as oxidized particles and together they have one of the highest capacities for oxidizing organic matter in marine sediments (Table 2). Reduced mineral phases of manganese and particularly iron occur extensively in sediments, so the oxidizing capacity is presumably used in sediments. The role of metal oxides in the oxidation of organic matter is poorly understood and rate measurements comparable to those reviewed here are not available. Depending on their physical availability, these oxides may be reduced inorganically (Stumm & Morgan 1981) and operate by cycling other reduced compounds. Organisms capable of reducing iron and manganese oxides have been cultured from soils and lake sediments. Their activities are summarized by Ehrlich (1981), but their importance in reducing iron and manganese oxides is not clear.

Sørensen (1982) measured Fe(III) reduction in slurries of marine

Table 3 Denitrification rates in sediments

Study/Location	Method	Rate		Comment
		m*M* yr^{-1}	μmole cm^{-2} yr^{-1}	
Bender et al (1977) Guinea Basin	NO_3^- profile, model	—	2.5 0.11	flux across sediment interface downward flux in sediments
Wilson (1978) Atlantic Ocean	N_2 excesses	—	0.5, 0.189 av	
Billen (1978) S. Bight, N. Sea	consumption rate of NO_3^- spike	23–205	—	
Sørensen (1978a,b)				
Randers Fjord (a)	C_2H_2 inhibition	12.7	—	
Limfjorden (b)	$^{15}NO_3^- \rightarrow N_2$	36–317	—	
Koike & Hattori (1978) Manguko-Ura	$^{15}NO_3^- \rightarrow N_2$	1.3–96	—	
Kaplan et al (1979) Great Sippewissett Marsh	N_2 production	—	0.88–1.77	seasonal study
Oren & Blackburn (1979) Kysing Fjord	$^{15}NO_3^- \rightarrow N_2$; corrected w/V_{max}, K_m	0.7–4.5	—	
Koike & Hattori (1979) Bering Sea shelf	$^{15}NO_3^- \rightarrow N_2$	10.5 (av)	—	

sediment along with denitrification and sulfate reduction. Ferric iron reduction was inhibited by additions of NO_3^- or NO_2^-, but resumed when the additions were depleted. Inhibition of sulfate reduction with molybdate did not affect Fe(III) reduction. Sørensen concluded that iron reduction was associated with facultative nitrate-reducing bacteria and that the process may be important in sediments at low NO_3^- concentrations.

Sulfate Reduction

Sulfate reduction has the largest organic matter oxidizing capacity of any process occurring in marine sediments (Table 2). The presence of large quantities of sulfate in marine systems leads to large differences in the sequences of reactions between marine systems and lakes.

Some of the most recent sulfate reduction rate measurements in marine sediments are summarized in Table 4. Jørgensen & Fenchel (1961) developed methods for the study of a model system and summarized some of the early tracer and incubation determinations of sulfate reduction rates. Goldhaber & Kaplan (1974, 1975) reviewed the sulfur cycle and factors controlling the sulfate reduction rate. Their work reports sulfate reduction rates, largely from model determinations.

Measurements of sulfate reduction rates in sediment have been demonstrated (Jørgensen 1978a,b) to be reliable and are used widely. Since $^{35}SO_4^=$ can be obtained carrier-free and sulfate pool sizes are generally large in marine sediments, these tracer measurements are true tracer experiments. Sulfate pool sizes can be measured easily in interstitial waters by using gravimetric or titrimetric (Reeburgh & Springer-Young 1983) methods. Adsorption has not been shown to be a problem. The only complication seems to be rapid pyrite formation (Howarth 1979), which occurs in salt marshes and has the effect of lowering the tracer-determined sulfate reduction rate.

Determination of the net sulfate reduction rate requires evaluation of the fate of reduced sulfur compounds. Such compounds may leave the sediments by entering the atmosphere (Hansen et al 1978) or through photosynthetic (Blackburn et al 1975, Jørgensen & Cohen 1977) or inorganic oxidation in the overlying water (Cline & Richards 1969).

Seasonal studies of sulfate reduction rates have been reported in a limited number of environments, namely Limfjorden (Jørgensen 1977), Colne Point salt marsh (Nedwell & Abram 1978), and Cape Lookout Bight (Klump 1980, Crill & Martens 1982). Because of the dominance of sulfate reduction and the ease of the rate determinations, sulfate reduction rates should probably be included as a part of any marine anoxic sediment rate study in the future.

The relationship between sulfate reduction rate and sedimentation rate

Table 4 Recent sulfate reduction rate determinations in marine sediments

Study/Location	Method	Rate ($mM\ yr^{-1}$)	Integrated rate (gross) ($mmole\ cm^{-2}\ yr^{-1}$)	Comment
Goldhaber et al (1977) Long Island Sound	jar experiment model	77 (surface) 2 (10 cm)	—	(FOAM site)
Jørgensen (1977) Limfjorden	$^{35}SO_4^=$ core injection	9–73 (surface) 0.2 (150 cm)	0.226 (10 cm)	2 yr study, sulfur budget determined
Murray et al (1978) Saanich Inlet	model	$K = 6 \times 10^{-9}\ s^{-1}$	—	
Nedwell & Abram (1978) Colne Point salt marsh	$^{35}SO_4^=$ core injection	—	0.44	1 yr study
Klump (1980), Crill & Martens (1982), Chanton & Martens (1982)	tube incubation $^{35}SO_4^=$	182–511 (1–4 cm) 0–55 (18–21 cm)	1.7 1.9	
Howarth & Teal (1980) Great Sippewissett Marsh	$^{35}SO_4^=$	—	7.5	
Devol & Ahmed (1981) Saanich Inlet	$^{35}SO_4^=$ core injection	52–78.8 (surface) 0.9 (30 cm)	0.48	max at 15 cm
Reeburgh (1980), Reeburgh & Alperin (unpublished) Skan Bay	$^{35}SO_4^=$ core injection	32 (surface) 9 (30 cm)	1.37 (1979), 0.43 (1980)	one core
Reeburgh & Alperin (unpublished) Chesapeake Bay	$^{35}SO_4^=$ core injection	150 (surface) 10 (55 cm)	2.26	one core

has been studied by Goldhaber & Kaplan (1975), Toth & Lerman (1977), and Berner (1978). Berner (1978) discusses a method for estimating sedimentation rates from the initial sulfate concentration gradient in marine sediments.

Methane Production

Methane production has received attention in rumen studies as a non-utilizable waste product and in digester studies as a desirable product. Methanogenesis has been reviewed by Wolfe (1971), Zeikus (1977), Mah et al (1977), and Bryant (1979). Reliable methane production rates in sediments have largely resulted from modeling, and several problems have emerged that appear to preclude tracer measurements of the methane production rate.

First, although a number of candidate reactions and mechanisms for methane production have been advanced, we still do not know which reaction is the dominant methane producer in marine sediments. Claypool & Kaplan (1974) used stable carbon isotope distributions ($\delta\ ^{13}CO_2$) and a Rayleigh distillation model to conclude that CO_2 reduction was the most important methane-producing reaction in sediments. Studies in Skan Bay (Shaw et al, unpublished) with $^{14}CO_2$ as tracer indicate that the CO_2 pool is so large and the resulting specific activity in a tracer experiment so low that unrealistically long incubations are necessary to produce detectable CH_4. Cappenberg (1974), Winfrey & Zeikus (1979), and Sansone & Martens (1981a) present evidence that acetate is an important precursor for methane. Recent turnover experiments on methionine (Zinder & Brock 1978, Phelps & Zeikus 1980) and methanol (Oremland et al 1982) suggest that both of these compounds may be methane precursors. Pool size measurements were not reported for either of these compounds.

The use of jar experiments to obtain methane production rates is complicated by competition between sulfate reducers and methanogens for hydrogen (Winfrey & Zeikus 1977, Oremland & Taylor 1978, Nedwell & Banat 1981), as well as by anaerobic methane oxidation. Martens & Berner (1974), Crill (1981), and Crill & Martens (1983) have made such measurements and observed no production of methane until sulfate was exhausted. As indicated earlier, these jar experiments are probably measuring net rates of methane production.

One method for determining the dominant reaction and the total amount of methane produced deals with determining a stable carbon isotope budget in sediments. Biogenic methane has a characteristic stable carbon isotope signature, and the carbon pool from which methane was produced should show an isotope "pull" equivalent to the "push" resulting from methane production. Previous work has involved only measurements of δ $^{13}CO_2$ and δ $^{13}CH_4$ (Claypool & Kaplan 1974, Doose 1980); by comparing pool sizes and isotope ratios in other carbon reservoirs, namely DOC (dissolved organic carbon), PIC (particulate inorganic carbon), and POC (particulate organic carbon), and by investigating specific classes of compounds the principal reaction can be determined from a stable isotope budget.

Stoessell & Byrne (1982) recently showed that methane does not adsorb in clay slurries. New solubility values for methane in seawater (Yamamoto et al 1976) have made determination of saturation more reliable.

Methane Oxidation

The most important sink for methane was believed until recently to be the atmosphere (Ehhalt 1974), where methane is ultimately oxidized to CO_2 in the troposphere by reaction with the OH radical. Two types of methane oxidation processes, aerobic (Rudd et al 1974, Rudd & Hamilton 1978) and anaerobic (Reeburgh & Heggie 1977, Reeburgh 1982), have been identified recently as important sinks for methane in freshwater systems, such as lakes and wetlands, and in marine sediments. Hanson (1980) and Rudd & Taylor (1980) have reviewed both processes. Water-column methane oxidation rates are summarized in Table 5 and sediment methane oxidation rates are summarized in Table 6.

Aerobic methane oxidation studies have been conducted in the water columns of lakes (Rudd et al 1974, Rudd & Hamilton 1978, Jannasch 1975, Harritts & Hanson 1980) and coastal waters (Sansone & Martens 1978). The rate determinations have involved measuring disappearance of methane in time-series incubations of water samples or labeling with $^{14}CH_4$ (Rudd et al 1974). These studies show that aerobic methane oxidation is confined to a thin depth interval in the water column by a lack of methane and inorganic nitrogen above and by a lack of oxygen below.

Anaerobic methane oxidation has been controversial ever since its occurrence in sediments was predicted by models (Reeburgh 1976, Barnes & Goldberg 1976, Martens & Berner 1977), but recent tracer experiments (Panganiban et al 1979, Reeburgh 1980, Iversen & Blackburn 1981, Devol, unpublished) and stable carbon isotope models (Reeburgh 1982) agree well with the predicted locations and magnitudes and indicate that the process does occur. Lidstrom (unpublished) has recently observed anaerobic methane oxidation in the anoxic water column of Framvaren fjord. The organisms responsible for anaerobic methane oxidation have not been isolated.

Laboratory evidence favoring (Davis & Yarbrough 1966) and disputing (Sorokin 1957) anaerobic methane oxidation has been presented. There are no known anaerobic organisms capable of using methane as the sole carbon

Table 5 Water-column methane oxidation rates

Study/Location	Method	Oxic/anoxic	Rate (μM yr^{-1})
Rudd et al (1974) Lake 120, ELA	tracer ($^{14}CH_4$)	oxic	1.4×10^3 (max)
Jannasch (1975) Lake Kivu	time series	oxic	15.6–325 175 av
Reeburgh (1976) Cariaco Trench	model	anoxic	$0.1–1.5 \times 10^{-2}$
Sansone & Martens (1978) Cape Lookout Bight	time series	oxic	3.6–76.6
Scranton & Brewer (1978) Ocean	apparent CH_4 utilization in dated water masses	oxic	1.5×10^{-4} (<150 yr water)
Panganiban et al (1979) Lake Mendota	tracer ($^{14}CH_4$)	anoxic	methane oxidation observed
Harrits & Hanson (1980) Lake Mendota	tracer ($^{14}CH_4$)	oxic	$10.5 \times 10^3–1.0 \times 10^7$ (max)
Lidstrom (unpublished) Framvaren Fjord	time series	anoxic	methane oxidation observed in 4 experiments from 150 to 177 m

source (Quayle 1972). Zehnder & Brock (1979, 1980) reported simultaneous production and oxidation of methane by nine methanogen strains, but the amounts oxidized (<1%) were too small to produce the net consumption observed in the low methane surface zone of marine sediments. Anaerobic methane oxidation is also confined to a narrow depth interval in sediments; the observed maximum rates lie at the bottom of the sulfate reduction zone, where sulfate is nearly exhausted. Since this process has not been observed in lake sediments and occurs only in a subsurface zone in marine sediments, it is probably connected with sulfate reduction.

These rates can be checked by comparing the integrated methane oxidation rate with the calculated upward flux of methane. These rates are nearly equal in recent Skan Bay work, indicating that methane diffusing upward into the low methane concentration zone undergoes net consumption, confirming the measurements. The integrated methane oxidation rate ranges between 5–20% of the integrated sulfate reduction rate in Skan Bay sediments. Devol & Ahmed (1981) proposed that a subsurface maximum in the sulfate reduction rate in Saanich Inlet was caused by anaerobic methane oxidation.

Table 6 Sediment methane oxidation rates

Study/Location	Method	Oxidation rate (mM yr^{-1})	Integrated oxidation rate (μmole cm^{-2} yr^{-1})
Reeburgh (1976) Cariaco Trench	model	E. basin 1.59 (10-cm zone) W. basin 0.55 (10-cm zone)	15.9 5.48
Barnes & Goldberg (1976) Santa Barbara Basin	model	0.232 (46-cm zone)	1.07×10^{-2}
Martens & Berner (1977) Long Island Sound	model	0.65 ($K_1 = 8 \times 10^{-9}$ s^{-1})	—
Kosiur & Warford (1979) Santa Barbara Basin	tracer (^{14}C-lactate, acetate) flask incubation	0.128 (av)	—
Bernard (1979) Gulf of Mexico	model	(K_1 = 14.4–1.8 $\times 10^{-10}$ s^{-1})	—
Reeburgh (1980) Skan Bay	tracer ($^{14}CH_4$) core injection	3.4 (max)	60 (measured)
Miller (1980) Guinea Basin, Eq. Atlantic	model ($\delta^{13}CO_2$ distribution)	—	8.8
Iversen & Blackburn (1981) Kysing Fjord	tracer ($^{14}CH_4$) core injection	0.017–0.102	0.12–0.63
Whiticar (1978, 1982) Eckernfördner Bay	model	sta. 2 0.095 (50-cm zone) sta. 4 0.046 (25-cm zone)	4.76 1.16
Devol (unpublished) Saanich Inlet	tracer ($^{14}CH_4$) core injection	6.6–12.4 (max)	25–71 (measured) 141 (calculated)
Reeburgh & Alperin (unpublished) Skan Bay	tracer ($^{14}CH_4$) core injection	2 (max)	cores 7, 8, 9 23.6 (measured) 23.6 (calculated)
Chesapeake Bay		10 (max)	220–360

Ammonium Production

Ammonium production rate measurements are summarized in Table 7. Ammonium is a product of the decomposition of organic nitrogen compounds in anoxic sediments and accumulates to m*M* concentrations in interstitial waters. Adsorption of ammonium has been found to be rapid and reversible in anoxic marine sediments (Rosenfeld 1979). Adsorbed ammonium was found to be predominantly associated with organic rather than mineral phases. Adsorption coefficients from several studies seem to be similar.

Although there are few direct measurements of ammonium production rates in marine sediments, two recent studies where measured and modeled rates agree well (Blackburn 1979, Rosenfeld 1981) suggest that the rate measurements and models are approaching the same point. Blackburn's stable isotope dilution method has the added advantage of providing a direct means of obtaining net and gross ammonium production.

Table 7 Ammonium production rates in sediments

Study/Location	Method	Adsorption coefficient	Rate (mM yr^{-1})	Comment
Berner (1974)	model	—		
Somes Sound			$K = 3.5 \times 10^{-9}$ s^{-1}	
Santa Barbara Basin			$K = 1.3 \times 10^{-11}$ s^{-1}	
Billen (1978) S. Bight, N. Sea	tube incubations	—	up to 78.8	
Murray et al (1978) Saanich Inlet	model	2	$K = 6.05 \times 10^{-9}$ s^{-1} (upper 15 cm) 4.17×10^{-10} s^{-1} (below 60 cm)	
Blackburn (1979) Limfjorden	$^{15}NH_4^+$		100. (net, 0–2 cm) 112. (total, 0–2 cm) 0.11 (12–14 cm)	
Klump (1980) Cape Lookout Bight	tube incubations	1.68	36.5 (0–2 cm) 7.3 (28–30)	1.9 mmole m^{-2} yr^{-1} over 32 cm
Rosenfeld (1981) Long Island Sound	model	1.6	0.44–0.57 (Sachem) 0.08–0.10 (FOAM)	
	jar expts. (sorption, temperature corrected)		0.65–0.48 (Sachem) 0.26–0.30 (FOAM)	

Table 8 Volatile fatty acid turnover rates in sediments

Study	Experiment type	Tracer	Sp. Act. (mCi/m mole)	Tracer conc. (μM)	Concentration (μM) (method)	Turnover rate constant (hr^{-1})	Turnover rate ($\mu M\ hr^{-1}$)	Comment
Cappenberg & Prins (1974) Lake Vechten	flask	U-^{14}C-L-lactate	45	< ambient	135	2.37	319.9	
	flask	U-^{14}C-acetate	57		95 (enzyme methods, Cappenberg 1974)	0.35	33.3	
Winfrey & Zeikus (1979) Lake Mendota	tube	U-^{14}C-acetate	56		2.7–4.5 (g.c. of HAc from porewater)	4.5	16	seasonal average
Ansbaek & Blackburn (1980) Limfjorden	core injection	U-^{14}C-acetate		10	0.1–6.0 (g.c. of HAc from porewater)	1.6–3.3; 2.1 (av)	6–12	nonexponential uptake suggesting multiple pools. Integrated acetate turnover 3 × integrated sulfate reduction
Sansone & Martens (1981a) Cape Lookout Bight	tube	1,2-^{14}C-acetate	53.5	< ambient	SO_4 zone			
					Jul 91	1.96	180	
					Jan 150	2.19	330	
					CH_4 zone			
					Jul 360	0.531	190	
					Jan 83	0.230	19	

Cape Lookout Bight	tube	1-^{14}C-propionate	56.7		SO_4 zone			
					Jul 5.08	1.40	7.11	
					Jan 12.1	0.223	2.70	
					CH_4 zone			
					Jul 13.2	0.462	6.10	
					Jan 2.1	0.330	0.70	
					(g.c. of methyl ester from whole sediment)			
Lovley & Klug (1982) Wintergreen Lake	syringe core injection	U-^{14}C-acetate	54	< ambient	27 Jun 110	3.11	342	
					4 Sep 100	1.59	159	
		2-^{14}C-propionate	55.7	< ambient	27 Jun 90	1.86	16.7	
					4 Sep 14	1.44	20	
		U-^{14}C-lactate	138.6	< ambient	4 Sep 1	2.76	3	
					(Bethge & Lindstrom 1974)			
Christensen & Blackburn (1982) Aarhus Bay, Danish coastal sediments	core injection	U-^{14}C-acetate	50	2	10–70 (vacuum distillation, g.c. of HAc from porewater)	3–7	3–13	exponential for short (8 min) incubations, 4 × – 10 × high relative to NH_4^+ production
Shaw, Alperin & Reeburgh (unpublished) Skan Bay	core injection	U-^{14}C-acetate	58.6	77	8–29 (modified Barcelona et al 1980)	0.2	1.6–5.8	

Volatile Fatty Acid Turnover

Table 8 summarizes the studies of volatile fatty acid turnover rates in sediments. These studies deal principally with acetate, but also include measurements on lactate and propionate. The tracer activities, tracer concentrations, and experimental conditions are reasonably similar for all of the studies summarized. Several authors (Ansbaek & Blackburn 1980, Christensen & Blackburn 1982) have noted that integration of the acetate turnover rates leads to values that are several times larger than the integrated rates of well-understood processes like sulfate reduction and ammonia production. This is unreasonable and requires an explanation.

Given the wide variety of environments and environmental conditions, the measured turnover rate constants for acetate seem to group remarkably well, corresponding to turnover times of about 20 minutes. Similar turnover times are observed for volatile fatty acids in the rumen. The lactate and acetate addition studies of Kosiur & Warford (1979), which were directed at methane production-consumption rates, show what appear to be much slower turnover times ($\sim$ days) and are reported in Table 6 (sediment methane consumption). This close grouping of measured acetate turnover rate constants may be explained in two ways: first, the studies are either correctly measuring the rate of a fundamental process that is relatively immune to environmental differences or second, the experimental conditions are so similar in all of the studies that, right or wrong, the results are the same.

The main source of variation in the turnover rates appears to be the pool size measurements. These determinations were made by a variety of analytical methods on whole sediment and interstitial waters collected by squeezing, centrifugation, and equilibration. The recent work of Christensen & Blackburn (1982) shows that tracer acetate was rapidly adsorbed into two pools, one permanently sorbed and one that can be released with excess acetate. They also presented evidence from gel filtration of interstitial water suggesting that a large portion of interstitial water acetate was complexed and unavailable. Thus the measured acetate pool size may be larger than the active or available pool.

Resolution of the questions about turnover rates of volatile fatty acids will require devising a way to measure or estimate the size of the active or available volatile fatty acid pools in sediments. Christensen & Blackburn (1982) indicate that 75–90% of the measured interstitial water acetate pool may be unavailable. They suggest investigations of availability involving size or other chromatographic separation, studies of the relative rates of remineralization of a tracer and pool constituent, and comparisons with some well-understood rate measurement.

Amino Acid Turnover

The studies of sediment amino acid turnover rates are summarized in Table 9. Two of these studies (Hanson & Gardner 1978, Henrichs et al 1982) were performed in salt marsh sediments. The results of these studies are probably not directly comparable to other sediment systems because of the presence of emergent plant root systems. Other rate measurements were not reported.

Christensen & Blackburn (1980) indicated that the alanine turnover rate in their studies exceeded the ammonium production rate. There are clearly too few comparable amino acid turnover rate measurements to draw firm conclusions, but it does appear that the situation with alanine and probably other amino acids is similar to that for volatile fatty acids—namely, the available pool is a fraction of the measured pool.

SUMMARY AND FUTURE WORK

This review has emphasized and summarized rate measurements of a number of processes important in remineralizing organic matter. It has also emphasized the necessity of showing that the rates of all processes occurring in a given sediment are internally consistent. Sulfate reduction, as indicated by Sørensen et al (1979), is the dominant reaction in anoxic systems in terms of its organic matter oxidizing capacity. The sulfate reduction rate is easily measured and gives results that agree well with diagenetic models. Ammonium production can also be measured reliably, although the analytical instrumentation used by Blackburn (1979) is probably less available. Since we are confident that both of these measurements give realistic results, either should be included in future turnover rate studies in anoxic sediments. These additional measurements allow determination of the importance of a reaction to the sediment system, and provide information that allows independent tests of the measured rates through diagenetic models or budgets. Where possible, detailed depth distributions of both concentrations and rates should be determined to make diagenetic modeling simpler. Determinations of fractional turnover rates or turnover experiments are a useful guide to future studies, but they give no information on system importance unless they are accompanied by pool size measurements.

Future measurements should preserve the integrity of the sediment studied as much as possible by minimizing preincubation manipulation, by adjusting tracer concentrations where possible to minimize perturbations, and by matching incubation temperatures and in situ temperatures. These measurements should also be directed toward determining the effects of

Table 9 Sediment amino acid turnover rates

Study/Location	Experiment type	Amino acid	Tracer conc. (n*M*)	Concentration (μM) (method)	Turnover rate constant (hr^{-1})	Turnover rate ($\mu M\ hr^{-1}$)	Comment
Hanson & Gardner (1978) Georgia salt marsh	tube	alanine	200–250	tall *Spartina* 30.3	0.11	3.25×10^{-3}	
				mud flat 14.3	0.023	0.33×10^{-3}	
				short *Spartina* 552	0.015	8.32×10^{-3}	
	tube	aspartic acid	150–200	tall *Spartina* 3.72	0.05	0.20×10^{-3}	
				mud flat 23.5	0.0094	0.22×10^{-3}	
				short *Spartina* 65.8 (amino acid analyzer, fluorometric detector, pore water by centrifugation)	0.0040	0.26×10^{-3}	
Christensen & Blackburn (1980) Limfjorden & Aarhus Bay	core injection	alanine	640	0.8 (HPLC—Lindroth & Mopper 1979 on centrifuged porewater)	9.60	3.13	in excess of NH_4^+ production
Henrichs et al (1982) Great Sippewissett Marsh	core injection	proline	750	27–59	0.58	15.6–34.2	
		alanine	1300	2–13	5.0	10–65	
		glutamic acid	900	11–50 (g.c.—Henrichs & Farrington 1979 on $NaNO_3$ or NaCl extracts)	2.22–2.85	31.4–142	

adsorption on pool sizes and rates. Development of methods capable of determining the active or effective pool size of intermediates is critical.

Where possible, seasonal rates and budgets like those available now for Limfjorden and Cape Lookout Bight should be determined to obtain a better understanding of how temperature, carbon input, and reaction rates vary over an annual cycle. This information will allow determination of the net rates of many of the reactions.

The work of Fenchel & Jørgensen (1977) was cited in the introduction of this review. This paper discussed detritus food chains and the role of bacteria, and laid down a broad framework. The Fenchel & Jørgensen paper considered element cycles, the sequence of degradation reactions, and the biomasses of bacteria and protozoans in sediment systems. While the perspective taken in this review is different, it should be pointed out that there are now numbers and rates associated with most of the processes discussed in that paper, and that we are approaching a point of understanding the rates and importance to the whole system of many of the processes responsible for degradation of organic detritus in aquatic sediments. Most of the rate measurements summarized here have been made in the past five years.

What does the future hold? What will future papers similar to Fenchel & Jørgensen (1977) and this one discuss? Analytical capabilities for small samples of organic molecules will improve and become more widespread, so we should be able to obtain agreement in our understanding of small molecules like amino acids and volatile fatty acids. The factors controlling the microbial availability of analytically determined substrates will receive attention. Methods capable of measuring biomass, physiological potential, metabolic activity, growth rate, and cell division rate of microbial populations will be refined and adapted to sediment studies. New organisms will be isolated. Developments in sediment traps will lead to a better understanding of the nature and flux of organic detritus to the sea floor. Studies of stable carbon isotopes in specific carbon pools and in classes of compounds will lead to a better understanding of sediment reactions and their extent. Many of these methods will be applied in hemipelagic and pelagic environments, resulting in a better understanding of denitrification and metal oxide reduction. Collaboration between microbiologists and geochemists in associations like FOAM (Friends of Anoxic Mud), $CH_4A \cdot O_2S$ (North Carolina), SKUM (Scandinavian Committee for Mud Research), and AARGH (Alaska Anoxic Research Group) should continue to make the study of anoxic mud scientifically exciting and rewarding.

ACKNOWLEDGMENTS

This work was supported in part by National Science Foundation grant OCE 81-17882 and funds from the state of Alaska. I thank Ruth Hand and Helen Stockholm for their care and patience in typing the manuscript. Mary Lidstrom and Allen Devol kindly supplied unpublished rate data. My colleagues Marc Alperin, Bob Barsdate, Ed Brown, Susan Henrichs, George Kipphut, and Dave Shaw discussed parts of the tables and text during preparation of the manuscript. Contribution number 518, Institute of Marine Science, University of Alaska.

Literature Cited

Aller, R. C. 1978. Experimental studies of changes produced by deposit feeders on pore water, sediment and overlying water chemistry. *Am. J. Sci.* 278:1185–1234

Aller, R. C., Yingst, J. Y. 1980. Relationship between microbial distributions and the anaerobic decomposition of organic matter in surface sediments of Long Island Sound, USA. *Mar. Biol.* 56:29–42

Anderson, J. J., Devol, A. H. 1973. Deep water renewal in Saanich Inlet, an intermittently anoxic basin. *Estuarine Coastal Mar. Sci.* 1:1–10

Ansbaek, J., Blackburn, T. H. 1980. A method for the analysis of acetate turnover in a coastal marine sediment. *Microb. Ecol.* 5:253–64

Barcelona, M. J., Liljestrand, H. M., Morgan, J. J. 1980. Determination of low molecular weight volatile fatty acids in aqueous samples. *Anal. Chem.* 52:321–25

Barnes, R. O., Goldberg, E. D. 1976. Methane production and consumption in anoxic marine sediments. *Geology* 4:297–300

Bender, M. L., Fanning, K. A., Froelich, P. N., Heath, G. R., Maynard, V. 1977. Interstitial nitrate profiles and oxidation of sedimentary organic matter in the eastern equatorial Atlantic. *Science* 198:605–9

Bernard, B. B. 1979. Methane in marine sediments. *Deep-Sea Res.* 26:429–43

Berner, R. A. 1964. An idealized model of dissolved sulfate distribution in recent sediments. *Geochim. Cosmochim. Acta* 28:1497–1503

Berner, R. A. 1971. *Principles of Chemical Sedimentology*. New York: McGraw-Hill. 240 pp.

Berner, R. A. 1974. Kinetic models for the early diagenesis of nitrogen, sulfur, phosphorus and silicon in anoxic marine sediments. In *The Sea*, ed. E. D. Goldberg, 5:427–49. New York: Wiley-Interscience. 895 pp.

Berner, R. A. 1976a. The benthic boundary layer from the viewpoint of a geochemist. In *The Benthic Boundary Layer*, ed. I. N. McCave, pp. 33–55. New York: Plenum. 323 pp.

Berner, R. A. 1976b. Inclusion of adsorption in the modelling of early diagenesis. *Earth Planet. Sci. Lett.* 29:333–40

Berner, R. A. 1978. Sulfate reduction and the rate of deposition of marine sediments. *Earth Planet. Sci. Lett.* 37:492–98

Berner, R. A. 1980. *Early Diagenesis: A Theoretical Approach.* Princeton, NJ: Princeton Univ. Press. 241 pp.

Berner, R. A. 1981. A rate model for organic matter decomposition during bacterial sulfate reduction in marine sediments. In *Biogéochimie de la Matière Organique à l'Interface Eau-Sediment Marin*, pp. 35–44. *Colloq. Int. C.N.R.S. No. 293*

Bethge, P. O., Lindstrom, K. 1974. Determination of organic acids of low molecular mass (C_1 to C_4) in dilute aqueous solution. *Analyst (London)* 99:137–42

Billen, G. 1978. A budget of nitrogen recycling in North Sea sediments off the Belgian coast. *Estuarine Coastal Mar. Sci.* 7:127–46

Blackburn, T. H. 1979. Method for measuring rates of NH_4^+ turnover in anoxic marine sediments, using a ^{15}N-NH_4^+ dilution technique. *Appl. Environ. Microbiol.* 37:760–65

Blackburn, T. H., Kleiber, P., Fenchel, T. 1975. Photosynthetic sulfide oxidation in marine sediments. *Oikos* 26:103–8

Breck, W. G. 1974. Redox levels in the sea. In *The Sea*, ed. E. D. Goldberg, 5:153–79. New York: Wiley-Interscience. 895 pp.

Bryant, M. P. 1979. Microbial methane production—theoretical aspects. *J. Anim. Sci.* 48:193–201

Cappenberg, Th. E. 1974. Interrelations between sulfate-reducing and methane-producing bacteria in bottom deposits of a fresh-water lake. II. Inhibition experiments. *Antonie van Leeuwenhoek J. Microbiol. Serol.* 40:297–306

Cappenberg, Th. E., Prins, R. A. 1974. Interrelations between sulfate-reducing and methane-producing bacteria in bottom deposits of a fresh-water lake. III. Experiments with ^{14}C-labeled substrates. *Antonie van Leeuwenhoek J. Microbiol. Serol.* 40:457–69

Chanton, J. P. 1979. *Lead-210 geochronology in a changing environment: Cape Lookout Bight, N.C.* MS thesis. Univ. N.C., Chapel Hill. 85 pp.

Chanton, J. P., Martens, C. S. 1982. The sulfur budget of an anoxic marine sediment. *Abstr. 45th Am. Soc. Limnol. Oceanogr. Meet.*

Christensen, D., Blackburn, T. H. 1980. Turnover of tracer (^{14}C, ^{3}H labelled) alanine in inshore marine sediments. *Mar. Biol.* 58:97–103

Christensen, D., Blackburn, T. H. 1982. Turnover of ^{14}C-labelled acetate in marine sediment. *Mar. Biol.* In press

Claypool, G. E., Kaplan, I. R. 1974. The origin and distribution of methane in marine sediments. In *Natural Gases in Marine Sediments*, ed. I. R. Kaplan, pp. 99–139. New York: Plenum. 324 pp.

Cline, J. D., Richards, F. A. 1969. Oxygenation of hydrogen sulfide at constant salinity, temperature and pH. *Environ. Sci. Technol.* 3:838–43

Crill, P. M. 1981. *Methane production and sulfate reduction in the anoxic, coastal marine sediment of Cape Lookout Bight, North Carolina.* MS thesis. Univ. N.C., Chapel Hill. 44 pp.

Crill, P. M., Martens, C. S. 1982. A comparison of methods for the determination of the rate of sulfate reduction in anoxic sediments. *Abstr. 45th Am. Soc. Limnol. Oceanogr. Meet.*

Crill, P. M., Martens, C. S. 1983. Spatial and temporal fluctuations of methane production in anoxic, coastal marine sediments. *Limnol. Oceanogr.* In press

Davis, J. B., Yarbrough, H. F. 1966. Anaerobic oxidation of hydrocarbons by *Desulfovibrio desulfuricans. Chem. Geol.* 1:137–44

Devol, A. H., Ahmed, S. I. 1981. Are high rates of sulphate reduction associated with anaerobic oxidation of methane? *Nature* 291:407–8

Doose, P. R. 1980. *The bacterial production of methane in marine sediments.* PhD Dissertation. Univ. Calif., Los Angeles. 240 pp.

Ehhalt, D. H. 1974. The atmospheric cycle of methane. *Tellus* 26:58–70

Ehrlich, H. L. 1981. *Geomicrobiology.* New York: Marcel Dekker. 393 pp.

Emerson, S., Jahnke, R., Bender, M., Froelich, P., Klinkhammer, G., Bowser, C., Setlock, G. 1980. Early diagenesis in sediments from the eastern equatorial Pacific. I. Pore water nutrient and carbonate results. *Earth Planet. Sci. Lett.* 49:57–80

Fanning, K. A., Manheim, F. T., eds. 1982. *The Dynamic Environment of the Ocean Floor.* Lexington, Mass.: Lexington Books. 502 pp.

Fenchel, T., Blackburn, T. H. 1979. *Bacteria and Mineral Cycling.* New York: Academic. 225 pp.

Fenchel, T. M., Jørgensen, B. B. 1977. Detritus food chains of aquatic ecosystems: The role of bacteria. *Adv. Microb. Ecol.* 1:1–58

Froelich, P. N., Klinkhammer, G. P., Bender, M. L., Luedtke, N. A., Heath, G. R., Cullin, D., Dauphin, P., Hammond, D., Hartman B., Maynard, V. 1979. Early oxidation of organic matter in pelagic sediments of the eastern equatorial Atlantic: suboxic diagenesis. *Geochim. Cosmochim. Acta* 43:1075–90

Gieskes, J. M. 1975. Chemistry of interstitial waters of marine sediments. *Ann. Rev. Earth Planet. Sci.* 3:433–53

Gieskes, J. M. 1981. Deep-sea drilling interstitial water studies: implications for chemical alteration of the oceanic crust, layers I and II. *Soc. Econ. Paleontol. Mineral. Spec. Publ. No. 32*, pp. 149–67

Glasby, G. P. 1973. Interstitial waters in marine and lacustrine sediments: a review. *J. R. Soc. N. Z.* 3:43–59

Goldberg, E. D., Bruland, K. 1974. Radioactive geochronologies. In *The Sea*, ed. E. D. Goldberg, 5:451–89. New York: Wiley-Interscience. 895 pp.

Goldhaber, M. B., Kaplan, I. R. 1974. The sulfur cycle. In *The Sea*, ed. E. D. Goldberg, 5:569–655. New York: Wiley-Interscience. 895 pp.

Goldhaber, M. B., Kaplan, I. R. 1975. Controls and consequences of sulfate reduction rates in recent marine sediments. *Soil Sci.* 119:42–55

Goldhaber, M. B., Aller, R. C., Cochran, J. K., Rosenfeld, J. K., Martens, C. S., Berner, R. A. 1977. Sulfate reduction, diffusion and bioturbation in Long Island Sound sediments: report of the FOAM group. *Am. J. Sci.* 277:193–237

Grundmanis, V., Murray, J. W. 1977. Nitrification and denitrification in marine sediments from Puget Sound. *Limnol. Oceanogr.* 22:804–13

Guinasso, N. L., Schink, D. R. 1975. Quantitative estimates of biological mixing rates in abyssal sediments. *J. Geophys. Res.* 80:3032–43

Hansen, M. H., Ingvorsen, K., Jørgensen, B. B. 1978. Mechanisms of hydrogen sulfide release from coastal marine sediments to the atmosphere. *Limnol. Oceanogr.* 23:68–76

Hanson, R. B., Gardner, W. S. 1978. Uptake and metabolism of two amino acids by anaerobic microorganisms in four diverse salt marsh soils. *Mar. Biol.* 46:101–7

Hanson, R. S. 1980. Ecology and diversity of methylotrophic organisms. *Adv. Appl. Microbiol.* 26:3–39

Harrits, S. M., Hanson, R. S. 1980. Stratification of aerobic methane-oxidizing organisms in Lake Mendota, Madison, Wisconsin. *Limnol. Oceanogr.* 25:412–21

Henrichs, S. M., Farrington, J. M. 1979. Amino acids in interstitial waters of marine sediments. *Nature* 272:319–22

Henrichs, S. M., Hobbie, J. M., Howarth, R. W., Helfrich, J., Kilham, P. 1982. Free amino acids in salt marsh sediments: concentrations and fluxes. *Limnol. Oceanogr.* In press

Hesslein, R. H. 1980. *In situ* measurements of pore water diffusion coefficients using tritiated water. *J. Fish. Res. Board Can.* 37:545–51

Hinga, K. R., Sieburth, J. McN., Heath, G. R. 1979. The supply and use of organic material at the deep-sea floor. *J. Mar. Res.* 37:557–79

Hobson, P. N., Bonsfield, S., Summers, R. 1974. Anaerobic digestion of organic matter. *Crit Rev. Environ. Control* 4:131–91

Howarth, R. W. 1979. Pyrite: its rapid formation in a salt marsh and its importance in ecosystem metabolism. *Science* 203:49–51

Howarth, R. W., Teal, J. M. 1980. Energy flow in a salt marsh ecosystem: the role of reduced inorganic sulfur compounds. *Am. Nat.* 116:862–72

Hungate, R. E. 1975. The rumen microbial system. *Ann. Rev. Ecol. Syst.* 6:39–66

Iversen, N., Blackburn, T. H. 1981. Seasonal rates of methane oxidation in anoxic marine sediments. *Appl. Environ. Microbiol.* 41:1295–1300

Jannasch, H. W. 1975. Methane oxidation in Lake Kivu. *Limnol. Oceanogr.* 20:860–64

Jørgensen, B. B. 1977. The sulfur cycle of a coastal marine sediment (Limfjorden, Denmark). *Limnol. Oceanogr.* 22:814–32

Jørgensen, B. B. 1978a. A comparison of methods for the quantification of bacterial sulfate reduction in coastal marine sediments. I. Measurement with radiotracer techniques. *Geomicrobiol. J.* 1:11–27

Jørgensen, B. B. 1978b. A comparison of methods for the quantification of bacterial sulfate reduction in coastal marine sediments. II. Calculations from mathematical models. *Geomicrobiol. J.* 1:29–47

Jørgensen, B. B. 1978c. A comparison of methods for the quantification of bacterial sulfate reduction in coastal marine sediments. III. Estimation from chemical and bacteriological field data. *Geomicrobiol. J.* 1:49–64

Jørgensen, B. B., Cohen, Y. 1977. Solar Lake (Sinai). 5. The sulfur cycle of the Benthic cyanobacterial mats. *Limnol. Oceanogr.* 22:657–66

Jørgensen, B. B., Fenchel, T. 1961. The sulfur cycle of a marine sediment model system. *Mar. Biol.* 24:189–201

Kaplan, I. R., ed. 1974. *Natural Gases in Marine Sediments.* New York: Plenum. 324 pp.

Kaplan, W. A., Valiela, I., Teal, J. M. 1979. Denitrification in a salt marsh ecosystem. *Limnol. Oceanogr.* 24:726–34

Karl, D. M. 1982. Microbial transformations of organic matter at oceanic interfaces: a review and prospectus. *EOS, Trans. Am. Geophys. Union* 63:138–40

Klump, J. V. 1980. *Benthic nutrient regeneration and the mechanisms of chemical sediment-water exchange in an organic-rich coastal marine sediment.* Dissertation. Univ. N.C., Chapel Hill. 160 pp.

Knowles, R. 1982. Denitrification. *Microbiol. Rev.* 46:43–70

Koike, I., Hattori, A. 1978. Denitrification and ammonia formation in anaerobic coastal sediment. *Appl. Environ. Microbiol.* 35:278–82

Koike, I., Hattori, A. 1979. Estimates of denitrification in sediments of the Bering Sea shelf. *Deep-Sea Res.* 26:409–16

Kosiur, D. R., Warford, A. L. 1979. Methane production and oxidation in Santa Barbara Basin sediments. *Estuarine Coastal Mar. Sci.* 8:379–85

Kriukov, P. A., Manheim, F. T. 1982. Extraction and investigative techniques for study of interstitial waters of unconsolidated sediments: a review. In *The Dynamic Environment of the Ocean Floor*, ed. K. A. Fanning, F. T. Manheim, pp. 3–26. Lexington, Mass.: Heath. 502 pp.

Krom, M. D., Berner, R. A. 1980a. The diffusion coefficients of sulfate, ammonium and phosphate ions in anoxic marine sediments. *Limnol. Oceanogr.* 25:327–37

Krom, M. D., Berner, R. A. 1980b. Adsorption of phosphate in anoxic marine sediments. *Limnol. Oceanogr.* 25:797–806

Krom, M. D., Sholkovitz, E. R. 1977. Nature and reactions of dissolved organic matter in the interstitial waters of marine sediments. *Geochim. Cosmochim. Acta* 41:1565–73

Krom, M. D., Westrich, J. T. 1981. Dissolved organic matter in the pore waters of recent marine sediments: a review. In *Biogéochimie de la Matière Organique à l'Interface Eau-Sediment Marin*, pp. 103–11. *Colloq. Int. C.N.R.S. No. 293*

Lehninger, A. L. 1975. *Biochemistry.* New York: Worth. 1055 pp. 2nd ed.

Li, Y. H., Gregory, S. 1974. Diffusion of ions in sea water and in deep-sea sediments. *Geochim. Cosmochim. Acta* 38:703–14

Lindroth, P., Mopper, K. 1979. High performance liquid chromatographic determination of subpicomole amounts of amino acids by precolumn fluorescence derivitization with o-pthaldialdehyde. *Anal. Chem.* 51:1667–74

Lindsay, W. L. 1979. *Chemical Equilibria in Soils*. New York: Wiley. 449 pp.

Lovley, D. R., Klug, M. J. 1982. Intermediary metabolism of organic matter in the sediments of a eutrophic lake. *Appl. Environ. Microbiol.* 43:552–60

Mah, R. A., Ward, D. M., Baresi, L., Glass, T. L. 1977. Biogenesis of methane. *Ann. Rev. Microbiol.* 31:309–42

Manheim, F. T. 1976. Interstitial waters of marine sediments. In *Chemical Oceanography*, ed. J. P. Riley, R. Chester, 6:115–85. London: Academic. 414 pp. 2nd ed.

Manheim, F. T., Sayles, F. L. 1974. Composition and origin of interstitial waters of marine sediments, based on deep sea drill cores. In *The Sea*, ed. E. D. Goldberg, 5:527–68. New York: Wiley-Interscience. 895 pp.

Martens, C. S., Berner, R. A. 1974. Methane production in the interstitial waters of sulfate-depleted marine sediments. *Science* 185:1167–69

Martens, C. S., Berner, R. A. 1977. Interstitial water chemistry of Long Island Sound sediments. I. Dissolved gases. *Limnol. Oceanogr.* 22:10–25

Martens, C. S., Klump, J. V. 1980. Biogeochemical cycling in an organic-rich coastal marine basin. I. Methane sediment-water exchange processes. *Geochim. Cosmochim. Acta* 44:471–90

McCave, I. N., ed. 1976. *The Benthic Boundary Layer*. New York: Plenum. 323 pp.

Miller, L. G. 1980. *Dissolved inorganic carbon isotope ratios in reducing marine sediments*. MS thesis. Univ. South. Calif., Los Angeles. 101 pp.

Murray, J. W., Grundmanis, V. 1980. Oxygen consumption in pelagic marine sediments. *Science* 209:1527–30

Murray, J. W., Grundmanis, V., Smethie, W. M. Jr. 1978. Interstitial water chemistry in the sediments of Saanich Inlet. *Geochim. Cosmochim. Acta* 42:1011–26

Nedwell, D. B., Abram, J. W. 1978. Bacterial sulfate reduction in relation to sulphur geochemistry in two contrasting areas of saltmarsh sediment. *Estuarine Coastal Mar. Sci.* 6:341–51

Nedwell, D. B., Banat, I. M. 1981. Hydrogen as an electron donor for sulfate-reducing bacteria in slurries of salt marsh sediment. *Microb. Ecol.* 7:305–13

Oremland, R. S., Taylor, B. F. 1978. Sulfate reduction and methanogenesis in marine sediments. *Geochim. Cosmochim. Acta* 42:209–14

Oremland, R. S., Marsh, L., Des Marais, D. J. 1982. Methanogenesis in Big Soda Lake, Nevada: an alkaline, moderately hypersaline desert lake. *Appl. Environ. Microbiol.* 43:462–68

Oren, A., Blackburn, T. H. 1979. Estimation of sediment denitrification rates at *in situ* nitrate concentrations. *Appl. Environ. Microbiol.* 37:174–76

Pamatmat, M. M. 1971. Oxygen consumption by the seabed. IV. Shipboard and laboratory experiments. *Limnol. Oceanogr.* 16:536–50

Panganiban, A. T., Patt, T. E., Hart, W., Hanson, R. S. 1979. Oxidation of methane in the absence of oxygen in lake water samples. *Appl. Environ. Microbiol.* 37:303–9

Phelps, T., Zeikus, J. G. 1980. Microbial ecology of anaerobic decomposition in Great Salt Lake. *Abstr. Ann. Meet. Am. Soc. Microbiol. 14*, p. 85

Quayle, J. R. 1972. The metabolism of one-carbon compounds by micro-organisms. *Adv. Microb. Physiol.* 7:119–203

Reeburgh, W. S. 1969. Observations of gases in Chesapeake Bay sediments. *Limnol. Oceanogr.* 14:368–75

Reeburgh, W. S. 1976. Methane consumption in Cariaco Trench waters and sediments. *Earth Planet. Sci. Lett.* 28:337–44

Reeburgh, W. S. 1980. Anaerobic methane oxidation: rate depth distributions in Skan Bay sediments. *Earth. Planet. Sci. Lett.* 47:345–52

Reeburgh, W. S. 1982. A major sink and flux control for methane in marine sediments: anaerobic consumption. In *The Dynamic Environment of the Ocean Floor*, ed. K. Fanning, F. T. Manheim, pp. 203–17. Lexington, Mass.: Heath

Reeburgh, W. S., Heggie, D. T. 1977. Microbial methane consumption reactions and their effect on methane distributions in freshwater and marine environments. *Limnol. Oceanogr.* 22:1–9

Reeburgh, W. S., Springer-Young, M. 1983. New measurements of sulfate and chlorinity in natural sea ice. *J. Geophys. Res.* In press

Revsbech, N. P., Jørgensen, B. B., Blackburn, T. H. 1980a. Oxygen in the sea bottom measured with a microelectrode. *Science* 207:1355–56

Revsbech, N. P., Sørensen, J., Blackburn, T. H., Lomhold, J. P. 1980b. Oxygen distribution in sediments measured with microelectrodes. *Limnol. Oceanogr.* 25:403–11

Rosenfeld, J. K. 1979. Ammonium adsorption in nearshore anoxic sediments. *Limnol. Oceanogr.* 24:356–64

Rosenfeld, J. K. 1981. Nitrogen diagenesis in Long Island Sound sediments. *Am. J. Sci.* 281:436–62

Rudd, J. W. M., Hamilton, R. D. 1978. Methane cycling in a eutrophic shield lake and its effects on whole lake metabolism. *Limnol. Oceanogr.* 23:337–48

Rudd, J. W. M., Taylor, C. D. 1980. Methane cycling in aquatic environments. *Adv. Aquatic Microbiol.* 2:77–150

Rudd, J. W. M., Hamilton, R. D., Campbell, N. E. R. 1974. Measurement of microbial oxidation of methane in lake water. *Limnol. Oceanogr.* 19:519–24

Sahores, J. J., Witherspoon, P. A. 1970. Diffusion of light paraffin hydrocarbons in water from 2°C to 80°C. In *Advances in Organic Geochemistry, 1966*, ed. G. D. Hobson, G. C. Spears, pp. 219–30. New York: Pergamon

Sansone, F. J., Martens, C. S. 1978. Methane oxidation in Cape Lookout Bight, North Carolina. *Limnol. Oceanogr.* 23:349–55

Sansone, F. J., Martens, C. S. 1981a. Methane production from acetate and associated methane fluxes from anoxic coastal sediments. *Science* 211:707–9

Sansone, F. J., Martens, C. S. 1981b. Determination of volatile fatty acid turnover rates in organic-rich marine sediments. *Mar. Chem.* 10:233–47

Scranton, M. I., Brewer, P. G. 1978. Consumption of dissolved methane in the deep ocean. *Limnol Oceanogr.* 23:1207–13

Sholkovitz, E. R. 1973. Interstitial water chemistry of the Santa Barbara Basin sediments. *Geochim. Cosmochim. Acta* 37:2043–73

Smith, K. L. Jr. 1978. Benthic community respiration in the N.W. Atlantic Ocean: *in situ* measurements from 40 to 5200 m. *Mar. Biol.* 47:337–47

Sørensen, J. 1978a. Denitrification rates in a marine sediment as measured by the acetylene inhibition technique. *Appl. Environ. Microbiol.* 36:139–43

Sørensen, J. 1978b. Capacity for denitrification and reduction of nitrate to ammonia in a coastal marine sediment. *Appl. Environ. Microbiol.* 35:301–5

Sørensen, J. 1982. Reduction of ferric iron in anaerobic marine sediment and interaction with reduction of nitrate and sulfate. *Appl. Environ. Microbiol.* 43:319–24

Sørensen, J., Jørgensen, B. B., Revsbech, N. P. 1979. A comparison of oxygen, nitrate and sulfate respiration in coastal marine sediments. *Microb. Ecol.* 5:105–15

Sorokin, Y. I. 1957. Ability of sulfate reducing bacteria to utilize methane for reduction of sulfate to hydrogen sulfide. *Mikrobiologiya* 115:816–18

Sorokin, Y. I. 1962. Experimental investigation of bacterial sulfate reduction in the Black Sea using ^{35}S. *Microbiology* 31:329–35 (English trans.)

Stoessell, R. K., Byrne, P. A. 1982. Methane solubilities in clay slurries. *Clays Clay Miner.* 30:67–72

Stumm, W., Morgan, J. J. 1981. *Aquatic Chemistry*. New York: Wiley-Interscience. 780 pp. 2nd ed.

Toth, D. J., Lerman, A. 1977. Organic matter reactivity and sedimentation rates in the ocean. *Am. J. Sci.* 277:465–85

Whiticar, M. J. 1978. *Relationships of interstitial gases and fluids during early diagenesis in some marine sediments.* Dissertation. Christian-Albrechts Univ. Kiel. 152 pp.

Whiticar, M. J. 1982. The presence of methane bubbles in the acoustically turbid sediments of Eckernfördner Bay, Baltic Sea. In *The Dynamic Environment of the Ocean Floor*, ed. K. A. Fanning, F. T. Manheim, pp. 219–35. Lexington, Mass.: Heath. 502 pp.

Wilson, T. R. S. 1978. Evidence for denitrification in aerobic pelagic sediments. *Nature* 274:354–56

Winfrey, M. R., Zeikus, J. G. 1977. Effect of sulfate on carbon flow during microbial methanogenesis in freshwater sediments. *Appl. Environ. Microbiol.* 33:275–81

Winfrey, M. R., Zeikus, J. G. 1979. Anaerobic metabolism of immediate methane precursors in Lake Mendota. *Appl. Environ. Microbiol.* 37:244–53

Wolfe, R. S. 1971. Microbial formation of methane. *Adv. Microb. Physiol.* 6:107–46

Wolin, M. J. 1979. The rumen fermentation: A model for microbial interactions in anaerobic ecosystems. *Adv. Microb. Ecol.* 3:49–77

Yamamoto, S., Alcauskas, J. B., Crozier, T. E. 1976. Solubility of methane in distilled water and seawater. *J. Chem. Eng. Data* 21:78–80.

Zehnder, A. J. B., Brock, T. D. 1979. Methane formation and methane oxidation by methanogenic bacteria. *J. Bacteriol.* 137:420–32

Zehnder, A. J. B., Brock, T. D. 1980. Anaerobic methane oxidation: occurrence and ecology. *Appl. Environ. Microbiol.* 39:194–204

Zeikus, J. G. 1977. The biology of methanogenic bacteria. *Bacteriol. Rev.* 41:514–41

Zilversmit, P. B. 1955. Meaning of turnover in biochemistry. *Nature* 175:863

Zinder, S. H., Brock, T. D. 1978. Methane, carbon dioxide and hydrogen sulfide production from the terminal methiol group of methionine by anaerobic lake sediments. *Appl. Environ. Microbiol.* 35:344–62

Ann. Rev. Earth Planet. Sci. 1983. 11: 299–327

METHANE AND OTHER HYDROCARBON GASES IN MARINE SEDIMENT[1]

George E. Claypool
US Geological Survey, Denver, Colorado 80225

Keith A. Kvenvolden
US Geological Survey, Menlo Park, California 94025

INTRODUCTION

Hydrocarbon gases are common in marine sediment accumulating in present-day oceans. Such gases originate from the decomposition of organic matter by biochemical and chemical processes. We consider seven hydrocarbon gases that occur in marine sediment (Table 1). In addition, inorganic gases such as nitrogen (N_2), argon (Ar), carbon dioxide (CO_2), and helium (He) are present, but usually as minor or trace components in natural gas. These will not be discussed here.

Methane (C_1) is almost always the dominant component of the natural gas mixtures. Usually accompanying C_1 are other hydrocarbon gases, including ethane (C_2), propane (C_3), isobutane (i-C_4), and normal butane (*n*-C_4), that are present in variable amounts from traces to 30–40 percent collectively. Marine sediments also contain volatile hydrocarbons of higher molecular weight, e.g. C_5 through at least C_7 (Hunt 1975), but our discussion is confined to permanent gas hydrocarbons C_1 through C_4. In addition to gaseous alkanes, the alkenes, ethene ($C_{2=}$) and propene ($C_{3=}$), are found also, but mainly in sediment near the seafloor and in the overlying water column.

There are three main stages of natural gas formation during the burial history of sediment. The earliest stage is biological C_1 formation, which

[1] The US Government has the right to retain a nonexclusive royalty-free license in and to any copyright covering this paper.

Table 1 Hydrocarbon gases in marine sediment

Name	Symbol	Molecular formula	Molecular weight
Methane	C_1	CH_4	16
Ethene	$C_{2=}$	C_2H_4	28
Ethane	C_2	C_2H_6	30
Propene	$C_{3=}$	C_3H_6	42
Propane	C_3	C_3H_8	44
Isobutane	i-C_4	C_4H_{10}	58
n-Butane	*n*-C_4	C_4H_{10}	58

occurs at low temperatures ($<50°C$) under certain environmental conditions. The next stage is early thermogenic (nonbiological) gas formation, in which the whole series of gaseous and liquid hydrocarbons are formed at rates that become geologically significant when burial temperatures are in the range of 80–120°C. Late thermogenic C_1-rich gas is produced during the last stage of gas formation, at temperatures higher than about 150°C at which previously formed heavier hydrocarbons are converted to C_1. Natural gas formed during each of these stages has a characteristic chemical and isotopic composition.

Cold bottom water ($\simeq 2°C$) and thin sediment cover (<1 km) over most of the deep ocean basins prevent attainment in the sediment of the temperatures required for significant nonbiological gas formation, except in unusual circumstances. Only at continental margins or at active ocean ridges are sediment temperatures achieved that permit significant nonbiological gas generation from high-temperature (80–150°C) decomposition of organic matter. As a consequence, the low-temperature biological decomposition of organic matter is the most important C_1-generating process operating in marine sediments accumulating in present-day oceans.

C_1 generation in nonmarine environments, e.g. "marsh gas," is an obvious consequence of organic decay under anaerobic conditions. However, prior to the advent of deep coring of the seafloor by the Deep Sea Drilling Project (DSDP), the extent of C_1 occurrence in the marine environment was not widely appreciated or understood (Atkinson & Richards 1967). Gassy marine sediments occasionally have been reported in coring operations in which penetration was limited to the upper 5 m of surface sediment (Revelle 1950, Emery & Hoggan 1958, Reeburgh 1969, 1976, 1980, Nissenbaum et al 1972, Rashid et al 1975, Martens & Berner 1977), but gassy sediments are more commonly found in DSDP rotary coring, at depths previously unattainable by conventional gravity or piston coring.

C_1 GENERATION AND CONSUMPTION PROCESSES

Sedimentary C_1 is formed by both biological and nonbiological decomposition of buried organic matter. Anaerobic microorganisms produce relatively pure C_1 enriched in the light isotope ^{12}C. C_1, as well as other alkane hydrocarbons, is produced by nonbiological, spontaneous decomposition of organic matter at rates that increase in an exponential fashion with increasing temperatures. At average Earth-surface temperatures (10–25°C), the rate of nonbiological hydrocarbon formation is extremely slow, and these hydrocarbons probably do not make a volumetrically significant contribution to economic gas occurrence, even when the process has continued over geological time periods. Only when organic matter is heated to temperatures of about 100°C (Frank et al 1974) does nonbiological gas formation occur at rates that are significant on the time-scale of the development period of sedimentary basins (10^8 yr).

Hydrocarbons (C_1 in particular) are subject to rapid biological oxidation in aqueous systems that contain dissolved O_2 (Rudd et al 1974, Patt et al 1974, Rudd & Hamilton 1975, 1978, Jannasch 1975, Rudd 1980). In addition, there is evidence that C_1 can be anaerobically oxidized in conjunction with active sulfate reduction (Reeburgh 1976, 1980, Reeburgh & Heggie 1977, Barnes & Goldberg 1976, Kosiur & Warford 1979). The processes of aerobic oxidation of hydrocarbons are well understood (McKenna & Kallio 1965), but the mechanisms for anaerobic degradation are still a matter of debate. In general, C_1 content in aerobic and sulfate-containing sediment is low ($< 10^{-4}$ standard volumes of gas per volume of sediment, v/v) and is most properly viewed as a result of the balance between C_1 production or influx, and C_1 consumption. In anoxic, sulfate-free, or low-sulfate sediment beneath a water column of sufficient depth to inhibit bubble formation, C_1 content can be high (0.05–2 v/v) and C_1 loss is by upward diffusion.

Biological Processes

The breakdown of organic matter in anoxic environments involves a complex sequence of processes symbolized in Figure 1, which is modified and expanded after Mah et al (1977). Dead organic tissue is dissaggregated and hydrolytically decomposed to yield the constituent biopolymers–cellulose, alginate, proteins, and lipids. The more labile of these biopolymers are hydrolyzed to yield the biomonomers (sugars, uronic acids, amino acids, fatty acids), which in turn are decomposed by fermentation and anaerobic oxidation reactions to give 2-, 3-, and 4-carbon organic compounds (acids, alcohols, aldehydes, etc). These short-chain organic compounds are further oxidized to CO_2, with consequent production of electrons or hydrogen. Continued metabolic activity depends on the removal of electrons by the biologically mediated reduction of certain inorganic compounds known as *electron acceptors*. A variety of multivalent elements (including O, Fe, Mn, N, S, C) can act as electron acceptors. In the interstitial waters of anoxic marine sediments, dissolved SO_4^{2-} and CO_2 are the most important electron acceptor compounds.

C_1 in low-temperature marine environments probably is produced from a limited range of substrates: H_2+CO_2, acetate, and formate (Mah et al 1977). Other compounds (methanol, trimethylamine, methionine, methyl mercaptan, and dimethyl sulfide) apparently can provide alternative carbon substrates for methanogenesis under unusual environmental conditions (Zinder & Brock 1978, Oremland et al 1982a). Of the more common substrates, formate is believed to be more readily converted to H_2+CO_2 than fermented to C_1 (Hungate et al 1970). With respect to the relative importance of acetate vs H_2+CO_2 as substrates for methanogenesis,

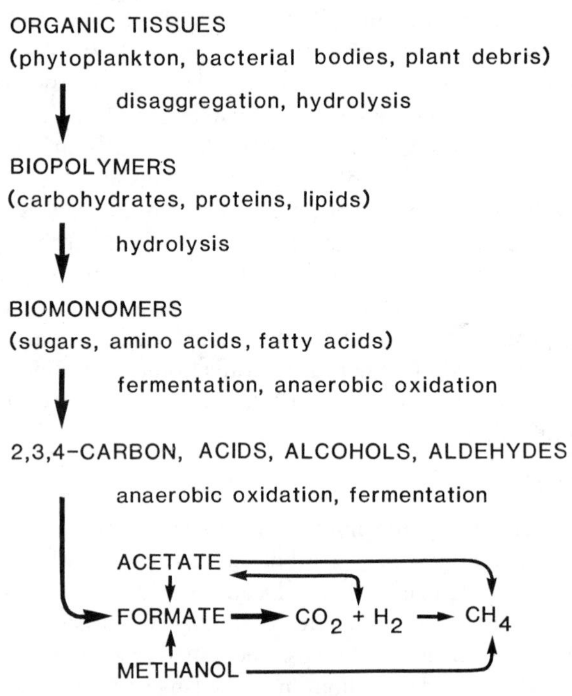

Figure 1 Sequence of organic matter degradation reactions in marine sediments.

environmental conditions also are important (Mah et al 1977, Mountfort & Asher 1978). Evidence for the importance of acetate fermentation in shallow-water, coastal lagoons is conflicting (Sørensen et al 1981, Sansone & Martens 1981). In deeper-water marine sediments, the evidence favors CO_2 reduction as the dominant methanogenic process; acetate fermentation is an energetically less favorable process (Zeikus 1977). Radiolabeling experiments have shown that acetate is anaerobically oxidized to CO_2 prior to C_1-generation in Santa Barbara Basin sediments (Warford et al 1979), and that 55–99% of C_1 in Baltic Sea sediments was formed via CO_2 reduction (Lein et al 1981).

Sulfate-reducing bacteria are believed to compete so favorably with CO_2-reducing microorganisms for available electron donors or hydrogen that C_1 production can be inhibited in the presence of dissolved SO_4^{2-} (Claypool & Kaplan 1974, Martens & Berner 1974, Winfrey & Zeikus 1977, Abram & Nedwell 1978, Sansone & Martens 1981). Apparently, biological C_1 production does not occur at optimum rates in most marine sediments until more than 80% of the dissolved SO_4^{2-} is depleted (Nikaido 1977). The depth of SO_4^{2-} depletion depends on the relative rates of biological processes and replenishment of SO_4^{2-} from overlying sea water, and varies from as little as a few centimeters to as much as 200 m. In marine environments, sediment immediately beneath the effective depth of SO_4^{2-} depletion invariably contains abundant C_1, and displays other geochemical changes indicating that rapid C_1 generation is taking place (Claypool & Kaplan 1974). Thus, a separation exists between sediment zones where C_1 production and SO_4^{2-} reduction are the dominant terminal processes. The two processes are not, however, mutually exclusive in the strictest sense (Oremland & Taylor 1978, Mountfort et al 1980, Oremland et al 1982b).

The apparent inhibition of methanogenesis by SO_4^{2-} is usually rationalized in terms of a competitive advantage conferred by greater relative free-energy yield for SO_4^{2-} reduction (Claypool & Kaplan 1974, Berner 1980). However, a more direct, mechanistic explanation involves the observed, more favorable kinetic parameters for H_2 utilization possessed by SO_4^{2-}-reducing bacteria compared with methanogens (Kristjansson et al 1982). In this manner, SO_4^{2-} concentration determines the dominant anaerobic respiration process, whereas the actual rates of either process are affected by other factors controlling substrate availability (Winfrey et al 1981).

Concentration And Isotopic Depth Profiles

One of the main lines of evidence for sequential SO_4^{2-}-reducing and C_1-producing anoxic diagenetic processes related to the most energetically efficient available electron acceptor comes from studies of interstitial waters from marine sediments (Berner 1980). In particular, the depth distribution of dissolved sulfur and carbon species and their stable isotope ratios clearly indicate distinct zones of diagenesis. An idealized depth plot of the concentration and δ-values[2] for some of the dissolved carbon and sulfur species in the porewater of typical anoxic marine sediments is shown as Figure 2. The depth scale of Figure 2 is arbitrary in the sense that the units of depth can vary from 10^{-1} to 10^2 m, depending on several factors, the most important of which is the rate of sediment accumulation. Figure 2 is a generalized composite of porewater data from several deep-water depositional environments, but draws most heavily on data from the Astoria Fan (Claypool 1974) and the Blake Outer Ridge (Claypool & Threlkeld 1983).

The depth profile of dissolved sulfate shows a regular decrease from seawater concentrations near the seafloor to zero at an arbitrary depth of 2 units. Over this same depth interval, the $\delta^{34}S$ of residual sulfate increases from $+20‰$ to about $+50‰$. Dissolved CO_2 concentration increases as sulfate concentration decreases, and may reach values that are twice the original sulfate concentration in seawater (56 millimolar). Because the increased $\sum CO_2$ is derived from oxidation of organic matter, the $\delta^{13}C$ of $\sum CO_2$ decreases rapidly with sediment depth from seawater values of $0‰$ near the seafloor to the $\delta^{13}C$ range of organic carbon ($-25 \pm 2‰$) and remains at this value down to a depth of about 2 units.

At depths greater than 2 units in Figure 2, the $\delta^{13}C$ of $\sum CO_2$ shifts abruptly to heavier values as a consequence of the preferential removal of ^{12}C-enriched CO_2 to form C_1. At the depth at which sulfate goes to zero the concentration of $\sum CO_2$ also decreases, reflecting the onset of rapid C_1 generation. In addition, this depth (2 units) is usually the shallowest depth of readily observable C_1, as marked by the appearance of gas pockets in DSDP cores, or as reflected by greatly increased C_1 concentrations in the few studies of deep marine sediments in which quantitative measurements were permitted by the sampling methods (Emery & Hoggan 1958). The C_1 with the most negative $\delta^{13}C$ (-100 to $-90‰$) also occurs at the depth of the transition from SO_4^{2-} reduction to CO_2 reduction, or at a depth of 2 units in Figure 2. Isotopically heavier C_1 (more positive $\delta^{13}C$) occurs at greater depths because of progressive depletion of ^{12}C in the $\sum CO_2$ reservoir from which the C_1 is formed. At depths shallower than 2 units in Figure 2, the small amounts of C_1 present have more positive $\delta^{13}C$ values, indicating partial oxidation in the zone of sulfate reduction (Doose & Kaplan 1981, Brooks et al 1983). This oxidation process selectively removes

[2] Stable isotope ratios for carbon and sulfur are reported in the standard δ-notation, where $\delta(‰) = [(R_{sample}/R_{standard}) - 1] \times 10^3$, $R = {}^{13}C/{}^{12}C$ and ${}^{34}S/{}^{32}S$; and the standards are PDB marine carbonate and Cañon Diablo meteoritic troilite, respectively.

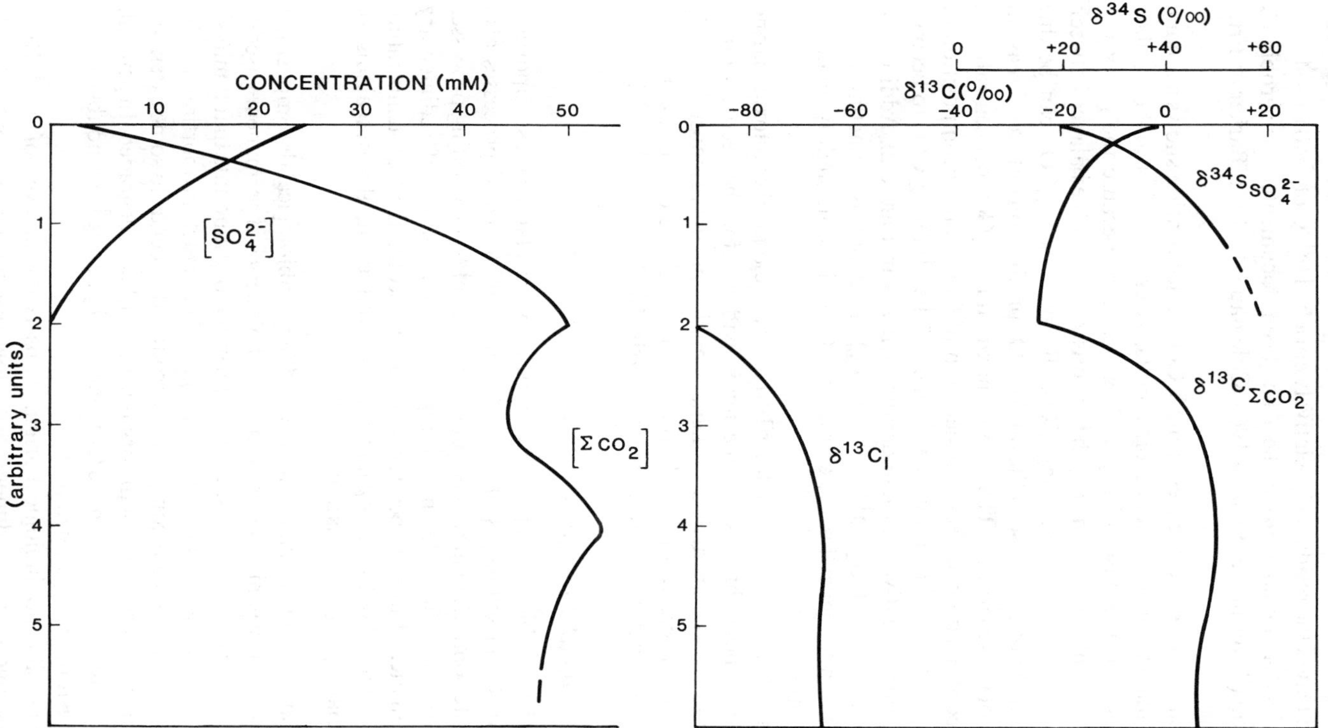

Figure 2 Generalized profiles of concentration and stable isotope ratio changes for dissolved sulfur and carbon species in anoxic marine sediments. Depth scale is arbitrary with depth units ranging from 10^{-1} to 10^2 m.

^{12}C and leaves the residual C_1 relatively enriched ($\delta^{13}C$ of -50 to $-40°/_{oo}$, or heavier). A similar isotope effect has been documented for aerobic C_1 oxidation (Silverman & Oyama 1968, Coleman et al 1981, Barker & Fritz 1981).

The curve in Figure 2 representing $\sum CO_2$ concentration shows a second deeper maximum at a depth of 4 units, after the cessation of sulfate reduction and the onset of rapid C_1 generation. Because the porewater $\sum CO_2$ concentration at any depth is a consequence of the balance between processes adding and removing CO_2, this increased $\sum CO_2$ can be due either to a relative increase in the rate of CO_2 production or to a decrease in the rate of C_1 generation. The second maximum in $\sum CO_2$ concentration is due to continued or increased generation of CO_2 from decomposition of organic matter, as shown by the tendency of the $\delta^{13}C$ of $\sum CO_2$ to become lighter (more negative) at these depths (below 4 in Figure 2). Addition of isotopically light CO_2 ($\delta^{13}C = -25°/_{oo}$) at depth is sometimes also reflected in a tendency for the cumulative C_1 to become slightly lighter at these depths.

The depth-related changes described above are typical of anoxic marine sediments. These changes are the main evidence for the recognition of depositional and diagenetic sedimentary environments responsible for C_1 generation. Wherever the conditions discussed above do not exist, C_1 is absent or present only at background levels (10^{-8}–10^{-5} v/v).

Nonbiological Processes

Although microbiological processes appear to be by far the most important source of C_1 in sediments of ocean basins, nonbiological processes also contribute some C_1. Three types of nonbiological C_1-generating processes are considered: first, the nonbiological decomposition of sedimentary organic matter at low temperatures ($<75°C$); second, the high-temperature ($>75°C$) degradation of organic matter; and third, the outgassing of primordial C_1 from the mantle.

LOW-TEMPERATURE NONBIOLOGICAL C_1 The nonbiological decomposition of sedimentary organic matter forms the whole series of permanent gaseous alkane hydrocarbons (C_1, C_2, C_3, C_4, etc) in relative amounts that diminish regularly with increasing molecular size. The exact proportion of the various hydrocarbon gas components produced depends to some extent on the chemical composition of the organic matter being degraded. In general, C_1 is 5–20 times as abundant as C_2 in gases originating from nonbiological decomposition of organic matter.

C_1 of possible nonbiological origin is observable at levels of 10^{-8}–10^{-5} v/v in pelagic marine sediments with low organic matter content (C_{org}

< 0.2%) beneath the zone of surface biological activity (Claypool 1983). Microbiological C_1 is absent from such sediments because of the lack of metabolizable organic matter and the presence of alternative electron acceptors. C_1 of microbiological origin typically is present in marine sediments of continental slopes and rises, with organic carbon contents in the range of 0.5–1%. Higher contents of nonbiological hydrocarbons also are present in methanogenic sediments, but the low-temperature nonbiological C_1 is completely masked by biological C_1. However, because most C_{2+} hydrocarbons are not a direct product of the microbiological decomposition of organic matter, the approximate content of nonbiological C_1 in gas of predominantly biological origin is about ten times the level of C_2.

C$_1$ OF HIGH-TEMPERATURE ORIGIN Most sediments accumulating in present-day ocean basins have not been exposed to high temperatures. Therefore, the limited occurrence in marine sediments of C_1 originating from the apparent high-temperature decomposition of organic matter (i.e. $\delta^{13}C_1$ of -50 to -30‰) can be explained in one of two ways: either the gas was generated at a more distant high-temperature location and migrated into cooler marine sediments, or instead the gas was generated in sediments subjected to high temperatures sometime during their burial history.

Large parts of the ocean basins are underlain by sediment thicknesses in excess of 3 km, especially near continental margins. Organic matter in such deeply buried sediments could have generated hydrocarbons that then migrated into the sediments within 1 km of the seafloor. DSDP holes drilled above salt domes in the deep Gulf of Mexico on Legs 1 and 10 found gas of apparent thermogenic origin, which probably migrated up from greater depths along with the salt diapir (Erdman et al 1969, Claypool et al 1973). Other DSDP holes have been drilled in regions with anomalously high thermal gradients (e.g. Gulf of California), and C_1 of high-temperature thermogenic origin was found at relatively shallow depths of burial (Galimov & Simoneit 1982). The chemical and isotopic composition of gases sampled in the DSDP cores is briefly reviewed in a following section.

PRIMORDIAL C$_1$ Interest in C_1 gas of primordial origin has increased recently as a result of the observation that primordial gases (^3He, C_1, H_2) are being injected into the deep oceans from hydrothermal systems operating at active mid-ocean ridges (Welhan & Craig 1979, Lupton & Craig 1981). This has led to speculation that gas from similar sources may have natural resource implications (Gold 1979, Gold & Soter 1980). The occurrence of primordial C_1 in marine sediments has not been documented, but is probably minor. Gas with composition indicative of likely mantle origin ($C_1/^3He = 3$–10×10^6, $\delta^{13}C_1 = -19‰$) has been most clearly identified in solution in water ($C_1 = 10^{-3}$ v/v) issuing from fresh basalt at 21°N on the East Pacific Rise (Welhan 1980). At spreading centers where hydrothermal systems are operating through appreciable thicknesses of organic-matter-rich sediment, as in Guaymas Basin in the Gulf of California, the hydrocarbon gases in the sediment and those being injected into the water column have a composition indicating origin from organic matter decomposition (Simoneit & Lonsdale 1982).

In general, the flux of C_1 into marine sediments from low-temperature, nonbiological organic-matter decomposition and from outgassing of the Earth is so slow that only a highly efficient concentrating and trapping mechanism could bring about a recognizable accumulation of such gas. In addition, the fact that such gas would be dispersed at low concentrations in marine sediment makes it extremely difficult to study. Gas of high-temperature origin is relatively rare in sediments of present-day ocean basins, but is probably much more important than C_1 gas from the other two nonbiological processes mentioned.

OCCURRENCE OF C_1 AND ORIGIN AND OCCURRENCE OF OTHER HYDROCARBON GASES

Trace or minor amounts of the other hydrocarbon gases (C_2, C_3, and C_4) commonly are found associated with C_1 where sensitive detection techniques have been used in gas analyses. Emery & Hoggan (1958) showed that near-surface marine sediment (<4 m subbottom) from Santa Barbara Basin off southern California contains mainly C_1, but also small amounts (10^{-7}–10^{-4} v/v) of C_2, $C_{2=}$, C_3, i-C_4, and n-C_4, as well as volatile hydrocarbons to C_7. Their sampling procedures involved collection of large-volume samples and transfer of the sediment from the seafloor to the laboratory without exposing the core to air. This sampling technique compensated for the small amounts of C_2 through C_4 hydrocarbons and the relatively insensitive detection techniques, which discouraged any immediate follow-up of these kinds of measurements.

Gases in Near-Surface Sediment

Published research on the occurrence of C_2 through C_4 hydrocarbons in near-surface marine sediments languished for almost twenty years until Bernard et al (1978) described the distribution of C_1, C_2, $C_{2=}$, C_3, and $C_{3=}$ in shelf and slope sediment (<2 m subbottom) of the Gulf of Mexico. Using similar headspace analysis techniques, Kvenvolden & Redden (1980) studied the occurrence and distribution of C_1, C_2, $C_{2=}$, C_3, $C_{3=}$, i-C_4, and n-C_4 in surface and near-surface sediment (<2.5 m subbottom) of the outer shelf, slope, and basin of the Bering Sea. For headspace analyses, sediment samples are obtained using standard gravity or piston-coring techniques. Recovered cores typically are less than 5 m long, and extracted gas is analyzed by gas chromatography. The methods of headspace analysis provide a rapid quantitative measure of a portion of the gases originally in the sediment.

Headspace analyses of samples from the Gulf of Mexico (Bernard et al 1978) and the Bering Sea (Kvenvolden & Redden 1980) provide a measure of concentrations of hydrocarbon gases in near-surface, partially oxic sediments of open-marine environments of the shelf, slope, and basin. For example, typical background concentrations of C_1 in these sediments range from about 10^{-6} to 10^{-4} standard volumes of gas per volume of interstitial water (v/v). These concentrations are as much as three to five orders of magnitude lower than the concentrations of C_1 found at shallow (<5 m) depths in some anoxic and low-sulfate sediments of restricted marine environments (Reeburgh 1969, 1976, 1980, Barnes & Goldberg 1976, Martens & Berner 1977, Kosiur & Warford 1979).

Generalizations also can be made about the occurrences of C_2 through C_4 hydrocarbons in sediments at shallow depth in open-marine environments. In the Gulf of Mexico and the Bering Sea, concentrations of C_2 are typically less than about 10^{-7} v/v, concentrations of C_3 are less than about 5×10^{-8} v/v, and concentrations of the sum of i-C_4 and n-C_4 are less than about 2.5×10^{-8} v/v. $C_{2=}$ and $C_{3=}$ concentrations are about the same order of magnitude as the concentrations of C_2 and C_3, with some variation depending on the environmental setting. Concentrations of C_2 through C_4 hydrocarbons do not appear to increase significantly with sediment depth within the uppermost 2 m; however, on a regional basis the average concentrations of C_2 through C_4 hydrocarbons do decrease with increasing depth of water from shelf to slope to basin.

Concentrations of C_2 through C_4 hydrocarbons may be much higher in sediments of restricted marine environments than in sediments of open-marine environments. For example, in sediments of Norton Sound, Alaska, concentrations of C_2+C_3 reach about 2×10^{-6} v/v and of i-C_4+n-C_4 about 2×10^{-7} v/v. However, C_1 in Norton Sound sediments reaches concentrations in excess of 2×10^{-2} standard volumes per volume of interstitial water at sediment depths of less than 3 m (Kvenvolden et al 1981a), so the concentration of C_{2+} hydrocarbons relative to C_1 actually can be less in restricted marine sediments, compared with open-marine sediments.

In considering the possible origin of hydrocarbons in near-surface, open-marine sediments, a comparison with hydrocarbons in seawater is instructive. Results of extensive studies of hydrocarbon gases in seawater have been summarized by Swinnerton & Lamontagne (1974). Table 2 shows that the hydrocarbon concentrations of near-surface marine sediment greatly exceed the average hydrocarbon concentration of open-ocean water. Thus, a water-column source for hydrocarbon gas in near-surface marine sediment is improbable.

Another possible explanation is that the hydrocarbon gases in the near-surface sediment of open-marine environments diffused or migrated upward from more deeply buried sediment. Although some movement from depth is certainly possible, the distribution of hydrocarbon gases in near-surface marine sediment, in general, does not provide compelling evidence for this source. Bernard (1979) concluded that upward diffusion from large accumulations of C_1 deeper than about 10 m would not be detectable in near-surface sediment. Profiles of concentrations of C_2 through C_4 hydrocarbons show little or no change with depth in near-surface sediments (Bernard et al 1978, Kvenvolden & Redden 1980). This kind of profile implies that the source of these hydrocarbons is probably not deeper in the sediment.

If the overlying water column and the underlying sediments are discounted as major sources of the hydrocarbons found in near-surface marine sediments, then these hydrocarbons must be generated in place. Microbial production (possibly in anoxic microenvironments) and consumption provide a reasonable explanation for the observed C_1 distributions in near-surface marine sediments of open-marine environments.

C_2, $C_{2=}$, C_3, and $C_{3=}$ hydrocarbons are known to be generated in connection with microbial processes (Davis & Squires 1954, Primrose & Dilworth 1976). Marine organisms produce both $C_{2=}$ and $C_{3=}$ (Hunt 1974), and studies of anoxic estuarine sediments indicate that small quantities of C_2 can be formed by certain methanogenic bacteria

Table 2 Concentrations of hydrocarbon gases in open-ocean water (Swinnerton & Lamontagne 1974) and near-surface sediment of open-marine environments (Bernard et al 1978, Kvenvolden & Redden 1980)

Gas	Open-ocean[a]	Near-surface sediment[a]
C_1	50	1000–100,000
C_2	0.5	20–100
$C_{2=}$	5	10–200
C_3	0.3	10–50
$C_{3=}$	1.4	5–100
C_4 (i-C_4 + n-C_4)	0.05	0–25

[a] Concentrations in 10^{-9} standard volumes of gas per volume of water (v/v).

(Oremland 1981). Prolonged incubation of these estuarine sediments also produce $C_{2=}$, C_3, $C_{3=}$, i-C_4, and *n*-C_4 (Vogel et al 1982). This kind of circumstantial evidence, combined with the common observation of small quantities of C_2 through C_4 hydrocarbons in recent marine sediments, has led some authors to suggest microbial processes as a reasonable source for these hydrocarbon gases (Emery & Hoggan 1958, Bernard et al 1978, Whelan et al 1980, Kvenvolden & Redden 1980).

Gases in Deeper Oceanic Sediment

Results obtained through the Deep Sea Drilling Project (DSDP) have provided an extensive record of hydrocarbon gases in deeper oceanic sediments, covering all of the oceans except the Arctic, and from sediment depths of a few meters to about 1500 m.

C_1 is the dominant hydrocarbon gas found in DSDP samples. Starting in about 1970 with DSDP Leg 10 in the Gulf of Mexico, gas samples have been collected routinely in evacuated containers (Gealy & Dubois 1971) or recovered from the headspace of sealed cans (McIver 1973) and analyzed by gas chromatography. Claypool et al (1973) summarize gas analyses of gas pockets in sediments collected on seven legs (10, 11, 13, 14, 15, 18 and 19) in the following areas: Gulf of Mexico, western Atlantic Ocean, Mediterranean Sea, Caribbean Sea, eastern and northern Pacific Ocean, and the Bering Sea. For all drilling sites except one, C_1 constituted more than 99.9 percent of the hydrocarbon gases detected. Utilizing gases in canned sediment samples, McIver (1975) showed that C_1 is virtually the only hydrocarbon gas present in 125 samples from 22 sites on eight DSDP legs (18, 19, 21, 23, 24, 27, 28, and 29) in the western and northern Pacific Ocean, Bering Sea, eastern Pacific Ocean, Arabian Sea, Red Sea, Gulf of Aden, Timor Sea, and the Antarctic Ocean. Thus the predominance of C_1 in marine sediments of the world's oceans is well established.

Although concentrations of C_2 are usually less than 0.1 percent of the hydrocarbon gases present in oceanic sediment, the ratio of C_2 to C_1 generally increases exponentially with depth in the sediment. This increasing ratio results mainly from increasing amounts of C_2 with depth, rather than from changes in the amounts of C_1. In Figure 3, the ratio of C_2 to C_1 is plotted as a function of depth of burial for a variety of depositional settings in the world's oceans, and the exponential increase in C_2 content with increasing depth is evident. At any given locality, the amount of C_2 appears to be proportional to the temperature and age of sediment (Rice & Claypool 1981).

Hydrocarbon gases larger than C_2, i.e. C_3, i-C_4, and *n*-C_4, also have been found in samples from DSDP drilling. $C_{2=}$ and $C_{3=}$ have not been reported. Routine detection and analysis of hydrocarbon gases larger than

C_2 required the application of a concentration technique to isolate the small amounts of the compounds usually present with C_1 from ocean sediment (Whelan 1979). Hunt & Whelan (1978) noted that C_1 was the dominant hydrocarbon gas in Black Sea sediment (Leg 42B), but that the gas mixture contained small amounts (10^{-6}–10^{-3} v/v) of C_2 through C_4 hydrocarbons. The concentrations of these hydrocarbons differed from each other by about an order of magnitude for each increasing carbon number; C_2 was most abundant. The trends of concentrations with depth for each of these hydrocarbons were similar.

Whelan (1979) observed trends of C_1 through C_4, as well as larger

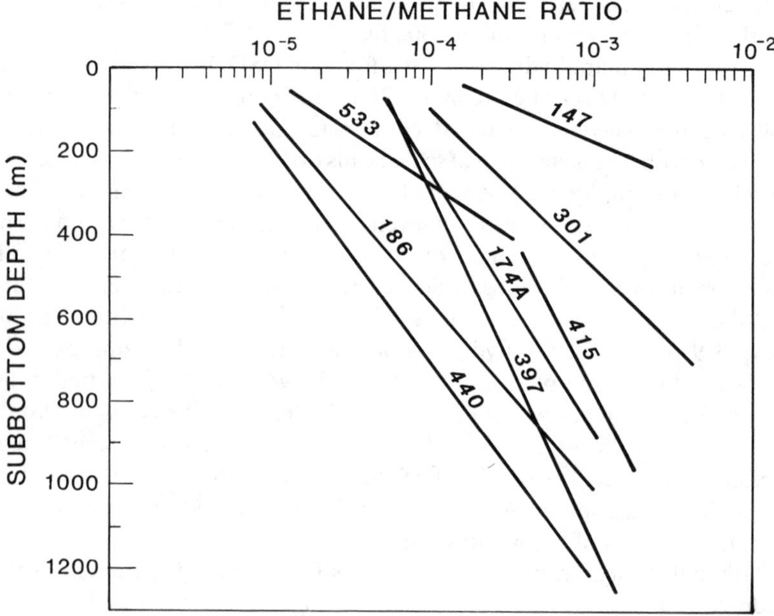

Figure 3 Ethane to methane ratios (C_2/C_1) vs depth in sediment from selected sites drilled by the Deep Sea Drilling Project.

Site	Leg	Location	Reference
147	15	Cariaco Trench, Caribbean	Rice & Claypool (1981)
174A	18	Astoria Fan, Pacific	Rice & Claypool (1981)
186	19	Aleutian Trench, Pacific	Rice & Claypool (1981)
397	47	African Margin, Atlantic	Whelan (1979)
415	50	Moroccan Basin, Atlantic	Galimov et al (1980)
440	57	Japan Trench, Pacific	Whelan & Sato (1980)
533	76	Blake Ridge, Atlantic	Kvenvolden & Barnard (1983a)

hydrocarbons, in continental margin sediments off northwest Africa that were similar to those observed for Black Sea sediments. Observations of C_2 through C_4 hydrocarbons in samples from DSDP drilling have now been augmented to include sediments from the Vigo Seamount (Whelan & Hunt 1979a), the Bay of Biscay (Whelan & Hunt 1979b), the Moroccan Basin (Galimov et al 1980, Whelan & Hunt 1980a), the Japan Trench (Whelan & Hunt 1980b), the California borderland (Whelan & Hunt 1981), and the Blake Outer Ridge (Kvenvolden & Barnard 1983a).

The C_2 through C_4 hydrocarbons accompaning C_1 in oceanic sediment probably have a complex history. They most likely represent products of a continuum of diagenetic processes, the first stages of which involve mainly microbiological processes operating near the sediment surface. These microbiological processes give way with sediment depth to processes that are thermally controlled at low temperatures. Thermal processes cause mild thermal decomposition of organic matter, resulting in the exponential increase with depth of C_2. Hydrocarbon gases C_3, i-C_4, and n-C_4 also tend, as a general rule, to increase slightly in concentration with depth. This increased concentration also may be attributed to low-temperature ($<50°C$) thermal reactions. At temperatures existing in sediment at depths greater than those sampled by DSDP for gas analyses (i.e. greater than about 1500 m in regions with average thermal gradients), the thermal processes become more intense and result in the generation of gas associated with petroleum formation.

On the basis of all these observations, the following generalizations can be made: (*a*) C_1 is the most abundant hydrocarbon gas in deeper oceanic sediments, commonly constituting more than 99% of the hydrocarbon gas mixture; (*b*) C_2 through C_4 hydrocarbons are ubiquitous but minor components of the gas mixture in oceanic sediment; (*c*) C_2 generally increases exponentially in concentration with depth of sediment; and (*d*) the trends of C_2 through C_4 concentrations with depth are usually similar.

The isotopic composition of C_1 in the hydrocarbon gas mixtures in oceanic sediments provides the best evidence of origin (Rice & Claypool 1981). Table 3 shows the range of isotopic compositions of C_1 found in sediments drilled by DSDP. In most of the examples, the $\delta^{13}C$ values fall in a range between about $-90‰$ and $-50‰$. This range is characteristic of microbiologically derived C_1 from natural sources, as summarized by Fuex (1977). C_1 with carbon isotopic compositions more positive than about $-55‰$ in natural gas of near-surface sediment may result from partial microbial oxidative processes, as discussed by Silverman & Oyama (1968), Coleman et al (1981), and Barker & Fritz (1981). Carbon isotopic compositions of C_1 more positive than $-55‰$ also may be attributed to high-temperature thermal processes giving rise to C_1 that has migrated

Table 3 Carbon isotope composition of C_1 recovered through DSDP drilling of oceanic sediment

Location	DSDP Leg	$\delta^{13}C_1$ (‰)	Reference
Gulf of Mexico	10	−84.0 to −48.7	Claypool et al (1973)
East Mediterranean Sea	13	−77.8 to −72.6	Claypool et al (1973)
Caribbean Sea	15	−81.3 to −69.1	Claypool et al (1973)
Carioca Trench	15	−76.3 to −59.6	Lyon (1973)
Astoria Fan	18	−88.7 to −75.9	Claypool et al (1973)
East Aleutian Trench	18	−80.8 to −72.6	Claypool et al (1973)
Bering Sea, North Pacific	19	−78.8 to −62.7	Claypool et al (1973)
Arabian Sea	23	−82.8 to −61.0	Claypool (unpublished)
South Red Sea	23	−76.8 to −60.9	Claypool (unpublished)
Gulf of Aden	24	−76.2 to −70.2	Claypool (unpublished)
Timor Trench	27	−77.0 to −58.6	Claypool (unpublished)
Ross Sea	28	−78.9 to −67.5	Claypool (unpublished)
Sea of Japan	31	−72.0 to −67.4	Jodele & Doose (unpublished)
Norwegian Sea	38	−87.3 to −71.2	Morris (1976)
Black Sea	42A	−72.0 to −63.0	Hunt & Whelan (1978)
Continental Rise, NW Africa	41	−73.9 to −51.7	Doose et al (1978)
Continental Rise, NW Africa	47	−80.0 to −60.0	Whelan (1979)
Moroccan Basin	50	−82.1 to −49.2	Galimov et al (1980)
Japan Trench	56, 57	−83.8 to −67.6	Whelan & Sato (1980)
California Borderland, Baja	63	−81.6 to −47.3	Claypool (unpublished)
Gulf of California	64	−79.2 to −40.4	Galimov & Simoneit (1982)
Blake Outer Ridge	11	−88.4 to −70.1	Claypool et al (1973)
Blake Outer Ridge	76	−93.8 to −65.8	Galimov & Kvenvolden (1983)

from deeper sources (Fuex 1977). Measurement of the H/D ratio of C_1 can resolve these alternative interpretations based on $\delta^{13}C$, because partially oxidized C_1 is even more enriched in D than ^{13}C (Coleman et al 1981).

At many DSDP sites, profiles of C_1 isotopic values can be constructed with depth of burial (Claypool et al 1973, Claypool & Kaplan 1974, Doose et al 1978, Whelan 1979, Galimov et al 1980, Whelan & Sato 1980, Galimov & Kvenvolden 1983). These profiles consistently show the trends seen in Figure 2, and are believed to result from a kinetic effect of biological methanogenesis (Rosenfeld & Silverman 1959).

Except for a few instances, the molecular and isotopic compositions of hydrocarbon gases in oceanic sediments cannot be explained satisfactorily by upward migration of thermogenically derived gases. Such migration would require diffusion of gas through unconsolidated and semiconsolidated oceanic sediment. That sufficient diffusion could take place over the distance required is unlikely, even if the gas were present in large amounts at depth (Bernard 1979).

IMPLICATIONS OF HYDROCARBON GASES IN MARINE SEDIMENT

Geochemical Prospecting

Although the hydrocarbon gases in the first 1000 m of oceanic sediment beneath the seafloor can, in general, be accounted for by in situ processes, situations exist where gases originating elsewhere are associated with sediment. As explained earlier, hydrocarbon gases may be generated by nonbiologic processes in more deeply buried sediment, where temperatures exceed 80°C and petroleum is being formed. These gases, derived from the thermal breakdown of organic matter during later diagenesis, may migrate along fractures and faults to the seafloor. There, as gas seeps, they may provide clues to the presence of their petroleum precursors in the subsurface.

Active gas seeps in the marine environment are manifest as gas bubbles venting from the seafloor and as anomalously high concentrations of gas dissolved in the water column. Anomalous concentrations of hydrocarbon gases in the water column can be identified by analyses of gases (*a*) recovered as bubbles (Bernard et al 1976, Reed & Kaplan 1977), (*b*) extracted from discrete water samples (Brooks et al 1973, Cline & Holmes 1977), and (*c*) detected by gas sniffers (Sigalove & Pearlman 1975, Sackett 1977, Reitsema et al 1978).

Hydrocarbon gas compositions and $\delta^{13}C$ values of C_1, measured in bubbles from natural seeps and underwater vents from offshore production operations, provide the basis for distinguishing biogenic (microbially derived) from thermogenic (petroleum-derived) gas (Bernard et al 1976). Biogenic gas has $C_1/(C_2+C_3)$ ratios greater than 1000, which means that the mixture is almost exclusively C_1. The isotopic composition of this C_1 is lighter (more negative) than $-60‰$. Thermogenic gas has significant quantities of low-molecular-weight hydrocarbons, with $C_1/(C_2+C_3)$ ratios ranging from 0 to 50. The isotopic composition of thermogenic C_1 is usually heavier (more positive) than $-50‰$. Thus, the ratio $C_1/(C_2+C_3)$ and the isotopic composition of C_1 can, in concert, be used as diagnostic parameters to ascertain the source of hydrocarbon gases in natural seeps. Figure 4 shows the molecular and isotopic compositions of gases in seeps and vents. Whereas vent gases clearly have a thermogenic source, only one seep (#5 in Figure 4) in the Gulf of Mexico has that source. Of the remaining thirteen sampled seeps in the Gulf of Mexico, eight have biogenic sources and five apparently have mixed origins (Bernard et al 1976). At Coal Oil Point and Carpinteria, offshore southern California, the composition of gas bubbling from submarine seeps indicates a thermogenic source (Reed & Kaplan 1977).

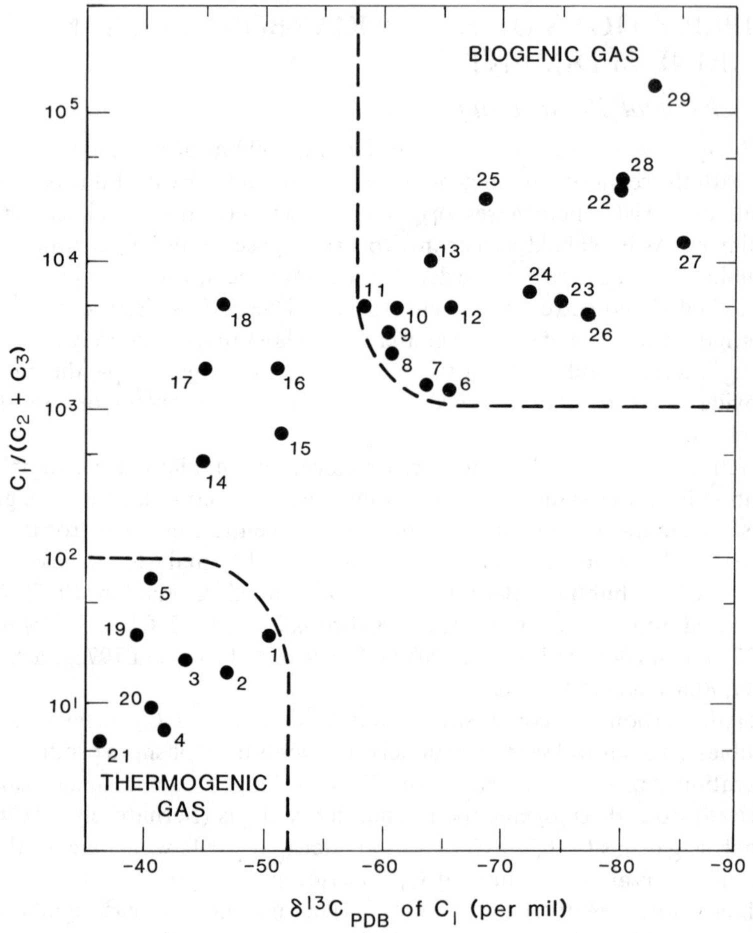

Figure 4 Relation between the logarithm of $C_1/(C_2+C_3)$ and the isotopic composition of C_1 for hydrocarbons in vents, seeps, and sediments.

Location	Numbers	Reference
Vents		
Gulf of Mexico	1–4	Bernard et al (1976)
Seeps		
Gulf of Mexico	5–18	Bernard et al (1976)
Offshore southern California	19–20	Reed & Kaplan (1977)
Norton Sound, Alaska	21–22	Kvenvolden et al (1979)
Norton Sound, Alaska	23–25	Kvenvolden et al (1981a)
Western Gulf of Alaska	26–29	Hampton & Kvenvolden (1981)

Before seeping into the water column, hydrocarbon gases may pass through unconsolidated surficial sediment at the seafloor. These hydrocarbons partially dissolve in the interstitial water and likely diffuse within the sediment. Therefore, in geochemical prospecting hydrocarbon gases in near-surface marine sediment might be examined for clues to the possible occurrence of petroleum at depth. Particularly informative results should be obtained from the analysis of sediments adjacent to active seeps. Carlisle et al (1975) examined sediments near an active seep and found a distinct hydrocarbon anomaly in which the $C_1/(C_2+C_3)$ ratio was less than 1, but no measurement was made of the isotopic composition of the C_1. In Norton Sound, Alaska, Kvenvolden et al (1979) measured the hydrocarbon gases in sediments associated with a CO_2 seep. A $C_1/(C_2+C_3)$ ratio of about 6 and a carbon isotopic composition of C_1 of $-36‰$ clearly indicate a thermogenic source of hydrocarbon gases (#21 in Figure 4). Even where active seepage is not present but where high concentrations of C_1 are present in sediment (exceeding about 10^{-3} standard volumes per volume of interstitial water), the molecular and isotopic gas compositions are diagnostic. For example, in Norton Sound (Kvenvolden et al 1981a) and the western Gulf of Alaska (Hampton & Kvenvolden 1981), the gas in near-surface, gas-charged sediment is from biogenic sources (Figure 4).

Although the $C_1/(C_2+C_3)$ ratio appears to be useful where gas concentrations are high, as in and near seeps and in gas-charged sediment, use of the ratio as an indicator of source is not yet clearly substantiated where low concentrations of hydrocarbon gases are dispersed within near-surface sediment (Kvenvolden & Redden 1980). Nevertheless, this ratio along with the ratio of $C_2/C_{2=}$ has been used to indicate the possible presence of thermogenic gas in near-surface sediment of St. George Basin in the southern Bering Sea (Kvenvolden et al 1981b), where two sites were identified with suspected thermogenic gas. This study also showed, however, that gas in surface grab samples has anomalously low $C_1/(C_2+C_3)$ ratios (less than 20). These low ratios probably do not signal the presence of thermogenic hydrocarbons, but rather indicate the preferential loss of C_1 owing to a balance between diffusion and selective microbial C_1 consumption.

Low concentrations of gas in near-surface sediment also limit the use of the isotopic composition of C_1 because the amount of C_1 collected usually is insufficient to obtain valid isotopic measurements. Where the carbon isotopic composition of C_1 can be determined for near-surface sediment, the C_1 may be anomalously heavy because the bacterial oxidation of C_1 leaves as a residual product an isotopically heavy C_1 (Coleman et al 1981). Thus, the $C_1/(C_2+C_3)$ ratios and the carbon isotopic compositions of C_1 in gases from near-surface sediment may have values that are characteristic of thermogenic sources, but the gases themselves may be the product of in situ processes strongly mediated by microbial activity. Application of these parameters as source indicators to the gases found in near-surface sediments is therefore limited, especially where low gas concentrations occur. Deeper sampling (to depths greater than about 2 m) may provide samples of gas that are not significantly affected by near-surface processes and thus are of potential use in geochemical prospecting for hydrocarbons.

Gas Hydrates

In general, solubilities of hydrocarbon gases in the interstitial waters of oceanic sediment increase with increasing depth of burial, but details regarding solubility have been worked out only for C_1 [see Rice & Claypool (1981) for a discussion]. Figure 5 shows the estimated solubility of C_1 as a function of depth. C_1 solubility increases with increasing depth of burial beneath water depths less than 1000 m. However, for deep-sea sediments at water depths greater than 1000 m, solubility initially decreases between sediment depths of 1000 and 2000 m before increasing at greater burial depths. Where water is saturated with C_1 at the depths indicated on Figure 5, the water can crystallize as a solid at temperatures above 0°C by the incorporation of C_1 into a clathrate, or three-dimensional framework of water molecules, that is stabilized by the included C_1 molecules. This water clathrate is commonly called a *gas hydrate*.

In gas hydrates, water crystallizes in the isometric system, rather than in the hexagonal system of normal ice. The isometric or cubic lattice contains voids, or cages, large enough to accommodate molecules of gas. Two structures of the cubic lattice are possible. In structure I, the cages are arranged in body-centered packing and include small hydrocarbon molecules such as C_1 and C_2 and nonhydrocarbons such as N_2, CO_2, and H_2S. In structure II, diamond packing is present; not only can C_1 and C_2 be included in the cages, but C_3 and i-C_4 also are needed to occupy some of the large cages in order to stabilize the structure. Apparently, gas molecules larger than i-C_4—for example, *n*-C_4, $C_{5's}$, etc—cannot be included in either structure I or II (Hand et al 1974, Hitchon 1974).

The pressure (depth)-temperature region in which pure C_1 forms a gas hydrate with pure water saturated with C_1 is shown in Figure 6. Gas extracted from oceanic sediment, however, is not pure C_1; it also has small concentrations of C_2 and hydrocarbon gases of higher molecular weight as well as CO_2. The presence of these additional components in the gas mixture causes the phase boundary (Figure 6) to shift to the right. Porewater in normal oceanic sediment is not pure but contains salts, particularly NaCl. The presence of salts in the water shifts the phase boundary to the left (Figure 6). For naturally occurring gases in oceanic

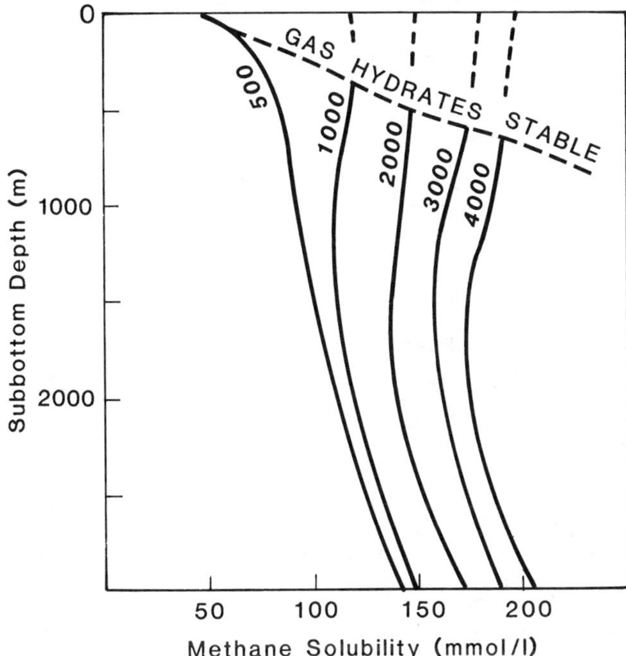

Figure 5 Estimated methane (C_1) solubility with depth of burial in oceanic sediment. The relation between C_1 solubility and the depth at which gas hydrates are stable is shown. Curves are for specific water depths (labeled in meters). The solubility data are from Culberson & McKetta (1951) and have been reduced 80% of the observed solubility of C_1 in pure water at a given temperature and depth to adjust for effects of 3.5% salinity. The series of curves indicate the changes of solubility at different pressure-temperature gradients associated with water depths from 0 to 4000 m. The pressure gradient is assumed to be hydrostatic (0.1 atm m^{-1}) and a geothermal gradient of 35°C km^{-1} is used. The temperature of sediment with no burial is assumed to be 20°C for a water depth of 0 m, 5°C for water depths of 500 m, and 2°C for water depths of 1000 m and greater. Figure is redrawn from Rice & Claypool (1981).

sediment, the shifts in the position of the phase boundary are of similar magnitude but in opposite directions. Thus, the effects approximately cancel each other, and the boundary for a pure-water and pure-methane system (Figure 6) provides a reasonable estimate of the pressure-temperature conditions under which natural gas hydrates, composed mainly of C_1, will be stable in oceanic sediment (Claypool & Kaplan 1974).

Because pressure is largely determined by water depth, the temperature established by the geothermal gradient mainly controls the lower depth limits within oceanic sediment at which gas hydrates are no longer stable and therefore decompose. The base of the gas hydrate zone follows a pressure-temperature surface that represents the maximum depth at which

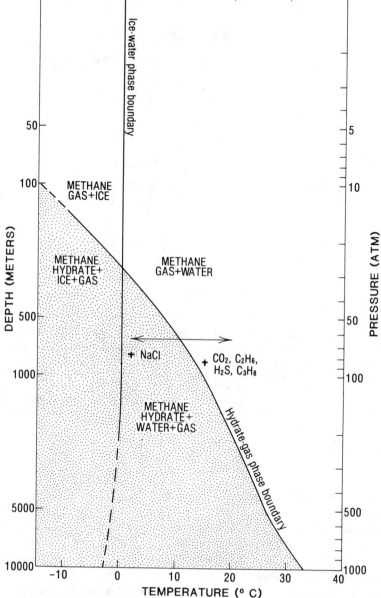

Figure 6 Phase diagram showing boundary between free methane gas (no pattern) and methane hydrate (pattern) for a pure-water and pure-methane system. Addition of NaCl to water shifts the curve to the left. Addition of CO_2, H_2S, C_2, and C_3 shifts the boundary to the right, thus increasing the area of the gas hydrate stability field. Depth scale assumes lithostatic and hydrostatic pressure gradients of 0.1 atm m^{-1}. Figure was constructed from Katz et al (1959) and redrawn from Kvenvolden & McMenamin (1980).

the gas hydrate is stable. Therefore, beneath the gas hydrate zone free gas may be present in gas-saturated sediments.

The base of the gas hydrate zone often correlates with anomalous acoustic reflectors in marine seismic data obtained from a number of areas on outer continental margins (Shipley et al 1979). The anomalous reflector approximately parallels the seafloor but deepens with increasing water depths. The depths at which this reflector occurs can be predicted based on considerations of the pressure-temperature stability field for gas hydrates and the geothermal gradient. The reflector commonly is called a *bottom-simulating reflector* (BSR). The BSR probably results from the velocity contrast between sediment cemented with gas hydrate and underlying sediment in which lower velocities occur because of the absence of gas hydrate and the possible presence of free gas.

Kvenvolden & McMenamin (1980) and Kvenvolden & Barnard (1983b) have summarized the geologic occurrences of natural gas hydrates and listed areas where the presence of marine gas hydrates can be inferred based on observations of BSRs on marine seismic records. Marine gas hydrates can be found in continental margin sediment of all the world's oceans. In spite of the apparent ubiquity of gas hydrates in continental margin sediment, solid samples of gas hydrate have been recovered from oceanic sediment in only three areas—the Black Sea in Russia (Yefremova & Zhizhchenko 1975), the Blake Outer Ridge in the Atlantic Ocean offshore from southern United States (Sheridan et al 1982), and the landward flank of the Middle America Trench in the Pacific Ocean offshore from Central America (Moore et al 1979, von Huene et al 1980, Aubouin et al 1982). Gas from these samples was mainly C_1 accompanied by minor amounts of the heavier hydrocarbon gases. The carbon isotopic composition of C_1 from sediments recovered from the Black Sea (Hunt & Whelan 1978) and from the Blake Outer Ridge (Claypool et al 1973, Galimov & Kvenvolden 1983) range from $-63‰$ to $-94‰$. These values fall in the range (-50 to $-90‰$) that is generally considered to indicate biogenic C_1. Thus, the C_1 in marine gas hydrates in these two areas likely results from the bacterial alteration of organic matter buried in the sediment.

Preliminary unpublished results of measurements of the molecular and isotopic compositions of gas from a massive gas hydrate recovered in sediments at one site from the Middle America Trench (DSDP Leg 84) indicate a possible thermogenic source for C_1 ($\delta^{13}C = -41‰$). However, this C_1 occurs with extremely ^{12}C-depleted $\sum CO_2$ ($+37‰$) and thus alternatively could be an unusual example of isotopically heavy biogenic C_1. Electric logging showed that the massive hydrate is about 3 m thick at a sediment depth of 250 m. The gas hydrate at this site apparently made room for itself during the crystal-forming process, resulting in a massive gas hydrate containing minor quantities of sediment.

Gas hydrates may act as a seal, trapping C_1 and other hydrocarbon gases beneath them (Hedberg 1980, Dillon et al 1980). If the interface between the base of the gas hydrate and the region of free gas intercepts porous and permeable beds, situations may be created where gas hydrates trap free gas in economically producible reservoirs. However, evidence for the widespread occurrence of such reservoirs associated with marine gas hydrates has not yet been clearly documented.

CONCLUSIONS

Hydrocarbon natural gas in oceanic sediment has multiple mechanisms of origin, although the most important one is C_1 generation by microbiological decomposition of organic matter. Profiles of $\delta^{13}C_1$, $\delta^{13}C$ of $\sum CO_2$, and C_2/C_1 ratios in the intermediate subbottom depth range (50–1000 m) indicate that microbiological C_1 generation is widespread in anoxic, sulfate-depleted ocean sediments. C_1 of thermogenic origin (along with C_2, etc) becomes a more important component of the gas mixture with increasing depth of burial. Thermogenic gas can be a major component locally where gas of a deeper, high-temperature ($>100°C$) origin migrates to shallower marine sediments. In oxidizing marine sediments at relatively shallow depths (<10 m) beneath the seafloor, the small amounts and variable composition of hydrocarbon gas are determined by both generation and consumption processes.

Attempts to locate buried petroleum accumulations in marine sediments by detecting gas seeps on the seafloor must take into account the effects on gas composition of both the mechanisms of origin and near-surface alteration. Where hydrocarbon gas occurs in concentrations sufficient to saturate the interstitial water of marine sediment, either a free gas phase will form or the gas in excess of porewater solubility will form solid gas hydrate, depending on pressure and temperature. Gas hydrates composed largely of microbiological C_1 appear to be widespread in marine sediments near continental margins, but gas of deeper origin also may contribute to gas hydrate formation in some circumstances.

ACKNOWLEDGMENT

Critical comments and constructive suggestions by R. S. Oremland and D. D. Rice significantly improved this review. D. R. Malone expertly processed the words.

Literature Cited

Abram, J. W., Nedwell, D. B. 1978. Inhibition of methanogenesis by sulfate-reducing bacteria competing for transferred hydrogen. *Arch. Microbiol.* 117:89–92

Atkinson, L. P., Richards, F. A. 1967. The occurrences and distribution of methane in the marine environment. *Deep-Sea Res.* 14:673–84

Aubouin, J., von Huene, R., Baltuck, M., Arnott, R., Bourgois, J., et al. 1982. Leg 84 of the Deep Sea Drilling Project, Subduction without accretion: Middle America Trench off Guatemala. *Nature* 297:458–60

Barker, J. F., Fritz, P. 1981. Carbon isotope fractionation during microbiol methane oxidation. *Nature* 293:289–91

Barnes, R. O., Goldberg, E. D. 1976. Methane production and consumption in anoxic marine sediments. *Geology* 4:297–300

Bernard, B. B. 1979. Methane in marine sediments. *Deep-Sea Res.* 26A:429–43

Bernard, B. B., Brooks, J. M., Sackett, W. M. 1976. Natural gas seepage in the Gulf of Mexico. *Earth Planet. Sci. Lett.* 31:48–54

Bernard, B. B., Brooks, J. M., Sackett, W. M. 1978. Light hydrocarbons in recent Texas continental shelf and slope sediments. *J. Geophys. Res.* 83:4053–61

Berner, R. A. 1980. *Early Diagensis: A Theoretical Approach.* Princeton, N.J.: Princeton Univ. Press. 241 pp.

Brooks, J. M., Frederick, A. D., Sackett, W. M., Swinnerton, J. W. 1973. Baseline concentrations of light hydrocarbons in Gulf of Mexico, *Environ. Sci. Technol.* 7:639–42

Brooks, J. M., Barnard, L. A., Wiesenberg, D. A., Kvenvolden, K. A. 1983. Molecular and isotopic compositions of gas in sediments at Site 533. In *Initial Reports of the Deep Sea Drilling Project*, R. E. Sheridan, F. Gradstein, et al, Vol. 76. Washington DC: USGPO. In press

Carlisle, C. T., Bayliss, G. S., VanDelinder, D. G. 1975. Distribution of light hydrocarbons in seafloor sediments: correlation between geochemistry, seismic structure, and possible reservoired oil and gas. *7th Offshore Technol. Conf.*, OTC 2341:65–70

Claypool, G. E. 1974. *Anoxic diagenesis and bacterial methane production in deep sea sediments.* PhD thesis. Univ. Calif., Los Angeles. 276 pp.

Claypool, G. E. 1983. Organic geochemistry of carbon-deficient sediments, Leg 78A, DSDP. In *Initial Reports of the Deep Sea Drilling Project*, J. C. Moore, B. Biju-Duval, et al, Vol. 78. Washington DC: USGPO. In press

Claypool, G. E., Kaplan, I. R. 1974. The origin and distribution of methane in marine sediment. In *Natural Gases in Marine Sediments*, ed. I. R. Kaplan, pp. 99–139. New York: Plenum. 324 pp.

Claypool, G. E., Threlkeld, C. N. 1983. Anoxic diagenesis and methane generation in sediments of the Blake Outer Ridge, DSDP Site 533, Leg 76. In *Initial Reports of the Deep Sea Drilling Project*, R. E. Sheridan, F. Gradstein, et al, Vol. 76. Washington DC: USGPO. In press

Claypool, G. E., Presley, B. J., Kaplan, I. R. 1973. Gas analyses in sediment samples from Legs 10, 11, 13, 14, 15, 18 and 19. In *Initial Reports of the Deep Sea Drilling Project*, J. S. Creager, D. W. Scholl, et al, 19:879–84. Washington DC: USGPO. 913 pp.

Cline, J. D., Holmes, M. L. 1977. Submarine seepage of natural gas in Norton Sound, Alaska. *Science* 198:1149–53

Coleman, D. D., Risatti, J. B., Schoell, M. 1981. Fractionation of carbon and hydrogen isotopes by methane-oxidizing bacteria. *Geochim. Cosmochim. Acta* 45:1033–37

Culberson, O. L., McKetta, J. J. Jr. 1951. The solubility of methane in water at pressures to 10,000 psia. *AIME Pet. Trans.* 192:223–26

Davis, J. B., Squires, R. M. 1954. Detection of microbially produced gaseous hydrocarbons other than methane. *Science* 119:381–82

Dillon, W. P., Grow, J. A., Paull, C. K. 1980. Unconventional gas hydrate seals may trap gas off southeast U.S. *Oil Gas J.* 78(1):124–30

Doose, P. R., Kaplan, I. R. 1981. Biogenetic control of gases in marine sediments of Santa Barbara basin, California. *Am. Assoc. Pet. Geol. Bull.* 65:919–20 (Abstr.)

Doose, P. R., Sandstrom, M. W., Jodele, R. F., Kaplan, I. R. 1978. Interstitial gas analysis of sediment samples from Site 368 and Hole 369A. In *Initial Reports of the Deep Sea Drilling Project*, Y. Lancelot, E. Seibold, et al, 41:861–63. Washington DC: USGPO. 1259 pp.

Emery, K. O., Hoggan, D. 1958. Gases in marine sediments. *Bull. Am. Assoc. Pet. Geol.* 42:2174–88

Erdman, J. G., Borst, R. L., Hines, W. J. 1969. Composition of gas sample 1 (core 5) by components (1.5.1). In *Initial Reports of the Deep Sea Drilling Project*, M. Ewing, J. L. Worzel, et al, 1:461–63. Washington DC: USGPO. 672 pp.

Frank, D. J., Gormly, J. R., Sackett, W. M. 1974. Revaluation of carbon-isotope composition of natural methanes. *Am. Assoc. Pet. Geol. Bull.* 58:2319–25

Fuex, A. N. 1977. The use of stable isotopes in hydrocarbon exploration. *J. Geochem. Explor.* 7:155–88

Galimov, E. M., Kvenvolden, K. A. 1983. Concentrations and carbon isotopic compositions of CH_4 and CO_2 in gas from sediments of the Blake Outer Ridge, DSDP Leg 76. In *Initial Reports of the Deep Sea Drilling Project*, R. E. Sheridan, F. Gradstein, et al, Vol. 76. Washington DC: USGPO. In press

Galimov, E. M., Simoneit, B. 1982. Geochemistry of interstitial gases in sedimentary deposits of the Gulf of California, Deep Sea Drilling Project Leg 64. In *Initial Reports of the Deep Sea Drilling Project*, J. R. Curray, D. G. Moore, et al, 64:781–87. Washington DC: USGPO. 1313 pp.

Galimov, E. M., Chinyanov, V. A., Ivanov, Ye. N. 1980. Isotopic composition of methane carbon and the relative content of gaseous hydrocarbons in the deposits of the Moroccan Basin of the Atlantic Ocean (Deep Sea Drilling Project Site 415 and 416). In *Initial Reports of the Deep Sea Drilling Project*, Y. Lancelot, E. L. Winterer, et al, 50:615–22. Washington DC: USGPO. 868 pp.

Gealy, E. L., Dubois, R. 1971. Shipboard geochemical analysis, Leg 7, *Glomar Challenger*. In *Initial Reports of the Deep Sea Drilling Project*, E. L. Winterer, W. R. Riedel, et al, 7:863–64. Washington DC: USGPO. 1757 pp.

Gold, T. 1979. Terrestrial sources of carbon and earthquake outgassing. *J. Pet. Geol.* 1:3–19

Gold, T., Soter, S. 1980. The deep earth gas hypothesis. *Sci. Am.* 242(5):154–61

Hampton, M. A., Kvenvolden, K. A. 1981. Geology and geochemistry of gas-charged sediment on Kodiak Shelf, Alaska. *Geo-Mar. Lett.* 1:141–47

Hand, J. H., Katz, D. L., Verma, V. K. 1974. Review of gas hydrates with implications for ocean sediments. In *Natural Gases in Marine Sediments*, ed. I. R. Kaplan, pp. 179–94. New York: Plenum. 324 pp.

Hedberg, H. D. 1980. Methane generation and petroleum migration. In *Problems of Petroleum Migration*, ed. W. H. Roberts III, R. J. Cordell, pp. 179–206, Tulsa, Okla: Am. Assoc. Pet. Geol. 273 pp.

Hitchon, B. 1974. Occurrence of natural gas hydrates in sedimentary basins. In *Natural Gases in Marine Sediments*, ed. I. R. Kaplan, pp. 195–225. New York: Plenum. 324 pp.

Hungate, R. E., Smith, W., Bauchop, T., Yu, I., Rabinowitz, J. C. 1970. Formate as an intermediate in the bovine rumen fermentation. *J. Bacteriol.* 102:389–97

Hunt, J. M. 1974. Hydrocarbon geochemistry of the Black Sea. In *The Black Sea—Geology, Chemistry, and Biology*, ed. E. T. Degens, D. A. Ross, Mem. 20:499–504. Tulsa, Okla: Am. Assoc. Pet. Geol. 633 pp.

Hunt, J. M. 1975. Origin of gasoline range alkanes in the deep sea. *Nature* 254:411–13

Hunt, J. M., Whelan, J. K. 1978. Dissolved gases in Black Sea sediments. In *Initial Reports of the Deep Sea Drilling Project*, D. A. Ross, Y. P. Neprochnov, et al, 42(2):661–65. Washington DC: USGPO. 1244 pp.

Jannasch, A. W. 1975. Methane oxidation in Lake Kivu (central Africa). *Limnol. Oceanogr.* 20:860–64

Katz, D. L., Cornell, D., Kobayashi, R., Poettmann, F. H., Vary, J. A., Elenblas, J. R., Weingug, C. F. 1959. *Handbook of Natural Gas Engineering.* New York: McGraw-Hill. 802 pp.

Kosiur, D. R., Warford, A. L. 1979. Methane production and oxidation in Santa Barbara basin sediments. *Estuarine Coastal Mar. Res.* 8:379–85

Kristjansson, J. K., Schonheit, P., Thauer, R. K. 1982. Different Ks values for hydrogen of methanogenic bacteria and sulfate-reducing bacteria: An explanation for the apparent inhibition of methanogenesis by sulfate. *Arch. Microbiol.* 131:278–82

Kvenvolden, K. A., Barnard, L. A. 1983a. Gas hydrates of the Blake Outer Ridge Site 533, DSDP/IPOD Leg 76. In *Initial Reports of the Deep Sea Drilling Project*, R. E. Sheridan, F. Gradstein, et al, Vol. 76. Washington DC: USGPO. In press

Kvenvolden, K. A., Barnard, L. A. 1983b. Hydrates of natural gas in continental margins. *Proc. Hedberg Conf., Am. Assoc. Pet. Geol. Mem. 34.* In press

Kvenvolden, K. A., McMenamin, M. A. 1980. Hydrates of natural gas: a review of their geologic occurrence. *US Geol. Surv. Circ. 825.* 11 pp.

Kvenvolden, K. A., Redden, G. D. 1980. Hydrocarbon gases in sediment of the shelf, slope, and basin of the Bering Sea. *Geochim. Cosmochim. Acta* 44:1145–50

Kvenvolden, K. A., Weliky, K., Nelson, C. H., DesMarais, D. J. 1979. Submarine seep of carbon dioxide in Norton Sound, Alaska. *Science* 205:1264–66

Kvenvolden, K. A., Redden, G. D., Thor, D. R., Nelson, C. H. 1981a. Hydrocarbon gases in near-surface sediment of the northern Bering Sea. In *The Eastern Bering Sea Shelf: Oceanography and Resources*, ed. D. W. Hood, J. A. Calder, 1:411–24, Seattle: Univ. Wash. Press. 625 pp.

Kvenvolden, K. A., Vogel, T. M., Gardner, J. V. 1981b. Geochemical prospecting for hydrocarbons in the outer continental shelf, southern Bering Sea, Alaska. *J. Geochem. Explor.* 14:209–19

Lein, A. Yu., Namsaraev, B. B., Trotsyuk, V. Ya., Ivanov, M. V. 1981. Bacterial methanogenesis in Holocene sediments of the Baltic Sea. *Geomicrobiol. J.* 2:299–315

Lupton, J. E., Craig, H. 1981. A major Helium-3 source at 15°S on the East Pacific Rise. *Science* 214:13–18

Lyon, G. L. 1973. Interstitial water studies, Leg 15, chemical and isotopic composition of gases from Cariaco Trench sediments. In *Initial Reports of the Deep Sea Drilling Project*, B. C. Heezen, I. D. MacGregor, et al, 20:773–74. Washington DC: USGPO. 958 pp.

Mah, R. A., Ward, D. M., Baresi, L., Glass, T. L. 1977. Biogenesis of methane. *Ann. Rev. Microbiol.* 31:309–41

Martens, C. S., Berner, R. A. 1974. Methane production in the interstitial waters of sulfate-depleted marine sediments. *Science* 185:1167–69

Martens, C. S., Berner, R. A. 1977. Interstitial water chemistry of anoxic Long Island Sound sediments. 1. Dissolved gases. *Limnol. Oceanogr.* 22:10–25

McIver, R. D. 1973. Appendix III. Hydrocarbon gases from canned core samples, sites 174A, 176, and 180. In *Initial Reports of the Deep Sea Drilling Project*, L. D. Kulm, R. von Huene, et al, 18:1013–14. Washington DC: USGPO. 1077 pp.

McIver, R. D. 1975. Hydrocarbon occurrences from JOIDES Deep Sea Drilling Project. *Proc. World Pet. Cong., 9th*, 2:269–80. London: Applied Sci. Publ. 376 pp.

McKenna, E. J., Kallio, R. E. 1965. The biology of hydrocarbons. *Ann. Rev. Microbiol.* 19:183–208

Moore, J. C., Watkins, J. S., et al. 1979. Middle American Trench. *Geotimes* 24(9):20–22

Morris, D. A. 1976. Organic diagenesis of Miocene sediments from Site 341, Vøring Plateau, Norway. In *Initial Reports of the Deep Sea Drilling Project*, M. Talwani, G. Udintsev, et al, 38:809–14. Washington DC: USGPO. 1256 pp.

Mountfort, D. O., Asher, R. A. 1978. Changes in proportions of acetate and carbon dioxide used as methane precursors during the anaerobic digestion of bovine waste. *Appl. Environ. Microbiol.* 35:648–54

Mountfort, D. O., Asher, R. A., Mays, E. L., Tiedje, J. M. 1980. Carbon and electron flow in mud and sandflat intertidal sediments at Deleware Inlet, Nelson, New Zealand. *Appl. Environ. Microbiol.* 39:686–94

Nikaido, M. 1977. On the relation between methane production and sulfate reduction in bottom muds containing sea water sulfate. *Geochem. J.* 11:199–206

Nissenbaum, A., Presley, B. J., Kaplan, I. R. 1972. Early diagenesis in a reducing fjord, Saanich Inlet, British Columbia—I. Chemical and isotopic changes in major components of interstitial water. *Geochim. Cosmochim. Acta* 36:1007–27

Oremland, R. S. 1981. Microbial formation of ethane in anoxic estuarine sediments. *Applied Environ. Microbiol.* 42:122–29

Oremland, R. S., Taylor, B. F. 1978. Sulfate reduction and methanogenesis in marine sediments. *Geochim. Cosmochim. Acta* 42:209–14

Oremland, R. S., Marsh, L., DesMarais, D. J. 1982a. Methanogenesis in Big Soda Lake, Nevada: an alkane, moderately hypersaline desert lake. *Appl. Environ. Microbiol.* 43:462–68

Oremland, R. S., Marsh, L. M., Polcin, S. 1982b. Methane production and simultaneous sulfate reduction in anoxic, salt marsh sediments. *Nature* 296:143–45

Patt, T. E., Cole, G. C., Bland, J., Hanson, R. S. 1974. Isolation and characterization of bacteria that grow on methane and organic compounds as sole source of carbon and energy. *J. Bacteriol.* 120:955–64

Primrose, S. B., Dilworth, M. J. 1976. Ethylene production by bacteria. *J. Gen. Microbiol.* 93:177–81

Rashid, M. A., Vilks, G., Leonard, J. D. 1975. Geological environment of a methane-rich recent sedimentary basin in the Gulf of St. Lawrence. *Chem. Geol.* 15:83–96

Reeburgh, W. S. 1969. Observations of gases in Chesapeake Bay sediments. *Limnol. Oceanogr.* 14:368–75

Reeburgh, W. S. 1976. Methane consumption in Cariaco Trench waters and sediments. *Earth Planet. Sci. Lett.* 28:337–44

Reeburgh, W. S. 1980. Anaerobic methane oxidation: rate depth distributions in Skan Bay sediments. *Earth Planet. Sci. Lett.* 47:345–52

Reeburgh, W. S., Heggie, D. T. 1977. Microbial methane consumption reactions and their effect on methane distributions in freshwater and marine environments. *Limnol. Oceanogr.* 22:1–9

Reed, W. E., Kaplan, I. R. 1977. The chemistry of marine petroleum seeps. *J. Geochem. Explor.* 7:255–93

Reitsema, R. H., Lindberg, F. A., Kaltenback, A. J. 1978. Light hydrocarbons in Gulf of Mexico water: sources and relation to structural highs. *J. Geochem. Explor.* 10:139–51

Revelle, R. 1950. Sedimentation and oceanography: survey of field observations. *Geol. Soc. Am. Mem. 43*, pt. 5. 6 pp.

Rice, D. D., Claypool, G. E. 1981. Generation, accumulation, and resource potential of biogenic gas. *Am. Assoc. Pet. Geol. Bull.* 65:5–25

Rosenfeld, W. D., Silverman, S. R. 1959. Carbon isotopic fractionation in bacterial production of methane. *Science* 130:1658–59

Rudd, J. W. M. 1980. Methane oxidation in Lake Tanganyika (East Africa). *Limnol. Oceanogr.* 25:958–63

Rudd, J. W. M., Hamilton, R. D. 1975. Factors controlling rates of methane oxidation and the distribution of the methane oxidizers in a small stratified lake. *Arch. Hydrobiol.* 75:522–38

Rudd, J. W. M., Hamilton, R. D. 1978. Methane cycling in a eutrophic shield lake and its effects on whole lake metabolism. *Limnol. Oceanogr.* 23:337–48

Rudd, J. W. M., Hamilton, R. D., Campbell, N. E. R. 1974. Measurement of microbial oxidation of methane in lake water. *Limnol. Oceanogr.* 19:519–24

Sackett, W. M. 1977. Use of hydrocarbon sniffing in offshore exploration. *J. Geochem. Explor.* 7:243–54

Sansone, F. J., Martens, C. S. 1981. Methane production from acetate and associated methane fluxes from anoxic coastal sediments. *Science* 211:707–9

Sheridan, R. E., Gradstein, F., et al. 1982. Early history of the Atlantic Ocean and gas hydrates in the Blake Outer Ridge: Results of the Deep Sea Drilling Project Leg 76. *Geol. Soc. Am. Bull.* 93:876–85

Shipley, T. H., Houston, M. H., Buffler, R. T., Shaub, F. J., McMillan, K. J., Ladd, J. W., Worzel, J. L. 1979. Seismic evidence for widespread possible gas hydrate horizons on continental slopes and rises. *Am. Assoc. Pet. Geol. Bull.* 63:2204–13

Sigalove, J. J., Pearlman, M. D. 1975. Geochemical seep detection for offshore oil and gas exploration. *7th Offshore Technol. Conf.*, OTC 2344:95–102

Silverman, M. P., Oyama, V. I. 1968. Automatic apparatus for sampling and preparing gases for mass spectral analysis in studies of carbon isotope fractionation during methane metabolism. *Anal. Chem.* 40:1833–37

Simoneit, B., Lonsdale, P. 1982. Hydrothermal petroleum in mineralized mounds at the sea bed of Guaymas Basin, Gulf of California. *Nature* 295:198–202

Sørensen, J., Christensen, D., Jørgensen, B. B. 1981. Volatile fatty acids and hydrogen as substrates for sulfate-reducing bacteria in anaerobic marine sediment. *Appl. Environ. Microbiol.* 42:5–11

Swinnerton, J. W., Lamontagne, R. A. 1974. Oceanic distribution of low-molecular-weight hydrocarbons—Baseline measurements. *Environ. Sci. Technol.* 8:657–63

Vogel, T. M., Oremland, R. S., Kvenvolden, K. A. 1982. Low-temperature formation of hydrocarbon gases in San Francisco Bay sediment. *Chem. Geol.* 37:289–98

von Huene, R., Aubouin, J., Azema, J., Blackinton, G., Carter, J. A., et al. 1980. Leg 67—the Deep Sea Drilling Project Mid-America Trench transect off Guatemala. *Geol. Soc. Am. Bull.* 91:421–32

Warford, A. L., Kosiur, D. R., Doose, P. R. 1979. Methane production in Santa Barbara Basin sediments. *Geomicrobiol. J.* 1:117–37

Welhan, J. A. 1980. Gas concentrations and isotope ratios at the 21°N EPR hydrothermal site. *EOS, Trans. Am. Geophys. Union* 61:996 (Abstr.)

Welhan, J. A., Craig, H. 1979. Methane and hydrogen in East Pacific Rise hydrothermal fluids. *Geophys. Res. Lett.* 6:829–31

Whelan, J. K. 1979. C_1–C_7 hydrocarbons from IPOD Holes 397 and 397A. In *Initial Reports of the Deep Sea Drilling Project*, U. von Rad, W. B. F. Ryan, et al, 47(1):531–39. Washington DC: USGPO. 835 pp.

Whelan, J. K., Hunt, J. M. 1979a. C_2 to C_7 hydrocarbons from IPOD Hole 398D. In *Initial Reports of the Deep Sea Drilling Project*, J.-C. Sibuet, W. B. F. Ryan, et al, 47(2):561–63. Washington DC: USGPO. 787 pp.

Whelan, J. K., Hunt, J. M. 1979b. Sediment C_1–C_7 hydrocarbons from IPOD Leg 48—Bay of Biscay. In *Initial Reports of the Deep Sea Drilling Project*, L. Montadert, D. G. Roberts, et al, 48:943–45, Washington DC: USGPO. 1183 pp.

Whelan, J. K., Hunt, J. M. 1980a. Sediment C_1–C_7 hydrocarbons from Deep Sea Drilling Project sites 415 and 416 (Moroccan Basin). In *Initial Reports of the Deep Sea Drilling Project*, Y. Lancelot, E. L. Winterer, et al, 50:623–24. Washington DC: USGPO. 868 pp.

Whelan, J. K., Hunt, J. M. 1980b. C_1–C_7 volatile organic compounds in sediments from Deep Sea Drilling Project Legs 56 and 57, Japan Trench. In *Initial Reports of the Deep Sea Drilling Project*, Scientific Party, 56–57:1349–65. Washington DC: USGPO. 1417 pp.

Whelan, J. K., Hunt, J. M. 1981. C_1–C_8 hydrocarbons in IPOD Leg 63 sediments from outer California and Baja California Borderlands. In *Initial Reports of the Deep Sea Drilling Project*, R. S. Yeats, B. U. Haq, et al, 63:775–84. Washington DC: USGPO. 967 pp.

Whelan, J. K., Sato, S. 1980. C_1–C_5 hydrocarbons from core gas pockets, Deep Sea Drilling Project Legs 56 and 57, Japan Trench Transect. In *Initial Reports of the Deep Sea Drilling Project*, Scientific Party, 56–57:1335–47. Washington DC: USGPO. 1417 pp.

Whelan, J. K., Hunt, J. M., Berman, J. 1980. Volatile C_1–C_7 organic compounds in surface sediments from Walvis Bay. *Geochim. Cosmochim. Acta* 44:1767–85

Winfrey, M. R., Zeikus, J. G. 1977. Effect of sulfate on carbon and electron flow during microbial methanogenesis in freshwater sediments. *Appl. Environ. Microbiol.* 33:275–81

Winfrey, M. R., Danielle, G. M., Bianchi, A. J. M., Ward, D. M. 1981. Vertical distribution of sulfate reduction, methane production, and bacteria in marine sediments. *Geomicrobiol. J.* 2:341–62

Yefremova, A. G., Zhizhchenko, B. P. 1975. Occurrence of crystal hydrates of gases in the sediments of modern marine basins. *Dokl. Acad. Sci. USSR Earth Sci. Sect.* 214:219–20

Zeikus, J. G. 1977. The biology of methanogenic bacteria. *Bacteriol. Rev.* 41:514–41

Zinder, S. H., Brock, T. D. 1978. Production of methane and carbon dioxide from methane thiol and dimethyl sulfide by anaerobic lake sediments. *Nature* 273:226–28

Ann. Rev. Earth Planet. Sci. 1983. 11: 329–58

IN SITU TRACE ELEMENT MICROANALYSIS[1]

D. S. Burnett
Division of Geological and Planetary Sciences, California Institute of Technology, Pasadena, California 91125

D. S. Woolum
Department of Physics, California State University, Fullerton, California 92634

1. INTRODUCTION

We define in situ trace element microanalysis as the measurement of concentrations of trace elements (< 1000 ppm) in individual mineral phases in polished sections, analogous to electron microprobe analysis for major elements. It is our opinion that such measurements are important and can provide a new dimension to petrology and geochemistry. This article is both a review of work already done and a summary of potential advances in the future. The advantages of in situ analyses, as opposed to methods involving physical and/or chemical mineral separations, are that the trace element data can be interpreted in a petrographic context and that ambiguities associated with the purity of the mineral separates can be avoided. With data on individual grains, comparisons of intergrain and intragrain (zoning) variations between major and trace elements can be made, and the importance of inclusions can be assessed, within the spatial resolution of the microanalysis technique.

Two general approaches to trace element microanalysis are covered here: techniques that capitalize on special nuclear properties, normally based on particle track radiography (discussed in Section 2); and techniques based on characteristic X-ray production (discussed in Section 3). Somewhat arbitrarily, we do not discuss in situ trace element microanalysis by secondary ion mass spectrometry (ion probes). This topic has recently

[1] Contribution No. 3808

0084–6597/83/0515–0329$02.00

been reviewed by Shimizu & Hart (1982). We consider the methods discussed here as complementary to the ion probe for the most part. The ability of the ion probe to study microscale isotopic variations is unique, and it would seem that such studies should be given priority for this instrument. The techniques discussed here are restricted to elemental analyses. In some cases they are less sensitive than the ion probe, and in other cases they are more sensitive. However, these techniques have the advantage that the basic physics, relating a detected signal to elemental concentration, is simpler and, at this stage, better understood.

Just as in conventional petrology, trace element microanalytical studies can be based on both natural and synthetic samples. From the point of view of microanalysis, synthetic samples have the great advantage that the bulk trace element level can be controlled, analytical interferences from other elements can be minimized, radioactive tracers can be added to permit autoradiographic analysis of the trace element distributions, and to some extent, the grain size can be optimized.

There is a major unresolved issue in experimental trace element geochemistry, however, concerning whether it is necessary to conduct studies at trace concentration levels using techniques like those discussed here (the so-called Henry's Law debate). Most work in this field is carried out to measure silicate crystal/silicate melt partition coefficients, D, and it is simpler in most cases to add 0.1–1% amounts of the "trace" element(s) being studied and to analyze the synthesized products by standard electron microprobe techniques. Most studies using this approach take care to demonstrate the constancy of D over some range of concentration above $\sim$0.1%, but the question remains whether the measured D is valid at ppm levels. Several mechanisms could cause the partition coefficient for element X to vary with the concentrations of X, e.g. (*a*) saturation of the liquid, causing a phase to precipitate with X as a stoichiometric constituent, (*b*) changes in the structure of the liquid due to high concentrations of X (compare Mysen & Virgo 1980), and (*c*) substitution into crystalline defects with saturation of defect sites at higher concentrations (Navrotsky 1978, Harrison & Wood 1980).

Mysen (1978) reviewed data indicating concentration-dependent D values and subsequently (Mysen 1979) published additional results for D_{Ni} (olivine), which showed about a factor of 2 decrease for olivine Ni concentrations above 3000 ppm. On the other hand, several other studies have been published indicating constant D from ppm to % levels [Drake & Holloway 1978 (Sm in plagioclase); Lindstrom & Weill 1978 (Ni in clinopyroxene, although see also Steele & Lindstrom 1981); Drake & Holloway 1981 (Ni in olivine)]. These authors have challenged some of the earlier work showing concentration variations in D. The case of D_{Ni}

(olivine) has been especially controversial with several studies (Hart & Davis 1978, Nabelek 1980, Drake & Holloway 1981) failing to find any decrease in D at high concentrations, as reported by Mysen (1979). The debate has focused on the possible effects of differences in techniques or in bulk composition, although the experiments of Drake & Holloway (1981) were designed to reproduce exactly some of those of Mysen (Mysen 1981, 1982, Nabelek 1981, Drake & Holloway 1982). In general it is more difficult to produce constancy from variability than vice versa, so the burden of proof in the D_{Ni} (olivine) case should lie with Mysen. Nevertheless, there are other examples of variation of D with concentration for different systems (Mysen 1978, Harrison & Wood 1980, Benjamin 1980), making it premature to conclude that D values measured at percent levels are *always* appliable at trace element levels. If possible, studies of trace element geochemistry should be done at ppm concentrations; however, the Harrison & Wood (1980) experiments indicate that the major complications may be at trace levels. These authors found systematic increases in D for Sm and Tm in garnet, with decreasing rare-earth concentrations below 10–100 ppm. They interpreted the variations as demonstrating the importance of defect substitution. Such effects cannot be ruled out as something important only at "very low" trace element concentrations. The possibility of defect substitution deserves much more study, as it possibly presents a great complication to the interpretation of trace element data in general.

2. PARTICLE TRACK RADIOGRAPHY

In this section we discuss methods of in situ trace element microanalysis based on the imaging of particles emitted from a polished rock section, either by decay of radioactivity contained in the sample (autoradiography) or by prompt radiation induced by an external bombardment of the sample (radiography). In applications discussed here, the image is formed in an external solid-state track detector (Fleischer et al 1975) or nuclear emulsion (Rogers 1979) affixed to the rock section.[2] As is discussed in more detail in the specific examples, radiography experiments have the disadvantage that the track detector must be irradiated with the sample.

The spatial resolution is set by the range of the detected particles and the thickness of the grain analyzed. In general, for quantitative analysis, grains are required that are large compared with the range (heavy charged particles) or half-thickness (electrons). The number of detected particles from an individual grain will depend on the concentration of the element

[2] Thorough discussions of relevant techniques are found in the books by Fleischer et al (1975) and Rogers (1979).

analyzed and the area of the grain; consequently, a convenient "figure of merit" for a given technique is the concentration-area product (ppm-μ^2) required to produce a given number of tracks from a single grain. To be specific, however, we will normally quote sensitivity in terms of the concentration required to produce 100 tracks in a specified grain size.

2.1 *Fission Track Radiography*

Nuclear fission occurs only with the heaviest elements, the actinide elements being the predominant fissionable elements. Consequently, this technique is, for practical purposes, restricted to Th and U in natural samples; however, these are geochemically important elements. The general aspects of detectors, track etching, track density measurements, and radiography ("element mapping"), including applications prior to 1973, are summarized in Fleischer et al (1975). Shorter reviews are given by Fleischer (1979) and Ahlen et al (1981).

URANIUM Uranium radiography is simple, unambiguous, and very sensitive; consequently, it has been widely used. The underlying reason is that ^{235}U is the only naturally occurring nucleus that is fissionable by thermal neutrons in a nuclear reactor. Fission of ^{232}Th requires a higher-energy bombardment, and the observed fission tracks will always contain a contribution from fission of ^{238}U and ^{235}U that must be subtracted out.

The average ranges of single fission fragments in silicate and phosphate minerals are 10–12 μ, and for simple quantitative analysis the grain radius and depth must be larger than this. If this criterion is satisfied, then the interior track density $\rho(\mathrm{cm}^{-2})$ observed in an external detector in direct contact with the grain will be uniform (assuming a uniform U distribution) and will be related to the U concentration by

$$\rho = \psi\sigma_f N \frac{R}{2}, \tag{1}$$

where N is the U concentration in atoms g^{-1}, σ_f is the thermal fission cross section for U of natural isotopic composition, ψ is the thermal neutron fluence, and R is the mean range (in g cm^{-2}) of a single fission fragment in the particular mineral. In practice, N is measured by irradiating the sample adjacent to a standard of known U content:

$$\rho/\rho_0 = \frac{N}{N_0}\frac{R}{R_0}, \tag{2}$$

where the zero subscripts refer to the standard. The range correction R/R_0 rarely differs from unity by more than $\pm 10\%$ and can be calculated with sufficient accuracy from existing fission-fragment range data (see Fleischer et al 1975) or semiempirical heavy-ion range tables (Northcliffe & Schilling 1970). Concentrations calculated from Equation (2) for U-bearing grains that are less than 20 μ and in a U-poor matrix will be systematically low [see Burnett et al (1971) for an alternative treatment of small grains], but qualitative or semiquantitative identification of small ($\lesssim 10$ μ) U-rich phases can be made for grains containing as few as 5×10^5 atoms. (The real problem in characterizing such "point sources" is that the probability of the grain being on the surface and "locatable" with an optical or scanning electron microscope, as opposed to being buried 5–10 μ below the surface, may be small).

Plastics and muscovite mica are the commonly used track detectors. Beautiful images of rock sections have been produced in Lexan plastic (Kleeman & Lovering 1967), presumably as a consequence of high densities of very short alpha tracks from (n, α) reactions on Li and B at grain boundaries (see Section 2.2). The images permit direct observation of the sources of U fission tracks; however, in order to form the images, irradiation temperatures must be kept low to prevent alpha track annealing (46°C for Hutcheon et al 1972). Fission track annealing in plastic track detectors is a potential problem, but Lexan has been used with neutron fluences up to at least 10^{17} cm^{-2} (D. Wark, private communication). Mica can be used for much higher fluences, but the limiting fluences for epoxy resins used in section preparation are $\lesssim 5 \times 10^{18}$ cm^{-2}. Polished meteorite slabs with mica detectors have been irradiated at $\sim 10^{19}$ cm^{-2} (Jones & Burnett 1979). At the limiting fluence for epoxy-mounted sections, the required U content to give 100 tracks (10% counting statistics) for a 100-μ grain is about 10 ppb. This is also approximately a practical limit, considering U contamination during section preparation and detector background. (Mica has 0.1–1000 ppb U.) For measurement of very low concentrations of U in larger ($\gtrsim 50$ μ) grains of nonopaque minerals, direct etching of induced fission tracks in annealed samples can be used (see, for example, Crozaz et al 1970, Jones & Burnett 1979). Concentrations of U in the 0.1–1 ppb range for individual grains of the major minerals of lunar mare basalts were measured in this way (Crozaz et al 1970).

THORIUM The thresholds for n-induced fission of ^{238}U and ^{232}Th are around 1 MeV, with cross sections increasing with energy and then leveling off above 15 MeV neutron energy. Fission cross sections at high energies are in general not well known, but at 14 MeV, $\sigma(^{232}\mathrm{Th})/\sigma(^{238}\mathrm{U}) = 0.32$. Limiting energies for charged-particle fission are set by the Coulomb barrier (~ 10 MeV for protons, 20 MeV for alpha particles), but above these energies fission cross sections of ^{238}U and ^{232}Th are comparable at ~ 1 barn. It is not possible to distinguish Th and U fission tracks, so the total track density in a high-energy irradiation is the sum of Th and U contributions. A separate low-energy irradiation is used to subtract out the U contribution. Quantitative Th/U *ratio* determinations are not limited to grains larger than the fission-fragment range, as are the absolute concentrations. U is more fissionable than Th, so σ_{Th}/σ_U will always be < 1, but it is clearly desirable to have this ratio as high as possible. If σ_{Th}/σ_U is known, then a single standard of known U content ($U_0 \gg Th_0$) is sufficient to determine both U and Th. For charged-particle irradiations where $\sigma^U \sim \sigma^{Th}$, the total track density is approximately proportional to Th+U.

Earlier work is thoroughly reviewed by Fleischer et al (1975). Subsequent studies using 30–40 MeV protons have been reported by Stapanian (1981) and Murrell & Burnett (1982a). The "fast" (fission spectrum) neutrons in a reactor are very unfavorable for Th determinations because of the high rate of ^{235}U fission by thermal and resonance energy neutrons. Even at very favorable locations (next to fuel rods), the best σ_{Th}/σ_U reported is ~ 0.08 (M. I. Stapanian, unpublished data). A very clean fission neutron spectrum from a thermal converter (Hughes 1953) has been successfully used by Crozaz (1974, 1979), who found $\sigma_{Th}/\sigma_U = 0.2$. Shirck (1975) proposed a clever technique for Th radiography using only thermal neutron irradiations based on 3 stages of irradiations. Fission-track measurements are made only during the first and third stages. The second stage irradiation is at a high fluence ($\sim 10^{20}$ cm^{-2}), converting ^{232}Th to ^{233}U following ^{233}Pa (27-d half-life) decay. This method has good sensitivity; 10 ppb Th can be measured to $\pm 25\%$ in 100-μ grains having Th/U = 4. The principal disadvantages are the low effective $\sigma_{Th}/\sigma_U = 0.1$ and the long time required for analysis. No investigations have exploited this technique. Murrell & Burnett (1982b), using the high-energy secondary neutron flux produced by stopping of the 675-MeV proton beam from the LAMPF accelerator, found $\sigma_{Th}/\sigma_U = 0.4$. (See also Haines et al 1976.) Values of σ_{Th}/σ_U as low as ~ 0.2 are useful because the typical Th/U abundance ratio is about 4.

The limiting factor in Th radiography appears to be the effect of radiation damage to the detector. Only mica has been used in studies to date. Light ions or neutrons do not produce tracks in mica, but nuclear collisions with the heavier (Al or greater) constituents of the mica can, with low efficiency, produce a recoil that will leave an etchable track. Most of these tracks are quite short ($< 2\ \mu$) but, at best, ruin the transparency of the detector and make reflected light or scanning electron microscope (SEM) observations impossible. The recoil track background can be reduced by annealing but cannot be totally removed (Haines et al 1976, Stapanian 1981). Some loss of fission tracks will also occur ($\sim 20\%$ for 2 h at 400°C; Stapanian 1981). For high-energy (> 50 MeV) irradiations, there is a residual population of 5–10 μ tracks, even after annealing (Stapanian 1981, Murrell & Burnett 1982b), that are indistinguishable from fission tracks. These set a background limit to Th analysis, which for LAMPF neutrons or 675-MeV protons would be ~1 ppm Th + U. These long recoils are not found in 30–40 MeV proton irradiations, and consequently, these irradiations appear to be best overall for Th radiography.

With an accelerator beam there is a trade off in the number of samples irradiated and the particle fluence for each sample. Using a 4×4 cm sample array (Woolum & Burnett 1981), fluences of up to 5×10^{17} protons cm^{-2} are practical. If the beam were focused on a single 1-cm^2 sample, fluences an order of magnitude larger, or even higher given unlimited accelerator time, could be obtained. The radiation damage effects on mica for such irradiations have not been checked. For a 5×10^{17} cm^{-2} fluence, 100 tracks will be obtained for individual 100-μ grains having 350 ppb U + Th. For other types of irradiations, track densities can be estimated from Equation (1).

OTHER ACTINIDES For practical purposes other actinides do not occur naturally, but ^{244}Pu was present in meteorites at the time of solar system formation. Synthetic studies of actinide geochemistry have been carried out for Pu as well as Th and U (Benjamin et al 1980, Jones 1981). These are summarized, in part, in the section on fission-track studies of chondrites. All actinide elements are readily fissionable, and spontaneous fission autoradiography is also feasible for some isotopes. In most cases, however, alpha autoradiography is more convenient (see Section 2.3), unless the slightly better spatial resolution of fission radiography is important. For example, ^{230}Th alpha autoradiography (Benjamin et al 1980) is more practical than ^{229}Th fission radiography.

NONACTINIDE FISSION Fission cross sections for elements lighter than Th decrease rapidly; even at bombarding energies of hundreds of MeV, the fission cross sections for Bi, Pb, Tl, Hg, and Au are about a factor of 5–10 less than those of the actinides. The geochemistry of these elements is very different from actinides; consequently, fission tracks from a high-energy irradiation associated with an actinide-poor phase (e.g. a sulfide or metal grain) might be plausibly associated with these elements. However, the cross sections of the 5 elements listed above are all within the same order of magnitude, and measurements of a specific element would be very difficult.

FISSION TRACK STUDIES OF CHONDRITES We consider these as an example. The actinide nuclei ^{232}Th, ^{235}U, ^{238}U, and ^{244}Pu are important for solar system chronology and for galactic "cosmochronology," since the average actinide solar system abundances depend on the properties of r-process nucleosynthesis in the Galaxy prior to the formation of the solar system (see, for example, Fowler 1977). Prior to 1970 it was assumed that chemical fractionation of actinide elements in meteorites would not be significant, especially for chondrites, which show relatively little fractionation of refractory lithophile elements on a whole rock basis. This assumption has been shown to be false, as discussed below.

Whole rock Th/U ratios are around 3.8 (Morgan & Lovering 1968). Pu/U whole rock measurements for chondrites based on ^{244}Pu fission Xe are difficult because of corrections for other Xe components. Although more data are needed, a consensus has recently developed, with Pu/U = 0.005 (Marti et al 1977, Hudson et al 1982, Jones 1982) as the best estimate of the solar system ratio. This is a factor of 3 lower than the previous estimate of Podosek (1970). As discussed below, track radiography and fossil fission track studies have produced extensive evidence for internal actinide fractionation within chondrites. Although not completely proved, it is very likely that the reason for Pu fractionation relative to either U or Th is that Pu is trivalent for the reducing conditions under which most meteorites have formed, whereas U and Th are tetravalent (Boynton 1978, Benjamin et al 1978). The fractionations between U and Th are less easily explained. Trivalent Pu has an ionic radius between that of Ce and Pr and thus, in terms of crystal chemistry, would be expected to behave as a light rare-earth element, and track radiography measurements of the relative partition coefficients for Pu (fission track) and Sm (beta track; see Section 2.4) for clinopyroxene (Jones 1981) are consistent with the rare-earth clinopyroxene partition coefficient pattern for a similar bulk composition (Grutzeck et al 1974). Similarly, the ratio Pu/Nd is impressively constant at 1.8×10^{-4}, by weight, among different meteorites, most spectacularly so between coexisting Ca-phosphate and clinopyroxene in the Angra dos Reis igneous meteorite (Marti et al 1977). It appears that meteorites or meteoritic minerals that show unfractionated Ce/Nd/Th/U abundances relative to Cl chondrites (usually regarded as the best approximation to nonvolatile solar composition) will also have the solar system Pu/Nd or Pu/U ratios (Benjamin et al 1978, Burnett et al 1982). Given this background, we now review what is known about actinide chemistry in the various types of chondritic materials.

Ca-Al-rich inclusions in carbonaceous chondrites The Ca-Al-rich inclusions found in carbonaceous chondrites such as Allende appear to be among the earliest-formed materials in the solar system. Their overall chemical and mineralogical composition is suggestive of a high-temperature fraction produced during the condensation of a gas of solar composition (Grossman & Larimer 1974, Grossman 1980), although they are very complex, even bewildering, objects when studied in detail. The inclusions are overall enriched in refractory lithophile elements, including the actinides and rare earths. Discussions of actinide condensation have assumed that perovskite is the most plausible actinide host mineral (Grossman & Larimer 1974, Boynton 1978). The predicted condensation sequence for a solar gas with decreasing temperature is Th, Pu, and then U, with Th and trivalent Pu condensation occurring at about the same temperatures as the light rare earths (Boynton 1978). The inclusion types most simply related to high-temperature condensation are *A* (melilite, spinel, and anorthite, with minor perovskite) and *B* (fassaitic clinopyroxene, melilite, spinel, and anorthite, with minor perovskite), with the *A* more likely to be a direct product of condensation (Grossman 1980). Many features of type *B* inclusions indicate melting. Both inclusion types have complex but mineralogically distinctive layered rims (Wark & Lovering 1977, 1980a,b).

Lovering et al (1976, 1979) and Wark & Lovering (1978) found perovskite was the sole U carrier in type *A* inclusions. Stapanian (1981) found that perovskite was the dominant U and Th host in two type *A* inclusions; in one of the two, U and Th were concentrated in a perovskite-rich rim, with typically U $\sim$ 500 ppb and a highly fractionated Th/U ratio of 20 ± 2. The total inclusion ($\sim 250\ \mu$) has about 100 ppb U and Th/U $\sim$ 10, dominated by the rims. Rim sampling is critical therefore, in any study of refractory lithophile elements in these inclusions. Ca-rich minerals such as Ca-phosphate or clinopyroxene in laboratory partitioning studies (Benjamin et al 1978, 1980) preferentially incorporate trivalent over tetravelent elements. Under all conditions Th is expected to be tetravalent. Thus, the high Th/U in perovskite may indicate that U is not trivalent in a high-temperature solar gas, because then Th/U ratios at or below the solar value would have been expected. The high Th/U in perovskite from *A* inclusions may reflect the volatility of tetravalent U. Alternatively, it is always possible that the *A* inclusions formed from a nonsolar parent material, already depleted in U.

In addition to perovskite, the *B* inclusions show a variety of micron-sized U and Th-rich accessory phases (Lovering et al 1976, 1979, Wark & Lovering 1978). Shirck (1975) showed that significant concentrations of U and Pu (by fossil fission tracks) are found in the major phases—clinopyroxene and melilite. However, the grain-to-grain variations in both U and Pu were large, with U concentrations ranging from 10 to 200 ppb in both phases. Both within and between grains, Pu fission tracks tended to correlate with U concentration. Spinel grains tended to be U-free but high surface U concentrations were observed around large spinels that could not be explained by mantling perovskite grains. Some enrichment of U in inclusion rims was found, but rims do not dominate as in the type *A* studied by Stapanian (1981). The very complex U and Pu distribution is in strong contrast to the apparently regular partitioning of rare earths between clinopyroxene and melilite mineral separates (Mason & Martin 1974, Nagasawa et al 1977). The actinide distributions presumably give us a better glimpse of reality. The complex U and Pu distributions might possibly be related to the prevalent secondary alteration found in type *B* inclusions (Grossman 1980).

Ordinary chondrites In the ordinary chondrites the Ca-phosphates—whitlockite (or merrillite; Dowty 1977) and chlorapatite—are the conspicuous actinide reservoirs. Cantelaube et al (1967) showed that U was concentrated to about 300 ppb in whitlockite from the St. Severin equilibrated (LL6) chondrite. Combining this with fission Xe measurements, Wasserburg et al (1969) and Lewis (1975) derived Pu/U = 0.04 for St. Severin whitlockite; however, the whitlockite alone did not account for the total rock U content. Podosek (1970) reported a total rock Pu/U for St. Severin of 0.015 (based on fission Xe measurements), and concluded that the Pu/U in whitlockite was enhanced by chemical fractionation, a conclusion that holds even stronger today, with the accepted whole rock Pu/U for St. Severin being 0.005 (Hudson et al 1982). Crozaz (1974) showed that Th/U = 10 in St. Severin whitlockite, a value that is also enhanced relative to the total rock value. Under reducing conditions in the laboratory, whitlockite preferentially incorporates Pu and Th relative to a silicate melt ($D_{Pu}/D_U \sim 6$ and $D_{Th}/D_U \sim 2$; Benjamin et al 1980).

In both the meteorites and synthetic samples the preference for Pu can be readily understood in terms of trivalent Pu relative to tetravalent U (or Th). The preference for Th over U is not so readily explained, but might possibly be accounted for by the crystal-liquid partition coefficient systematics developed by Matsui et al (1977) for different elements partitioning into a given mineral. It was observed that the maximum *D* value among elements with a given valence occurred for an optimum ionic radius. If this is true, $D_{Pu} < D_U$ would be expected for tetravalent Pu, and this is compatible with existing partition coefficients (Benjamin 1980).

The enhanced Pu/U and Th/U ratios in equilibrated chondrite whitlockite are due primarily to incorporation of a lower fraction of the total rock U ($\sim 10\%$, Jones & Burnett 1979). Thus, whitlockite is not the major U host phase; however, none of the characterizable phases of St. Severin, including clinopyroxene and apatite, can account for the whole rock U, which is interstitially located on grain boundaries. Chlorapatite has much higher U concentrations than whitlockite (ap/whit ~ 20 in St. Severin, and somewhat lower in other equilibrated chondrites; Pellas & Storzer 1975, Crozaz 1974, 1979, Jones & Burnett 1979), and it is very inhomogenously distributed; however, even in apatite-rich sections, this phase cannot account for the total U, unless much of the apatite occurs as grains of micron-size or less. This latter possibility cannot be excluded and would, in fact, explain the apparent lack of a chlorine material balance in St. Severin. It is possible that all of the ^{244}Pu in St. Severin is in whitlockite (compare Alaerts et al 1979), but because of the spread in the total rock fission Xe^{136} contents in the literature (Jones 1982), it may be that only about one half is in this phase (Jones & Burnett 1979).

Adopting the Nd concentration in St. Severin whitlockite from Chen & Wasserburg (1981), a Pu/Nd ratio of 1.9×10^{-4} is obtained, similar to values of Marti et al (1977) and supporting the close coherence of Pu and light rare earths. However, Ebihara & Honda (preprint, 1982) find $[Nd]_{ap} \sim 1.5[Nd]_{whit}$ for Bruderheim (L6), whereas Pellas & Storzer (1981) report $[Pu]_{ap} = 0.15\text{–}0.2\ [Pu]_{whit}$ for H6 chondrites. Conceivably Pu-rare-earth coherence might break down internally within apatite-rich chondrites.

St. Severin has been extensively recrystallized by thermal metamorphism, with almost total obliteration of the characteristic chondrule texture (see, for example, Wasson 1974). Fission track and Xe studies on less equilibrated chondrites shed some light on the more extensive St. Severin data. Pu/U ratios for whitlockites from a variety of chondrites (Kirsten et al 1978) show a surprisingly large spread, with values up to 0.12 in Nadiabondi (H5) compared with 0.04 in St. Severin. In Nadiabondi the higher ratio is due to a low U content for Nadiabondi whitlockite [50 ppb, Chen & Wasserburg (1981), Murrell & Burnett (1982c); 80 ppb, Pellas et al (1979)]. The Pu/Nd ratio in Nadiabondi whitlockite is similar to St. Severin (Chen & Wasserburg 1981). In Nadiabondi and several H3 chondrites, U is instead concentrated in chondrule glass, and in H3 chondrites U is less than 20 ppb in whitlockite. Thus U in chondrites moves during metamorphism from chondrules into phosphates; however, the transfer may be incomplete, even in well-equilibrated meteorites like St. Severin. In Bremervörde (H3) the whitlockites have ^{244}Pu track densities comparable to St. Severin and no tracks in olivine adjacent to chondrule glass, implying that Pu incorporation in whitlockite occurs very early, perhaps even prior to concentration of rare earths in phosphates (Murrell & Burnett 1982c). All in all it appears difficult to use the Pu/U ratio in whitlockites for chonological purposes, although Pu concentrations may be interpretable. The extreme internal fractionation of Pu and U in ordinary chondrites is a complication to the measurement of reliable solar system actinide abundances from these meteorites.

Enstatite chondrites The highly reduced enstatite chondrites (Wasson 1974) appear simpler in that U and Th, like many other elements, deviate from lithophile character and appear to be totally concentrated in oldhamite (CaS) in the equilibrated (E6) meteorites (Furst et al 1982, Murrell & Burnett 1982a). Albandite (MnS) is a minor U reservoir ($\sim 1/20$ that of CaS). Niningerite [(Mg, Mn)S] is the major ($\sim 75\%$) U-bearing phase in the less equilibrated Abee (E4). Fossil tracks, presumably ^{244}Pu, were found in one enstatite-oldhamite grain boundary in Hvittis (E6), suggestive of ^{244}Pu concentration in oldhamite as well. It may be that the actinides and rare earths are quantitatively concentrated in oldhamite, making the E6 meteorites seem very favorable for Pu/Xe measurements; however, Kennedy (1981) found no strong fission Xe enrichments in stepwise-heating Xe release measurements of bulk E chondrites.

2.2 *Alpha Radiography Using Nuclear Reactions*

This technique is the analogue of fission radiography except that a plastic track detector sensitive to alpha particles is attached to the polished section during an irradiation. At high bombarding energies ($\lesssim 10$ MeV), alpha emission is not specific to any one element. Several light nuclei that would be of geochemical interest (e.g. ^{15}N, ^{19}F) exhibit exothermic (p, α) reactions; however, there are competing reactions on major constituents (e.g. ^{18}O), and it has not been demonstrated that the plastic track detectors could survive useful proton doses. Only thermal neutron (n, α) reactions appear useful, and these are dominated by the high cross sections of ^{10}B (3840 barns) and ^{6}Li (950 barns). Background from ^{17}O(n, α), and possibly ^{14}N(n, p), in both detectors and sample set an ultimate limit of about 0.2 ppm B or ~ 1 ppm Li. The relative track production rate (per unit weight) of Li to B is 0.16 (Furst 1979). Cellulose-based plastics, most commonly cellulose nitrate, provide no simple means for distinguishing B from Li, so applications to date have only involved samples (e.g. fossil materials) for which it was known that the track density from Li would be much smaller than that from B (Furst et al 1976, Furst 1981). Boron contamination is a very serious problem, and analyses below 20 ppm require great care (Weller et al 1978). Sensitivity is not a problem, given the limits set by ^{17}O background and contamination, and fluences of only 10^{12}–10^{13} n cm^{-2} are required. However, irradiations are best done with a well-thermalized flux to eliminate fast neutron recoil tracks and, at low temperatures, to prevent track annealing. The spatial resolution, set by the alpha ranges, is good (3–5 μ in rock). For cellulose nitrate there is a practical track density upper limit of $\sim 10^8$ cm^{-2} for counting because of electron beam damage at the required scanning electron microscope magnifications, but this still permits B- or Li-rich phases of $\sim 2\,\mu$ to be located or quantitative analyses of ~ 10-μ grains. In addition to the alpha particles, the ^{7}Li recoils from the ^{10}B$(n, \alpha)^{7}$Li reaction will produce etchable tracks, although the empirically determined (Furst 1979) net efficiency for total track etching and counting in cellulose nitrate leads to track production rates roughly comparable to those expected from the alphas *alone*. Thus Equation (1) can be used to estimate track densities if the factor $R/2$ is replaced by $R/4$ ($R \sim 10^{-3}$ g cm^{-2} for silicate material). The full energy tritons from ^{6}Li$(n, \alpha)^3$H do not register tracks in cellulose nitrate but, in principle, should register in CR-39 plastic (Ahlen et al 1981, Cassou & Benton 1978, Fleischer 1979). Using a thin alpha absorber, this would permit distinction between Li and B, with good sensitivity and reasonable spatial resolution, but this has not been demonstrated; the only relevant study indicates only low-energy triton registration (Pilione & Carpenter 1981). If distinctions between low-energy protons and alphas are possible, N-mapping through ^{14}N(n, p) might be feasible.

In addition to obvious possibilities for laboratory studies of Li and B geochemistry, use of ^{17}O-enriched materials would permit ^{17}O(n, α) radiography, which would seem attractive for kinetic studies (e.g. O diffusion or exchange rates).

2.3 *Autoradiography by Alpha Radioactivity*

Elemental microdistributions can be determined for naturally occurring alpha emitters and for those elements that are suitable targets for producing induced alpha activity. In the latter case, nuclear reactions are employed, but, as opposed to work discussed in the previous section, the alpha emission is not prompt. This has the advantage that the alpha-particle detector is not irradiated. The critical factors with regard to feasibility are the alpha-particle half-lives, activation cross sections, grain size, alpha-particle range, and elemental concentration.

For the case of the natural alpha emitters, the half-lives are very long ($\geqslant 10^9$ yr). The alpha activity of a grain is proportional to the product of the concentration divided by the half-life. With such low activity, the product of the concentration and grain area must be relatively large, with, for example, percent levels of U required to produce 100 tracks from $\sim$100-μ grains with 10^6 s exposures.

For synthetic samples, ^{230}Th ($T_{1/2} = 7.8 \times 10^4$ yr) alpha radiography has been used to measure Th partition coefficients for Ca-phosphates and clinopyroxene (Benjamin et al 1980). Th, U, and Sm analyses in the same sample are possible using alpha, fission, and beta tracks, respectively (see below). In addition, ^{148}Gd ($T_{1/2} = 93$ yr) is potentially important, and by using freshly purified ^{210}Pb ($T_{1/2} = 22$ yr) tracer, experimental Pb geochemistry should be possible by recording the alpha decay of the ^{210}Po ($T_{1/2} = 138$ d) daughter.

In the case of induced activity, the half-life cannot be much shorter than

$\sim$1 h to allow time for the assembly of samples and detectors after the activation irradiation. Neither can it be excessively long ($\geqslant$100 d), in order to avoid unreasonably long activation irradiations and detector exposures. In any given activation scheme, several alpha emitters may be produced. Since the alpha-particle energies from a thick target vary from zero up to the decay energy, significant differences in the half-lives must exist in order to avoid, or to be able to correct for, interferences. The nuclei observed must also be uniquely ascribable to a specific target element, which in practice means the element must be produced by a fairly simple nuclear reaction on a major isotope with a relatively high cross section. Only ^{211}At-^{211}Po (7.2 h) and ^{210}Po (138 d) appear usable, and the production reactions that have been demonstrated are ^{209}Bi$(\alpha, 2n)$ ^{211}At (Woolum et al 1976) and ^{208}Pb$(\alpha, 2n)$ ^{210}Po (Hamilton 1971). Use of ^{209}Bi$(n, \gamma\beta)$ ^{210}Po appears feasible. In principle, these same nuclei can be produced by heavy ion (Li, B, C) reactions on other Pb isotopes, Tl, and Hg, but unique production from a given element does not occur. Ahlen et al (1981) have suggested ^{206}Pb(^{3}He, $3n$) ^{206}Po. The alpha branching ratio in ^{206}Po is only 5%, which makes it less sensitive than those given above (which have 100% alpha decays); however, for higher Pb concentrations it would be more convenient because of the relatively long ^{210}Po half-life. Because of inappropriate half-lives or low alpha/electron capture branching ratios, rare-earth alpha emitters are not useful for autoradiography on natural samples.

At ^{4}He bombarding energies of 30 MeV, essentially the total reaction cross section for nuclei having atomic mass numbers $\sim$200 ($\sim$1 barn) appears as the $(\alpha, 2n)$ reaction and can be used to efficiently activate ^{211}At and ^{210}Po from ^{209}Bi and ^{208}Pb, respectively (Woolum et al 1976, Hamilton 1971). Using cellulose nitrate, the full alpha energies (6–7 MeV) do not leave etchable tracks. Therefore, only slowed-down alphas from subsurface decays produce etchable tracks, but this is somewhat of an advantage in minimizing contributions from Pb surface contamination. The actual depth response and spatial resolution are somewhat complicated (Woolum et al 1979), but, in general, grains larger than about 35 μ are required for quantitative analysis. Experimental track production rates (assuming complete decay) of $\sim 8 \times 10^5$/cm^2-ppm for both ^{208}Pb and Bi have been achieved. For practical exposures, about 1 ppm ^{208}Pb or Bi is required for relatively precise ($\pm$10%) analyses of individual 100-μ grains. In 0.1–0.3 mm chondrite metal grains it has been possible to measure down to 0.03 ppm Bi at lower precision (Woolum & Burnett 1981). This technique has been used to demonstrate a previously unrecognized siderophile character for both Bi and Pb in L chondrites (Woolum & Burnett 1981), although it had been anticipated in nebular condensation calculations by Larimer (1973). In Khohar, Bi and Pb are concentrated in a population of kamacite (Ni-poor Fe-metal phase) grains that appear to be finely polycrystalline, so there is the possibility that the Bi and Pb are adsorbed on metal surfaces rather than dissolved. The incorporation of Pb and Bi in these metals is, however, more likely to have occurred during a reheating event on a parent body (probably shock-induced) and not in the solar nebula. Contrary to iron meteorites, ^{208}Pb in chondrites shows little tendency to concentrate in troilite. Larimer's (1973) calculations predict that $\sim 95\%$ of the Bi and Pb would condense as pure phases. Bi/Pb microdistribution studies of Allende, enstatite chondrites, and ordinary chondrites show no evidence for these. While the track distributions qualitatively appear uniform for Allende and the enstatite meteorites, detailed track-by-track mapping for the meteorites (Woolum et al 1979) indicate that Bi is not totally randomly distributed; the data are consistent with a model in which 90% of the Bi resides in $\sim 10^{-16}$ g Bi "point" sources and the rest in $\sim 10^{-14}$ g "point" sources, which, if pure Bi, are $\sim 10^2$ Å and $\sim 10^3$ Å in size.

2.4 *Beta Autoradiography*

In this case a nuclear emulsion is exposed to the polished section and Ag grains produced by the stopping of betas emitted by the sample are analyzed by optical counting or Ag X rays. Although this method is widely used for biological work (e.g. Rogers 1979), recent applications to geochemical studies have been as a consequence of Mysen & Seitz (1975). This technique has no apparent value for natural samples, e.g. activated by a thermal neutron irradiation, because the beta activity of a given mineral is not specific to an element and tends to be dominated by major element activities such as ^{45}Ca. The required solid angles for good spatial resolution for a micro-gamma-ray spectrometer lead to vanishingly small detection efficiencies.

Beta radiography has important applications for synthetic samples using radioactive tracers, although only one element can be analyzed at a time. Spatial resolution is better for low-energy betas (less than a few hundred KeV); however, unlike heavy charged particles, betas do not have a well-defined range, because of the continuous energy spectrum of beta particles and their complex trajectories (due to a high probability of scattering). Definitions of beta ranges are given in nuclear chemistry texts. Empirically, betas show an approximately exponential attenuation in intensity in passing through absorbing material. Consequently, a useful measure of spatial resolution is the half thickness, $D_{1/2}$, the amount of material required to decrease the beta counting rate by a factor of 2. Empirically (Evans 1955),

$$D_{1/2} \cong 40\, E^{1.14},$$

where $D_{1/2}$ is in mg cm^{-2} (thickness times density) and E is the maximum beta energy in MeV. The equation was obtained for Al but depends only weakly on absorber composition. Table 1, similar to Mysen & Seitz (1975), gives nuclei readily suited for beta radiography in terms of having maximum energies less than about 0.35 MeV ($D_{1/2} \lesssim 40\ \mu$) and convenient half-lives (35 d to 10^4 yr).

Table 1 consists of nuclei that could be easily utilized by copying the relatively simple procedures in the literature. These should be regarded as the minimum number of useful nuclei. There are many possibilities that have not been demonstrated, e.g. use of low energy Auger or internal conversion electrons following electron capture decay. The electron capture decay of ^{55}Fe has biological applications (Rogers 1979). Alternatively, beta emitters of any energy can be used for uniform thin-section samples by adjusting the sample thickness to control the spatial resolution. Micron-scale resolution using ^{14}C ($D_{1/2} \sim 17\ \mu$ in minerals) has been demonstrated in biological work (Rogers 1979). Using the exponential attenuation approximation, track densities for simple exposure geometries are given by Jones (1981). In this approximation, the maximum track production rate $\dot{\rho}_{max}$ (betas/cm^2-s) for an emulsion in contact with

Table 1 Radioactive nuclei potentially important for beta autoradiography of synthetic samples

Nucleus	Half-life[a]	Maximum energy (MeV)	Half-thickness[b] (mg cm^{-2})
^{95}Nb	35 d	0.16	5.0
^{103}Ru	40	0.23	7.3
^{203}Hg	47	0.21	6.8
^{46}Sc	84	0.36	12
^{35}S	88	0.17	5.2
^{121m}Te[c]	154	$\leqslant 0.05$	$\leqslant 1.3$
^{65}Zn	244	0.33	11
^{155}Eu	1.8 yr	0.15	8.2
^{171}Tm	1.9	0.097	2.8
^{147}Pm	2.6	0.23	7.3
^{3}H	12	0.018	0.41
^{151}Sm	90	0.076	2.1
^{63}Ni	92	0.066	1.8
^{166m}Ho	1200	0.075	2.1
^{14}C	5730	0.156	4.8

[a] Nuclei listed in order of increasing half-life, in days and years.

[b] Approximately composition-independent; divide by density to convert to linear dimensions.

[c] Internal conversion electrons.

an "infinitely thick" sample is related to the specific beta activity I (decays $g^{-1}\ s^{-1}$) by

$$\dot{\rho}_{max} = I\lambda/4, \tag{3}$$

where λ is the mean attenuation length ($\lambda = D_{1/2}/\ln 2$). A simple case, useful for estimating required crystal sizes for quantitative analysis, is that of the track density at the center of a hemispherical grain of radius r:

$$\rho/\rho_{max} = (1 - e^{-r/\lambda}). \tag{4}$$

For example, for ^{151}Sm ($\lambda \cong 7\ \mu$ in minerals), the track density at the center of a 30 μ diameter grain in a Sm-free matrix is still about 12% less than that for an infinitely large grain. In general, quantitative analysis should be restricted to grain diameters large compared with $D_{1/2}$.

Quantitative analyses of the nuclear emulsion images have been made by optical reflected-light counting (Mysen & Seitz 1975) or by electron-induced Ag X-ray counting (Holloway & Drake 1977, Benjamin et al 1977, Jones & Burnett 1981). Both methods have complications. Optical counting for very low energy ($E_{max} \sim 0.2$ MeV) betas is based on individual Ag grains, and an individual beta will produce more than one grain. At higher track densities, coalesced or juxtaposed grains may be counted as a single grain; thus the observed grain density from a given source with increasing exposure time will tend to be nonlinear, i.e. saturate. Mysen & Seitz report linear "track" densities vs exposure time below $6 \times 10^7\ cm^{-2}$ for ^{14}C and infer higher maximum track densities for ^{151}Sm. We have been unable to confirm linearity with ^{151}Sm for grain densities as low as $1 \times 10^7\ cm^{-2}$, but we cannot rule out that better results can be obtained with greater practice and patience. It is clear, however, that for optical counting, concentrations and exposure times should be adjusted so that all measurements are done with similar grain densities.

Holloway & Drake (1977) (see also Drake & Holloway 1981) used an electron microprobe to count Ag L X-rays on the developed emulsion; however, Jones & Burnett (1981) have shown that selective volatilization of the organic constituents of the emulsion relative to Ag occurs essentially instantaneously upon exposure to the microprobe electron beam (15 KV, 5 nA, 30-μ spot). The amount of volatization is less for regions of higher Ag counting rate; thus errors in measured relative counting rates (e.g. in D values) will result. The lower current densities (typically by a factor of 40) for an SEM electron beam produce much less damage, but still give acceptable counting rates. Other complications with Ag X-ray counting are latent image fading and intermittent nonlinear response (Jones & Burnett 1981); these probably arise because only the outermost emulsion layers are being analyzed. Beta autoradiography should not be used without a parallel program of tests and controls on the emulsion batch used and the counting techniques employed.

Beta autoradiography has primarily been used for trace element partition coefficient studies and is the basis for most of the "Henry's Law" studies discussed in Section 1. It has also been used by Jones (1981) to show that Pu and Nd do not have equal partition coefficients despite the apparent lack of fractionation observed in many meteoritic samples (Marti et al 1977).

3. X-RAY FLOURESCENCE TECHNIQUES

X-ray flourescence techniques have the advantage of a much broader repertoire of detectable elements in natural samples than the particle track techniques discussed in Section 2. The sensitivities are a continuous and generally slowly varying function of the atomic number of the target elements. Elemental analyses are insensitive to chemical speciation, and the physics involved in X-ray production and detection (cross sections, absorption coefficients) is well understood, making quantitative analysis straightforward, in principle. We discuss in situ trace element microdistribution studies using heavy charged particles and photons. We recognize that in special circumstances, e.g. Ni in forsterite or K in FeS, good sensitivity (~ 100 ppm) is possible with electron microprobes for relatively light elements (Z less than about 30). We assume, however, that most readers are familiar with these capabilities.

3.1 *Charged-Particle-Induced X-Rays*

COMPARISON OF ELECTRONS, PROTONS, AND HEAVIER IONS Relative to incident electrons, incident heavy charged particles, e.g. protons, produce orders of magnitude less continuous radiation (bremsstrahlung) due to their much greater mass. The cross sections for characteristic X-ray line emission are similar for 1–5 MeV protons and 10–50 KeV electrons. Thus, the X-ray signal to beam-induced background is much higher for heavy charged particles compared with electrons. This allows for improvements over electron probe sensitivities. The major background source for ion bombardments is the secondary bremsstrahlung generated by the ionization produced in the sample by the incident ion (Folkmann et al 1974). Because the production of characteristic X rays and the production of secondary electron bremsstrahlung are both consequences of the primary ionization, all charged particles give the same signal to secondary bremsstrahlung background. Thus ions heavier than protons offer no advantages and have the following disadvantages. (*a*) For comparable X-ray yields, the heavier ions produce greater gamma-ray backgrounds.

(*b*) Incident ions heavier than protons are harder to focus for microanalysis. (*c*) Heavy ions, due to their much larger energy loss, produce greater heating and radiation damage to the sample. Minimum detection limits are about a factor of 10 higher for alpha particles, relative to protons (Cooper 1973). Consequently, we consider only protons.

PROTON-INDUCED X-RAY EMISSION (PIXE) PIXE, or proton-probe, work has had wide application in environmental and biological studies, and many review articles are available (see, for example, Cahill 1980, Johansson & Johansson 1976, Johansson 1977, and references within these papers). Most studies are on large area samples, but a number of facilities have developed intense microbeams from 2 to ~50 μ in size, some by using collimators (Horowitz & Grodzins 1975, Shroy et al 1978) and others by using solely magnetic or electrostatic focusing [see Cahill (1980) for a tabulation through November 1979; also, Maggiore (1982), Koyama-Ito & Grodzins (1980)].

X-ray production drops rapidly below 1-MeV incident proton energy. Significant gamma-ray production for nuclear reactions begins above ~5 MeV, so 2–4 MeV are common energies for proton-induced X-ray studies. For 2, 3, and 4 MeV, the depths of X-ray production in a mineral are roughly 25, 60, and 100 μ, respectively. For comparison, the half-thicknesses for photon absorption at 5, 8, and 14 KeV (the X-ray region of interest) are roughly 10, 50, and 200 μ. Thus, the depth of sample analyzed may be set by X-ray production (proton-stopping), by X-ray absorption, or by the thickness of the sample. In general, it seems unnecessary to require a beam spot size much smaller than the analysis depth; consequently, proton beam sizes of 10–20 μ are required for optimal spatial resolution in typical rock sections. For better spatial resolution, ultrathin samples mounted on inert backings are required.

Pinpointing the position of the microbeam on a multiphase sample is important in PIXE microprobe work. Optical viewing or elemental X-ray imaging is typically used. However, Younger & Cookson (1979) recently reported a secondary electron imaging system, similar to those of SEMs.

High sensitivities for trace element analyses are quoted in many reviews, but these are based on the implicit assumption of a low Z (e.g. organic) matrix, where significant interference effects from the major elements occurring in geological materials can be neglected. Because the cross sections for K X-ray production drop rapidly with increasing Z (a factor of ~100 between Ca and Mo at 2 MeV), and because the L/K production ratio is >100 for $Z \gtrsim 30$, L lines are used for the heavier elements. But with a Si(Li) detector with typically ≥ 150 eV resolution there are unresolved or marginally resolved overlaps between the heavy trace element L lines and the major element K lines (e.g. the rare earths and the first-series transition metals). High counting rates from major element lines cause severe pulse pileup and deadtime effects. Absorbers can be used to some extent to attenuate the major element X rays, but care must be exercised in choosing the composition and thickness of the absorber to achieve a favorable *relative* attenuation of major element to trace element X rays. Also, secondary flourescence in the absorber should not introduce a peak in the energy region of interest.

It is unlikely that the conventional use of PIXE, employing a Si(Li) detector, will ever be a broadband general technique for natural samples. However, analyses are possible for cases where the trace element lines have much higher energy than any major element line, or for synthetic samples where the trace element concentration levels and matrix composition can be controlled. An example of such a case (Woolum & Boisseau 1981) is shown in Figure 1 for a synthetic glass (nominally Fe-free) containing 150–

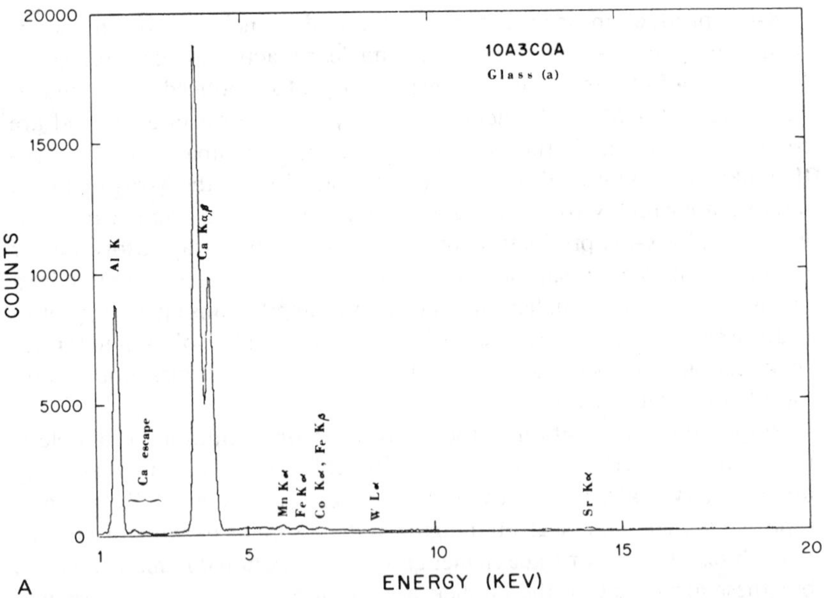

Figure 1 PIXE spectrum of a nominally Fe-free, Th-doped, synthetic glass sample, using a 2.0 MeV, ~100 μ diameter proton beam and a Si(Li) detector with a 2 mil Al absorber to attenuate the major element X rays. The analysis time was about 20 min. (*a*) The full spectrum. The Al peak is primarily due to secondary fluorescence in the absorber, as well as from Al X rays from the sample; (*b*) the spectrum above the Ca K peak, showing transition metals, W, and Pt peaks riding on the broad secondary bremsstrahlung background, as well as Th and Sr peaks at higher energy; (*c*) the Th and Sr peak region at expanded scale, showing a 5:1 peak-to-background ratio obtained with 150–200 ppm present in the sample.

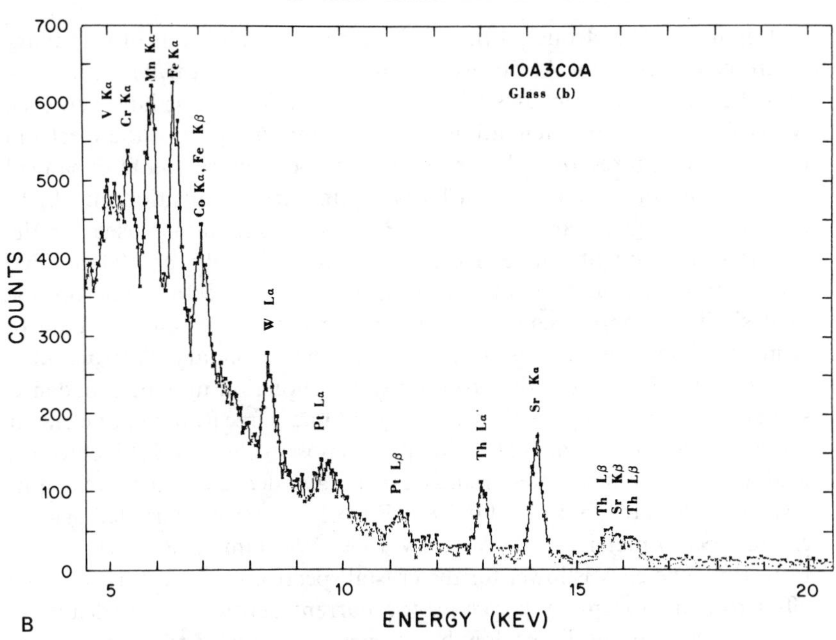
10A3COA
Glass (b)
V Kα
Cr Kα
Mn Kα
Fe Kα
Co Kα, Fe Kβ
W Lα
Pt Lα
Pt Lβ
Th Lα
Sr Kα
Th Lβ
Sr Kβ
Th Lβ
COUNTS
ENERGY (KEV)

B

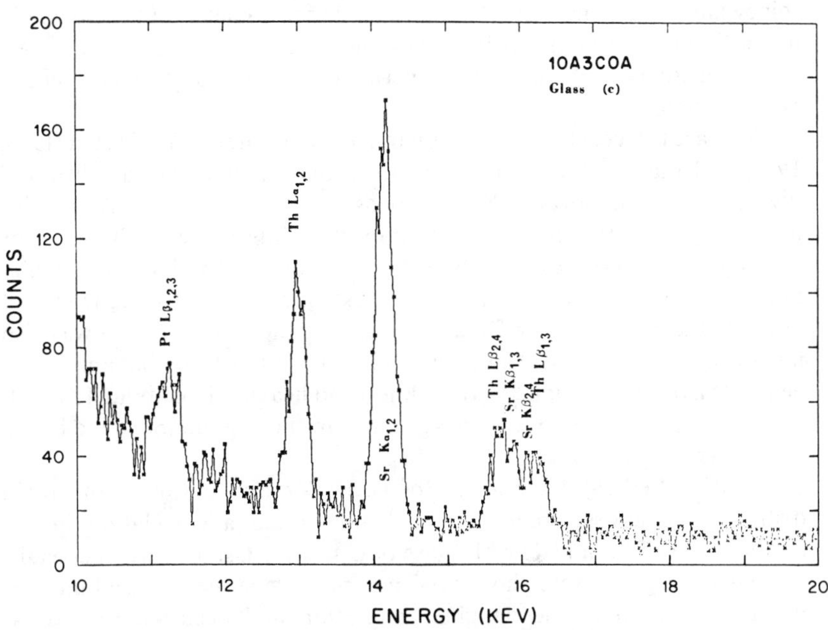
10A3COA
Glass (c)
Pt Lβ1,2,3
Th Lα1,2
Sr Kα1,2
Th Lβ2,4
Sr Kβ1,3
Sr Kβ2,4
Th Lβ1,3
COUNTS
ENERGY (KEV)

C

200 ppm Th. The strong Al peak in Figure 1*a* is the result of secondary flourescence from an Al absorber window on the Si(Li) detector. This absorber strongly attenuates all major elements K X-rays; however, even with a factor of $\sim 10^3$ attenuation, Ca is the dominant peak in the spectrum (Figure 1*a*). Figures 1*b* and 1*c* show the high-energy region at expanded scales. In addition to a clear Th Lα peak, lines from transition metals, W and Sr in the glass, and a trace of Pt from the container used for the synthesis are evident. These elements are probably between 100 and 500 ppm, amounts not normally detected in routine electron microprobe analysis. The pronounced continuous background so evident in Figure 1*b* is mostly secondary bremsstrahlung from the secondary electrons produced in the stopping of the proton beam, although it may be broadened somewhat by pulse pileup. If a trace element line is free from major element interferences, analysis at ~ 50 ppm appears possible with a Si(Li) detector. Major/minor element interferences can be avoided using a wavelength-dispersive detector, but very little PIXE work has been attempted using a crystal spectrometer (Poole & Shaw 1969, Woolum & Boisseau 1981). Detection efficiency is lower for the crystal spectrometer but this could be offset to some extent by larger proton current densities, since deadtime problems are absent. [Very low beam currents are used to maintain low deadtimes for the Si(Li).] Preliminary estimates indicate comparable or somewhat better sensitivites for the crystal spectrometer, but these would not be limited, in general, to only favorable matrix conditions. The ultimate PIXE sensitivity probably will be determined by available beam current or sample stability.

There are few geochemical proton microprobe studies. Bosch et al (1978, 1980) and Blank et al (1981) have shown that ilmenite, armalcolite, and ulvöspinel are important Y host phases in Apollo 17 mare basalts, indicating that either ilmenite is an important lunar rare-earth carrier or that significant Y-REE fractionation has occurred. The latter possibility might explain small "Y anomalies" seen in some mare basalts (e.g. Wakita et al 1970). Partitioning of Sr, Zr, and Nb among the opaque oxides was also measured by Blank et al. This study and Figure 1 illustrate that analysis of elements from Se through Te (via K lines) and heavier than about Ta (via L lines) is favorable due to low backgrounds in this region above the broad secondary bremsstrahlung peak.

Clark et al (1979) determined Pb, Tl, Hf, Zr, and U (by fission track) distributions across zircons. They demonstrated a correlation in the distributions of Pb and Tl with those of U in zoned, single zircon crystals. The U-rich regions were "hot spots" and not a separate included mineral phase. Apparently, little daughter migration had occurred relative to parent in these crystals. Pb and Tl were undetected with the electron probe, but ~100 ppm levels of Pb were determined with PIXE, despite the unfavorable location of the Pb L lines below the dominating Zr K lines and despite interferences between Pb Lα, Tl, and Hf. A RbCl wafer was used to critically absorb the Zr X rays and to improve the Pb(L)/Zr(K) ratio by a factor of ~11; the secondary flourescence (Rb K) peak from the absorber was sufficiently higher in energy to make quantitative work using the Pb Lβ line possible.

Gentry et al (1976) used PIXE microprobe analysis of monazite inclusions surrounded by giant pleochroic halos as evidence for primordial superheavy elements; however, a number of similar studies were negative (Bosch et al 1977, Cahill et al 1978, Cookson et al 1978). These studies point out the importance of considering potential contributions from nuclear reaction gamma rays in an X-ray spectrum.

Wark et al (1980) have published spectra from perovskite in Allende Ca-Al-rich inclusions, showing clearly identifiable peaks from Sr, Zr, Y, and, in some cases, Zn and As. The Sr was interpreted as originating from melilite underlying the small (10–20 μ) perovskite grains. An attempt was made to classify the rare-earth patterns relative to bulk inclusion REE patterns based on "continuum" levels in the heavy REE region above the Fe K lines, but the preliminary analysis was not convincing. However, there are clear differences in the Y levels, which possibly could be used for classification purposes.

Enstatite chondrite oldhamite, in which U and Th are unfractionated, is potentially important for ^{244}Pu chronology (Murrell & Burnett 1982a); however, preliminary PIXE analyses of Ce and Nd at the 20–30 ppm level (Woolum et al 1982) indicated definite U-REE, and thus possible Pu-U, fractionation in oldhamite from the Peña Blanca Spring aubrite.

3.2 *Synchrotron Radiation*

Synchrotron radiation is the magnetic analogue of bremsstrahlung, i.e. electromagnetic radiation emitted by charged particles accelerated by deflection in a magnetic field. Relativistic electrons in circular orbits provide intense sources of such radiation, and synchrotron radiation has been widely applied in solid-state research (e.g. Kunz 1979, Winick & Doniach 1980). Winick (1980) provides a concise summary of synchrotron radiation properties. In the laboratory frame, the radiation pattern is strongly peaked in the (forward) direction of motion of the electrons, and the vertical angular divergence of the radiation, tangential to the orbit, is small, typically $\sim\gamma^{-1}$ for relativistic electron energies, where γ is the electron energy divided by its rest mass energy ($\gamma = E/mc^2 \gg 1$). The synchrotron radiation is predominantly polarized with the electric vector in the orbital plane (parallel to the electron's radial acceleration). Photons are emitted over a broad spectral range. For electron energies in the 1–10 GeV range, peak fluxes in the 1–100 KeV range can be obtained. Despite the broad energy distribution of synchrotron radiation, the intensities are sufficiently large that it is practical to consider both selecting a relatively narrow energy region with a monochromator and focusing to a small spot with curved crystal spectrometers or "mirrors," while still having enough intensity remaining to excite detectable yields of fluorescent X rays. For example, a synchrotron microprobe is being proposed for the electron storage ring at the Brookhaven National Synchrotron Light Source (NSLS). Tunable beams up to ~ 50 KeV containing $\geqslant 10^8$ photons/s/μm^2 with an energy bandwidth $\leqslant 3\%$ are predicted (B. Gordon, private communication). A detailed review of the potential of X-ray flourescence microprobes using synchrotron radiation is given by Sparks (1980). There are several advantages over characteristic X-ray production using electrons or protons. (1.) The flourescence cross sections are 10–1000 times higher than for charged particles and, for a given energy photon, increase with increasing Z. (2.) The photon energy can be tuned to optimize analysis for a given range of elements. Since flourescent X-ray production is primarily by the photoelectric effect, critical absorption can be exploited. This is a distinct advantage when simultaneous multielement analysis is not the primary concern. For example, with 40–60 KeV photons, K lines for the heavy rare earths can be excited with cross sections $\sim 10^2$ times higher than Fe-group K lines. Use of trace element K lines would eliminate most of the interference problems between trace element L lines and major element K lines when a Si(Li) detector is used, but deadtime and sum peak limitations from major element lines may still require the use of crystal spectrometer detectors for flourescent X rays, assuming sufficient intensity. (3.) Because the primary mode of interaction for ~ 50 KeV photons in geological materials is the photoelectric effect, and because the photoelectrons have very low energies, negligible secondary bremsstrahlung background is produced. The primary background is photons from either coherent (elastic) or Compton scattering. Background can be reduced with a detector in the plane of the electron orbit at 90° to the primary photon beam because of the reduced yields of coherent scattering of the polarized synchrotron radiation in this direction. Compton-scattered photons are produced both in the sample and in the detector, with the latter being the most important source of background beneath the flourescent X-ray lines in many applications (Sparks 1980). The energy spectrum for secondary Compton photons peaks near the primary energy but drops off rapidly at lower energies. For this reason, flourescence by monochromatic synchrotron radiation produces much lower background than by continuum synchrotron radiation. For $Z = 20$ to 40 in organic matrices, Cooper (1973) found similar signal-to-background ratios for 2–4 MeV protons and 5–15 KeV monochromatic photons from X-ray tubes and electron capture sources. Sparks (1980) calculated comparable signal/background ratios for monochromatic synchrotron radiation and protons for $20 < Z < 40$, but for $Z > 45$ the relative signal/background ratios were ~2–10, in favor of photons. (4.) Unlike electrons or protons, most photons that do not interact with an inner-shell electron do not damage the sample with heat or radiation. Also, the photon cross sections are much higher than the charged-particle cross sections. Thus heating rates for photons are about a factor of 10^2–10^6 less than for charged particles (Sparks 1980) for the same net X-ray counting rate.

For primary photon energies of 10–50 KeV, attenuation half-thicknesses in geological materials are 0.1–10 mm; consequently, analysis depth will be determined by sample thickness or by flourescent X-ray attenuation. As with proton-induced X rays, a practical spatial resolution (i.e. beam-spot size) of 10–20 μ is indicated. Extraordinary design efforts to achieve micron-size beam spots do not appear to be justified unless a parallel effort to make uniform ultrathin sections on this scale is also undertaken.

The thorough calculations of Sparks (1980) indicate that, for 10-μ grains and the proper choice of monochromatic photon beams, analyses could be made of any trace element heavier than Na at ppm levels. Facilities are not yet available for work under optimum conditions, and very few measurements have been carried out on relevant materials. The most relevant analysis was of a monazite grain with 37 KeV photons as part of the search for superheavy elements in giant pleochroic halos (Sparks et al 1977, 1978). A conservative limit of ~1–2 ppm (~5×10^9 atoms/inclusion) of superheavy elements was set. The photon energy was selected to eliminate K lines from rare earths (major elements in monazite). Relative to PIXE spectra of similar monazite grains (Gentry et al 1976, Bosch et al 1977), significantly improved (up to factor of ~10) peak-to-background ratios were demonstrated by Sparks et al (1978).

4. SUMMARY

Synchrotron X-ray flourescence on paper appears to be the best broad-based technique for analysis of most trace elements on a microscale, although in achieving its greatest sensitivity, monochromatic photons will have to be used, which will reduce simultaneous multielement analysis capabilities. However, it is still a technique of the future, undemonstrated and with only one dedicated facility (NSLS) proposed for the United States. It may also be that in the future, high-resolution ion microprobes will be much more available and the secondary ion yield-concentration relationships better understood so that these can be the dominant instruments for in situ trace element microanalysis. But for the present, the other techniques discussed here are all that is available. For natural samples they have definite limitations: restriction to specific elements in the case of track radiography, and relatively lower sensitivity for proton-induced heavy element ($Z \gtrsim 40$) X rays. Despite the broad-based nature of the PIXE technique, with a Si(Li) detector the problems must be well chosen so that major element interferences (peak overlap, and pileup and deadtime effects) are absent, although efforts to incorporate wavelength-dispersive analysis for PIXE may improve this situation. On the other hand, these techniques have the advantage that, within these limitations, they have demonstrated applicability. In addition, they are compatible with relatively restricted research budgets. This assumes the continued existence and availability of nuclear research facilities. However, there are presently many small accelerators, which are potentially capable of proton microbeam X-ray fluorescence studies. For experimental trace element geochemistry the beta autoradiography and/or proton-induced X-ray techniques offer opportunities for relatively low-cost studies, and we would anticipate that it is in this area that the greatest contributions of the techniques discussed here can be made. For all these techniques, special sample preparation methods and backings are necessary to minimize contamination, heating, and background. It is unreasonable to believe that one will ever be able to mail in standard petrographic thin sections for trace element analysis.

Acknowledgment

Much of the research discussed in this paper was supported by NASA through grants NSG 7202 (DSB) and NSG 7314 (DSW) and by NSF grant EAR 8121381 (DSB).

Literature Cited

Ahlen, S. P., Price, B., Tarle, G. 1981. Track-recording solids. *Phys. Today* 34:32–39

Alaerts, L., Lewis, R. S., Anders, E. 1979. Isotopic anomalies of noble gases in meteorites and their origins—III. LL—chondrites. *Geochim. Cosmochim. Acta* 43:1399–415

Benjamin, T. M. 1980. *Experimental actinide element partitioning between whitlockite, apatite, diopsidic clinopyroxene, and anhydrous melt at one bar and 20 kilobars pressure*. PhD thesis. Calif. Inst. Technol., Pasadena

Benjamin, T. M., Arndt, N. T., Holloway, J. R. 1977. Instrumental technique for beta-track mapping. *Carnegie Inst. Washington Yearb.* 76:658–60

Benjamin, T. M., Heuser, W. R., Burnett, D. S. 1978. Laboratory studies of actinide partioning relevant to ^{244}Pu chronometry. *Proc. Lunar Planet. Sci. Conf., 9th*, pp. 1393–406

Benjamin, T. M., Heuser, W. R., Burnett, D. S., Seitz, M. G. 1980. Actinide crystal-liquid partitioning for clinopyroxene and $Ca_3(PO_4)_2$. *Geochim. Cosmochim. Acta* 44:1251–64

Blank, H., Nobiling, R., Traxel, K., El Goresy, A. 1981. Partitioning of trace elements among coexisting opaque oxides in lunar basalts using a proton probe microanalyzer. *Meteoritics* 16:294–95

Bosch, F., El Goresy, A., Krätschmer, W., Martin, B., Povh, B., Nabiling, R., Traxel,

K., Schwalm, D. 1977. On the possible existence of superheavy elements in monazite. *Z. Phys. A* 280:39–44

Bosch, F., El Goresy, A., Martin, B., Povh, B., Nobiling, R., Schwalm, D., Traxel, K. 1978. The proton microprobe: a powerful tool for nondestructive trace element analysis. *Science* 199:765–68

Bosch, F., El Goresy, A., Herth, W., Martin, B., Nobiling, R., Povh, B., Reiss, H. D., Traxel, K. 1980. The Heidelberg proton microprobe. *Nucl. Sci. Appl.* 1:1–39

Boynton, W. V. 1978. Fractionation in the solar nebula. II. Condensation of Th, U, Pu, and Cm. *Earth Planet. Sci. Lett.* 40:63–70

Burnett, D., Monnin, M., Seitz, M., Walker, R., Yuhas, D. 1971. Lunar astrology—U-Th distributions and fission-track dating of lunar samples. *Proc. Lunar Sci. Conf., 2nd*, pp. 1503–19

Burnett, D. S., Stapanian, M. I., Jones, J. H. 1982. Meteorite actinide chemistry and cosmochronology. In *Essays in Nuclear Astrophysics*, ed. D. Clayton, D. Schramm, pp. 141–58. Cambridge Univ. Press

Cahill, T. A. 1980. Proton microprobes and particle-induced X-ray analytical systems. *Ann. Rev. Nucl. Part. Sci.* 30:211–52

Cahill, T. A., Fletcher, N. R., Medsker, L. R., Nelson, J. W., Kaufmann, H., Flocchini, R. G. 1978. Proton-induced X-ray analysis of monozite inclusions possessing pleochroic halos. *Phys. Rev. C* 17:1183–95

Cantelaube, Y., Maurette, M., Pellas, P. 1967. Traces d'ions lourds dons les mineroux de la chondrite de St. Severin. In *Radioactive Dating and Methods of Low Level Counting*, pp. 215–29. Vienna: IAEA

Cassou, R. M., Benton, E. V. 1978. Properties and applications of Cr-39 polymeric nuclear track detector. *Nucl. Track Detect.* 2:169–71

Chen, J. H., Wasserburg, G. J. 1981. Precise isotopic analysis of uranium in picomole and subpicomole quantities. *Anal. Chem.* 53:2060

Clark, G. J., Gulson, B. L., Cookson, J. A. 1979. Pb, U, Tl, Hf and Zr distributions in zircons determined by proton microprobe and fission track techniques. *Geochim. Cosmochim. Acta* 43:905–18

Cookson, J. A., Fletcher, N. R., Kemper, K. W., Medsker, L. R., Cahill, T. A. 1978. Analysis of a giant halo monazite inclusion on the Harwell proton microprobe. *Proc. Int. Symp. Superheavy Elements*, ed. M. A. K. Lodhi, pp. 164–69. New York: Pergamon

Cooper, J. A. 1973. Comparison of particle and photon excited X-ray flourescence applied to trace element measurements of environmental samples. *Nucl. Instrum. Methods* 106:525–38

Crozaz, G. 1974. U, Th and extinct ^{244}Pu in the phosphates of the St. Severin meteorite. *Earth Planet. Sci. Lett.* 23:164–69

Crozaz, G. 1979. Uranium and thorium microdistributions in stony meteorites. *Geochim. Cosmochim. Acta* 43:127–36

Crozaz, G., Haack, U., Hair, M., Maurette, M., Walker, R., Woolum, D. 1970. Nuclear track studies of ancient solar radiations and dynamic lunar surface processes. *Proc. Apollo 11 Lunar Sci. Conf.*, pp. 2051–80

Dowty, E. 1977. Phosphate in Angra dos Reis: structure and composition of the $Ca_3(PO_4)_2$ minerals. *Earth Planet. Sci. Lett.* 35:347–51

Drake, M. J., Holloway, J. R. 1978. 'Henry's Law' behaviour of Sm in a natural plagioclase/melt system. *Geochim. Cosmochim. Acta* 42:679–84

Drake, M. J., Holloway, J. R. 1981. Partitioning of Ni between olivine and silicate melt: the 'Henry's Law problem' reexamined. *Geochim. Cosmochim. Acta* 45:431–37

Drake, M. J., Holloway, J. R. 1982. Partitioning of Ni between olivine and silicate melt: the Henry's Law problem reexamined: reply to discussion by B. Mysen. *Geochim. Cosmochim. Acta* 46:299

Evans, R. D. 1955. *The Atomic Nucleus*. New York/Toronto/London: McGraw-Hill. 972 pp.

Fleischer, R. L. 1979. Where do particle tracks lead? *Am. Sci.* 67:194–203

Fleischer, R. L., Price, P. B., Walker, R. M. 1975. *Nuclear Tracks in Solids: Principles and Applications*. Berkeley: Univ. Calif. Press. 605 pp.

Folkmann, F., Gaarde, C., Huus, T., Kemp, K. 1974. Photon-induced X-ray emission as a tool for trace element analysis. *Nucl. Instrum. Methods* 116:487–99

Fowler, W. A. 1977. Nuclear cosmochronology. *Proc. Robert A. Welch Conf. Chem. Res., 21st, Houston, Cosmochem.*, pp. 61–133

Furst, M. J. 1979. *The use of boron concentrations in fossil materials as a paleosalinity indicator*. PhD thesis. Calif. Inst. Technol., Pasadena. 187 pp.

Furst, M. J. 1981. Boron in siliceous materials as a paleosalinity indicator. *Geochim. Cosmochim. Acta* 45:1–13

Furst, M., Lowenstam, H. A., Burnett, D. S. 1976. Radiographic study of the distribution of boron in recent mollusk shells. *Geochim. Cosmochim. Acta* 40:1381–86

Furst, M. J., Stapanian, M. I., Burnett, D. S. 1982. Observation of non-lithophile behavior for U. *Geophys. Res. Lett.* 9:41–44

Gentry, R. V., Cahill, T. A., Fletcher, N. R., Kaufmann, H. C., Medsker, L. R., Nelson,

J. W., Flocchini, R. G. 1976. Evidence for primordial superheavy elements. *Phys. Rev. Lett.* 37:11–15

Grossman, L. 1980. Refractory inclusions in the Allende meteorite. *Ann. Rev. Earth Planet. Sci.* 8:559–608

Grossman, L., Larimer, J. W. 1974. Early chemical history of the solar system. *Revs. Geophys. Space Phys.* 12:71–101

Grutzeck, M., Kridelbaugh, S., Weill, D. 1974. The distribution of Sr and REE between diopside and silicate liquid. *Geophys. Res. Lett.* 1:273–75

Haines, E., Weiss, J. R., Burnett, D. S., Woolum, D. S. 1976. Th-U fission radiography. *Nucl. Instrum. Methods* 135:125–31

Hamilton, E. I. 1971. Pb analysis by alpha particle tracks. *Nature* 231:524–25

Harrison, W. J., Wood, B. J. 1980. An experimental investigation of the partitioning of REE between garnet and liquid with reference to the role of defect equilibria. *Contrib. Mineral. Petrol.* 72:145–55

Hart, S. R., Davis, K. E. 1978. Nickel partitioning between olivine and silicate melt. *Earth Planet. Sci. Lett.* 40:203–19

Holloway, J. R., Drake, M. J. 1977. Quantitative microautoradiography by X-ray emission microanalysis. *Geochim. Cosmochim. Acta* 41:1395–97

Horowitz, P., Grodzins, L. 1975. Scanning, proton-induced X-ray microspectrometry in an atmospheric environment. *Science* 189:795–97

Hudson, B., Hohenberg, C. M., Kennedy, B. M., Podosek, F. A. 1982. ^{244}Pu in the early solar system. *Lunar Planet. Sci. XIII*, pp. 346–47.

Hughes, D. J. 1953. *Pile Neutron Research.* Cambridge, Mass: Addison-Wesley

Hutcheon, I. D., Phakey, P. P., Price, P. B. 1972. Studies bearing on the history of lunar breccias. *Proc. Lunar Sci. Conf., 3rd*, pp. 2845–65

Johansson, S. A. E., ed. 1977. Proc. Int. Conf. Part. Induced X-Ray Emission. Its Anal. Appl. *Nucl. Instrum. Methods* 142:1–338

Johansson, S. A. E., Johansson, T. B. 1976. Analytical application of particle induced X-ray emission. *Nucl. Instrum. Methods* 137:473–516

Jones, J. H. 1981. *Studies of the geochemical similarity of plutonium and samarium and their implications for the abundance of ^{244}Pu in the early solar system.* PhD thesis. Calif. Inst. Technol., Pasadena. 197 pp.

Jones, J. H. 1982. The geochemical coherence of Pu and Nd and the ^{244}Pu/^{238}U ratio of the early solar system. *Geochim. Cosmochim. Acta* 46:1793–1804

Jones, J. H., Burnett, D. S. 1979. The distribution of U and Pu in the St. Severin chondrite. *Geochim. Cosmochim. Acta* 43:1895–1905

Jones, J. H., Burnett, D. S. 1981. Quantitative radiography using Ag X-rays. *Nucl. Instrum. Methods* 180:625–33

Kennedy, B. M. 1981. *Potassium-argon and iodine-xenon gas retention ages of enstatite chondrite meteorites.* PhD thesis. Washington Univ., St. Louis. 298 pp.

Kirsten, T., Jordan, J., Richter, H., Pellas, P., Storzer, D. 1978. Plutonium and uranium distribution patterns in phosphates from ten ordinary chondrites. *Proc. Int. Conf. Geochron. Cosmochron. Isotope Geol., 4th*, ed. R. Zartman, pp. 215–19. *US Geol. Surv. Rep. 78-701*

Kleeman, J. D., Lovering, J. F. 1967. Uranium distribution in rocks by fission-track registration in lexan plastic. *Science* 156:512–13

Koyama-Ito, H., Grodzins, L. 1980. Computer calculations of solenoidal focussing of MeV proton beams to one micrometer diameter. *Nucl. Instrum. Methods* 174:331–40

Kunz, C., ed. 1979. *Synchrotron Radiation Techniques and Applications.* Berlin: Springer-Verlag. 442 pp.

Larimer, J. W. 1973. Chemical fractionations in meteorites—VII. Cosmothermometry and cosmobarometry. *Geochim. Cosmochim. Acta* 37:1603–23

Lewis, R. S. 1975. Rare gases in separated whitlockite from the St. Severin chondrite: Xe and Kr from fission of extinct ^{249}Pu. *Geochim. Cosmochim. Acta* 39:417–32

Lindstrom, D. J., Weill, D. F. 1978. Partitioning of transition metals between diopside and coexisting silicate liquids—I. Nickel, cobalt, and manganese. *Geochim. Cosmochim. Acta* 42:817–32

Lovering, J. F., Hinthorne, J. R., Conrad, R. L. 1976. Direct ^{207}Pb/^{206}Pb dating by ion microprobe of uranium-thorium rich phases in Allende calcium-aluminum rich clasts (CARC's). In *Lunar Sci. VII*, pp. 504–6 (Abstr.)

Lovering, J. F., Wark, D. A., Sewell, D. K. B. 1979. Refractory oxide, titanate, niobate and silicate accessory mineralogy of some Type B Ca-Al-rich inclusions in the Allende meteorite. In *Lunar Planet. Sci. X*, pp. 745–47 (Abstr.)

Maggiore, C. J. 1982. The Los Alamos nuclear microprobe with a superconducting solenoid final lens. *Nucl. Instrum. Methods* 191:199–203

Marti, K., Lugmair, G. W., Scheinin, N. B. 1977. Sm-Nd-Pu systematics in the early solar system. In *Lunar Planet. Sci. VIII*, pp. 619–21

Mason, B., Martin, P. M. 1974. Minor and trace element distribution in melilite and pyroxene from the Allende meteorite. *Earth Planet. Sci. Lett.* 22:141–44

Matsui, Y., Onuma, N., Nagasawa, H., Higuchi, H., Banno, S. 1977. Crystal structure control in trace element partition between crystal and magma. *Bull. Soc. Fr. Mineral. Cristallogr.* 100:315–24

Morgan, J. W., Lovering, J. F. 1968. U and Th abundances in chondritic meteorites. *Talenta* 15:1079–95

Murrell, M. T., Burnett, D. S. 1982a. Actinide microdistributions in the enstatite meteorites. *Geochim. Cosmochim. Acta.* 46:2453–60

Murrell, M. T., Burnett, D. S. 1982b. Thorium-uranium fission radiography II. *Nucl. Instrum. Methods* 199:617–21

Murrell, M. T., Burnett, D. S. 1982c. Actinide chemistry in ordinary chondrites. *Meteorit. Soc. Meet., 45th, St. Louis* (Abstr.)

Mysen, B. O. 1978. Limits of solution of trace elements in minerals according to Henry's Law: Review of experimental data. *Geochim. Cosmochim. Acta* 42:871–86

Mysen, B. O. 1979. Nickel partitioning between olivine and silicate melt: Henry's Law revisited. *Am. Mineral.* 64:1107–14

Mysen, B. O. 1981. Nickel partitioning between olivine and liquid in natural basalt: Henry's Law behavior—comment on a paper by P. I. Nabelek. *Earth Planet. Sci. Lett.* 52:222–24

Mysen, B. O. 1982. Partitioning of Ni between olivine and silicate melt: the "Henry's law problem" reexamined: discussion. *Geochim. Cosmochim. Acta* 46:297–98

Mysen, B. O., Seitz, M. G. 1975. Trace element partitioning determined by beta track mapping: An experimental study using carbon and samarium as examples. *J. Geophys. Res.* 80:2627–35

Mysen, B. O., Virgo, D. 1980. Trace element partitioning and melt structure: an experimental study at 1 atm pressure. *Geochim. Cosmochim. Acta* 44:1917–30

Nabelek, P. I. 1980. Nickel partitioning between olivine and liquid in natural basalts: Henry's Law behavior. *Earth Planet. Sci. Lett.* 48:293–302

Nabelek, P. I. 1981. Nickel partitioning between olivine and liquid in natural basalts: Henry's Law behavior—Reply to B. O. Mysen. *Earth Planet. Sci. Lett.* 52:225–26

Nagasawa, H., Blanchard, D. P., Jacobs, J. W., Brannon, J. C., Philpotts, J. A., Onuma, N. 1977. Trace element distributions in mineral separates of the Allende inclusions and their genetic implications. *Geochim. Cosmochim. Acta* 41:1587–1600

Navrotsky, A. 1978. Thermodynamics of element partitioning: (1) Systematics of transition metals in crystalline and molten silicates and (2) Defect chemistry and "the Henry's Law problem." *Geochim. Cosmochim. Acta* 42:887–902

Northcliffe, L. C., Schilling, R. F. 1970. Range and stopping-power tables for heavy ions. *Nucl. Data Tables* A7(3–4):233–463

Pellas, P., Storzer, D. 1975. Uranium and plutonium in chondritic phosphates. *Meteoritics* 10:471–72

Pellas, P., Storzer, D. 1981. ^{244}Pu fission track thermometry and its application to stony meteorites. *Proc. R. Soc. London Ser. A* 374:253–70

Pellas, P., Storzer, D., Kirsten, T., Jordan, J., Richter, H. 1979. 244-Pu/U-238 ratios in whitlockites of ordinary chondrites: A possible chronological tool. In *Lunar Planet. Sci. X*, pp. 969–71

Pilione, L. J., Carpenter, B. S. 1981. Nuclear track determination of Li and B in various matrices. *Nucl. Instrum. Methods* 188:639–46

Podosek, F. A. 1970. Dating of meteorites by the high-temperature release of iodine correlated ^{129}Xe. *Geochim. Cosmochim. Acta* 34:341–65

Poole, D. M., Shaw, J. L. 1969. In *Cong. X-Ray Optics Microanal.*, ed. J. Mollenstedt, K. H. Gaukler, p. 319. Berlin: Springer-Verlag

Rogers, A. W. 1979. *Techniques of Autoradiography*. Amsterdam: Elsevier/North-Holland Biomedical Press. 429 pp.

Shimizu, N., Hart, S. R. 1982. Applications of the ion microprobe to geochemistry and cosmochemistry. *Ann. Rev. Earth Planet. Sci.* 10:483–526

Shirck, J. R. 1975. *Fission track studies of Pu-244 in white inclusions of the Allende meteorite*. PhD thesis. Washington Univ., St. Louis. 175 pp.

Shroy, R. E., Kraner, H. W., Jones, K. W. 1978. Proton microprobe with windowless-exit port. *Nucl. Instrum. Methods* 157:163–68

Sparks, C. J. Jr. 1980. X-ray flourescence microprobe for chemical analysis. In *Synchrotron Radiation Research*, ed. H. Winick, S. Doniach, 14:459–512. New York: Plenum

Sparks, C. J. Jr., Raman, S., Yaken, H. L., Gentry, R. V., Krause, M. O. 1977. Search with synchrotron radiation for superheavy elements in giant-halo inclusions. *Phys. Rev. Lett.* 38:205–8

Sparks, C. J. Jr., Raman, S., Ricci, E., Gentry, R. V., Krause, M. O. 1978. Evidence against superheavy elements in giant-halo inclusions re-examined with synchrotron radiation. *Phys. Rev. Lett.* 40:507–11

Stapanian, M. I. 1981. *Induced fission track measurements of carbonaceous chondrite*

Th/U ratios and Th/U microdistributions in Allende inclusions. PhD thesis. Calif. Inst. Technol., Pasadena. 282 pp.

Steele, I. M., Lindstrom, D. J. 1981. Ni partitioning between diopside and silicate melt. *Geochim. Cosmochim. Acta* 45:2177–83

Wakita, H., Schmitt, R. A., Rey, P. 1970. Elemental abundances of major, minor and trace elements in Apollo 11 rocks, soil and core samples. *Proc. Apollo 11 Lunar Sci. Conf.*, pp. 1685–717

Wark, D. A., Lovering, J. F. 1977. Marker events in the early evolution of the solar system: Evidence from rims on Ca-Al-rich inclusions in carbonaceous chondrites. *Proc. Lunar Sci. Conf., 8th*, pp. 95–112

Wark, D. A., Lovering, J. F. 1978. Classification of Allende coarse-grained Ca-Al-rich inclusions. *Lunar Planet. Sci. IX*, pp. 1211–13 (Abstr.)

Wark, D. A., Lovering, J. F. 1980a. More early solary system stratigraphy: Coarse-grained CAI's. *Lunar Planet. Sci. XI*, pp. 1208–10 (Abstr.)

Wark, D. A., Lovering, J. F. 1980b. Second thoughts about rims. *Lunar Planet. Sci. XI*, pp. 1211–13 (Abstr.)

Wark, D. A., Hughes, T. C., Lucas, M., McKenzie, C. D. 1980. Proton microprobe analysis of perovskite in Allende Ca-Al-rich inclusions. *Lunar Planet. Sci. XI*, pp. 1205–7

Wasserburg, G. J., Huneke, J. C., Burnett, D. S. 1969. Correlation between fission tracks and fission-type xenon in meteoritic whitlockite. *J. Geophys. Res.* 74:4221–32

Wasson, J. T. 1974. *Meteorites: Classification and Properties.* New York/Berlin/Heidelberg: Springer-Verlag. 316 pp.

Weller, M. R., Furst, M., Tombrello, T., Burnett, D. S. 1978. Boron concentrations in carbonaceous chondrites. *Geochim. Cosmochim. Acta* 42:999–1009

Winick, H. 1980. Properties of synchrotron radiation. In *Synchrotron Radiation Research*, ed. H. Winick, S. Doniach, 2:11–25. New York: Plenum

Winick, H., Doniach, S., eds. 1980. *Synchrotron Radiation Research.* New York: Plenum. 754 pp.

Woolum, D., Boisseau, P. 1981. Proton-probe analyses of planetary materials. *Lunar Planet. Sci. XII*, pp. 1206–8

Woolum, D. S., Burnett, D. S. 1981. Metal and Pb/Bi microdistribution studies of an L3 chondrite: Their implications for a meteorite parent body. *Geochim. Cosmochim. Acta* 45:1619–32

Woolum, D. S., Burnett, D. S., August, L. S. 1976. Lead-bismuth radiography. *Nucl. Instrum. Methods* 138:655–62

Woolum, D. S., Bies-Horn, L., Burnett, D. S. August, L. S. 1979. Bismuth and ^{208}Pb microdistributions in enstatite chondrites. *Geochim. Cosmochim. Acta* 43:1819–28

Woolum, D. S., Burnett, D. S., Murrell, M. T., Benjamin, T., Maggiore, C. J., Rogers, P. 1982. Trace elements in Peña Blanca Springs Oldhamite. *Meteoretical Soc. Meet., 45th, St. Louis* (Abstr.)

Younger, P. A., Cookson, J. A. 1979. A secondary electron imaging system for a nuclear microprobe. *Nucl. Instrum. Methods* 158:193–98

Ann. Rev. Earth Planet. Sci. 1983. 11: 359–69

SEISMIC GAPS IN SPACE AND TIME[1]

Karen C. McNally
Charles F. Richter Seismological Laboratory, Earth Sciences Board, University of California, Santa Cruz, California 95064

INTRODUCTION

Large, shallow earthquakes along plate boundaries account for the largest release of seismic energy worldwide. Studies of seismic gaps provide valuable insight into this process and are also the cornerstone of earthquake prediction research. Not only have the locations and sizes of earthquakes been successfully identified prior to their occurrence, but these same locations have also provided a focus for fundamental research regarding the earthquake process. In addition, the identification of seismic gaps provides an opportunity for public planning to mitigate future disasters caused by earthquakes.

The concept of seismic gap originated with empirical observations in Japan in the 1920s, and was later provided with a physical basis by the theory of plate tectonics in the 1960s. Today, studies of seismic gaps involve detailed analyses of the sizes and source rupture processes of earthquakes, as well as the characteristics of fault zones (e.g. fault geometry, inferred heterogeneities, seismic vs aseismic slip along fault zones) and detailed geologic studies documenting the repeat times of fault rupture in recent geologic history.

One of the simplest and clearest definitions of a seismic gap is given by Aki (1978):

> ... the aftershock areas of large earthquakes in the past tens to hundreds of years tend to fill up a seismic zone without significant overlap. In the context of plate tectonics, this implies that the relative motion of plates is accomplished by a series of ruptures along the plate margin, each associated with a major earthquake. Then, it is natural to anticipate the next earthquake in a gap of aftershock zones unfilled for more than the expected recurrence time of major earthquakes for the seismic zone. Thus, the gap is important for a long range prediction of major earthquakes in the plate margin ...

[1] Portions of this article reprinted by permission of the Seismological Society of America.

0084–6597/83/0515–0359$02.00

This review discusses three main aspects of Aki's apparently simple definition. The question of (*a*) the physical nature of gaps (what, where, and how big?) follows a historical outline of the development of the concept. Then, (*b*) the relationship between seismic gaps and plate tectonics is reviewed, followed by (*c*) a discussion of work to discern when gaps break. The references cited are not intended as a complete list of all works related to this rather broad topic. However, by referring to works cited in the present overview, the reader can obtain a detailed listing of references for any particular aspect. I apologize to those authors whose work is not mentioned specifically.

HISTORY

Apparently the earliest effort to use a concept like that of seismic gap to forecast a future large earthquake was by A. Imamura prior to the devastating Tokyo earthquake of 1923, which caused 140,000 deaths, largely due to fire (Aki 1980). Unfortunately, as a result of both fears of causing social unrest and a lack of confidence in the scientific basis of his forecast, other scientists did not endorse Imamura's warning to take preventative measures against this disaster. [Aki (1980) provides an interesting perspective of this incident in his Presidential Address to the Seismological Society of America.]

Kanamori (1981) discusses the work of Imamura (1928), who suggested that the seismic potential at that time (1928) was relatively high at a particular location in the Tokaido-Nankaido region in southwest Japan. This conclusion was based on the observation that large earthquakes repeated in nearly the same location and that, as a consequence, average "repeat-times" could be determined for earthquakes on discrete fault segments. Imamura obtained a repeat-time value of 100 to 150 years for the area southeast of the Kii Peninsula and Shikoku Island, and he pointed out that 70 years had elapsed with no significant earthquake in this region. On this basis, he suggested that a large earthquake should be expected. Such an earthquake did in fact occur in 1944 (Tonankai, $M_S = 8.0$) and 1946 (Nankaido, $M_S = 8.2$) (Kanamori 1981).

Fedotov (1965) conducted a major study of great ($M_S \geq 7.75$), shallow earthquakes in the Kamchatka, Kurile, and Japan regions. He noted that these earthquakes ruptured discrete fault segments, defined by aftershock zones, and he identified several locations where a long time had passed without major rupture. On this basis, he suggested several areas as probable locations of future large earthquakes. Since 1965, several major earthquakes have occurred in these zones, thus "filling" the seismic gaps. This work has been described recently by Kanamori (1981), Kasahara (1981), Kelleher et al (1973), and McCann et al (1979).

Since these initial studies, the seismic gap concept has been widely applied to a number of areas worldwide, including studies by Mogi (1968), Utsu (1970), Sykes (1971), Kelleher (1972), and Kelleher et al (1973). McCann et al (1979) provided an extensive review of previous work, as well as an updated categorization of all seismic gaps worldwide. Successful long-term forecasts of eight major earthquakes using the seismic gap method are reported.

In two notable cases, a seismic-gap-type concept has been accepted for long-term forecasts of major earthquakes in the future. As a consequence, two large population centers—the Tokai region (located between Tokyo and Nagoya in central Japan) and Los Angeles, California—are taking preventive measures against disaster. A comprehensive discussion of the earthquake prediction program in Japan is available in English by Mogi (1981b). On the basis of the seismic gap concept, the Tokai region has been indicated as a potential region for a great, shallow earthquake of magnitude ~ 8. Numerous measurements of the region are telemetered to the center of the Japan Meteorological Agency (JMA) in Tokyo and are monitored continuously. When anomalous behavior is observed, the Earthquake Prediction Council makes an evaluation that is reported to the Director General of JMA, who reports to the Prime Minister. Extensive public preparations at many levels are ongoing.

In the Los Angeles area, the public has begun efforts to withstand a repeat of the great 1857 earthquake along the southern San Andreas Fault. The most comprehensive data regarding the recurrence times of great earthquakes along the San Andreas are the result of geological studies by Sieh (1978). McCann et al (1979) assign this location to category 1, which represents a gap of the highest seismic potential. In contrast to Japan, the earthquake prediction program of the US Geological Survey is more oriented toward learning how to predict earthquakes, although extensive monitoring efforts are also being conducted in the Los Angeles area (see, for example, Wesson & Filson 1981).

Numerous studies to refine the understanding of the nature of seismic gaps are underway, and are testimony to the general success of this concept in describing the earthquake process. Several of these studies are described below.

THE PHYSICAL NATURE OF SEISMIC GAPS: WHAT, WHERE, AND HOW BIG?

It is reasonable to expect that structural features of fault zones might be related to the sizes and locations of aftershock zones, if it is true that motion along a fault zone is accommodated by repeated breaks of discrete, nonoverlapping fault segments. The sizes of large earthquakes, and hence the sizes of seismic gaps, were examined by Kelleher et al (1974), who studied variations in interface geometry along five major island arcs and found that great earthquakes with very long rupture lengths (greater than 400 km) occur where gently dipping slabs of lithosphere have a broad contact zone against the overthrust slabs (1957 Aleutian, 1964 Alaskan, 1960 Chilean, 1952 Kamchatka, and 1965 Rat Island earthquakes). In contrast, moderately large earthquakes (maximum rupture length less than 150 km) tend to occur at locations along the Japan-Kurile-Kamchatka, New Hebrides, and Middle America seismic zones, where the downgoing slabs dip more steeply, with a narrow zone of contact with the overriding lithosphere. In earlier work, Isacks et al (1968) suggested that the upper limit to the sizes of earthquakes might be controlled by the area of lithospheric contact. Numerous studies, including those by Kanamori (1971, 1977a), Uyeda & Kanamori (1979), and Ruff & Kanamori (1980), have compared the features of subduction zones worldwide (such as the variation in coupling between plates, the convergence rates, the age of the downgoing plate, the maximum depth to which the continuous Benioff zone penetrates, and back-arc spreading) with the sizes of the largest characteristic earthquakes in order to correlate the physical behavior of each. Based on results of the earlier works relating the coupling between downgoing and upper plates to the characteristic earthquake size ($7.0 \leq M_W \leq 9.5$), Ruff & Kanamori (1980) have recently correlated the degree of coupling with convergence rate and lithospheric age. They find that the largest earthquakes are found where the oceanic lithosphere is younger and convergence rates are faster; smaller earthquakes are found where the lithosphere is older and convergence rates are slower. (The results for Peru and Central America are exceptions to this trend, however.) These general results are useful for estimating the characteristic dimensions of the largest seismic gaps in each area, and are also of practical value for public planning. However, a common problem is that the known seismic history is quite short relative to the repeat times for the largest earthquakes.

A different problem for understanding the physical nature of seismic gaps is why, in general, do aftershock zones abut along discrete segments, rather than overlap? Carr & Stoiber (1977) suggested that geological and geophysical data indicate breaks transverse to the trench that occur in the underthrust slab and in the overriding lithosphere and limit the rupture lengths of great earthquakes in Central America. Mogi (1969) concluded that transverse structural features, such as faults, trenches, and ridges, separated adjacent ruptures in Japan. Kelleher (1970) and Sykes (1971) made similar observations for the Aleutians. Aki (1978, 1979) studied the initiation and stopping of earthquake rupture at boundaries between neighboring rupture zones. These boundaries, or barriers, could be caused by fault geometry, such as bends, steps, or corners (see also Segall & Pollard 1980, Bakun et al 1980, Mavko 1982), or by crustal inhomogeneities inferred from velocity anomalies. In Aki's model, these barriers act as stress concentrators and delimit the earthquake rupture zones. Aki's studies emphasized the rupture during the 1966 ($M_S = 6.5$) earthquake along the San Andreas Fault near Parkfield, California. Although this event is smaller than those discussed previously, the physical models are probably applicable to larger earthquakes in the consideration of nonuniform distribution of strength over a fault plane. Lindh & Boore (1981) also studied this event and concluded that rupture began at a fault bend (which appears to continue at depth) and stopped at an en-echelon offset of the fault. Between these barriers, they envision a "stuck patch" on the fault (Wesson et al 1973). When strain that has been accumulating at the bend in the fault causes failure at that point, rupture progresses through the "stuck patch," but is stopped by the barrier produced by the en-echelon fault offset.

Kanamori (1977b) suggested that a fault zone could be represented by a random distribution of stress concentrations on asperities of different scales that might be due to geometry or heterogeneities of the frictional strength or both. Stress concentrates near the strongest asperity, which is at the hypocenter of the future mainshock. When this asperity breaks, the entire fault plane ruptures (see, for example, Lay & Kanamori 1980, Stewart et al 1981, Chael & Stewart 1982).

In some cases the failure of one large asperity could trigger the failure of another large asperity nearby (Lay & Kanamori 1980, 1981, Kanamori & McNally 1982). Such triggering would be influenced by various factors (e.g. the level of stress at the adjacent asperities, the existence of barriers segmenting the fault zone, etc). This situation is problematic for the use of seismic gaps in predicting earthquakes: at times, several asperities may rupture simultaneously and produce a very large event; at other times, the same asperities may break singly as smaller, separate events (Kanamori & McNally 1982).

It may not be the case, however, that rupture in the mainshock always occurs with the failure of the strongest asperity. Recent works by Ebel & Helmberger (1982) and Hartzell & Helmberger (1982) found that in one case fault rupture initiated with the failure of a strong asperity (1968 Borrego Mountain, California, earthquake, $M_L = 6.4$), but in the other case (1979 Imperial Valley, California, earthquake, $M_L = 6.6$) the strongest asperities were broken as rupture progressed, following the initial break. Both works cited supporting evidence for the existence of asperities in the geologic, seismic, and/or geodetic data.

A future direction for the study of seismic gaps would be to continue the investigation of fault rupture processes and to correlate these results with other data, such as patterns of small earthquakes and geological, geophysical, and geodetic observations. With such work, the knowledge of the physical nature of seismic gaps might be refined toward greater applicability for earthquake prediction. An obvious goal is to identify fault asperities, "stuck patches," and barriers and to observe associated behavior at these locations before failure in major earthquakes. A useful approach to studying the effects of both the asperity and barrier models has been developed by Rudnicki & Kanamori (1981).

PLATE TECTONICS AND SEISMIC GAPS

The plate tectonic model clearly provides the physical framework of strain buildup and release, which is fundamental to the general success of the seismic gap methods in anticipating future large earthquakes. Davies & Brune (1971) compared seismic slip rates with plate convergence rates and found reasonable agreement. However, more refined comparisons have been attempted as more data have become available. An important contribution is the calculation of present-day plate motions by Minster & Jordan (1978). These results provide the basis for examining the detailed mechanisms of subduction and earthquake recurrence along plate boundaries. Unfortunately, plate convergence rates do not predict when the next seismic gap will break. Sykes & Quittmeyer (1981) point out that large earthquakes in two regions where the rates of plate convergence are comparable, i.e. the Middle America Trench offshore Mexico and parts of the Aleutians, have average repeat times of about 35 and 60 years, respectively. Also, they note that even for a particular point along a given plate boundary, repeat times may vary by at least a factor of 1.5 to 2. The average rates of plate convergence for the last 3 million years are not necessarily applicable to the last 100 or so years of seismic data. In addition, the plate tectonic models assume that plates are rigid and that boundaries are continuous. The ratio of the seismic slip rate to the rate of plate motion varies among subduction zones (Kanamori 1977a, Sykes & Quittmeyer 1981), as well as from one earthquake source region to another along the same subduction zone (e.g. McNally & Minster 1981). It has been found that the seismic slip rate can match the plate rate locally, but elsewhere may be suppressed as a result of the subduction of seafloor topography, or may be accommodated wholly or in part by aseismic slip. As discussed above, Kanamori and others have attempted to quantify the relative amounts of aseismic slip for various subduction zones through a model of gradual decoupling between the downgoing and upper plates. This work suggests the important role of aseismic slip in the subduction process and in the evaluation of the seismic potential of seismic gaps. Sykes & Quittmeyer

(1981) have attempted to quantify the average component of inferred aseismic slip for various subduction zones worldwide. Conclusive assessments are difficult because of the relatively short history of earthquakes compared with geologic time, e.g. few data are available for earthquakes that repeat at 500–1000 year intervals. The best "earthquake catalogs" for these cases may come from geologic methods of dating fault disturbances. A complete review of this work has been provided by Sieh (1981).

SEISMIC GAPS AND GREAT EARTHQUAKES: WHEN?

McCann et al (1979) used the seismic gap technique to estimate the sizes and locations of future large ($M_S \geq 7$), shallow earthquakes, but the time of occurrence was only forecast to within tens of years. One should refer to these authors for a discussion of seismic gaps for specific regions. In their definition, seismic gaps are any regions along active plate boundaries that have not experienced a large thrust or strike-slip earthquake for more than 30 years. The category of highest seismic potential is assigned to regions having experienced at least one large earthquake in the historic past, with the most recent event occurring more than 100 years ago. While application of standard criteria is quite useful as a first approximation, this approach does not take into account the variations in rates of plate motion, strain accumulation, or characteristic repeat times of earthquakes along the various fault zones. McNally & Minster (1981) extended the work of Kelleher et al (1973) and McCann et al (1979) for large, shallow earthquakes along the Middle America Trench, where the repeat times of earthquakes average only 35 years. They differenced the earthquake moment release as a function of time and distance along the trench in order to identify locations where the seismic potential is currently high because of a deficiency in seismic slip.

The extent to which the time of the next earthquake can be described deterministically vs probabilistically is unknown because of the lack of a definitive theory of the physical process. A useful contribution is the work to develop statistical models of earthquake occurrence. Vere-Jones (1978) and Vere-Jones & Ozaki (1982) have studied data for historical strong earthquakes affecting the Kamakura, Japan, region (compiled by Kawasumi 1970) and have outlined a general method that includes specification of a model through a suitable form of conditional risk or intensity function, estimation of parameters through a likelihood maximization procedure, and application of likelihood ratio tests to evaluate the effect of increasing the complexity of the model (Vere-Jones 1978). Brillinger (1982) reviews a number of contemporary statistical methods that are relevant to estimating the probabilities that seismic events take place at a particular location during specified future time periods. His preliminary study of Sieh's (1978) geologic data on dates of earthquakes along the San Andreas Fault near Pallett Creek, California, over the last 1400 years indicates that a Weibull distribution with shape parameter 2 reasonably describes the distribution of times between successive events. (The Weibull distribution is a generalization of the exponential distribution that allows for a power dependence of the hazard on time.) A hazard function, which corresponds to hazard increasing linearly with time since the most recent events, is used to estimate that the probability of the next event within ten years is 0.073, with an estimated standard error of 0.025.

A model of the physical process of earthquakes repeating on the same fault segment has recently been suggested by Shimazaki & Nakata (1980) and has met with considerable interest in the United States and Japan (see, for example, Mogi 1981a, Sykes & Quittmeyer 1981). They presented three models of stress accumulation and release that might be used to infer the future slip. In the first ("strictly periodic") case, both the initial and final stresses are constant, so each earthquake duplicates its predecessor. In this case, one can predict at any time both the time and size of the next event. In the second ("time predictable") case, the rupture will always occur when the stress builds up to a constant breaking stress, but the final stress following rupture varies with time, so that the time, but not the size, can be predicted. Finally, in the third ("slip predictable") case, the stress level after the event is constant (and presumably quite small), but the initial stress varies with time, enabling the size of a hypothetical earthquake at some given time, but not the time itself, to be known. Results comparing the "time predictable" model with real data appear encouraging (e.g. Mogi 1981a).

FUTURE WORK

Several advances in the studies of seismic gaps are anticipated within the next 5 years. New analog and digital data from the World-Wide Standard Seismograph Network operated by the US Geological Survey (USGS) will facilitate studies of earthquake source parameters, resulting in an increased understanding of the physical nature of seismic gaps and the earthquake rupture process. In addition, a program by the USGS to archive and distribute copies of seismogram data for large historic earthquakes will allow re-analysis of old records using current techniques of waveform modeling for studies of the "repeatability" of earthquake rupture patterns.

A fundamental goal for earth scientists is to develop new and existing models of strain accumulation and failure to the point at which they can be definitively tested using available data. Near-source information on prior seismicity, mainshock and aftershock earthquakes using temporary seismographic and strong ground motion networks is needed to constrain the results from the global network data for testing such models. The near-source data can also be used to upgrade the earthquake locations and hence the resolution capability of the global network. This will allow better integration of results from seismological studies with other geophysical and geologic data, enhancing understanding of the physical process of earthquake generation.

It is now possible to determine the source parameters of earthquakes as small as $M_S = 5$ efficiently and reliably with the new global digital network data. This provides a comprehensive sampling of earthquake faulting mechanisms along plate boundaries, which can also be used to refine the determinations of present-day plate motions.

Finally, techniques for objectively describing seismicity data are urgently needed. For this, a new emphasis must be placed on statistical studies integrated with a knowledge of the physics of earthquakes. A definitive physical theory, as well as extensive or conclusive data that are directly relevant, is lacking. Despite these limitations, physical models of the earthquake process can be incorporated into statistical models for quantitative testing of earthquake catalogs. Geologic studies to find the dates of previous fault ruptures and careful reconstructions of historic reports of damage in large earthquakes are of fundamental importance to extend the earthquake occurrence history. Every effort should be made to uniformly report and determine source parameters of modern earthquakes.

SUMMARY

The seismic gap concept has proven very successful in forecasting the sizes and locations of large, shallow earthquakes. The new seismological, geological, and statistical studies described above should lead to important insights regarding the physical nature of seismic gaps and the earthquake process.

Acknowledgments

This work was performed under US Geological Survey Contract Number 14-08-0001-20546. I would also like to thank G. Beroza, E. Brown, and D. Harlow for critical review.

Literature Cited

Aki, K. 1978. Origin of the seismic gap: what initiates and stops a rupture propogation along a plate boundary? *Proc. Conf. VI, Methodology for Identifying Seismic Gaps and Soon-to-Break Gaps*, pp. 3–46. *US Geol. Surv. Open-File Rep. 78-943*

Aki, K. 1979. Characterization of barriers on an earthquake fault. *J. Geophys. Res.* 84:6140–48

Aki, K. 1980. Presidential address: possibilities of seismology in the 1980s. *Bull. Seismol. Soc. Am.* 70:1969–76

Bakun, W. H., Stewart, R. M., Bufe, C. G., Marks, S. M. 1980. Implication of seismicity for failure of a section of the San Andreas Fault. *Bull. Seismol. Soc. Am.* 70:185–201

Brillinger, D. R. 1982. Seismic risk assessment: some statistical aspects. *Earthquake Predict. Res.* 1:2

Carr, M. J., Stoiber, R. E. 1977. Geologic setting of some destructive earthquakes in Central America. *Geol. Soc. Am. Bull.* 88:151–56

Chael, E. P., Stewart, G. S. 1982. Recent large earthquakes along the Middle America Trench and their implications for the subduction process. *J. Geophys. Res.* 87:329–38

Davies, G., Brune, J. 1971. Regional and global fault slip rate from seismicity. *Nature Phys. Sci.* 229:101–7

Ebel, J. E., Helmberger, D. V. 1982. *P*-wave complexity and fault asperities: the Borrego Mountain, California, earthquake of 1968. *Bull. Seismol. Soc. Am.* 72:413–37

Fedotov, S. A. 1965. Regularities of the distribution of strong earthquakes in Kamchatka, the Kuril Islands, and northeast Japan. *Trudy Inst. Fiz Zemli Acad. Nauk SSSR* 36:66–93

Hartzell, S., Helmberger, D. V. 1982. Strong-motion modeling of the Imperial Valley earthquake of 1979. *Bull. Seismol. Soc. Am.* 72:571–96

Imamura, A. 1928. On the seismic activity of central Japan. *Jpn. J. Astron. Geophys.* 6:119–37

Isacks, B., Oliver, J., Sykes, L. R. 1968. Seismology and the new global tectonics. *J. Geophys. Res.* 73:5855–99

Kanamori, H. 1971. Great earthquakes at island arcs and the lithosphere. *Tectonophysics* 12:187–98

Kanamori, H. 1977a. Seismic and aseismic slip along subduction zones and their tectonic implications. In *Island Arcs, Deep Sea Trenches and Back-Arc Basins, Maurice Ewing Ser.*, ed. W. C. Pitman III, M. Talwani, 1:163–74. Washington DC: Am. Geophys. Union. 470 pp.

Kanamori, H. 1977b. Use of seismic radiation to infer source parameters. *Proc. Conf. III, Fault Mechanics and Its Relation to Earthquake Prediction*, pp. 283–317. *US Geol. Survey Open-File Rep. 78-380*

Kanamori, H. 1981. The nature of seismicity patterns before large earthquakes. In *Earthquake Prediction: An International Review, Maurice Ewing Ser.*, ed. P. G. Richards, D. W. Simpson, 4:1–19. Washington DC: Am. Geophys. Union. 680 pp.

Kanamori, H., McNally, K. C. 1982. Variable rupture mode of the subduction zone along the Ecuador-Columbia coast. *Bull. Seismol. Soc. Am.* 72:1241–53

Kasahara, K. 1981. *Earthquake Mechanics.* Cambridge/New York/Melbourne: Cambridge Univ. Press. 248 pp.

Kawasumi, H. 1970. Proofs of 69 years periodicity and imminence of destructive earthquake in Southern Kwanto district and problems in the countermeasures thereof. *Chigaku Zasshi* 79:115–37

Kelleher, J. A. 1970. Space-time seismicity of the Alaska-Aleutian seismic zone. *J. Geophys. Res.* 75:5745–56

Kelleher, J. A. 1972. Rupture zones of large South American earthquakes and some predictions. *J. Geophys. Res.* 77:2087–2103

Kelleher, J. A., Savino, J., Rowlett, H., McCann, W. R. 1974. Why and where great thrust earthquakes occur along island arcs. *J. Geophys. Res.* 79:4889–99

Kelleher, J. A., Sykes, L. R., Oliver, J. 1973. Possible criteria for predicting earthquake locations and their applications to major plate boundaries of the Pacific and Caribbean. *J. Geophys. Res.* 78:2547–85

Lay, T., Kanamori, H. 1980. Earthquake doublets in the Solomon Islands. *Phys. Earth Planet. Inter.* 21:238–304

Lay, T., Kanamori, H. 1981. An asperity model of large earthquake sequences. In *Earthquake Prediction: An International Review, Maurice Ewing Ser.*, ed. P. G. Richards, D. W. Simpson, 4:579–92. Washington DC: Am. Geophys. Union. 680 pp.

Lindh, A. D., Boore, D. M. 1981. Control of rupture by fault geometry during the 1966 Parkfield earthquake. *Bull. Seismol. Soc. Am.* 71:95–116

Mavko, G. M. 1982. Fault interaction near Hollister, California. *J. Geophys. Res.* 87:7807–16

McCann, W. R., Nishenko, S. P., Sykes, L. R., Krause, J. 1979. Seismic gaps and plate tectonics: seismic potential for major boundaries. *Pageoph* 117:1083–1147

McNally, K. C., Minster, J. B. 1981.

Nonuniform seismic slip rates along the Middle America Trench. *J. Geophys. Res.* 86:4949–59

Minster, J. B., Jordan, T. H. 1978. Present-day plate motions. *J. Geophys. Res.* 83:5331–34

Mogi, K. 1968. Sequential occurrences of recent great earthquakes. *J. Phys. Earth* 16:30–36

Mogi, K. 1969. Some features of recent seismic activity in and near Japan (2). *Bull. Earthquake Res. Inst. Univ. Tokyo* 47:395–417

Mogi, K. 1981a. Seismicity in western Japan and long-term earthquake forecasting. In *Earthquake Prediction: An International Review, Maurice Ewing Ser.*, ed. P. G. Richards, D. W. Simpson, 4:43–51. Washington DC: Am. Geophys. Union. 680 pp.

Mogi, K. 1981b. Earthquake prediction program in Japan. In *Earthquake Prediction: An International Review, Maurice Ewing Ser.*, ed. P. G. Richards, D. W. Simpson, 4:635–66. Washington DC: Am. Geophys. Union. 680 pp.

Rudnicki, J. W., Kanamori, H. 1981. Effects of fault interaction on moment, stress drop, and strain energy release. *J. Geophys. Res.* 86:1785–93

Ruff, L., Kanamori, H. 1980. Seismicity and the subduction process. *Phys. Earth Planet. Inter.* 23:240–52

Segall, P., Pollard, D. D. 1980. Mechanics of discontinuous faults. *J. Geophys. Res.* 85:4337–50

Shimazaki, K., Nakata, T. 1980. Time-predictable recurrence model for large earthquakes. *Geophys. Res. Lett.* 7:279–82

Sieh, K. E. 1978. Prehistoric large earthquakes produced by slip on the San Andreas Fault at Pallett Creek, California. *J. Geophys. Res.* 83:3907–39

Sieh, K. E. 1981. A review of geological evidence for recurrence times of large earthquakes. In *Earthquake Prediction: An International Review, Maurice Ewing Ser.*, ed. P. G. Richards, D. W. Simpson, 4:181–207. Washington DC: Am. Geophys. Union. 680 pp.

Stewart, G. S., Chael, E. P., McNally, K. C. 1981. The November 29, 1978, Oaxaca, Mexico, earthquake: a large simple event. *J. Geophys. Res.* 86:5053–60

Sykes, L. R. 1971. Aftershock zones of great earthquakes, seismicity gaps, and earthquake prediction for Alaska and the Aleutians. *J. Geophys. Res.* 79:8021

Sykes, L. R., Quittmeyer, R. C. 1981. Repeat times of great earthquakes along simple plate boundaries. In *Earthquake Prediction: An International Review, Maurice Ewing Ser.*, ed. P. G. Richards, D. W. Simpson, 4:217–47. Washington DC: Am. Geophys. Union. 680 pp.

Utsu, T. 1970. Seismic activity and seismic observation in Hokkaido in recent years. *Rep. Coord. Comm. Earthquake Predict.* 2:1–2 (In Japanese)

Uyeda, S., Kanamori, H. 1979. Back-arc opening and the mode of subduction. *J. Geophys. Res.* 84:1049–61

Vere-Jones, D. 1978. Earthquake prediction—a statistician's view. *J. Phys. Earth* 26:129–46

Vere-Jones, D., Ozaki, T. 1982. Some examples of statistical estimation applied to earthquake data. I. Cyclic Poisson and self-exciting models. *Ann. Inst. Stat. Math.* 34:189–207

Wesson, R. L., Burford, R. O., Ellsworth, W. L. 1973. Relationship between seismicity, fault creep and crustal loading along the central San Andreas Fault. *Proc. Conf. Tectonic Probl. San Andreas Fault Syst., Stanford Univ. Publ. Geol. Sci.* 13:303–21

Wesson, R. L., Filson, J. R. 1981. Development and strategy of the earthquake prediction program in the United States. In *Earthquake Prediction: An International Review, Maurice Ewing Ser.*, ed. P. G. Richards, D. W. Simpson, 4:671–80. Washington DC: Am. Geophys. Union. 680 pp.

Ann. Rev. Earth Planet. Sci. 1983. 11 : 371–414

TERRESTRIAL INERT GASES: Isotope Tracer Studies and Clues to Primordial Components in the Mantle

John E. Lupton
Marine Science Institute and Department of Geological Sciences, University of California, Santa Barbara, California 93106

INTRODUCTION

The discovery of primitive rare-gas isotopes in the interior of the Earth, namely ^{3}He and ^{129}Xe, has initiated an important set of inquiries into the noble gas content of the Earth's mantle, crust, and atmosphere, and the relationship between these terrestrial rare-gas components and those observed in extraterrestrial samples such as meteorites and the solar wind. These studies have been guided by ^{3}He, since excess ^{3}He has become the unambiguous trademark of primordial components from the Earth's mantle. For this reason, this paper focuses on helium isotope variations in terrestrial systems, with occasional excursions into the discussions of heavier rare-gas isotopes.

The primary purpose of this paper is twofold: first, to identify the basic components or end members that, when mixed together, fractionated, and perturbed, produce the rare-gas isotope variations that we observe in terrestrial systems; and second, to discuss a few specific applications of rare-gas isotopes to studies of ocean circulation, mid-ocean ridge tectonic and hydrothermal processes, and continental zones of geothermal activity. This article emphasizes *isotopic* signatures—rare-gas abundance patterns and Earth degassing models are not treated here. The complexities of heavy rare-gas isotopes are discussed only insofar as they help to identify basic terrestrial components, especially primordial mantle reservoirs. The reader is referred to review articles by Ozima & Alexander (1976), Bernatowicz & Podosek (1978), Tolstikhin (1978), Craig & Lupton (1981), Fisher (1981), and to two forthcoming books (Mamyrin & Tolstikhin 1983, Ozima & Podosek 1983) for additional discussions of terrestrial inert gases.

0084–6597/83/0515–0371$02.00

HELIUM ISOTOPE GEOCHEMISTRY

Basic Components

Helium isotope ratios vary by over four orders of magnitude in terrestrial samples: from $^3He/^4He \simeq 10^{-9}$ in uranium minerals (Shukolyukov 1970) up to $^3He/^4He \simeq 5 \times 10^{-5}$ in Hawaiian lavas (Kaneoka & Takaoka 1980). Although these are extreme examples, $^3He/^4He$ variations of two or even three orders of magnitude are frequently observed, even within a single geographic zone. With few exceptions, most terrestrial $^3He/^4He$ variations can be explained in terms of mixing between three major components or end members: atmospheric helium with $^3He/^4He = 1.4 \times 10^{-6}$, radiogenic helium with $^3He/^4He \simeq 10^{-7}$, and mantle helium with $^3He/^4He \simeq 10^{-5}$. We now elaborate on the definition of these major helium components.

ATMOSPHERIC HELIUM The Earth's atmosphere contains helium in a concentration of 5.24×10^{-6} by volume, with an isotopic ratio of $^3He/^4He = (1.39 \pm 0.01) \times 10^{-6}$, which is an average of the two detailed isotopic determinations by Mamyrin et al (1970) and Clarke et al (1976). The isotopic ratio of air helium is quite constant over the globe due to the rapid mixing time of the atmosphere (~ 10 yr) compared to the residence time for helium ($\sim 10^6$ yr). Because of its uniform isotopic composition, air helium is used in most laboratories as a convenient helium isotopic standard, and it has become the convention to express helium isotope variations relative to the $^3He/^4He$ ratio in air. In this paper, we use the units R/R_A, where $R = {}^3He/^4He$ and $R_A = ({}^3He/^4He)_{AIR}$.

RADIOGENIC HELIUM Helium in well gas and in typical continental rocks has $R/R_A \simeq 0.01$–0.1, significantly depleted in ^3He or enriched in ^4He relative to atmospheric helium. This helium, which is sometimes called "crustal" or "continental," is produced by the radioactive decay of U and Th series elements. Although radiogenic helium is most abundant in the continental crust, where U and Th are most concentrated, radiogenic helium production also occurs in oceanic basalts and presumably in the mantle itself. Thus the term "continental" is really a misnomer, since radiogenic helium is found in all reservoirs. The ^4He in radiogenic helium is produced directly by the alpha particle decays, while the ^3He is produced by the nuclear reaction $^6Li\,(n, \alpha)\,{}^3H\,(\beta^-)\,{}^3He$. Although the exact proportions of ^3He and ^4He generated depend on the relative concentrations of Li, U, and Th, it has been shown that typical crustal rocks produce helium with $R/R_A \simeq 0.01$–0.1 (Morrison & Pine 1955, Gerling et al 1971). As expected, much higher $^3He/^4He$ ratios (up to $R/R_A \simeq 9$) are found in lithium-rich minerals such as spodumene (Aldrich & Nier 1948), and somewhat lower ratios ($R/R_A \simeq 10^{-3}$) in uranium minerals (Shukolyukov 1970, Clarke & Kugler 1973). Tolstikhin (1978) has reviewed the subject of terrestrial radiogenic rare gases in some detail.

MANTLE HELIUM Until about 15 years ago, there was little evidence that significant quantities of primitive or "original" gases were trapped in the interior of the Earth. Since helium escapes from the atmosphere on a time scale of $\sim 10^6$ years, this implied that all present-day terrestrial helium is either radiogenic or cosmogenic (produced by cosmic-ray interactions). The first indication that this was incorrect came when Clarke et al (1969) reported slightly elevated $^3He/^4He$ ratios ($R/R_A \simeq 1.22$) in helium dissolved in deep Pacific Ocean water. This ^3He excess was attributed to the leakage of primordial ^3He into the oceans at active spreading centers such as the East Pacific Rise. This idea was supported by the work of Dymond & Hogan (1973) and Fisher (1974), who found nonatmospheric noble gas abundance patterns in oceanic basalt glasses, which they interpreted in terms of the abundance patterns of primordial rare gases in meteorites. The discovery by Krylov et al (1974) and Lupton & Craig (1975) of elevated $^3He/^4He$ ratios ($R/R_A \simeq 9$) in oceanic basalt glasses confirmed that an anomalous rare-gas component is trapped in the Earth's interior and that the excess ^3He in the oceans is due to the flux of this "primordial" helium from the mantle into the oceans and in turn into the Earth's atmosphere. At about the same time, Mamyrin et al (1969) and Kamenskii et al (1971) reported similar $^3He/^4He$ enrichments (up to $R/R_A = 10$) in volcanic gases. Subsequent studies have shown that essentially all fresh mid-ocean ridge basalts (MORB's) contain excess ^3He in their glassy margins, and that similar ^3He enrichments ($R/R_A = 5$–30) are found in many other tectonic environments, including subduction zones, hot spots and oceanic islands, and continental geothermal areas, which are presumably also sites of mantle helium leakage or degassing.

The excess ^3He in oceanic basalts is explained as a primordial component by default: no radiogenic or nucleogenic mechanism has been found that could produce such high $^3He/^4He$ ratios over broad regions of the Earth's mantle. Figure 1 compares ^3He, ^4He, and Ne in terrestrial components (air, crustal or radiogenic gases, and mantle-derived gases) with extraterrestrial components such as the solar wind (Geiss et al 1972) and primordial rare gases in meteorites (PRGM, Pepin & Signer 1965). MORB's have a surprisingly uniform helium isotopic composition ($R/R_A \simeq 9$) but exhibit large variations in $^3He/Ne$; hot spots, such as Hawaii, have even higher $^3He/^4He$ ratios presumably because they are tapping a deeper part of the mantle more enriched in ^3He (Craig & Lupton 1976). The addition of pure radiogenic helium ($R/R_A \simeq 0.01$, $^3He/Ne = \infty$) to the primordial gas component on the figure would cause the $^3He/^4He$ ratio to decrease without affecting the $^3He/Ne$ ratio. Thus hot spot and MORB-type helium could be made from the solar wind or PRGM component by dilution with radiogenic ^4He from the decay of U and Th in the mantle over geologic time. The variations in $^3He/Ne$ in the hot spot and MORB fields can be explained by varying amounts of dilution with atmospheric-type gases, possibly introduced to the mantle at subduction zones. Although this simple idea is difficult to confirm, no strong evidence has been presented to suggest that the ^3He-rich component in the Earth's interior, which we assume is a remnant from the time of the Earth's formation, is different from the primordial gases in meteorites. Arguments as to whether the Earth contains "solar" or "planetary"-type primordial rare gases may never be resolved because the primitive terrestrial rare-gas signature has been obscured by the addition of radiogenic and atmospheric components. For example, Craig & Lupton (1981) point out that if the mantle initially had $R/R_A \simeq 200$ as measured in the gas-rich meteorites, then the ^4He in the oceanic basalts is 5% primordial and 95% radiogenic. Because higher

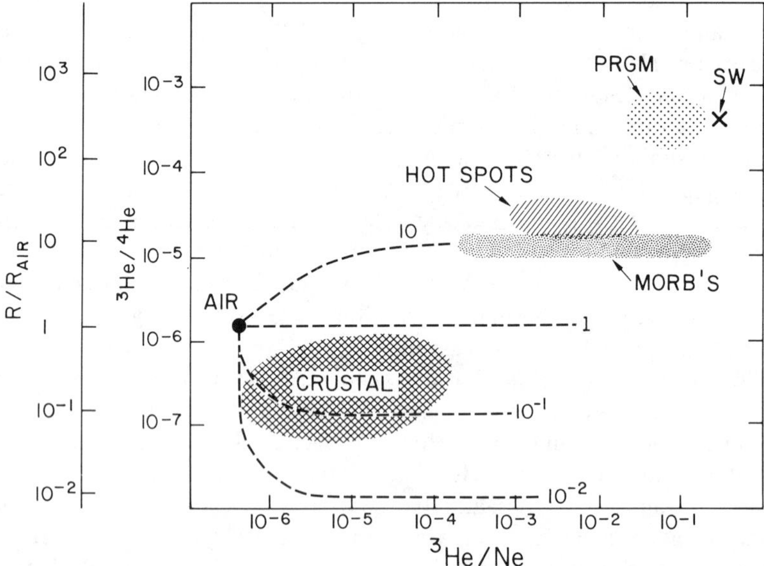

Figure 1 $^3He/^4He$ versus $^3He/Ne$ for various terrestrial components, including air, crustal gases, mid-ocean ridge basalts (MORB's), hot spots, primordial rare gases in meteorites (PRGM), and solar wind (SW). The dashed lines depict mixing lines between air and a helium-rich component with $R/R_A = 10$, 1, 10^{-1}, and 10^{-2}. Addition of radiogenic helium to extraterrestrial components (PRGM or SW) causes a decrease in $^3He/^4He$ with little change in $^3He/Ne$. Thus the MORB and hot-spot fields could be generated by dilution of primordial gases with radiogenic helium and subsequent mixing with atmospheric gases.

$^3He/^4He$ ratios imply a more primitive sample less diluted with radiogenic helium, considerable effort has been devoted to the search for high ^3He enrichments which may have originated in the deep mantle. We discuss the rare-gas signature of the deep mantle in a later section.

OTHER COMPONENTS In addition to the three major types of terrestrial helium (i.e. atmospheric, radiogenic, and primordial), there are other sources that are significant in certain areas. The most important of these is the "tritiugenic" helium (Jenkins & Clarke 1976) that is produced by tritium (^3H), which decays to ^3He with an 18-year mean life. This ^3He component is only important in the oceans or in natural continental waters where significant concentrations of nuclear-era tritium occur. Measurements of tritium and dissolved ^3He have been used to calculate "tritium-^3He ages" for Atlantic Ocean waters (Jenkins et al 1972, Jenkins & Clarke 1976, Jenkins 1980), and in limnological studies (Torgersen et al 1977).

Another component that must be considered is the generation of ^3He by natural reactors such as the Oklo Mine (Cowan 1976). Drozd et al (1974) have reported spectra for xenon isotopes that confirm natural spontaneous fission at the Oklo Mine, and thus it is possible that anomalous helium could also be generated by an intense natural reactor. Although the total effect of Oklo-type reactors is too small to be detected in atmospheric rare gases (Drozd 1974, Bernatowicz & Podosek 1978), natural reactors must be considered whenever isolated isotopic anomalies are encountered.

The Terrestrial Helium Budget

Helium is unique among the terrestrial rare gases in that both helium isotopes escape continuously from the atmosphere, and thus a significant fraction of the Earth's helium inventory has been lost into interplanetary space during geologic time. The concentration of ^3He and ^4He in air is controlled by the dynamic balance between the separate sources and sinks for these two isotopes. Early attempts to construct a steady-state helium balance for the Earth were not very successful because the only important ^3He sources were assumed to be production by cosmic-ray interactions in the atmosphere, and crustal production by neutrons on ^6Li (Morrison & Pine 1955, Bates & McDowell 1957, Kockarts & Nicolet 1962). These ^3He sources were insufficient to balance the thermal escape mechanism, which favors ^3He over ^4He by approximately 100 : 1.

Table 1 summarizes our present knowledge of the terrestrial budget of ^3He and ^4He. The budget has been considerably modified by the discovery of a nonthermal escape mechanism for both ^3He and ^4He (Axford 1968, Patterson 1968) and of several new ^3He sources, including the flux of primordial helium from the interior of the Earth (Clarke et al 1969, Craig et

Table 1 Estimates of terrestrial sources and sinks for ^3He and ^4He, after Buhler et al (1976)

	^3He flux (atoms cm^{-2} s^{-1})	^4He flux (10^6 atoms cm^{-2} s^{-1})	References
Sources			
Auroral precipitation (solar wind)	~5	—	Buhler et al (1976)
Interstellar gas	~5	—	Holzer & Axford (1971)
Primordial flux—oceans	4–5	—	Craig et al (1975), Craig & Lupton (1981)
Primordial flux—subaerial	1–10?	—	Craig & Lupton (1971)
Cosmic rays—direct accretion	~1	—	Lupton (1973)
Cosmic dust and meteorites	<1	—	Johnson (1971)
Galactic cosmic-ray interactions	~0.6	—	Johnson & Axford (1969)
Radiogenic—continental crust	~0.1	3	Morrison & Pine (1955), Craig & Lupton (1981)
Radiogenic—ocean crust	—	0.3	Craig & Lupton (1981)
Total	16?	~3	
Losses			
Thermal escape	6	0.1?	Johnson & Axford (1969), Axford (1968), Kockarts (1973)
Nonthermal escape	1	2–8	Johnson & Axford (1969), Banks & Holzer (1969)
Total	7	2–8	

al 1975), auroral precipitation of solar-wind helium (Axford et al 1972, Buhler et al 1976), interstellar neutral helium (Holzer & Axford 1971), and direct accretion of solar cosmic rays (Lupton 1973). Thus it is now estimated that most of the atmospheric ^{3}He is derived from the primordial flux and from the solar wind and interstellar gas; and that most ^{3}He leaves by thermal escape, whereas ^{4}He is lost mainly by nonthermal mechanisms. It should be emphasized that most of these fluxes are very poorly known. For example, the rate of thermal escape depends critically on the thermospheric temperature, and the theoretical details of nonthermal escape are still under discussion (Donahue 1971). Although the primordial flux of ^{3}He through the oceans is fairly well established at 5 atoms cm^{-2}s^{-1} (Craig et al 1975), the subaerial flux from areas such as Yellowstone Park is unknown. Craig & Lupton (1981) estimate that the generation and loss rates for ^{3}He are approximately in balance at a flux of 10 atoms cm^{-2}s^{-1}. The reader is referred to articles by Johnson & Axford (1969), Kockarts (1973), and Buhler et al (1976) for discussions of helium fluxes.

Although the foregoing has assumed that a steady-state balance exists, the residence time for atmospheric helium ($\sim 10^{6}$ yr) is quite short compared to the age of the Earth, and it is probable that substantial variations in the atmospheric content of ^{3}He and ^{4}He have occurred during geologic time. Sheldon & Kern (1972) have proposed that the dominant mechanism of helium loss is ion-pumping by the solar wind during geomagnetic field reversals, rather than any steady loss mechanism. On the other hand, Lupton & Craig (1975) have suggested that the atmospheric ^{3}He/^{4}He ratio would increase during a geomagnetic reversal or null due to greater penetration of cosmic radiation and the solar wind. Whatever the result, it is clear that changes in the total inventory of ^{3}He and ^{4}He in air may accompany geomagnetic reversals, and it is of interest to search for evidence of such effects in the geologic record. Craig & Chou (1982) have reported low ^{3}He/^{4}He ratios in Greenland ice cores ($R/R_{A} = 0.82$–0.90), but they attribute this ^{3}He depletion to fractionation during diffusive loss of gases from the ice.

HELIUM ISOTOPE VARIATIONS IN DIFFERENT TECTONIC ENVIRONMENTS

The excess ^{3}He in oceanic basalts and in other mantle-derived samples is the one primordial component that can be clearly identified among the terrestrial noble gases. For this reason, the study of terrestrial ^{3}He/^{4}He variations has received considerable attention during the last decade. Of course one can argue that studies based on only two rare-gas isotopes cannot delineate very many components or end members. In a strict sense

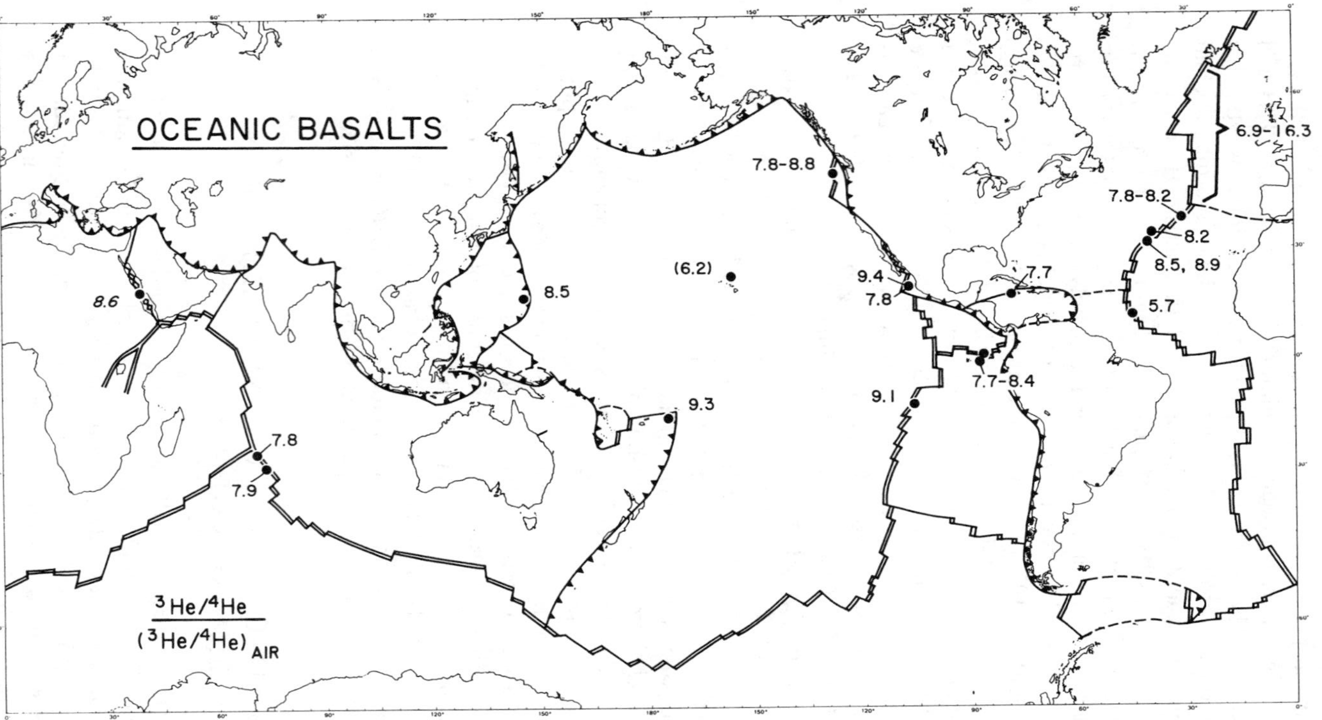

Figure 2 Helium isotope ratios R/R_A for oceanic basalt glasses. The major plate boundaries are shown for comparison. All samples are young tholeiitic basalt glass from ocean spreading centers or back-arc basins, with the exception of a $^3He/^4He$ ratio for the Red Sea brines (assumed to have been extracted from basalts of the Red Sea rift), and one value for an older basalt from a location near Hawaii (shown in parentheses). Data from Krylov et al (1974), Lupton & Craig (1975), Craig & Lupton (1976), Lupton et al (1977), Poreda & Craig (1979), Poreda et al (1980), Kurz & Jenkins (1981), Kurz et al (1982b), and Lupton (1982).

this is true. However, these studies have shown that the $^3He/^4He$ ratio of the "mantle helium" leaking from the interior of the Earth is not uniform, and that distinct $^3He/^4He$ signatures are found in association with different tectonic environments. Thus it seems that "mantle helium" is not a single component, but is instead composed of several subcomponents with distinct signatures. The clearest expressions of these subcomponents are seen in oceanic systems where the influence of radiogenic helium is the smallest. We now discuss the different "types" of mantle helium associated with various tectonic regimes, with an aim toward addressing the question of mantle heterogeneity.

Oceanic Basalts

The $^3He/^4He$ ratio in young oceanic basalts is very uniform for samples collected at different plate boundaries and with differing He concentrations and rare-gas abundance patterns. This finding was first reported by Lupton & Craig (1975) and has since been verified by several other studies (see Figure 2). The average basalt $^3He/^4He$ ratio ($R/R_A = 9$) is generally accepted as the helium signature for the upper mantle, and the constancy of this ratio is taken as evidence that the upper mantle is well mixed on a global scale. The uniformity of the basalt $^3He/^4He$ ratio is even more surprising when one considers that the helium isotopes in basalts probably came from separate sources: the ^3He is primordial while the ^4He is radiogenic. Thus the uniform $^3He/^4He$ may imply not that the mantle is well mixed, but rather that the ratio of U and Th to primordial ^3He in the upper mantle is constant.

GASES IN GLASS, HOLOCRYSTALLINE BASALT, AND VESICLES The fact that mantle gases are trapped in oceanic basalts is due to several favorable effects: rapid quenching of the magma by contact with seawater; hydrostatic pressure of seawater, which reduces outgassing of the magma; and low silica content, which is responsible for the slow helium diffusion rate through basalt glass. For example, rhyolite lavas erupted on the continents have very poor helium retention compared to oceanic basalts. Thus primordial rare gases are found only in the glassy rinds that comprise the outer 1 cm or so of oceanic basalt pillows—the crystalline interiors contain rare gases with seawater abundance patterns and with $^3He/^4He$ close to the atmospheric ratio (Dymond & Hogan 1973, Lupton & Craig 1975). This is attributed to two effects: (*a*) rare gases have an affinity for the melt relative to crystals and the rare gases originally dissolved in the melt occupy sites along the grain boundaries when crystallization is complete, and (*b*) gas exchange occurs between holocrystalline basalt and seawater. It should be mentioned that Kirsten et al (1981), in apparent contradiction to the above discussion, find that a systematic difference in the rare-gas concentration patterns in glasses vs crystalline basalts exists only for helium.

The $^3He/^4He$ ratio is the only rare-gas parameter that is relatively constant in oceanic basalts: the absolute concentrations and relative abundances of the rare gases exhibit large variations even between samples from the same pillow (Craig & Lupton 1981). In particular, the absolute concentrations of helium in basalt glasses vary by over a factor of 100, and the average basalt ^3He concentration is too low to account for the estimated ^3He flux through the oceans (Craig et al 1975), assuming 150 km^3 of basalt is degassed each year. Lupton & Craig (1975) suggested that this flux discrepancy as well as the high variability of basalt He concentrations is due to partitioning of helium between the melt and vesicles. Kirsten (1968) showed that helium and other rare gases partition favorably into the gas phase from an enstatite melt. If a basalt contains a significant fraction of vesicles by volume, then the gases retained in the glass phase are residuals remaining after exsolution into vesicles. Kurz & Jenkins (1981) have studied gas-melt partitioning by comparing helium released from basalt glass by crushing in vacuo vs melting. Although their results may have been affected by loss of gas from the vesicles before analysis, they found a good correlation between the volume fraction of helium released by crushing vs the volume fraction of vesicles, corresponding to a gas-melt partition coefficient for helium of 1.6×10^{-3} cm^3 STP g^{-1} atm^{-1}. Kurz & Jenkins (1981) also found that the helium released by crushing is isotopically indistinguishable from that contained in the glass. This is contradicted by the recent work of Rison & Craig (1981), who find different $^3He/^4He$ ratios in vesicles vs glass for basalts from Loihi seamount. Additional evidence that basalt glass and vesicles can contain different rare-gas signatures has been reported by Marty et al (1983), who have distinguished two noble gas components in a Mid-Atlantic Ridge basalt on the basis of $^{40}Ar/^{36}Ar$ ratios measured during stepwise heating. These studies suggest that the measurement of rare gases in basalt glass is not so simple, and that the subject of melt-vesicle partitioning merits further study.

HELIUM DIFFUSION From the foregoing discussion, it is clear that most of the original helium contained in a ocean ridge crest magma is lost during eruption and that only a small remnant is captured in the glass. How then does the helium trapped in the basalt glass evolve with time? Craig & Lupton (1976) calculated that the $^3He/^4He$ ratio and the helium content of basalt glass should decrease with time because of diffusive loss and dilution by in-situ radiogenic ^4He production. Craig & Lupton (1981) speculated that it may be possible to establish a chronology for a sequence of basalts at a given spreading center based on the reduction in the $^3He/^4He$ ratio by diffusion loss. However, the specifics of this $^3He/^4He$ depletion depend on several factors that are difficult to determine, including the initial $^3He/^4He$ signature, the He diffusion coefficient, the thickness of the glass rind, and the U and Th content. Estimates for the diffusion coefficient of helium in basaltic glass at ocean floor temperatures (0°C) range over several orders of magnitude: 1×10^{-17} cm^2 s^{-1} (Kurz & Jenkins 1981), 5×10^{-15} cm^2 s^{-1} (Craig & Lupton 1976), and 4×10^{-14} cm^2 s^{-1} (Jambon & Shelby 1979). On the basis of their low determination for the diffusion coefficient, Kurz & Jenkins (1981) conclude that no significant diffusion effects will take place in less than 100 m.y. for a typical basalt glass rim. However, significant depletions in $^3He/^4He$ (down to $R/R_A = 3$) and in absolute He concentrations have been reported for basalt glasses with estimated ages between 4 and 10 m.y. (Lupton & Craig 1975, Craig & Lupton 1976, Takaoka & Nagao 1980, Kirsten et al 1982). Although it is possible that some process other than diffusion is responsible, these data nonetheless indicate that helium loss is important for basalt glasses on a time scale of $\sim 10^7$ years.

HELIUM SIGNATURE VS BASALT TYPE Figure 2 summarizes measurements of helium isotope ratios for tholeiitic basalts of the type that make up the bulk of the oceanic crust. How do these typical ridge-crest tholeiitic basalts compare with other basalt types such as alkali basalts or basalts erupted in different tectonic environments such as back-arc basins? Lupton & Craig (1975) found low helium ratios ($R/R_A = 5.6$ and $R/R_A < 0.4$) for two alkali basalts (not shown in Figure 2), but these results are difficult to compare with typical MORB's because of the greater age, higher U and Th contents, and shallower eruption depths of these alkali basalt samples. Kurz & Jenkins (1981) have measured helium isotopes in a variety of MORB's from different spreading centers, but report no systematic $^3He/^4He$ variations. Tholeiitic and alkali basalts from Loihi seamount exhibit quite different rare-gas signatures (Kurz et al 1981, 1982a, Rison & Craig 1981), but as we discuss later, Loihi is a hotspot and therefore quite different from an ocean ridge crest environment. Figure 2 includes results for two back-arc basin tholeiites from the Lau Basin (Lupton & Craig 1975) and the Mariana Trough (Poreda & Craig 1979), and one sample from the Mid-Cayman Rise (Kurz & Jenkins 1981), which is a small spreading-center segment bordered by a transform fault. The helium in these samples is isotopically indistinguishable from typical MORB helium, indicating that these zones of crustal extension extract helium from the upper mantle in the same way as normal ocean ridge spreading centers.

VARIATIONS ALONG A SPREADING-CENTER AXIS Thus far two detailed studies have been undertaken to delineate isotopic variations of helium,

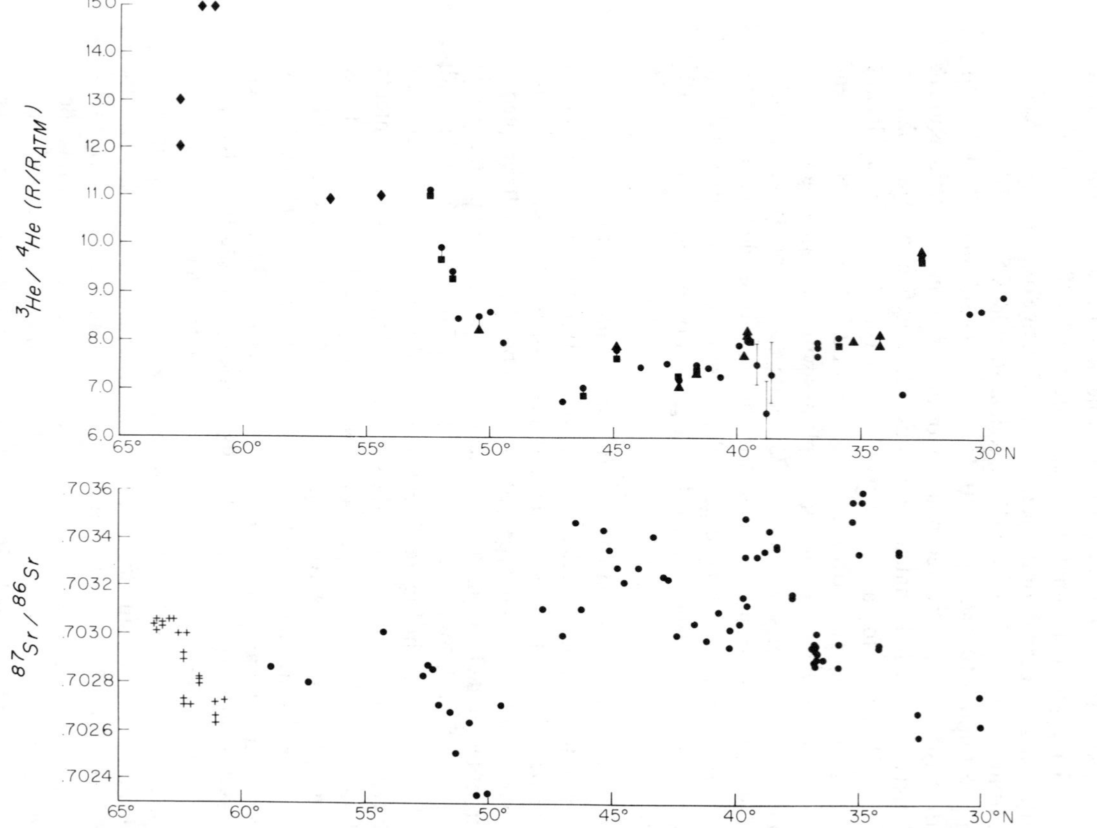

Figure 3 Helium isotope ratio R/R_A and $^{87}Sr/^{86}Sr$ versus latitude for basalt glass from the crest of the Mid-Atlantic Ridge (from Kurz et al 1982b). Reprinted by permission from the authors and from Elsevier Scientific Publishing Company. Helium isotope data from Poreda et al (1980) (points shown as diamonds) and from Kurz et al (1982b). Strontium data from White et al (1976), White & Schilling (1978), and Hart et al (1973).

strontium, etc, along the axis of a spreading center. One of these is a study (still in progress) of basalts from a 400-km section of the Juan de Fuca Ridge from 44°N to 48°N. Preliminary results for helium (Lupton 1982) and heavier rare-gas isotopes (Hart et al 1982) indicate a systematic variation in $^{3}He/^{4}He$ and $^{4}He/^{40}Ar$ from north to south along the ridge axis. The Juan de Fuca Ridge characteristically erupts Fe- and Ti-rich basalts (Delaney et al 1981), and the $^{3}He/^{4}He$ ratio appears to correlate with the degree of Fe-Ti enrichment. These along-axis variations may be caused by differences in the differentiation of the magma, by the influence of the Cobb hot spot, which is situated near the ridge crest at ~46°N, or by seawater interaction with the melt.

The second detailed study of ridge crest basalts has involved a suite of samples from the Mid-Atlantic Ridge collected by J. G. Schilling and co-workers. These samples provide detailed coverage of the ridge beginning south of the Azores Fracture Zone at 28°N, across the Azores Platform, and along the axis of the Reykjanes Ridge to Iceland at 65°N. As shown in Figure 3, $^{87}Sr/^{86}Sr$ ratios reach a minimum at 50°N, and then increase along the Reykjanes Ridge toward Iceland (White et al 1976, White & Schilling 1978, Hart et al 1973). Similar trends have been reported for La/Sm, K, Rb, Cs, F, Br, Cl, and Pb isotopes for this suite of samples (White & Schilling 1978, Schilling et al 1980, Schilling 1975, Sun et al 1975). One interpretation of these variations is mixing between depleted MORB-type mantle reservoirs and undepleted plume-type reservoirs—the regions of high $^{87}Sr/^{86}Sr$ are then explained as "contamination" with more radiogenic Sr from plumes at the Azores and Iceland. However, the helium isotope results for the region from 27°N to 57°N (Kurz et al 1982b) and for the Reykjanes Ridge (Poreda et al 1980) do not mirror the strontium isotope trends (see Figure 3). When $^{3}He/^{4}He$ is plotted versus $^{87}Sr/^{86}Sr$, the results fall into two distinct provinces: the Azores Platform basalts from 33°N to 50°N form one group, and the basalts from south of 33°N and from north of 50°N and the Reykjanes Ridge form a separate field (see Figure 6). Very high ^{3}He enrichments ($R/R_A = 18$–23) have been found in the geothermal gases of Iceland (Polak et al 1975), and thus elevated $^{3}He/^{4}He$ ratios might be expected for Reykjanes Ridge basalts due to influence of the Iceland hot spot. On the other hand, the $^{3}He/^{4}He$ signature lower than MORB for the Azores Platform is surprising, since one would argue on the basis of the strontium signature that the Azores plume is derived from a mantle source less depleted than the source region for the Iceland plume. Thus one might expect more primitive helium with higher $^{3}He/^{4}He$ at the Azores. These helium and strontium variations can thus be explained via mixing between three reservoirs: (*a*) the Iceland plume with high $^{3}He/^{4}He$ and intermediate $^{87}Sr/^{86}Sr$, (*b*) the Azores plume with low $^{3}He/^{4}He$ and high $^{87}Sr/^{86}Sr$, and

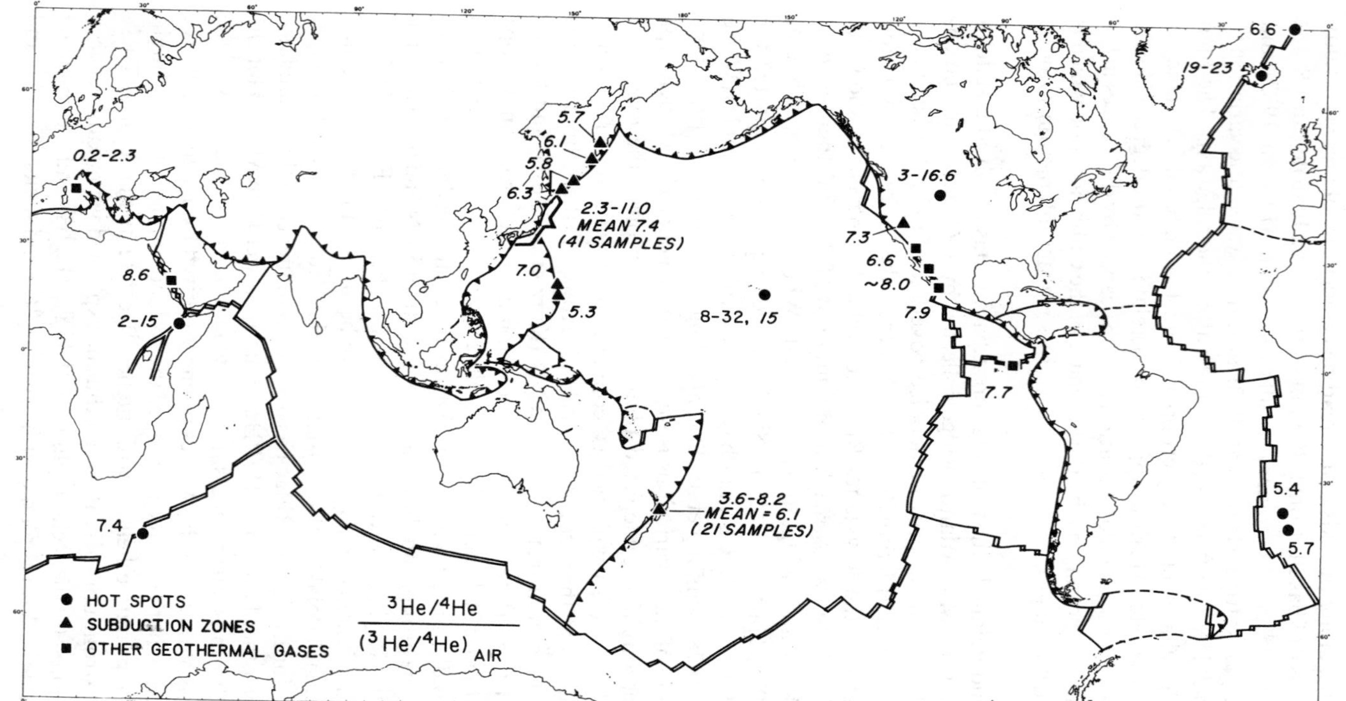

Figure 4 Helium isotope ratios R/R_A for hotspots (*circles*), subduction zones (*triangles*), and other geothermal sites (*squares*). Ratios for gases and hydrothermal fluids are shown in italics, other values are for rocks. Data from Baskov et al (1973), Polak et al (1975), Craig & Lupton (1976, 1978), Lupton et al (1977), Craig et al (1978a,b), Welhan et al (1979), Lupton et al (1980), Jenkins et al (1978), Lupton (1979), Nagao et al (1981), Kurz et al (1982a), Torgersen et al (1982), and Bertrami et al (1982).

(*c*) a MORB reservoir with intermediate $^3He/^4He$ and low $^{87}Sr/^{86}Sr$ (Kurz et al 1982b). Note that the very high helium isotope ratios (R/R_A up to 16) reported by Poreda et al (1980) for Reykjanes Ridge basalts are a striking violation of the normal ridge crest helium signature. Curiously, the maximum $^3He/^4He$ ratios are found at 62°N on the Reykjanes Ridge and somewhat lower ratios are observed nearer Iceland.

Hot Spots

The first indication of a mantle helium component distinct from MORB helium was the $R/R_A = 15$ signature reported for the fumarole gas from Kilauea volcano, Hawaii (Craig & Lupton 1976). This helium isotope ratio, significantly higher than MORB, was taken as evidence that the Hawaiian hot spot is tapping a deeper region of the mantle more enriched in ^3He than the MORB reservoir. As shown in Figure 4, extreme $^3He/^4He$ ratios have been observed at several other supposed hot spots such as Iceland (Polak et al 1975, Poreda et al 1980, Hooker et al 1982), Yellowstone Park (Craig et al 1978b), and the Ethiopian Rift Valley (Craig & Lupton 1978, Craig et al 1977). In all of these hot spot zones, the $^3He/^4He$ ratios vary considerably and only a fraction of the samples exhibit the highest ratios, which are assumed to represent the actual mantle plume component. These fluctuations are probably due to the incorporation of various amounts of crustal helium in the plume signal—in the case of Hawaii, this crustal component might be the gases retained in aged oceanic crust.

If we assume that the elevated $^3He/^4He$ observed in mantle-derived samples is a remnant of extraterrestrial helium with $R/R_A \simeq 200$ as depicted in Figure 1, then higher $^3He/^4He$ signatures should indicate more primitive terrestrial components that have suffered less dilution with radiogenic helium. The highest terrestrial $^3He/^4He$ ratios are those reported by Kaneoka & Takaoka (1980) for olivine and clinopyroxine phenocrysts in ankaramite from Haleakala volcano, Maui ($R/R_A = 34$–37), and by Saito et al (1978) for a kaersutite from Kakanui, New Zealand ($R/R_A = 35$). The kaersutite result is apparently questionable, since Poreda & Craig (private communication) found $R/R_A = 6$ for the same sample. However, several very high ($R/R_A = 27$–32) helium ratios have been reported in Hawaiian samples from Haleakala volcano and from Loihi seamount (Kurz et al 1981, Rison & Craig 1981, Kurz 1982a, Kyser & Rison 1982).

Thus the Hawaiian hot spot has the most primitive terrestrial helium measured thus far, and consequently rare-gas studies of Hawaiian lavas and volcanic gases are being actively pursued by several research groups. The helium isotope measurements reported to date are summarized in Figure 5. Much of the current work has focused on Loihi seamount, which is the southeasternmost active volcano in the Hawaiian-Emperor chain and probably marks the present-day location of the Hawaiian hot spot. The presence of both alkalic and tholeiitic basalts on Loihi is of interest because it was previously supposed that only tholeiitic lavas were erupted during the early stages of evolution of a Hawaiian volcano (Clague et al 1981, Moore et al 1982). Kurz et al (1981, 1982a) were the first to verify that very primitive volatiles are present in the Loihi lavas, finding helium isotope ratios of $R/R_A = 23.1$ and 31.9 in tholeiites and 24.1 in an alkali basalt. Rison & Craig (1981) have confirmed this signature, reporting $R/R_A = 27.2$ for a tholeiite from Loihi and $R/R_A = 5.4$ and 24.6 for Loihi alkali basalts. It is curious that Kilauea, which is located between Loihi and Maui, has consistently lower $^3He/^4He$ ratios in the volcanic gases ($R/R_A = 15$) and in the lavas ($R/R_A = 9$–18) (Craig & Lupton 1976, Kaneoka & Takaoka 1978, Kurz et al 1982a, Rison & Craig 1981). Kaneoka & Takaoka (1978, 1980) have found MORB-type helium ratios ($R/R_A = 7.5$–10.6) in two Hualalai dunites and in ultramafic nodules from Salt Lake Crater, Oahu, suggesting that the helium in these samples was derived from the MORB reservoir rather than the deep mantle. On the

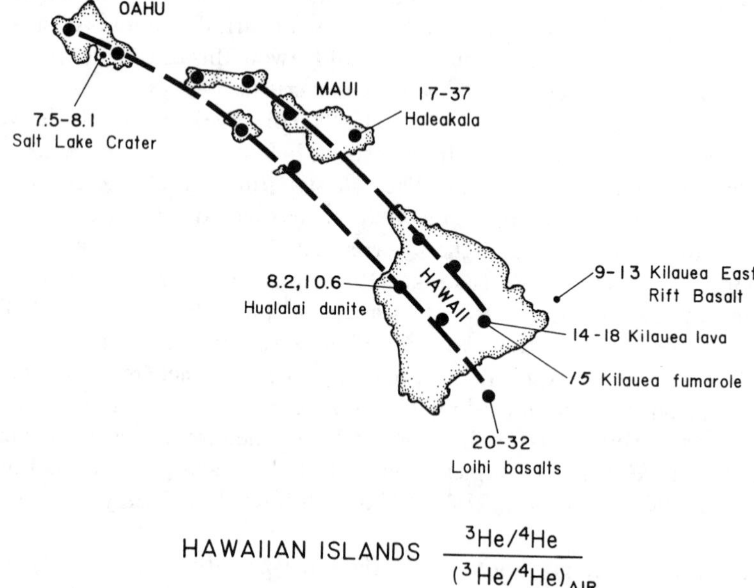

Figure 5 Helium isotope ratios R/R_A for samples from the Hawaiian Islands and Loihi seamount. Gas sample from Kilauea fumarole shown in italics, other samples are rocks. Map of Hawaiian volcanic centers after Moore et al (1982). Helium data from Craig & Lupton (1976), Kaneoka & Takaoka (1978, 1980), Rison & Craig (1981), and Kurz et al (1982a).

other hand, helium measurements at Haleakala and Kilauea suggest that phenocrysts are perfectly good "bottles" for studying volatiles from the deep mantle (Kaneoka & Takaoka 1980, Kurz et al 1982a, Kyser & Rison 1982). A study of all the eruptive phases of a single Hawaiian volcano would be very useful in understanding the diverse rare-gas signatures for the various Hawaiian volcanic centers.

$^3He/^4He$–$^{87}Sr/^{86}Sr$ Correlations: The Tristan Component

When $^{143}Nd/^{144}Nd$ values for MORB and for various hot spots and oceanic islands are plotted vs $^{87}Sr/^{86}Sr$, a negative correlation is observed that is explained via mixing between depleted mantle (MORB) and undepleted reservoirs (O'Nions et al 1977, DePaolo & Wasserburg 1976). In this model, the low $^{143}Nd/^{144}Nd$ and high $^{87}Sr/^{86}Sr$ ratios for Tristan da Cunha and Gough indicate that these oceanic islands are derived from the least depleted regions of the mantle, close to "bulk earth" composition. The distinction between MORB and hot spot $^3He/^4He$ signatures is also consistent with this simple two-reservoir model. However, Pb isotopes show more complicated trends, and a third component, such as recycled oceanic crust or sediments, must be invoked in order to explain the isotopic variations of Sr, Nd, and Pb (Sun 1980).

Kurz et al (1982a) published the first combined study of $^3He/^4He$ and $^{87}Sr/^{86}Sr$ variations, and their results show that more than two mantle components or reservoirs are required to explain the helium-strontium systematics of oceanic islands. The results reported by Kurz et al (1982a) in this initial study fall in two separate trends. One trend is defined by Reykjanes Ridge basalts, Iceland, and Loihi, and is consistent with mixing between the MORB reservoir and a primordial mantle reservoir. Presumably other high $^3He/^4He$ hot spots such as Yellowstone and the Ethiopian Rift Valley would also fall on this trend if uncontaminated strontium signatures could be obtained in continental areas. The second trend is defined by a series of oceanic islands that have $^3He/^4He$ lower than MORB, including Prince Edward, Jan Mayen, Tristan da Cunha, and Gough (see Figures 4 and 6). Kurz et al (1982a) explain these variations in terms of mixing between three mantle reservoirs: (*a*) MORB, (*b*) primitive or undepleted mantle (epitomized by Loihi), and (*c*) a reservoir with low $^3He/^4He$ and high $^{87}Sr/^{86}Sr$, which might be called the "Tristan component."

The interpretation of these results is of course difficult, because helium and strontium have much different affinities and might be expected to behave quite differently during partial melting events. Kurz et al (1982a) suggest that the Tristan component originated in a region of the mantle contaminated with subducted oceanic crust or sediments—this interpretation is consistent with the low $^3He/^4He$ and high $^{87}Sr/^{86}Sr$ for Tristan. However, this signature could also be generated by multiple melting events in which the mantle beneath Tristan da Cunha and Gough has lost helium by degassing while retaining U and Th. The $^3He/^4He$ ratio would then decrease as a result of high $U/^3He$ and $Th/^3He$ ratios (Kurz et al 1982a).

Because subaerial basalts are highly degassed, Kurz et al (1982a) analyzed phenocrysts from Tristan da Cunha, Gough, Jan Mayen, and Prince Edward in order to obtain consistent $^3He/^4He$ signatures. This raises the question as to whether such phenocrysts contain helium isotopes representative of the mantle source region. As mentioned earlier, comparison of $^3He/^4He$ in phenocrysts vs basalt glass at Kilauea suggests that

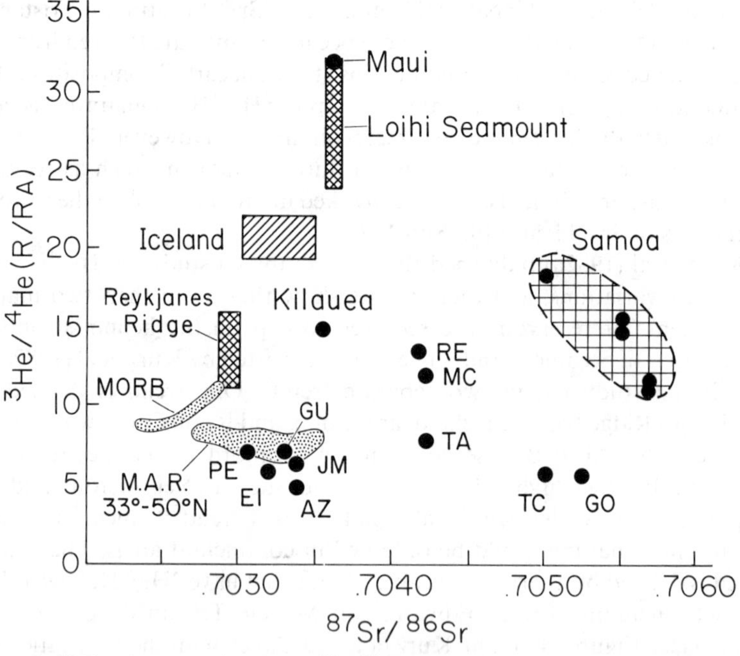

Figure 6 Helium isotope ratio R/R_A vs $^{87}Sr/^{86}Sr$ for mid-ocean ridge basalts (MORB's) and volcanic rocks from various oceanic islands [after a similar figure by Kurz et al (1982a)]. These variations require mixing between at least three mantle reservoirs. Some of the islands have been abbreviated as follows: PE = Prince Edward, EI = Easter Island, AZ = Azores, GU = Guadalupe, JM = Jan Mayen, RE = Reunion, MC = Macdonald seamount, TA = Tahiti, TC = Tristan da Cunha, GO = Gough. Helium isotope data from Craig & Lupton (1978) Poreda et al (1980), Kurz et al (1982a,b), Craig & Rison (1982), Kaneoka & Takaoka (1982), and Rison & Craig (1982). Strontium isotope data from McDougall & Compston (1965), Hedge et al (1972), Hart (1973), Hart et al (1973), Hedge et al (1973), O'Nions & Pankhurst (1974), Batiza (1977), O'Nions et al (1977), Hedge (1978), Zindler et al (1980), and White & Hofmann (1982).

the phenocrysts contain helium isotopically indistinguishable from the bulk lava (Kaneoka & Takaoka 1978). Another problem concerns the signatures for Loihi vs Kilauea. These two Hawaiian volcanoes have similar $^{87}Sr/^{86}Sr$ ratios, and yet their $^{3}He/^{4}He$ signatures are quite different ($R/R_A = 30$ and 15, respectively). This discrepancy may be resolved when the helium isotope systematics of the Hawaiian islands have been studied in more detail.

More recently, helium isotope ratios have been reported for several oceanic islands not included in the study by Kurz et al (1982a), including Easter Island, the Azores, Guadalupe, Grand Comore, Reunion, Macdonald seamount, Tahiti, and Samoa (Craig & Rison 1982, Kaneoka & Takaoka 1982, Rison & Craig 1982). As shown in Figure 6, Easter Island, Guadalupe, the Azores, and Tahiti all fall approximately on the low $^{3}He/^{4}He$ mixing trend previously defined by MORB, Jan Mayen, Prince Edward, Tristan, and Gough. The curious helium and strontium pattern observed in Mid-Atlantic Ridge basalts from the Azores Platform (Figure 3) is thus consistent with the signature for the Azores plume itself. Only six oceanic hot spots have $^{3}He/^{4}He$ ratios higher than MORB: Hawaii, Iceland, Reykjanes Ridge, Reunion, Macdonald seamount, and Samoa. Rison & Craig (1982) discovered that Samoa is a striking exception to the Tristan and Loihi trends: xenoliths in Samoan lavas have "enriched" strontium with $^{87}Sr/^{86}Sr \simeq 0.7050$–$0.7057$, similar to Tristan and Gough, but have $^{3}He/^{4}He$ ratios up to $R/R_A = 18$, much higher than Tristan. These new data support the "multiple melting" explanation for Tristan and other oceanic islands with low $^{3}He/^{4}He$, since the high $^{3}He/^{4}He$ signature for Samoa precludes significant contribution from subducted crust.

Subduction Zones

In addition to hot spots and ocean spreading centers, a third type of volcanic area of some importance is the island arc and continental margin province, which is generally characterized by more acidic volcanic rocks. Baskov et al (1973) and Craig et al (1978a) found elevated $^{3}He/^{4}He$ ratios in volcanic gases from Kamchatka ($R/R_A = 5.7$), the Kuriles ($R/R_A = 5.4$–6.3), the Marianas ($R/R_A = 5.3$, 7.0), Hakone Volcano, Japan ($R/R_A = 6.2$), and Lassen Park, California ($R/R_A = 7.7$, 8.1), thus proving that mantle volatiles are present in convergent margin systems. Subsequent studies of geothermal gases in the Japanese Islands (Nagao et al 1981, Sano et al 1983) and in New Zealand (Torgersen et al 1982) have confirmed that subduction zones are characterized by a surprisingly uniform $^{3}He/^{4}He$ signature averaging $R/R_A = 6$ (Figure 4).

Craig et al (1978a) pointed out that the existence of such high $^{3}He/^{4}He$ ratios in subduction zones places strong constraints on the relative amounts of remelted subducted oceanic crust vs new mantle material involved in the generation of andesitic and more acidic lavas at convergent plate margins. In oceanic crust with 0.3 ppm uranium, the original MORB helium ratio ($R/R_A = 9$) will have decreased to $R/R_A \simeq 2$ after 100 m.y. as a result of the addition of radiogenic ^{4}He. Assuming, in addition, that the crust has been almost completely outgassed, the ^{3}He/^{4}He ratio in downgoing oceanic slab must be even more radiogenic (say $R/R_A = 0.1$). Thus essentially all of the ^{3}He in subduction zone gases must be derived from primary mantle material. Although it is impossible to calculate a magmatic recipe for subduction zone lavas, approximately one half of the subduction zone *helium* must be supplied directly from the mantle. Studies of the helium trapped in the arc-type and orogenic lavas themselves (rather than the volcanic gases) may help to further define the subduction zone volcanic process. If the remelting of subducted crust is important, then it should be possible to see correlations between the ^{3}He/^{4}He ratio in the arc volcanics and the age of the crust being subducted.

Other Mantle Helium Sources

DIAMONDS Because of their supposed mantle origin, chemical inertness, and high temperature stability, diamonds should be ideal for studying the rare-gas content of the Earth's interior. However, the analysis of gases in diamonds presents severe technical problems because of the very low gas concentrations (3×10^{-8} cc He total for a typical 0.3 g diamond). Takaoka & Ozima (1978) analyzed the rare gases in two batches of diamonds from the Kimberley mines, South Africa, and found a clear mantle helium signature ($R/R_A = 6, 14$) and ^{40}Ar/^{36}Ar above the atmospheric ratio. In a recent paper, Ozima & Zashu (1983a) report highly variable helium isotope results for 13 South African diamonds—the ^{3}He/^{4}He ratios range from $R/R_A = 0.03$ to $R/R_A = 168$ and 226 in two diamonds with very low helium concentrations. Ozima & Zashu were unable to identify any radioactive or nuclear mechanisms that could generate such elevated ^{3}He/^{4}He ratios in diamond, and conclude that these high ratios are due to very primitive helium captured early in the Earth's history. This would require diamond with very low U and Th contents (say <0.1 ppb) in order to preserve this helium signature for several b.y. The authors attribute the lower ^{3}He/^{4}He observed in most of the diamonds to a greater radiogenic ^{4}He component; this explanation is supported by the fact that low ^{3}He/^{4}He diamonds have higher absolute He concentrations. If it represents a primitive terrestrial signature, then this elevated ^{3}He/^{4}He signature in diamonds has important implications for the early history of the Earth, since this ratio is higher than planetary helium ($R/R_A = 100$) and closer to solar helium ($R/R_A \simeq 280$). It is obvious that more diamonds should be examined in order to completely understand the wide range of $^{3}He/^{4}He$ ratios that are observed. Ozima & Zashu suggest that an important test of their interpretation would be to date diamonds in order to correlate the $^{3}He/^{4}He$ signature with the estimated times of formation.

OTHER MANTLE XENOLITHS Kaneoka et al (1978) have examined several ultramafic nodules from the South African kimberlite region with the idea of studying rare gases from the Earth's interior. Their results for $^{3}He/^{4}He$ are highly variable, ranging from $R/R_A < 0.4$ up to $R/R_A = 4.5$. Kyser & Rison (1982) have analyzed rare gases in xenoliths from South Africa, France, and Arizona with similar variable results ($R/R_A < 0.3$ up to $R/R_A = 9$). Although such mantle fragments and crystalline precipitates should reflect the rare-gas signature of the subcontinental mantle, the results are difficult to interpret because it is not known how well such samples preserve the record of their mantle environment (Ozima & Podosek 1983).

JOSEPHINITE Bird & Weathers (1975) proposed a deep-mantle origin for josephinite, an iron-nickel alloy-bearing rock from southwestern Oregon, suggesting that josephinite is a specimen from the Earth's core. When excess ^{3}He, ^{21}Ne, and ^{129}Xe were reported for josephinite (Downing et al 1977, Bochsler et al 1978), this was taken as verification of the core hypothesis for its origin. In particular, Bochsler et al (1978) inferred a $^{3}He/^{4}He$ ratio in josephinite of $R/R_A \simeq 350$, higher than any other terrestrial sample. However, when the La Jolla rare-gas group (Craig et al 1979) attempted to duplicate this result using a different gas extraction technique, they found much lower $^{3}He/^{4}He$ ratios ($R/R_A = 5.4$) and no appreciable ^{21}Ne excess in josephinite, and even lower ratios ($R/R_A = 0.25$) in a native iron from Disko Island, Greenland. Bernatowicz et al (1979) measured Ne and heavier rare gases in josephinite and found elemental abundance patterns resembling surface waters, and no large ^{21}Ne and ^{129}Xe anomalies. Thus Craig et al (1979) and Bernatowicz et al (1979) find no evidence requiring a core origin for josephinite, and suggest that the previously reported isotopic anomalies were experimental artifacts.

Summary of Mantle Helium Components

After all this discussion, how many separate subcomponents can be clearly identified among the various mantle helium sources? The answer is two: MORB helium and undepleted mantle (such as Loihi). All other components can be produced from these two mantle components by dilution with old oceanic crust and sediments, or by degassing and subsequent addition of radiogenic ^{4}He. However, with the help of strontium isotopes as an additional tracer, it is possible to list four different helium components that consistently occur in association with certain tectonic features:

1. MORB helium ($R/R_A \simeq 9$, $^{87}Sr/^{86}Sr \simeq 0.7025$) is characterized by a very uniform isotopic signature. This component, which is also found in back-arc basins, is thought to represent the upper (depleted) mantle.
2. Hot spot helium ($R/R_A = 15$–30, $^{87}Sr/^{86}Sr = 0.7030$–0.7035) is derived from the deep (undepleted) mantle, and represents the most primitive terrestrial rare-gas component.
3. The Tristan component ($R/R_A \simeq 6$–7, $^{87}Sr/^{86}Sr = 0.7040$–0.7050), discovered by Kurz et al (1982a) in basalts from Tristan da Cunha and Gough Island, may be generated in a region of the mantle contaminated with subducted oceanic crust and sediments. On the other hand, these ^3He-depleted oceanic islands may be the result of mantle metasomatism, and thus may not represent a distinct mantle reservoir.
4. Subduction zone helium ($R/R_A \simeq 5$–8) probably does not qualify as a distinct mantle component, and yet $^3He/^4He$ ratios in geothermal gases at convergent plate margins fall consistently near $R/R_A = 6$, indicating that uniform proportions of remelted downgoing slab and primary mantle material are involved in the generation of subduction zone volcanics.

A TWO-LAYERED MANTLE: $^3He/^4He$ AND $^{40}Ar/^{36}Ar$ VARIATIONS

Helium, strontium, and neodymium isotopic data from ocean ridge crests and hot spots require a heterogeneous mantle with separate reservoirs for MORB-type and plume-type components. One approach has been to assume a two-layered mantle in which the upper depleted layer gives rise to the MORB component, and the deeper layer, which is relatively fertile and undepleted, is the source region for the mantle plume components observed at Hawaii and Iceland. Kaneoka & Takaoka (1980) and Kaneoka (1981) have discussed the systematics of $^3He/^4He$ and $^{40}Ar/^{36}Ar$ ratios in oceanic samples, and conclude that these rare-gas indicators are compatible with the two-layered mantle idea. As illustrated in Figure 7, most of the helium and argon isotopic variations for MORB's and oceanic hot spots can be explained by mixing between three end members: (*a*) plume-type or undepleted source material with $R/R_A \simeq 40$, $^{40}Ar/^{36}Ar = 400$; (*b*) MORB-type or depleted material with $R/R_A = 9$, $^{40}Ar/^{36}Ar \simeq 16{,}000$; and (*c*) air with $R/R_A = 1$, $^{40}Ar/^{36}Ar = 295$. The most important support for this idea comes from the inverse correlation between $^3He/^4He$ and $^{40}Ar/^{36}Ar$ observed for the most uncontaminated samples from the Hawaiian hot spot, corresponding to two-component mixing between plume-type and MORB-type end members. Argon isotope results for ultramafic nodules also suggest that the $^{40}Ar/^{36}Ar$ ratio in the mantle decreases with depth (Kaneoka et al 1978, Kyser & Rison 1982).

Hart et al (1979) have proposed a mantle degassing model that is consistent with these helium and argon systematics. They argue that only the upper depleted layer of the mantle has been degassed, while the lower mantle has retained its primordial gases. The combined system of atmosphere plus sialic crust has thus been formed from the upper mantle; and the $^{40}Ar/^{36}Ar = 400$ signature in primordial gases from the deep mantle is similar to the ratio in air because the atmosphere-crust-upper mantle system has behaved as a closed reservoir for argon (but not for helium), with the same proportions of ^{36}Ar and K as the lower undegassed mantle. The upper mantle then exhibits high $^{40}Ar/^{36}Ar$ ratios because it has preferentially lost argon relative to potassium. Of course, the similarity between deep-mantle argon and air argon could be explained by other mechanisms such as the cycling of atmospheric components through the deep mantle, but these alternative models seem less plausible than the

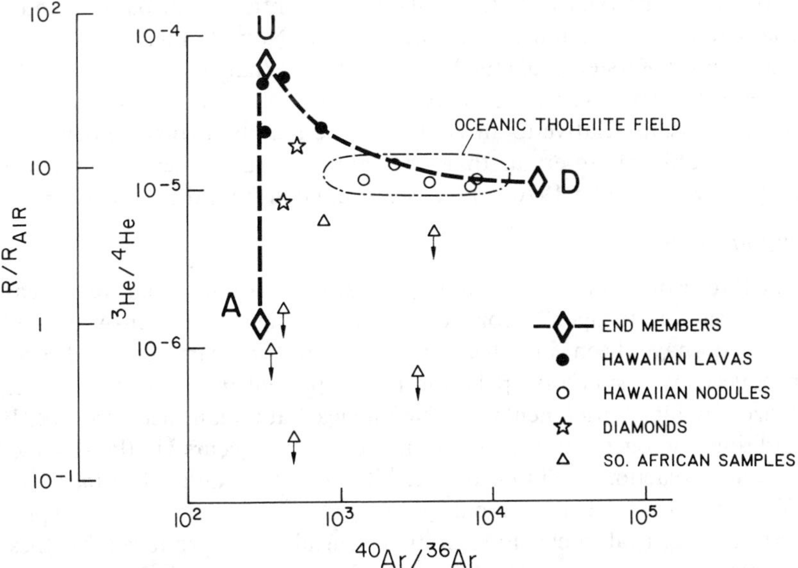

Figure 7 $^{3}He/^{4}He$ versus $^{40}Ar/^{36}Ar$ for various mantle-derived samples [after similar plots by Kaneoka & Takaoka (1980) and Kaneoka (1981)]. Arrows indicate upper limits for $^{3}He/^{4}He$ determinations. Heavy dashed lines indicate mixing lines between air (A), undepleted mantle (U) and depleted mantle (D) reservoirs. Data from Kaneoka & Takaoka (1978, 1980), Kaneoka et al (1978), Takaoka & Ozima (1978), and Kyser & Rison (1982).

upper-mantle degassing model, which also provides an explanation for the lower $^{3}He/^{4}He$ and higher $^{40}Ar/^{36}Ar$ ratios in MORB relative to mantle plumes.

For completeness, it should be mentioned that the ideas of Kaneoka & Takaoka (1980) and Hart et al (1979) are not universally accepted. One can argue that the relatively low $^{40}Ar/^{36}Ar$ ratios observed in certain mantle-derived samples are due to atmospheric contamination and do not represent the argon signature of the undepleted mantle. For example, Fisher (1981) assumes $^{40}Ar/^{36}Ar = 1.5 \times 10^{4}$ for the mantle as a whole. However, an explanation in terms of atmospheric contamination is difficult to reconcile with the observed correlation of low $^{40}Ar/^{36}Ar$ with high $^{3}He/^{4}He$ as depicted in Figure 7.

OTHER INERT GAS SIGNATURES

Excess Neon

Studies of the elemental abundance pattern of rare gases in MORB glasses have shown a persistent excess of neon relative to other noble gases (Dymond & Hogan 1973, Kirsten et al 1981, Ozima & Zashu 1983b). This anomaly is observed in both Pacific and Atlantic Ocean basalts, and is characterized by a hundredfold enrichment in $Ne/^{36}Ar$ relative to atmospheric gases. Kirsten et al (1982) have suggested that, apart from ^{3}He, this Ne excess can be taken as an independent tracer of mantle gases. The excess neon in basalts relative to air could be generated by two-stage degassing from a noble gas reservoir that originally had atmospheric abundances (Ozima & Alexander 1976), or by escape of neon from the atmosphere.

Neon Isotopes

The three isotopes of neon exhibit quite large variations in nature. At least five separate "trapped" neon components, designated Neon-A through Neon-E, have been identified in extraterrestrial samples; and neon produced by cosmic-ray spallation is also present in meteorites, characterized by high enrichments of ^{21}Ne, the least abundant neon isotope. In addition, radiogenic production of neon isotopes occurs via the so-called Wetherill reactions: $^{17}O\ (\alpha, n)^{20}Ne$, $^{18}O\ (\alpha, n)^{21}Ne$, and $^{25}Mg\ (n, \alpha)^{22}Ne$ (Wetherill 1954). The abundances of ^{17}O, ^{18}O, and ^{25}Mg in typical terrestrial crustal rocks is such that crustal neon produced by these reactions is enriched in $^{21}Ne/^{22}Ne$ and depleted in $^{20}Ne/^{22}Ne$ relative to atmospheric neon. Significant ^{21}Ne enrichments attributed to radiogenic production have been observed in a variety of terrestrial samples, including helium-rich natural gases (Emerson et al 1966), and minerals such as anorthosite, quartz, beryl, and tourmaline (Zadnick 1981, Zadnick & Jeffery 1982).

The isotopic composition of neon in the present-day terrestrial atmosphere does not resemble any of the neon components identified in extraterrestrial samples, which raises questions concerning the origin of terrestrial neon. However, the presence of excess ^{3}He in the Earth's mantle suggests that planetary- or solar-type neon might be found in terrestrial samples having elevated $^{3}He/^{4}He$ ratios. Craig & Lupton (1976) found excess $^{20}Ne/^{22}Ne$ in Kilauea fumarole gas and in MORB glasses (up to 5% $^{20}Ne/^{22}Ne$ excess for Kilauea gas), which they interpreted as evidence for a small solar neon component (solar-wind neon has $^{20}Ne/^{22}Ne = 13.7$, about 1.4 times the atmospheric ratio; see Geiss et al 1972). The existence of excess $^{20}Ne/^{22}Ne$ and $^{21}Ne/^{22}Ne$ in mantle-derived lavas and xenoliths has been verified by other experimenters (Kyser & Rison 1982, Zadnick 1981, Zadnick & Jeffery 1982), although these are no longer attributed to extraterrestrial components. The correlation of excess ^{21}Ne with ^{4}He suggests that the excess ^{21}Ne is nucleogenic, while the observed $^{20}Ne/^{22}Ne$ enrichments are assumed to be due to simple mass fractionation (Kyser & Rison 1982). Studies of volcanic gases and rocks have demonstrated that mass-fractionation effects can produce the $^{20}Ne/^{22}Ne$ enrichments reported for mantle-derived samples (Kaneoka 1980). The most persuasive evidence has come from temporal changes in the gases from Showa-shinzan volcano: from 1958 to 1977 the $^{3}He/^{4}He$ ratio decreased from $R/R_A = 5.3$ to 2.0 while the $^{20}Ne/^{22}Ne$ ratio increased from 9.84 to 10.36 (Nagao et al 1980, Nagao et al 1981). Thus the current wisdom says that the neon in present-day air is primordial terrestrial neon plus a small radiogenic ^{21}Ne component, and that the deviations from atmospheric composition observed in some terrestrial samples can be explained by nucleogenic or mass-fractionation mechanisms and do not require extraterrestrial neon components (Ozima & Podosek 1983).

Xenon Isotopes

Like neon, the isotopic composition of atmospheric xenon cannot be easily explained in terms of xenon components observed elsewhere in the solar system. Furthermore, as judged by comparison with extraterrestrial xenon, atmospheric xenon contains about 7% excess ^{129}Xe relative to the isotope pattern predicted for primordial terrestrial xenon (Bernatowicz & Podosek 1978). This excess ^{129}Xe is attributed to the decay of now extinct ^{129}I (half-life = 17 m.y.), implying that condensation of Earth materials occurred early in the solar system's history, before extinction of ^{129}I (Ozima & Podosek 1983).

Another type of excess ^{129}Xe is observed in certain terrestrial samples—namely a ^{129}Xe excess relative to the atmospheric xenon abundance. The classic example is the 8% enrichment of ^{129}Xe discovered in CO_2 well gas from Harding County, New Mexico (Butler et al 1963). This excess, which is also attributed to ^{129}I decay, demonstrates the existence of reservoirs that have not mixed with atmospheric xenon since the time when the Earth's original inventory of ^{129}I had not completely decayed, i.e. within say 50 m.y. of the time of the Earth's formation. Thus, like excess ^3He from the mantle, excess ^{129}Xe is an unambiguous marker for primitive terrestrial gases.

The Harding County gas has been reanalyzed several times since the original work by Butler et al, each time with more refined techniques and increased accuracy (Boulos & Manuel 1971, Hennecke & Manuel 1975a, Phinney et al 1978, Smith & Reynolds 1981). These studies have shown that, in addition to the ^{129}Xe excess, the CO_2 well gas contains excess ^{131}Xe through ^{136}Xe as a result of spontaneous fission of ^{238}U and ^{244}Pu; a mantle helium component ($R/R_A = 3$); slight enrichments of ^{20}Ne and ^{21}Ne; and radiogenic argon ($^{40}Ar/^{36}Ar \simeq 7000$). Thus although the excess ^3He and ^{129}Xe are clearly primitive "inherited" components, the Harding County gas contains several other components: the CO_2 is probably derived from crustal carbonates, the relatively low $^3He/^4He$ ratio and high $^{40}Ar/^{36}Ar$ indicate significant radiogenic addition of ^4He and ^{40}Ar, and the excess ^{20}Ne and ^{21}Ne are probably due to mass fractionation.

Excess ^{129}Xe has been reported in a number of mantle-derived samples, such as Hawaiian nodules (Hennecke & Manuel 1975b, Kaneoka & Takaoka 1978, 1980, Kyser & Rison 1982) and kimberlites (Kaneoka et al 1978), although the amount of the ^{129}Xe excess is much smaller than in the CO_2 well gas. Considering the many occurrences of excess ^3He from the mantle, it is surprising that excess ^{129}Xe is not more prevalent. For example, isotopic anomalies are generally absent in the krypton and xenon in MORB glasses (Kirsten et al 1981, Ozima & Zashu 1983b). However, small enrichments of ^{129}Xe have now been reported for MORB's from the Mid-Atlantic Ridge and the East Pacific Rise (Staudacher & Allegre 1981, Kirsten et al 1981). If a distinct mantle xenon component exists, it would presumably be characterized by excess ^{129}Xe plus a fission xenon component. The variable occurrence of anomalous xenon in mantle-derived samples may be due to mixing between the mantle end member (perhaps similar to the CO_2 well gas xenon) and atmospheric xenon.

APPLICATIONS

We now discuss the application of ^3He and other noble gas isotope tracers to natural processes such as ocean mixing, ocean ridge crest tectonism, and continental geothermal systems. These three examples of rare-gas applications, which are by necessity treated only briefly here, were chosen from a much more extensive list—topics not discussed include earthquake prediction, volcano hazards monitoring, uranium prospecting, and hydrologic and limnologic studies, to name a few. The reader is referred to Craig & Lupton (1981) for a more extensive review of ^{3}He and rare gases in oceanic systems.

Ocean Tracer Studies

Because it is injected into the oceans at mid-depth via localized sources at the major spreading centers, ^{3}He is an excellent stable-conservative tracer for studying deep-ocean mixing and circulation. Although the tectonic activity along ocean ridge crests is smoothly varying when averaged over geologic time scales, the injection of ^{3}He and other mantle volatiles is probably episodic, since this injection involves cycles of magma intrusion and extraction of volatiles via hydrothermal circulation of seawater through the ridge crest basalts (Macdonald 1982). This idea is supported by measurements of ^{3}He in the oceans: well-defined plumes of ^{3}He-rich water have been observed emanating from several localized ridge crest sites, such as 21°N on the East Pacific Rise (EPR) (Lupton et al 1980), the Galapagos Rift (Lupton et al 1980), the Guaymas Basin in the Gulf of California (Lupton 1979), the 15°S region of the EPR (Lupton & Craig 1981), the Juan de Fuca Ridge (Lupton et al 1981, Normark et al 1982), and the Mariana Trough back-arc basin (Horibe et al 1982). Because seawater contains a large component of dissolved atmospheric helium, the injection of mantle helium produces relatively small increases in the ocean water ^{3}He/^{4}He ratio relative to the ratio in air. For this reason, oceanic ^{3}He/^{4}He variations are usually expressed as a percent deviation in units of $\delta(^3\text{He})$, where $\delta(^3\text{He}) = 100\,(R/R_A - 1)$. Plumes of ^{3}He emanating from active oceanic injection sites typically exhibit maximum ^{3}He/^{4}He ratios of $\delta(^3\text{He}) = 50\%$ (or $R/R_A = 1.50$), which corresponds to an approximate 1 to 15 mixture of extracted MORB helium ($R/R_A = 9$) and dissolved air helium ($R/R_A = 1.0$).

As expected, the largest flux of ^{3}He occurs on those ridge crest segments with the highest spreading rates, such as the EPR crest between the equator and 30°S latitude. Figure 8 shows ^{3}He/^{4}He ratios in seawater for an east-west section spanning the EPR crest at 15°S latitude. This $\delta(^3\text{He})$ section delineates one of the most dramatic geochemical features observed in the deep ocean: a huge asymmetric plume of ^{3}He centered at 2500 m depth over the EPR crest (Lupton & Craig 1981). Although the plume extends several thousand kilometers to the west of the EPR crest, it is absent in the profiles to the east of the spreading-center axis. This marked asymmetry in the ^{3}He distribution, which has not yet been measured in any other geochemical

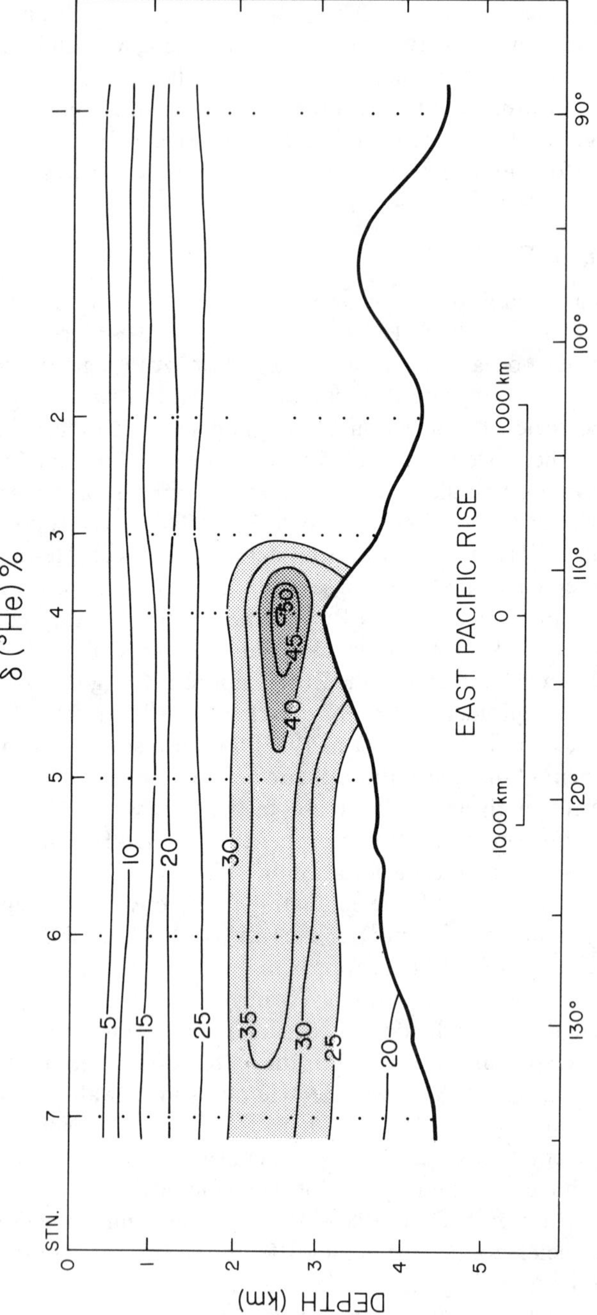

Figure 8 Contours of $\delta(^3He)$ in section view over the East Pacific Rise at 15°S, showing a huge plume of ^3He-rich water emanating from the ridge crest. Data from Lupton & Craig (1981), reprinted by permission from the American Association for the Advancement of Science.

property (including temperature, radium, or silica) suggests that the deep circulation in this region is from east to west in apparent contradiction to the classic deep-circulation pattern of Stommel & Arons (1960). Using the ^{3}He plume as a guide, Reid (1982) has located small temperature excesses in the deep southeastern Pacific that are probably due to heat extracted from the EPR by the same hydrothermal systems responsible for the ^{3}He injection. Stommel (1982) has proposed that the 15°S ^{3}He plume is not passive, but is dynamically active due to the driving force of heat added on the ridge crest.

In addition to ^{3}He, other hydrothermally injected species, such as manganese, hydrogen, and methane, can be detected in the water column near active spreading centers. One of the most striking examples is the vertical profile of manganese in the water column at the 21°N EPR site, which almost exactly duplicates the ^{3}He profile (Lupton et al 1980). Both ^{3}He and Mn show a sharp maximum at 2400 m depth, indicating a layer of water enriched in ^{3}He and Mn extracted from the 21°N ridge crest basalts via hydrothermal activity. Unlike ^{3}He, the other hydrothermal tracers are nonconservative: H_2 and CH_4 are consumed by biological activity (Welhan & Craig 1979) and Mn is scavenged from the water column by particulate fluxes (Weiss 1977). Thus residence times must be determined before the resulting tracer distributions can be interpreted. Lupton & Craig (1981) have suggested that if the deep-ocean consumption rates can be established, then the decrease in these tracers along the axis of a ^{3}He plume will provide an internal clock or "speedometer" for the abyssal circulation in that region. Furthermore, the scavenging of Mn from such plumes will produce a "manganese shadow" in the underlying sediments, which will provide a record of the patterns of circulation and ridge crest injection in the past.

Because the mixing time of the oceans ($\sim 10^3$ yr) is approximately the same as the oceanic residence time for mantle helium, the localized inputs of ^{3}He produce strong vertical and lateral gradients in ^{3}He/^{4}He, even in areas of the ocean distant from spreading-center activity. Such helium isotope variations in seawater have been successfully used to identify and trace major water masses. For example, in both the Atlantic and the Pacific, the Antarctic Bottom Water component can be easily traced by its distinctive ^{3}He/^{4}He signature compared with the overlying deep water (Jenkins & Clarke 1976, Lupton & Craig 1981). Figure 9 shows a typical ^{3}He/^{4}He profile in the southwest Pacific in which three different water masses can be clearly identified: (*a*) a shallow component with low $\delta(^3\text{He})$ that is approximately in equilibrium with air, (*b*) Pacific Deep Water (depth 1500–2700 m) with elevated ^{3}He/^{4}He [$\delta(^3\text{He}) \simeq 20\%$] due to ^{3}He injection on the East Pacific Rise system, and (*c*) Antarctic Bottom Water (depth ≥ 3000 m)

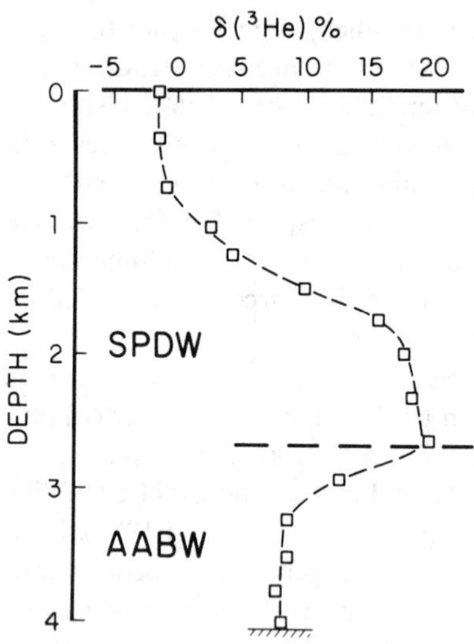

Figure 9 Profile of $\delta(^3He)$ vs water depth at GEOSECS station 278 in the southwest Pacific (36.5°S, 179.6°W). The dashed line indicates the boundary between South Pacific Deep Water (SPDW) and Antarctic Bottom Water (AABW). Data from Lupton & Craig (1981), reprinted by permission from the American Association for the Advancement of Science.

marked by $\delta(^3He) \simeq 8\%$ that originated in the circumpolar region to the south and is intruding beneath the overlying Pacific Deep Water (Lupton & Craig 1981).

The two helium isotopes are the only rare-gas components in seawater affected by input of gases extracted from oceanic basalts—the other rare gases are swamped by the dissolved atmospheric component. Measurements of so-called gas saturation anomalies, i.e. deviations of the observed concentrations of Ne, Ar, N_2, Ar, and Xe from the value for air-saturated seawater, have been used to identify effects such as changes in sea level atmospheric pressure, nonadiabatic temperature changes after gas equilibration, and air injection via bubbles, thereby allowing more accurate estimates of the excess ^{4}He in the ocean (Bieri et al 1966, Bieri 1971, Craig & Weiss 1971). Recent work on rare gases in marine sediments suggests that sediment fluxes of ^{3}He and ^{4}He must also be considered in the oceanic helium budget (Barnes & Clarke 1980, Sayles & Jenkins 1982). In addition to dissolved air and input of mantle helium, ^{3}He produced by the decay of anthropogenic tritium also contributes to oceanic helium isotopic variations. As mentioned earlier, measurements of the ^{3}H-^{3}He radioactive-stable tracer pair can be used to "date" water masses in areas unaffected by input of primordial ^{3}He (Jenkins & Clarke 1976, Jenkins 1980), and of course tritium itself is a powerful oceanic tracer (Östlund et al 1974).

The monumental task of mapping the three-dimensional distribution of ^{3}He, ^{3}H, and other tracers in the major oceans has been underway for over a decade. Although some of the results have been published separately (Craig et al 1975, Jenkins & Clarke 1976, Lupton 1976, Lupton & Craig 1981), the bulk of these data will be described in the forthcoming GEOSECS Expedition Atlas (1983). Unfortunately, the GEOSECS Expedition (Craig & Turekian 1976) did not provide detailed coverage of the eastern Pacific, and thus much of our knowledge of ^{3}He injection on the East Pacific Rise is based on the results of smaller oceanographic studies, such as the work shown in Figure 8.

Submarine Hydrothermal Activity

Deep-sea hydrothermal vents were discovered at the Galapagos Rift, where plumes of warm ^{3}He-rich hydrothermal water detected with an unmanned vehicle (Weiss et al 1977, Lupton et al 1977) led the way to direct observation and collection of submarine hot-spring water with the manned, submersible ALVIN (Corliss et al 1979). Subsequent expeditions found similar hydrothermal systems at several other spreading centers, leading to the conclusion that such systems are common on active oceanic ridges, and that the hydrothermal circulation of seawater through ridge crest basalts is the principal mechanism by which helium and other mantle volatiles are extracted from fresh oceanic crust. Fyfe & Lonsdale (1981) have reviewed this topic, and list over 20 sites of seawater hydrothermal activity. However, *first-hand* observations of submarine hot-water vents have only been reported for a handful of locations, including the Galapagos Rift (Corliss et al 1979), the East Pacific Rise at 21°N latitude (RISE Project Group 1980), the Guaymas Basin (Lonsdale 1980), the Juan de Fuca Ridge (Normark et al 1981, 1982), the EPR at 20°S (Craig 1981, Ballard et al 1981), and the Red Sea geothermal brines in the axial rift of the Red Sea (Craig 1969, Lupton et al 1976, Backer & Schoell 1972). A second list can be compiled that includes sites where hydrothermal activity can be inferred from indirect geochemical or geophysical evidence: 13°N on the EPR crest (Merlivat et al 1981), 3.5°S on the EPR (Lonsdale 1977), 15°S on the EPR (Lupton & Craig 1981), and the TAG Hydrothermal Field at 26°N on the Mid-Atlantic Ridge (Rona et al 1975, Jenkins et al 1980). Although most ocean floor hydrothermal systems discovered to date are associated with the intrusional zone of an oceanic ridge crest, submarine vents have also been observed on young seamounts (Lonsdale et al 1982) and in back-arc basin extensional zones such as the Mariana Trough (Horibe et al 1982).

^{3}He has played a key role in geochemical studies of submarine hot springs for two reasons: (*a*) it is an unambiguous tag for mantle volatiles extracted from the basalts, and (*b*) since the ^{3}He flux through the oceans is fairly well known, ^{3}He concentrations can be used to estimate global oceanic hydrothermal fluxes of other chemicals. Measurements of ^{3}He, ^{4}He, heat, and other properties in pure vent fluids collected at the

Galapagos Rift (up to ~12°C) and at the 21°N EPR site (up to 350°C) have established that the heat/^{3}He ratio in these fluids is constant at 7.6×10^{-8} cal atom^{-1}, suggesting that helium and heat are extracted from MORB's at the same rate by hydrothermal circulation (Jenkins et al 1978, Craig et al 1980). The global convective heat flux from mid-ocean ridge hydrothermal activity can be calculated using the vent fluid heat/^{3}He ratio plus the oceanic ^{3}He flux (Craig et al 1975), assuming that all the excess ^{3}He in the oceans is injected by submarine hydrothermal activity. The resulting estimate (4.9×10^{19} cal yr^{-1}) is in remarkably good agreement with independent estimates based on geophysical considerations (Jenkins et al 1978). Interestingly, a deep-water temperature excess has been detected at the 21°N EPR site (Lupton et al 1980) and in the eastern Pacific between 10°S and 20°S (Reid 1982) in zones that coincide with plumes of excess ^{3}He. The magnitude of the temperature anomaly in each case is nearly identical to that predicted on the basis of the ^{3}He excess, using the heat/^{3}He ratio observed in undiluted hydrothermal vent fluids. This suggests that the regions of temperature excess can be attributed to input of hot vent water, and confirms that ^{3}He and heat are conservative tracers in the deep sea.

Pure 350°C vent water from the 21°N EPR site contains ^{3}He in concentrations about 2200 times that in saturated seawater (Lupton et al 1979), a striking confirmation that the ^{3}He plumes emanating from mid-ocean ridges actually originate at submarine hot springs. Moreover, the ^{3}He/^{4}He ratio of this hydrothermal vent helium is $R/R_A = 8.0$, nearly identical to the helium signature for basalt glasses from this area (Lupton et al 1980). Similar agreement is observed between vent helium (Jenkins et al 1978) and basalt helium (Poreda & Craig 1979, Kurz & Jenkins 1981) at the Galapagos Rift, confirming that submarine hydrothermal circulation is directly extracting MORB-type helium from the local basalts. The pure vent fluids also contain high concentrations of Mn, H_2, CH_4, Si, and other species extracted from the basalts, although in some cases the concentrations vary relative to ^{3}He. For example, ^{3}He, heat, Si, and CO_2 appear to covary exactly in the 21°N and Galapagos fluids, while Mn, Li, Ca, and Ba behave in a "nonconservative" way and exhibit varying ratios to ^{3}He and the other "conservative" properties (Corliss et al 1979, Lupton et al 1980).

Welhan and co-workers have studied methane dissolved in the 21°N EPR vent fluids and claim to have demonstrated the presence of "abiogenic" or inorganically produced methane in these fluids; this claim is based on the uniform CH_4/^{3}He ratio, the unique isotopic signature of the methane ($\delta^{13}C = -15$ to -18‰), and the absence of significant sedimentary organic carbon sources on the EPR crest (Welhan & Craig 1979, Welhan 1981, Welhan et al 1981). The CH_4/^{3}He ratio is apparently diagnostic of the origin of the methane: low $CH_4/^3He$ ratios (5×10^6) observed in the 21°N fluids are characteristic of mantle methane, while ratios 10–1000 times higher are observed in continental geothermal systems containing thermocatalytic methane. By estimating the ^3He content of a hypothetical undissipated atmosphere and the total carbon content of the Earth, Kamenskii et al (1976) calculated that the mantle should contain primitive carbon and ^3He in the ratio $C/^3He \simeq 10^7$, surprisingly close to the value observed in 21°N hydrothermal fluids. These findings are particularly important in light of recent claims by Gold & Soter (1980) that the Earth's interior contains significant quantities of primitive deep-source natural gas.

An interesting sidelight to hydrothermal vent studies has been the discovery that hot water emitted by the vents rises for hundreds of meters as buoyant plumes before reaching density equilibrium and beginning to spread laterally. This effect was very clearly demonstrated by the ^3He- and Mn-rich layer in the water column 200 m above the 21°N EPR vent field (Lupton et al 1980). Ironically, this layer at 21°N can be easily detected only with tracers such as ^3He, Mn, or CH_4—the corresponding temperature anomaly is only 0.020°C. Studies of the discharge rates from the 21°N EPR "black-smoker" vents are now in progress (Macdonald et al 1980, Converse et al 1982). When accurate estimates of the fluxes of heat and water are available, it should be possible to construct reasonable physical models of these buoyant plumes so that the height reached by the ^3He-rich vent effluent can be related to the average temperature excess and flow rate of the vents themselves.

Continental Geothermal Areas

Most of our knowledge of primordial rare-gas components has come from studies of oceanic systems, where the crust can be easily penetrated by mantle plumes, and dilution with radiogenic components is small owing to relatively low concentrations of K, U, and Th. However, very clear expressions of mantle rare-gas components are also found on the continents. One of the best examples is Yellowstone Park, Wyoming, where hot spot helium marked by $R/R_A = 16.5$ penetrates U- and Th-rich continental crust (Craig et al 1978b). In fact, all of the continental geothermal areas that have been surveyed using helium isotopes have elevated $^3He/^4He$ ratios, indicating mantle helium input. Besides Yellowstone, the list includes Cerro Prieto, Mexico (Welhan et al 1979), the Imperial Valley in southern California (Welhan et al 1978, Lupton et al 1983), the Geysers (Torgersen & Jenkins 1982), the Ethiopian Rift Valley (Craig et al 1977), and the Taupo Volcanic Zone in New Zealand (Torgersen et al 1982).

Helium isotope ratios in geothermal fluids seem to show regional trends

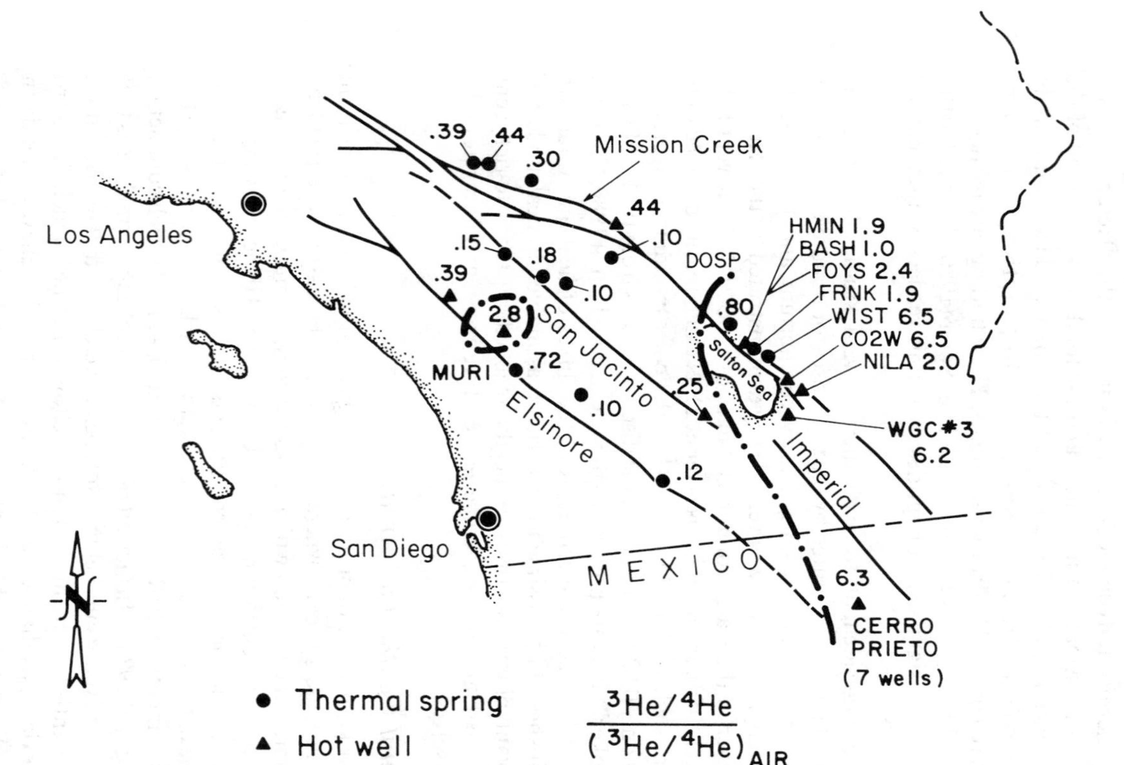

Figure 10 Helium isotope ratios R/R_A for hot springs and hot wells along southern California fault zones. Heavy dashed lines mark boundaries of zones with strong mantle helium input, as indicated by elevated $^3He/^4He$ ratios. Some of the areas sampled have been abbreviated as follows: DOSP = Dos Palmos Springs, HMIN = Hot Mineral Well, BASH = Bashford's Baths, FOYS = Fountain of Youth, FRNK = Frink Spring, WIST = Wister Mud Pots, CO2W = CO_2 Wells, NILA = Niland Slabs, WGC #3 = Western Geothermal Sinclair Well #3, MURI = Murietta Hot Springs. Data from Welhan et al (1978, 1979) and Lupton et al (1983).

even when absolute helium concentrations and other parameters do not, indicating that the $^3He/^4He$ signature gives information about the subsurface plumbing. As an example, the $^3He/^4He$ variations in hot springs and hot wells along southern California fault zones are shown in Figure 10. With one exception, $^3He/^4He$ ratios higher than air are only found in the Imperial Valley–Salton Trough region, indicating that significant input of mantle helium is confined to this zone. Furthermore, the close correspondence of the helium signatures for Cerro Prieto, WGC Well #3, Wister Mud Pots, and CO_2 Wells suggests that these sites are all connected to a single aquifer or reservoir extending from the southern end of the Salton Sea down to Cerro Prieto. This trend is not seen in the more conventional geothermal indicators, such as water chemistry, gas composition, or stable isotopes (D/H, $\delta^{18}O$, etc).

One of the differences between expressions of mantle volatiles observed in oceanic areas versus the continents is that much greater variations in the $^3He/^4He$ ratio are observed in continental geothermal zones. For example, at Yellowstone Park helium signatures down to $R/R_A = 2.8$ are found within the caldera boundary, and only a small fraction of the springs and geysers have the $R/R_A = 16.5$ signature that is thought to characterize the Yellowstone hot spot. These variations are attributed to two-component mixing between a mantle end member with high $^3He/^4He$ and radiogenic helium depleted in ^3He. (Atmospheric and tritium-produced helium are usually not important.) As shown in Figure 11, helium and argon isotope measurements by Torgersen et al (1982) in the natural springs and geothermal boreholes of New Zealand's North Island support this explanation: a strong correlation is observed between $^3He/^4He$ on the one hand and absolute He concentrations and $^{40}Ar/^{36}Ar$ on the other, again indicating mixing between a magmatic component (with high $^3He/^4He$, low $^{40}Ar/^{36}Ar$, and relatively low He) and a radiogenic component (with low $^3He/^4He$, high absolute He, and high $^{40}Ar/^{36}Ar$).

The dominant mantle component in New Zealand is the subduction zone helium characteristic of convergent plate margins, but as shown in Figure 11 (*top*), significant deviations from the average $R/R_A = 6$ signature are found within the Taupo Volcanic Zone. Surprisingly, significant helium isotope variations are even observed among boreholes samples in the Wairakei and Broadlands geothermal fields (Figure 12). This finding is in sharp contrast to the helium results at Cerro Prieto, Mexico, which showed uniform $^3He/^4He$ ratios over the borehole field (Welhan et al 1979). This difference may be related to the evolution of the rare-gas signatures during the lifetime of a geothermal field; thus the helium ratios at Cerro Prieto may be due to the relatively young age of that field. This hypothesis is now being tested by additional helium measurements at Wairakei.

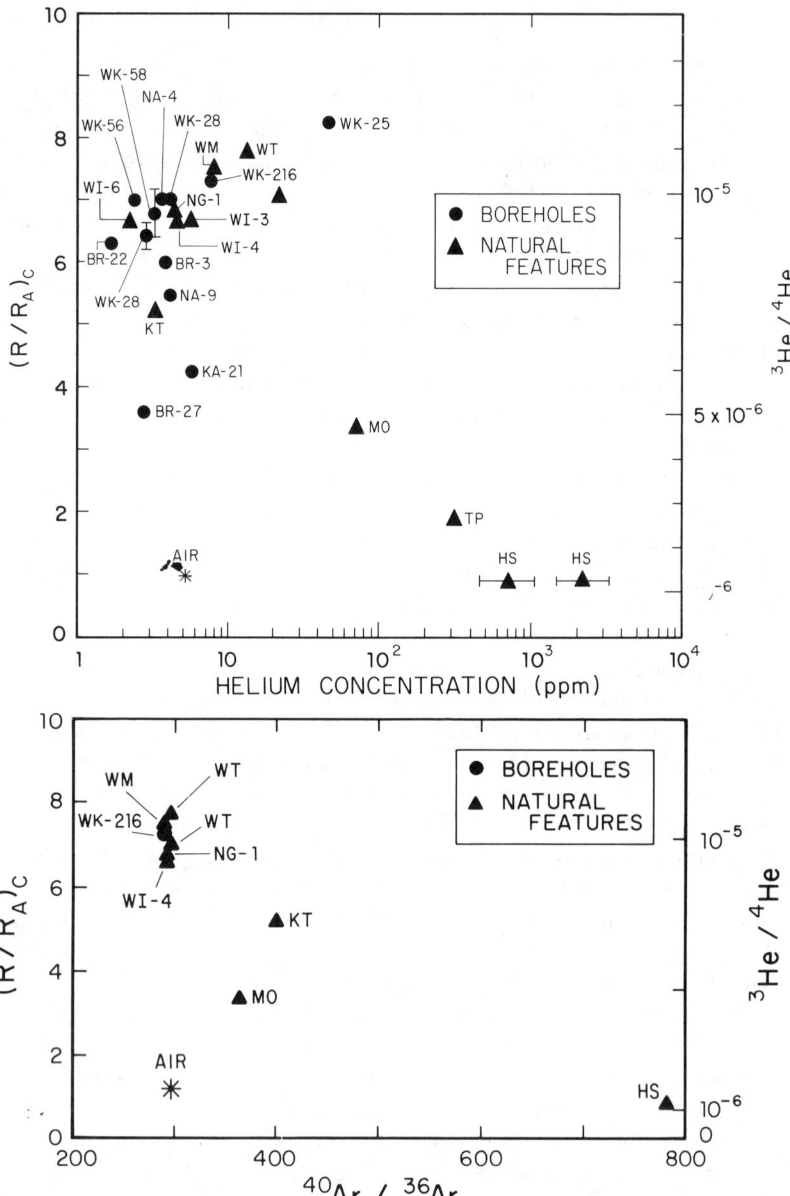

Figure 11 (*top*) Helium isotope ratio R/R_A, corrected for admixture of air, plotted versus helium concentration (in ppm) for geothermal boreholes (*circles*) and natural features (*triangles*) in New Zealand. (*bottom*) Corrected helium isotope ratio R/R_A versus $^{40}Ar/^{36}Ar$ for samples from New Zealand. Areas sampled have been abbreviated as follows: WK = Wairakei, BR = Broadlands, KA = Kawerau, NA = Ngawha, NG = Ngauruhoe, KT = Ketetahi, WT = Waiotapu, WM = Waimangu, WI = White Island, MO = Morere, TP = Te Puia, HS = Hanmer Spring. Data from Torgersen et al (1982), reprinted by permission from Elsevier Scientific Publishing Company.

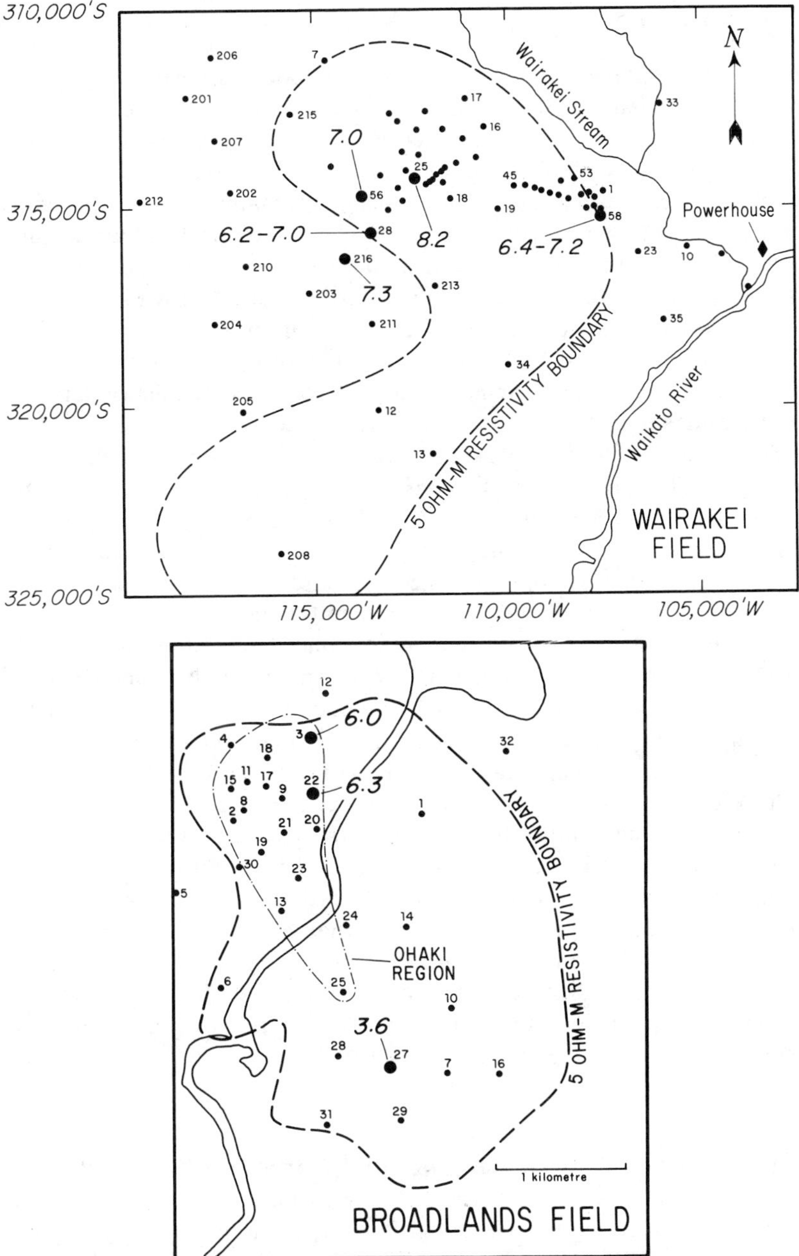

Figure 12 Helium isotope variations within the Wairakei and Broadlands geothermal fields on the North Island, New Zealand. Large numbers indicate $^3He/^4He$ ratios, R/R_A, corrected for admixture of air; small numbers denote the borehole number. Data from Torgersen et al (1982).

CONCLUSION

In this article I have outlined the separate components, particularly mantle components, that we "think" can be distinguished from the complicated array of isotope variations in terrestrial rare gases. Some of these components, particularly those produced by radioactive or nuclear mechanisms, are easily identified. Others, such as signatures from depleted vs undepleted mantle reservoirs, are not as clear, and one could argue quite convincingly that we have a great deal to learn about both the original and the present-day rare-gas content of the Earth's interior. However, we have learned where to look and what tools to use in order to probe the deep Earth: the most useful information about the isotope geochemistry of the mantle has come from combined studies of oceanic spreading centers, hot spots, and subduction areas, using $^{3}He/^{4}He$, $^{40}Ar/^{36}Ar$, $^{129}Xe/^{132}Xe$, $^{87}Sr/^{86}Sr$, and $^{143}Nd/^{144}Nd$ as tracers. Studies of diamonds, xenoliths, and other mantle-derived samples are also quite promising, but the results are more variable and therefore more difficult to interpret.

I have also discussed some examples of the application of rare gases, particularly helium isotopes, as tracers in natural systems. In some ways this has been more successful than our attempts to use inert gases to identify primitive terrestrial reservoirs and address questions concerning the early history of the Earth. For example, whether or not we understand the origin and significance of the helium in the mantle, when this ^{3}He-rich helium leaks into the oceans at active spreading centers it becomes an extremely powerful tracer for mapping ocean circulation patterns and for studying ridge crest tectonic and hydrothermal processes. Helium isotope studies have thus become an integral part of the current effort in ocean science, and this will continue to be an important application in the future. As applied to continental areas, the usefulness of rare-gas isotopes as tracers is not very well established. However, we do know that helium isotopes give completely different information from the more conventional tracers used in geothermal and hydrologic prospecting. Future studies of continental thermal areas will help to define the role of rare gases in probing for geothermal and deep-source gas reservoirs.

ACKNOWLEDGMENTS

This paper profited from discussions with I. Kaneoka, M. Ozima, R. Hart, M. Garcia, K. Marti, G. Wetherill, H. Wakita, I. N. Tolstikhin, F. A. Podosek, Y. Horibe, and H. Craig. I thank Jill Marson for typing the manuscript and Dave Crouch for drafting the figures. Mark Kurz and Bill Jenkins kindly provided figures from their manuscripts. My research on ^{3}He and rare gases has been funded by the US National Science Foundation and by the US Geological Survey.

Literature Cited

Aldrich, L. T., Nier, A. O. 1948. The occurrence of He^3 in natural sources of helium. *Phys. Rev.* 74:1590–94

Axford, W. I. 1968. The polar wind and the terrestrial helium budget. *Geophys. Res.* 73:6855–59

Axford, W. I., Buhler, F., Chivers, H. J. A., Eberhardt, P., Geiss, J. 1972. Auroral helium precipitation. *J. Geophys. Res.* 77:6724–30

Backer, H., Schoell, M. 1972. New deeps with brines and metalliferous sediments in the Red Sea. *Nature Phys. Sci.* 240:153–58

Ballard, R. D., Morton, J., Francheteau, J. 1981. Geology and high temperature hydrothermal circulation of ultra-fast spreading ridge: East Pacific Rise at 20°S. *EOS, Trans. Am. Geophys. Union* 62:912 (Abstr.)

Banks, P. M., Holzer, T. E. 1969. High-latitude plasma transport: the polar wind. *J. Geophys. Res.* 26:6317–32

Barnes, R. O., Clarke, W. B. 1980. $^3He/^4He$, He, Ne, and Ar concentrations in pore fluids of the DSDP sites 398 to 438. *EOS, Trans. Am. Geophys. Union* 61:996 (Abstr.)

Baskov, Y., Vetshteyn, V., Surikov, S., Tolstikhin, I., Malyuk, G., Mishina, T. 1973. Isotopic composition of H, O, C, Ar, and He in hot springs and gases in the Kurile-Kamchatka volcanic region as indicators of formation conditions. *Geochem. Int.* 10:130–38

Bates, D. R., McDowell, M. R. C. 1957. Atmospheric helium. *J. Atmos. Terr. Phys.* 11:200–8

Batiza, R. 1977. Petrology and chemistry of Guadalupe Island: An alkalic seamount on a fossil ridge crest. *Geology* 5:760–74

Bernatowicz, T. J., Podosek, F. A. 1978. Nuclear components in the atmosphere. In *Terrestrial Rare Gases*, ed. E. C. Alexander, M. Ozima, *Adv. Earth Planet. Sci.* 3:99–135. Tokyo: Jpn. Sci. Soc. Press

Bernatowicz, T. J., Goettel, K. A., Hohenberg, C. M., Podosek, F. A. 1979. Anomalous noble gases in josephinite and associated rocks? *Earth Planet. Sci. Lett.* 43:368–84

Bertrami, R., Lombardi, S., Hooker, P. J., McRae, T. A., O'Nions, R. K., Oxburgh, E. R. 1982. Helium isotope variation in geothermal areas of northern Italy. *EOS, Trans. Am. Geophys. Union* 63:457 (Abstr.)

Bieri, R. H. 1971. Dissolved noble gases in marine waters. *Earth Planet. Sci. Lett.* 10:329–33

Bieri, R. H., Koide, M., Goldberg, E. D. 1966. The noble gas contents of Pacific seawaters. *J. Geophys. Res.* 71:5243–65

Bird, J. M., Weathers, M. S. 1975. Josephinite: specimens from the Earth's core? *Earth Planet. Sci. Lett.* 28:51–64

Bochsler, P., Stettler, A., Bird, J. M., Weathers, M. S. 1978. Excess ^3He and ^{21}Ne in josephinite. *Earth Planet. Sci. Lett.* 39:67–74

Boulos, M. S., Manuel, O. K. 1971. The xenon record of extinct radioactivities in the earth. *Science* 174:1334–36

Buhler, F., Axford, W. I., Chivers, H. J. A., Marti, K. 1976. Helium isotopes in an aurora. *J. Geophys. Res.* 81:111–15

Butler, W. A., Jeffrey, P. M., Reynolds, J. H., Wasserburg, G. J. 1963. Isotopic variations in terrestrial xenon. *J. Geophys. Res.* 68:3283–91

Clague, D. A., Moore, J. G., Normark, W. R. 1981. Loihi seamount: early alkalic eruptive phase. *EOS, Trans. Am. Geophys. Union* 62:1082 (Abstr.)

Clarke, W. B., Kugler, G. 1973. Dissolved helium in groundwater: a possible method for uranium and thorium prospecting. *Econ. Geol.* 68:243–51

Clarke, W. B., Beg, M. A., Craig, H. 1969. Excess ^3He in the sea: evidence for terrestrial primordial helium. *Earth Planet. Sci. Lett.* 6:213–20

Clarke, W. B., Jenkins, W. J., Top, Z. 1976. Determination of tritium by mass spectrometric measurement of ^3He. *Int. J. App. Radiat. Isot.* 27:515–22

Converse, D. R., Holland, H. D., Edmond, J. M. 1982. Hydrothermal flow rates at 21°N. *EOS, Trans. Am. Geophys. Union* 63:472 (Abstr.)

Corliss, J. B., Dymond, J., Gordon, L. I., Edmond, J. M., von Herzen, R. P., Ballard, R. D., Green, K., Williams, D., Bainbridge, A., Crane, K., van Andel, T. H. 1979. Submarine thermal springs on the Galapagos Rift. *Science* 203:1073–83

Cowan, G. A. 1976. A natural fission reactor. *Sci. Am.* 235:36–47

Craig, H. 1969. Geochemistry and origin of the Red Sea brines. In *Hot Brines and Recent Heavy Metal Deposits in the Red Sea*, ed. E. T. Degens, D. A. Ross, pp. 208–42. Berlin: Springer-Verlag

Craig, H. 1981. Hydrothermal plumes and tracer circulation along the East Pacific Rise: 20°N to 20°S. *EOS, Trans. Am. Geophys. Union* 62:893 (Abstr.)

Craig, H., Chou, C. C. 1982. Helium isotopes and gases in Dye 3 ice cores. *EOS, Trans. Am. Geophys. Union* 63:298 (Abstr.)

Craig, H., Lupton, J. E. 1976. Primordial neon, helium, and hydrogen in oceanic basalts. *Earth Planet. Sci. Lett.* 31:369–85

Craig, H., Lupton, J. E. 1978. Helium isotope variations: evidence for mantle plumes at Yellowstone, Kilauea, and the Ethiopian Rift Valley. *EOS, Trans. Am. Geophys. Union* 59:1194 (Abstr.)

Craig, H., Lupton, J. E. 1981. Helium-3 and

mantle volatiles in the ocean and the oceanic crust. In *The Sea*, ed. C. Emiliani, 7:391–428. New York: Wiley

Craig, H., Rison, W. 1982. Helium-3: Indian Ocean hot spots and the East African rift. *EOS, Trans. Am. Geophys. Union* 63:1144 (Abstr.)

Craig, H., Turekian, K. K. 1976. The GEOSECS Program: 1973–1976. *Earth Planet. Sci. Lett.* 32:217–19

Craig, H., Weiss, R. F. 1971. Dissolved gas saturation anomalies and excess helium in the ocean. *Earth Planet. Sci. Lett.* 10:289–96

Craig, H., Clarke, W. B., Beg, M. A. 1975. Excess ^{3}He in deep water on the East Pacific Rise. *Earth Planet Sci. Lett.* 26:125–32

Craig, H., Lupton, J. E., Horowitz, R. M. 1977. Isotopic geochemistry and hydrology of geothermal waters in the Ethiopian Rift Valley. *Scripps Inst. Oceanogr. Ref. No. 77-14*, Univ. Calif., San Diego

Craig, H., Lupton, J. E., Horibe, Y. 1978a. A mantle helium component in circum-Pacific volcanic gases: Hakone, the Marianas, and Mt. Lassen. In *Terrestrial Rare Gases*, ed. E. C. Alexander, M. Ozima. Tokyo: Jpn. Sci. Soc. Press; *Adv. Earth Planet. Sci.* 3:3–16

Craig, H., Lupton, J. E., Welhan, J. A., Poreda, R. 1978b. Helium isotope ratios in Yellowstone and Lassen Park volcanic gases. *Geophys. Res. Lett.* 5:897–900

Craig, H., Poreda, R., Lupton, J. E., Marti, K., Regnier, S. 1979. Rare gases and hydrogen in josephinite. *EOS, Trans. Am. Geophys. Union* 60:970 (Abstr.)

Craig, H., Welhan, J. A., Kim, K., Poreda, R., Lupton, J. E. 1980. Geochemical studies of the 21°N EPR hydrothermal fluids. *EOS, Trans. Am. Geophys. Union* 61:992 (Abstr.)

Delaney, J. R., Johnson, H. P., Karsten, J. L. 1981. The Juan de Fuca Ridge—hot spot—propagating rift system: new tectonic, geochemical, and magnetic data. *J. Geophys. Res.* 86:11747–50

DePaolo, D. J., Wasserburg, G. J. 1976. Nd isotopic variations and petrogenic models. *Geophys. Res. Lett.* 3:249–52

Donahue, T. M. 1971. Polar ion flow: wind or breeze? *Rev. Geophys. Space Phys.* 9: 1–9

Downing, R. G., Hennecke, E. W., Manuel, O. K. 1977. Josephinite: a terrestrial alloy with radiogenic xenon-129 and the noble gas imprint of iron meteorites. *Geochem. J.* 11:219–29

Drozd, R. J. 1974. *Krypton and xenon in lunar and terrestrial samples.* PhD thesis. Washington Univ., St. Louis, Mo.

Drozd, R. J., Hohenberg, C. M., Morgan, C. J. 1974. Heavy rare gases from Rabbit Lake (Canada) and the Oklo mine (Gabon): natural spontaneous chain reactions in old uranium deposits. *Earth Planet. Sci. Lett.* 23:28–33

Dymond, J., Hogan, L. 1973. Noble gas abundance patterns in deep-sea basalts—primordial gases from the mantle. *Earth Planet. Sci. Lett.* 20:131–39

Emerson, D. E., Stroud, L., Meyer, T. O. 1966. The isotopic abundance of neon from helium-bearing natural gases. *Geochim. Cosmochim. Acta* 30:847–54

Fisher, D. E. 1974. The planetary primordial component of rare gases in the deep Earth. *Geophys. Res. Lett.* 1:161–64

Fisher, D. E. 1981. Heavy and radiogenic rare gases trapped in deep-sea basalts. In *The Sea*, ed. C. Emiliani, 7:429–42. New York: Wiley

Fyfe, W. S., Lonsdale, P. 1981. Ocean floor hydrothermal activity. In *The Sea*, ed. C. Emiliani, 7:589–638. New York: Wiley

Geiss, J., Buhler, F., Cerutti, H., Eberhardt, P., Filleux, C. H. 1972. Solar wind composition experiment, Apollo 16 Preliminary Science Report. *NASA Spec. Publ. SP-315*, pp. 14.1–14.10

GEOSECS Expedition Atlas. 1983. *Shorebased Data.* Washington DC: USGPO

Gerling, E. K., Mamyrin, B. A., Tolstikhin, I. N., Yakovleva, S. S. 1971. Helium isotope composition in some rocks. *Geochem. Int.* 8:755–61

Gold, T., Soter, S. 1980. The deep-Earth-gas hypothesis. *Sci. Am.* 242:154–61

Hart, R., Dymond, J., Hogan, L. 1979. Preferential formation of the atmosphere-sialic crust system from the upper mantle. *Nature* 278:156–59

Hart, R., Dymond, J., Hogan, L. 1982. Noble gas evidence for sea water interaction with residual melts along the Juan de Fuca Ridge. *EOS, Trans. Am. Geophys. Union* 63:1154 (Abstr.)

Hart, S. R. 1973. Submarine basalts from Kilauea Rift, Hawaii: nondependence of trace element composition on extrusion depth. *Earth Planet. Sci. Lett.* 20:201–3

Hart, S. R., Schilling, J. G., Powell, J. L. 1973. Basalts from Iceland and along the Reykjanes Ridge: Sr isotope geochemistry. *Nature* 246:104–7

Hedge, C. E. 1978. Strontium isotopes in basalts from the Pacific Ocean basin. *Earth Planet. Sci. Lett.* 38:88–94

Hedge, C. E., Peterman, Z. E., Dickinson, W. R. 1972. Petrogenesis of lavas from western Samoa. *Geol. Soc. Am. Bull.* 83:2709–14

Hedge, C. E., Watkins, N. D., Hildreth, R. A., Doering, W. P. 1973. ^{87}Sr/^{86}Sr ratios in basalts from islands in the Indian Ocean. *Earth Planet. Sci. Lett.* 21:29–34

Hennecke, E. W., Manuel, O. K. 1975a. Noble gases in CO_2 well gas, Harding County, New Mexico. *Earth Planet. Sci. Lett.* 27:346–55

Hennecke, E. W., Manuel, O. K. 1975b. Noble gases in a Hawaiian xenolith. *Nature* 257:778–80

Holzer, T. E., Axford, W. I. 1971. Interaction between interstellar helium and the solar wind. *J. Geophys. Res.* 76:6965–70

Hooker, P. J., Condomines, M., O'Nions, R. K., Oxburgh, E. R. 1982. Helium and other isotopic variations in the Krafka rifting zone, Iceland. *EOS, Trans. Am. Geophys. Union* 63:457 (Abstr.)

Horibe, Y., Kim, K. R., Craig, H. 1982. Deep ocean hydrothermal vents in the Mariana Trough. *5th Int. Conf. Geochron. Cosmochron. Isotope Geol., Nikko Park, Jpn.*, p. 154 (Abstr.)

Jambon, A., Shelby, J. E. 1979. Helium diffusion and solubility in obsidians and basaltic glass in the range 200–300°C. *Earth Planet. Sci. Lett.* 51:206–14

Jenkins, W. J. 1980. Tritium and ^{3}He in the Sargasso Sea. *J. Mar. Res.* 38:533–69

Jenkins, W. J., Clarke, W. B. 1976. The distribution of ^{3}He in the western Atlantic Ocean. *Deep-Sea Res.* 23:481–94

Jenkins, W. J., Beg, M. A., Clarke, W. B., Wangersky, P. J., Craig, H. 1972. Excess ^{3}He in the Atlantic Ocean. *Earth Planet. Sci. Lett.* 16:122–26

Jenkins, W. J., Edmond, J. M., Corliss, J. B. 1978. Excess ^{3}He and ^{4}He in Galapagos submarine hydrothermal waters. *Nature* 272:156–58

Jenkins, W. J., Rona, P. A., Edmond, J. M. 1980. Excess ^{3}He in deep-water over the Mid-Atlantic Ridge at 26°N: evidence of hydrothermal activity. *Earth Planet. Sci. Lett.* 49:39–44

Johnson, H. E. 1971. *Astrophysical flow problems*. PhD thesis. Univ. Calif., San Diego

Johnson, H. E., Axford, W. I. 1969. Production and loss of ^{3}He in the earth's atmosphere. *J. Geophys. Res.* 74:2433–38

Kamenskii, I. L., Yakutseni, V. P., Mamyrin, B. A., Anufriyev, S. G., Tolstikhin, I. N. 1971. Helium isotopes in nature. *Geokhimiya* 8:914–31

Kamenskii, I. L., Lobkov, V. A., Prasolov, E. M., Beskrovny, N. S., Kudryavtseva, E. I., Anufriev, G. S., Pavlov, V. B. 1976. The components of the upper mantle of the Earth in gases of Kamchatka. *Geokhimiya* 5:682–95

Kaneoka, I. 1980. Rare gas isotopes and mass fractionation: an indicator of gas transport into or from a magma. *Earth Planet. Sci. Lett.* 48:284–92

Kaneoka, I. 1981. Noble gas constraints on the layered structure of the mantle. *Rock Magn. Paleogeophys.* 8:94–99

Kaneoka, I., Takaoka, N. 1978. Excess ^{129}Xe and high ^{3}He/^{4}He ratios in olivine phenocrysts of Kapuho lava and xenolithic dunites from Hawaii. *Earth Planet. Sci. Lett.* 39:382–86

Kaneoka, I., Takaoka, N. 1980. Rare gas isotopes in Hawaiian ultramafic nodules and volcanic rocks: constraints on genetic relationships. *Science* 208:1366–68

Kaneoka, I., Takaoka, N. 1982. Noble gas state in the Earth's interior—the significance of noble gas data on samples from the hot spot area. *5th Int. Conf. Geochron. Cosmochron. Isotope Geol., Nikko Park, Jpn.*, pp. 177–78 (Abstr.)

Kaneoka, I., Takaoka, N., Aoki, K. 1978. Rare gases in mantle-derived rocks and minerals. In *Terrestrial Rare Gases*, ed. E. C. Alexander, M. Ozima. Tokyo: Jpn. Sci. Soc. Press; *Adv. Earth Planet. Sci.* 3:71–84

Kirsten, T. 1968. Incorporation of rare gases in solidifying enstatite melts. *J. Geophys. Res.* 73:2807–10

Kirsten, T., Richter, H., Storzer, D. 1981. Abundance patterns of rare gases in submarine basalts and glasses. *Meteoritics* 16:341 (Abstr.)

Kirsten, T., Richter, H., Storzer, D. 1982. Rare gas isotope systematics of submarine basalts and glasses. *5th Int. Conf. Geochron. Cosmochron. Isotope Geol., Nikko Park, Jpn.*, pp. 183–84 (Abstr.)

Kockarts, G. 1973. Helium in the terrestrial atmosphere. *Space Sci. Rev.* 14:723–57

Kockarts, G., Nicolet, M. 1962. Le problème aéronomique de l'hélium et de l'hydrogène neutres. *Ann. Geophys.* 18:269–90

Krylov, A. Ya., Mamyrin, B. A., Khabarin, L. A., Mazina, T. I., Silin, Yu. I. 1974. Helium isotopes in ocean floor bedrock. *Geokhimiya* 8:1220–25

Kurz, M. D., Jenkins, W. J. 1981. The distribution of helium in oceanic basalt glasses. *Earth Planet Sci. Lett.* 53:41–54

Kurz, M. D., Jenkins, W. J., Hart, S. R., Clague, D. 1981. Helium isotopic systematics of oceanic islands and Loihi seamount. *EOS, Trans. Am. Geophys. Union* 62:1083 (Abstr.)

Kurz, M. D., Jenkins, W. J., Hart, S. R. 1982a. Helium isotopic systematics of oceanic islands and mantle heterogeneity. *Nature* 297:43–47

Kurz, M. D., Jenkins, W. J., Schilling, J. G., Hart, S. R. 1982b. Helium isotopic variations in the mantle beneath the central North Atlantic Ocean. *Earth Planet. Sci. Lett.* 58:1–14

Kyser, T. K., Rison, W. 1982. Systematics of rare gas isotopes in basic lavas and ultramafic xenoliths. *J. Geophys. Res.* 87:5611–30

Lonsdale, P. 1977. Blustering of suspension-feeding macrobenthos near abyssal hydrothermal vents at oceanic spreading centers. *Deep-Sea Res.* 24:852–63

Lonsdale, P. 1980. Hydrothermal plumes and baritic sulfide mounds at a Gulf of California spreading center. *EOS, Trans. Am. Geophys. Union* 61:995 (Abstr.)

Lonsdale, P., Batiza, R., Simkin, T., Vanko, D. 1982. Submersible study of active hydrothermal vents and massive sulphide deposits on young off-ridge volcanoes near the East Pacific Rise at 20°48′N. *EOS, Trans. Am. Geophys. Union* 63:472 (Abstr.)

Lupton, J. E. 1973. Direct accretion of ^{3}He and ^{3}H from cosmic rays. *J. Geophys. Res.* 78:8330–37

Lupton, J. E. 1976. The ^{3}He distribution in deep water over the Mid-Atlantic Ridge. *Earth Planet. Sci. Lett.* 32:371–74

Lupton, J. E. 1979. Helium-3 in the Guaymas Basin: evidence for injection of mantle volatiles in the Gulf of California. *J. Geophys. Res.* 84:7446–52

Lupton, J. E. 1982. Helium isotope variations in Juan de Fuca Ridge basalts. *EOS, Trans. Am. Geophys. Union* 63:1147 (Abstr.)

Lupton, J. E., Craig, H. 1975. Excess ^{3}He in oceanic basalts: evidence for terrestrial primordial helium. *Earth Planet. Sci. Lett.* 26:133–39

Lupton, J. E., Craig, H. 1981. A major ^{3}He source on the East Pacific Rise. *Science* 214:13–18

Lupton, J. E., Weiss, R. F., Craig, H. 1976. Mantle helium in the Red Sea brines. *Nature* 266:244–46

Lupton, J. E., Weiss, R. F., Craig, H. 1977. Mantle helium in hydrothermal plumes in the Galapagos Rift. *Nature* 267:603–4

Lupton, J. E., Craig, H., Klinkhammer, G. 1979. Helium-3 and manganese at the 21°N EPR hydrothermal site. *EOS, Trans. Am. Geophys. Union* 60:864 (Abstr.)

Lupton, J. E., Klinkhammer, G. P., Normark, W. R., Haymon, R., Macdonald, K. C., Weiss, R. F., Craig, H. 1980. Helium-3 and manganese at the 21°N East Pacific Rise hydrothermal site. *Earth Planet. Sci. Lett.* 50:115–27

Lupton, J. E., Johnson, H. P., Delaney, J. H. 1981. Hydrothermal ^{3}He on the Juan de Fuca Ridge *EOS, Trans. Am. Geophys. Union* 62:913 (Abstr.)

Lupton, J. E., Craig, H., Poreda, R., Welhan, J., Horowitz, R. 1983. Helium isotope variation along southern California fault zones. In preparation

Macdonald, K. C. 1982. Mid ocean ridges: Fine scale tectonic, volcanic and hydrothermal processes within the plate boundary zone. *Ann. Rev. Earth Planet. Sci.* 10:155–90

Macdonald, K. C., Becker, K., Spiess, F. N., Ballard, R. D. 1980. Hydrothermal heat flux of the "black smoker" vents on the East Pacific Rise. *Earth Planet Sci. Lett.* 48:1–7

Mamyrin, B. A., Tolstikhin, I. N. 1983. *Helium Isotopes in Nature*. Amsterdam: Elsevier. In press

Mamyrin, B. A., Tolstikhin, I. N., Anufriyev, G. S., Kamenskii, I. L. 1969. Abnormal isotopic composition of helium in volcanic gases. *Dokl. Akad. Nauk SSSR* 184:191

Mamyrin, B. A., Anufriev, G. S., Kamenskii, I. L., Tolstikhin, I. N. 1970. Determination of the isotopic composition of atmospheric helium. *Geochem. Int.* 7:498–505

Marty, B., Zashu, S., Ozima, M. 1983. Two noble gas components in a Mid-Atlantic Ridge basalt. *Nature*. Submitted for publication

McDougall, I., Compston, W. 1965. Strontium isotope composition and potassium-rubidium ratios in some rocks from Réunion and Rodriquez, Indian Ocean. *Nature* 207:252–53

Merlivat, L., Boulegue, J., Dimon, B. 1981. Helium isotopes and manganese distribution in the water column at 13°N on the East Pacific Rise. *EOS, Trans. Am. Geophys. Union* 62:913 (Abstr.)

Moore, J. G., Clague, D. A., Normark, W. R. 1982. Diverse basalt types from Loihi seamount, Hawaii. *Geology* 10:88–92

Morrison, P., Pine, J. 1955. Radiogenic origin of the helium isotopes in rock. *Ann. NY Acad. Sci.* 62:71–92

Nagao, K., Takaoka, N., Matsuo, S., Mizutani, Y., Matsubayashi, O. 1980. Change in rare gas composition of the fumarolic gases from the Showa-shinzan volcano. *Geochem. J.* 14:139–43.

Nagao, K., Takaoka, N., Matsubayashi, O. 1981. Rare gas isotopic compositions in natural gases of Japan. *Earth Planet. Sci. Lett.* 53:175–88

Normark, W. R., Delaney, J. R., Morton, J. L., Koski, R., Barnes, I., Stevenson, A., Hayba, D., Bargar, K., Johnson, H. P., Clague, D. 1981. Hydrothermal vents and sulfide deposits on the southern Juan de Fuca Ridge. *EOS, Trans. Am. Geophys. Union* 62:913 (Abstr.)

Normark, W. R., Lupton, J. E., Murray, J. W., Koski, R. A., Clague, D. A., Morton, J. L., Delaney, J. R., Johnson, H. P. 1982. Polymetallic sulfide deposits and water column tracers of active hydrothermal vents on the southern Juan de Fuca Ridge. *Mar. Technol. Soc. J.* 16:46–53

O'Nions, R. K., Pankhurst, R. J. 1974. Petrogenetic significance of isotope and trace element variation in volcanic rocks from the Mid-Atlantic. *J. Petrol.* 15:603–34

O'Nions, R. K., Hamilton, P. J., Evensen, N. M. 1977. Variations in $^{143}Nd/^{144}Nd$ and $^{87}Sr/^{86}Sr$ ratios in oceanic basalts. *Earth Planet. Sci. Lett.* 34:13–22

Östlund, H. G., Dorsey, H. G., Rooth, C. G. 1974. GEOSECS North Atlantic radiocarbon and tritium results. *Earth Planet. Sci. Lett.* 23:69–86

Ozima, M., Alexander, E. C. 1976. Rare gas fractionation patterns in terrestrial samples and the earth-atmosphere evolution model. *Rev. Geophys. Space Phys.* 14:385–90

Ozima, M., Podosek, F. A. 1983. *Geochemistry of Noble Gases.* Cambridge Univ. Press. In press

Ozima, M., Zashu, S. 1983a. Primitive He in diamonds. *Science.* In press

Ozima, M., Zashu, S. 1983b. Noble gases in submarine pillow volcanic glasses. *Earth Planet. Sci. Lett.* 62:24–40

Patterson, T. N. L. 1968. Escape of atmospheric helium by nonthermal processes. *Rev. Geophys. Space Phys.* 6:553–57

Pepin, R. O., Signer, P. 1965. Primordal rare gases in meteorites. *Science* 149:253–65

Phinney, D., Tennyson, J., Frick, U. 1978. Xenon in CO_2 well gas revisited. *J. Geophys. Res.* 83:2312–19

Polak, B. G., Kononov, V. I., Tolstikhin, I. N., Mamyrin, B. A., Khabarin, L. 1975. The helium isotopes in thermal fluids. In *Thermal and Chemical Problems of Thermal Waters*, ed. A. T. Johnson, 119:15–29. Grenoble: Int. Assoc. Hydrol. Sci. Publ.

Poreda, R., Craig, H. 1979. Helium and neon in oceanic volcanic rocks. *EOS, Trans. Am. Geophys. Union* 60:969 (Abstr.)

Poreda, R., Craig, H., Schilling, J. G. 1980. $^3He/^4He$ variations along the Reykjanes Ridge. *EOS, Trans. Am. Geophys. Union* 61:1158 (Abstr.)

Reid, J. L. 1982. Evidence of an effect of heat flux from the East Pacific Rise upon the characteristics of mid-depth waters. *Geophys. Res. Lett.* 9:381–84

RISE Project Group. 1980. East Pacific Rise: hot springs and geophysical experiments. *Science* 207:1421–33

Rison, W., Craig, H. 1981. Loihi seamount: mantle volatiles in the basalts. *EOS, Trans. Am. Geophys. Union* 62:1083 (Abstr.)

Rison, W., Craig, H. 1982. Helium-3: coming of age in Samoa. *EOS, Trans. Am. Geophys. Union* 63:1144 (Abstr.)

Rona, P. A., McGregor, B. A., Betzer, P. R., Bolger, G. W., Krause, D. C. 1975. Anomalous water temperatures over the Mid-Atlantic Ridge crest at 26° north latitude. *Deep-Sea Res.* 22:611–18

Saito, K., Basu, A. R., Alexander, E. C. 1978. Planetary-type rare gases in an upper mantle-derived amphibole. *Earth Planet. Sci. Lett.* 39:274–80

Sano, Y., Tominaga, T., Nakamura, Y., Wakita, H. 1983. $^3He/^4He$ ratios of methane-rich natural gases in Japan. *Geochem. J.* Submitted for publication

Sayles, F. L., Jenkins, W. J. 1982. Advection of pore fluids through sediments in the equatorial east Pacific. *Science* 217:245–47

Schilling, J. G. 1975. Azores mantle blob: rare earth evidence. *Earth Planet. Sci. Lett.* 25:103–15

Schilling, J. G., Bergeron, M. B., Evans, R. 1980. Halogens in the mantle beneath the North Atlantic. *Philos. Trans. R. Soc. London Ser. A* 297:147–78

Sheldon, W. R., Kern, J. W. 1972. Atmospheric helium and geomagnetic field reversals. *J. Geophys. Res.* 77:6194–201

Shukolyukov, Y. A. 1970. Nuclear fission of uranium in nature. *Atomizdat, Moscow*, p. 270

Smith, S. P., Reynolds, J. H. 1981. Excess ^{129}Xe in a terrestrial sample as measured in a pristine system. *Earth Planet. Sci. Lett.* 54:236–38

Staudacher, Th., Allegre, C. J. 1981. ^{129}Xe excess in typical MORB from the Pacific Ocean: consequences about the mean age of the atmosphere and the structure of the earth's mantle. *EOS, Trans. Am. Geophys. Union* 62:420 (Abstr.)

Stommel, H. 1982. Is the south Pacific Helium-3 plume dynamically active? *Earth Planet. Sci. Lett.* 61:63–67

Stommel, H., Arons, A. B. 1960. On the abyssal circulation of the world ocean. Part II—An idealized model of circulation pattern and amplitude in oceanic basins. *Deep-Sea Res.* 6:217–33

Sun, S.-S. 1980. Lead isotopic study of young volcanic rocks from mid-ocean ridges, oceanic islands and island arcs. *Philos. Trans. R. Soc. London Ser. A* 297:409–45

Sun, S.-S., Tatsumoto, M., Schilling, J. G. 1975. Mantle plume mixing along the Reykjanes Ridge axis: lead isotopic evidence. *Science* 190:143–47

Takaoka, N., Nagao, K. 1980. Rare-gas studies of Cretaceous deep-sea basalts. In *Initial Reports of Deep Sea Drilling Project*, 51–53:1121–26. Washington DC: USGPO

Takaoka, N., Ozima, M. 1978. Rare-gas isotopic composition in diamonds. In *Terrestrial Rare Gases*, ed. E. C. Alexander, M. Ozima. Tokyo: Jpn. Sci. Soc. Press; *Adv. Earth Planet. Sci.* 3:65–70

Tolstikhin, I. N. 1978. A review: some recent advances in isotope geochemistry of light rare gases. In *Terrestrial Rare Gases*, ed. E. C. Alexander, M. Ozima. Tokyo: Jpn. Sci. Soc. Press; *Adv. Earth Planet. Sci.* 3:33–62

Torgersen, T., Jenkins, W. J. 1982. Helium isotopes in geothermal systems: Iceland, The Geysers, Raft River, and Steamboat Springs. *Geochim. Cosmochim. Acta* 46:739–48

Torgersen, T., Top, Z., Clarke, W. B., Jenkins, W. J., Broecker, W. S. 1977. A new method for physical limnology—tritium-helium-3 ages—results for Lakes Erie, Huron, and Ontario. *Limnol. Oceanogr.* 22:181–93

Torgersen, T., Lupton, J. E., Sheppard, D., Giggenbach, W. 1982. Helium isotope variations in the thermal areas of New Zealand. *J. Volcanol. Geotherm. Res.* 12:283–98

Weiss, R. F. 1977. Hydrothermal manganese in the deep sea: scavenging residence time and Mn/^{3}He relationships. *Earth Planet. Sci. Lett.* 37:257–62

Weiss, R. F., Lonsdale, P., Lupton, J. E., Bainbridge, A. E., Craig, H. 1977. Hydrothermal plumes in the Galapagos Rift. *Nature* 267:600–3

Welhan, J. A. 1981. *Carbon and hydrogen gases in hydrothermal systems: the search for a mantle source.* PhD thesis. Univ. Calif., San Diego. 194 pp.

Welhan, J. A., Craig, H. 1979. Methane and hydrogen in East Pacific Rise hydrothermal fluids. *Geophys. Res. Lett.* 6:829–31

Welhan, J., Lupton, J., Craig, H. 1978. Helium isotope ratios in southern California fault zones. *EOS, Trans. Am. Geophys. Union* 59:1197 (Abstr.)

Welhan, J. A., Poreda, R., Lupton, J. E., Craig, H. 1979. Gas chemistry and helium isotopes at Cerro Prieto. *Geothermics* 8:241–44

Welhan, J., Kim, K., Craig, H. 1981. Hydrocarbons in 21°N hydrothermal fluids. *EOS, Trans. Am. Geophys. Union* 62:913 (Abstr.)

Wetherill, G. W. 1954. Variations in the isotopic abundances of neon and argon extracted from radioactive minerals. *Phys. Rev.* 96:679–83

White, W. M., Hofmann, A. W. 1982. Sr and Nd isotope geochemistry of oceanic basalts and mantle evolution. *Nature* 296:821–25

White, W. M., Schilling, J. G. 1978. The nature and origin of geochemical variations in Mid-Atlantic Ridge basalts from the central North Atlantic. *Geochim. Cosmochim. Acta* 42:1501–17

White, W. M., Schilling, J. G., Hart, S. R. 1976. Evidence for the Azores mantle plume from strontium isotope geochemistry of the central North Atlantic. *Nature* 263:659–63

Zadnick, M. G. 1981. *Variations in the isotopic composition of neon in terrestrial matter.* PhD thesis. Univ. West. Aust., Nedlands. 231 pp.

Zadnick, M. G., Jeffery, P. M. 1982. Observations on the isotopic composition of terrestrial neon. *5th Int. Conf. Geochron. Cosmochron. Isotope Geol., Nikko Park, Jpn.*, pp. 407–8 (Abstr.)

Zindler, A., Hart, S., Frey, F., Jakobbson, S. 1980. Nd and Sr isotope ratios and rare earth element abundances in Reykjanes Peninsula basalts: evidence for mantle heterogeneity beneath Iceland. *Earth Planet. Sci. Lett.* 45:249–62

Ann. Rev. Earth Planet. Sci. 1983. 11: 415–59

THE ATMOSPHERES OF THE OUTER PLANETS[1]

Garry E. Hunt
Laboratory for Planetary Atmospheres, University College London, London WC1E 6BT, England[2]

INTRODUCTION

One of the fundamental goals of the research program for solar system exploration is to provide a quantitative explanation for the extreme differences between the terrestrial planets and the major planets—Jupiter, Saturn, Uranus, and Neptune. The latter are huge, rapidly rotating, low-density objects, with optically reducing atmospheres. They contain more than 99% of the planetary mass of the solar system. The low density of these objects suggests that, like the stars, they are entirely composed of light elements—hydrogen, helium, carbon, and nitrogen—whereas silicates—iron and nickel mainly—constitute the cores of the inner planets. Since hydrogen and helium are thought to be the principal constituents of the solar nebulae, understanding the origin and evolution of these giant planets may hold important clues to the formation of the solar system.

In addition to the major planets, there are important satellite atmospheres too. Titan possesses a substantial atmosphere and its origin and evolution may provide important clues to the development of the atmospheres of the terrestrial planets. The tenuous atmosphere of the Jovian satellite, Io, results primarily from the intense interactions between the charged-particle population of the magnetosphere in which the body is embedded, with additional material injected from the active volcanoes that are present over the entire surface of the satellite.

For centuries, studies of planetary atmospheres have been carried out, primarily by astronomers. Their telescopic observations, of limited spatial,

[1] Contribution 103 of the Laboratory for Planetary Atmospheres, University College London.

[2] Present address: Atmospheric Physics Group, Blackett Laboratory, Imperial College, London SW7 2BZ, England.

0084–6597/83/0515–0415$02.00

temporal, and spectral resolution, have provided a catalog of planetary characteristics that have formed the foundation of our understanding of these distant objects. In many cases, these observations only allowed a qualitative description of the measured phenomena.

In the past decade, a considerable increase in our understanding of the planets has been made through the analysis of observations made by space probes, such as Pioneer 10 and 11 and Voyagers 1 and 2, including new insights into the atmospheric processes of Jupiter, Saturn, and Titan. These data have also been complemented by observations made by Earth-orbiting satellites such as the International Ultraviolet Explorer (IUE). With observations made by a wide range of instruments, using the entire electromagnetic spectrum, the subject of planetary atmospheres has become a logical extension of the terrestrially based subjects of cloud physics, atmospheric physics, and atmospheric chemistry. Advances in image-processing systems [for example, at Imperial College (Hunt et al 1983), at the Jet Propulsion Laboratory (Jepsen et al 1980), and at the University of Wisconsin (Limaye et al 1982)] have enabled the first quantitative analyses of meteorological phenomena to be made. Thus, investigations can now be made into the driving mechanisms that cause the visible appearances of the planets. Furthermore, data have been used in the construction of more realistic numerical models to simulate the observations and thereby assist in quantifying the interpretations. This natural extension of the terrestrially based disciplines now allows a thorough investigation to be made of physical, chemical, and dynamical processes with varied boundary conditions. There is no doubt that studies of planetary atmospheres provide insight into the behavior of processes under varying conditions. This work is essential for a complete understanding of the Earth.

In this paper, I review the current understanding of the atmospheres of Jupiter, Saturn, and Titan. I first discuss the compositions of these atmospheres and analyze observations for both cosmogonical and cosmological investigations. This is followed by a discussion of the visible and thermal structures of these atmospheres, which relate closely to their compositions. A detailed discussion is then given on the meteorologies of these planets, particularly for Jupiter and Saturn, which can now be discussed in a quantitative manner.

This review includes the results of studies carried out up to June 1982.

JUPITER AND SATURN

Atmospheric Composition

The advances in infrared astronomy during the past few years are primarily responsible for the rapid increase in our knowledge of the composition of the atmospheres of the distant, giant planets. For Jupiter, infrared spectroscopy is responsible for the discovery of most of the minor molecules. As recently as 1970, only H_2, CH_4, and NH_3 had been positively identified as constituents in the Jovian atmosphere. Since then, more than a dozen minor constituents have been detected. Most of these molecules have been observed in the wavelength range 1–3 μm or in the far-infrared range of $\lambda > 5$ μm. Observations made from Earth-based telescopes utilize the terrestrial atmospheric windows, which occur at wavelengths centered on 5, 10, and 20 μm. Then, by means of very sensitive detectors, it is possible to obtain high signal-to-noise ratios to produce extremely high spectral resolution on the brighter planets. With Fourier transform spectrometers, the resolving power can reach 10^5 in the near-infrared (Maillard et al 1973, Lecacheux et al 1976) and 10^4 at 5 μm (Larson 1980) and at 10 μm (Tokunaga et al 1979). This type of instrument has also been flown on the Voyager spacecraft to obtain observations of the entire infrared spectrum at a resolution of 4.2 cm^{-1} (Hanel et al 1979).

In Figure 1*A*, the infrared spectrum of Jupiter is compared with those of Earth and Mars for the region 100–2300 cm^{-1}. In this region of the spectrum, many molecules exhibit strong vibration-rotation bands, whose structure is relatively simple. In many cases, these bands have been extensively studied in the laboratory, which greatly assists in their identification and interpretation.

Molecular hydrogen is symmetric and therefore does not have a permanent dipole moment. However, a weak collision-induced dipole spectrum exists, which creates significant absorption for the very long pathlengths of about 40 km encountered in the Jovian and Saturnian atmospheres above the 1 bar level. The presence of hydrogen in the Jovian spectrum in the region 100–750 cm^{-1} is clearly seen in Figure 1*A*. In Figure 1*B*, we compare the Jovian spectrum with similar observations of Saturn made by the IRIS instrument during the Voyager encounters (Hanel et al 1981b). The brightness temperature of the midlatitude spectrum of Saturn is about 30–40 K lower than Jupiter. The S_0 and S_1 lines of molecular hydrogen are evident at 350 and 600 cm^{-1}, but the detailed shapes appear to differ in the two planetary spectra. These differences may, however, be affected by the viewing geometry of the observations.

The spectral features of tropospheric ammonia, which are prominent in the Jovian spectrum (Figures 1*A*,*B*), are relatively weak in the Saturn observations (Figure 1*B*), where the NH_3 is strongly depleted by condensation in the upper troposphere as a consequence of the low temperatures. With the reduced NH_3 concentration, the 1000 cm^{-1} region of the Saturnian spectrum is dominated by the presence of PH_3. Phosphine, which is a nonequilibrium product, should be found at levels where the temperature is $\sim$2000 K, rather than at the cloud tops, where it is $\sim$90 K.

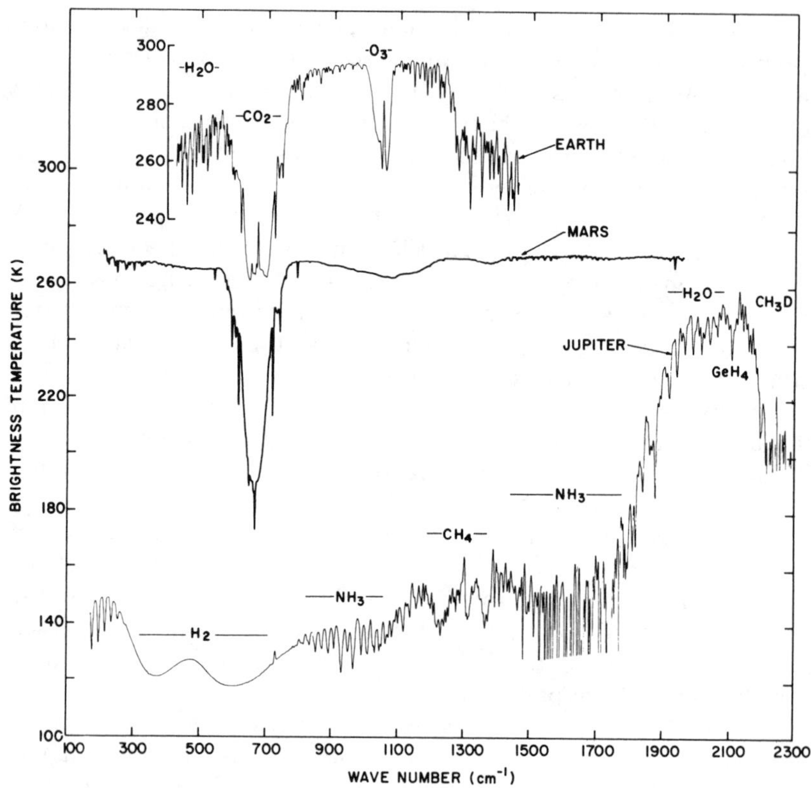

Figure 1A The brightness temperatures of the Earth, Mars, and Jupiter, obtained by IRIS instruments mounted on NIMBUS 4 (Earth), Mariner 9 (Mars), and Voyager 1 (Jupiter) spacecraft (after R. Hanel, private communication).

The evidence of the domination of phosphine in the Saturnian spectrum is the first suggestion of a vigorous meteorology that possesses stronger vertical motions.

Helium is the second most abundant element in the Sun and is probably an important constituent of the outer planets too. However, it is extremely difficult to detect in planetary atmospheres. The first positive detection of helium in these atmospheres was made by Carlson & Judge (1974) from the observation of the He I line at 58.4 nm, although there are many uncertainties in interpreting this observation owing to the lack of neighboring spectral observations of hydrogen. A more precise result can be determined through the influence of helium on the far-infrared thermal emission spectrum. Trafton (1967) demonstrated that the pressure-induced absorption due to collision between hydrogen molecules and hydrogen and helium is responsible for a large fraction of the far-infrared opacity of Jupiter. Gautier & Grossman (1972) developed a method for inferring the helium abundance in the 300–700 cm^{-1} region that uses the different spectral characteristics of the H_2-H_2 and H_2-He absorption coefficients. Using the Voyager observations of Hanel et al (1981a), Gautier et al (1981) have derived values using two methods. The first scheme uses only the IRIS spectra from selected locations on the planet, while the second method uses a thermal profile independently derived from radio occultation measure-

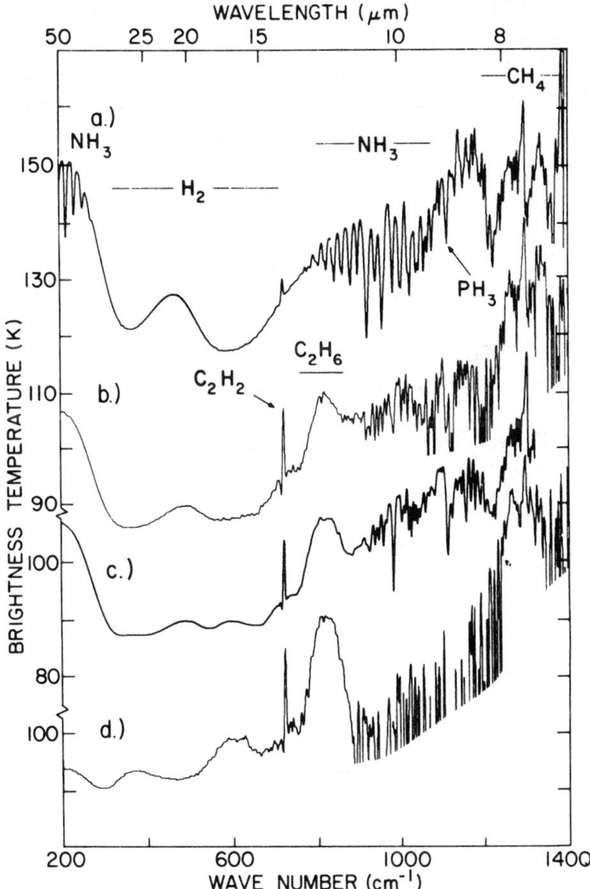

Figure 1B Saturnian and Jovian spectra: (*a*) spectrum of the North Equatorial Belt of Jupiter recorded by Voyager 1 IRIS in March 1979; (*b*) large average of Saturn spectra taken between 15° and 50° north and south; (*c*) synthesized Saturn spectra using only molecular absorption and a midlatitude temperature profile; (*d*) south polar spectrum of Saturn recorded at an emission angle of 76°.

ments and infrared spectra recorded near the occultation point. Gautier et al (1981) obtain a hydrogen mole fraction of 0.897 ± 0.03 by the first method and a value of 0.88 ± 0.036 by the second. These correspond to helium mass fractions of 0.19 ± 0.05 and 0.21 ± 0.06, respectively.

A similar approach has been applied to Saturn, and Hanel et al (1981b) find values for H_2 of ~ 0.94 and for He of $\sim 0.068 \pm 0.025$. When compared with Jupiter, these results suggest that Saturn has a depletion in helium that must be significant relative to the internal structure of the planet and its early history.

Table 1 lists the molecules that have been detected in the atmospheres of Jupiter and Saturn in the period extending to June 1982. The region of the spectrum where these molecules are detected is extremely significant, since there is considerable variation in the level of line formation with wavelength.

Clearly, in the troposphere, the interpretation of the measurements will be affected by clouds. The basic microphysical properties and spatial variations of these clouds are not accurately known, which puts some uncertainty into the derived abundances (Hunt 1978).

All the molecular identifications, apart from those of He and HD, are the result of infrared observations made with resolving power better than 10^3. On Jupiter, H_2, $^{13}CH_4$, CH_3D, C_2H_2, C_2H_6, CO, PH_3, and $^{15}NH_3$ have been detected from the ground, while H_2 and GeH_4 were observed from the

Table 1 Observed molecules in atmospheres of Jupiter and Saturn

Molecule	Spectral range/μm Jupiter	Saturn
He	0.0584	—
HD	0.746	0.6064
H_2	0.8, 2.5, 1.25, 4.285	0.8, 1.25
CH_4	0.8, 1.1	0.8, 1.1
$^{13}CH_4$	1.1	1.1
CH_3D	5	5
NH_3	1–2, 10, 50–200	0.645
$^{15}NH_3$	10	—
H_2O	5	—
CO	5	—
GeH_4	5	—
PH_3	2, 5, 10	3, 5, 10
C_2H_2	$\sim$0.17, 13	$\sim$0.17
C_2H_6	12	12
C_3H_4	—	$\sim$15
C_3H_8	—	$\sim$13

Kuiper Airborne Observatory. On Saturn, Hanel et al (1981b) detected C_3H_4 for the first time; these species have also been found in the atmosphere of Titan.

The detection of C_2H_2 and C_2H_6 in emission in the infrared spectrum of Jupiter by Ridgway (1974) was the first observational evidence for CH_4 photodissociation by the solar ultraviolet radiation in the upper Jovian atmosphere. Hanel et al (1981a) estimate mixing ratios of 3×10^{-8} for C_2H_2 and 5×10^{-6} for C_2H_6. However, these values, which refer to stratospheric levels, do show some latitudinal and hemispheric variations. The abundance of ethane relative to acetylene in Jupiter's atmosphere appears to be about three times larger in the polar regions than at lower latitudes. Furthermore, there is an overall increase in the abundance ratio by a factor of 1.7 between the Voyager encounters. Obviously, it is not possible to account for this large variation through photodissociation processes. Barocki et al (1983) believe that this observation is more consistent with the suggestion by Bar-Nun (1979) that some C_2H_2 is generated by lightning discharges. Cook et al (1979a) have shown that the Jovian lightning coincides with areas of convective activity. Barocki et al (1983) suggest that cloud structures are consistent with more convective cloud systems in the equatorial region than at the poles, which appears to agree with this hypothesis.

Kunde et al (1982) have recently carried out a detailed analysis of the composition of Jupiter's atmosphere in the region of the North Equatorial Belt, which, at the time of the Voyager 1 encounter, was one of the convective regions. They find the NH_3 profile is close to saturation in the 0.2–0.4 bar region, reaching depletion by a factor of about 3 at the 0.7 bar level. In the 0.8–1 bar region, they estimate a mole fraction of $\sim 1.78 \pm 0.89 \times 10^{-4}$, which is in agreement with Lambert (1978). The distribution of PH_3 is reduced in the region 0.4–0.7 bar, which is expected from the coupled photochemistry of PH_3 and NH_3 (Strobel 1977). The PH_3 abundance in the 1–4 bar region is found to be $\sim 6 \pm 2 \times 10^{-7}$, which is similar to the solar value of Cameron (1973). The mole fraction of CH_3D, determined from the 1120–1220 cm^{-1} and 2100–2200 cm^{-1} spectral range, is $\sim 3.5^{+1.0}_{-1.3} \times 10^{-7}$, while that of GeH_4 is $7 \pm 2 \times 10^{-10}$ in the 2–5 bar region, which is a factor of 10 lower than the solar value. The abundance of water vapor is difficult to estimate because of possible contamination of the line structure from H_2O-H_2 broadening. However, Kunde et al (1982) estimate a mole fraction of 1×10^{-6} at the 2.5 bar level, increasing to 3×10^{-5} at the 4 bar level.

The spectrum of Saturn (Figure 1*B*) also shows emission by stratospheric CH_4 and its photochemical derivatives C_2H_2 and C_2H_6. Hanel et al (1981b) estimate mole fractions of 2×10^{-8} for C_2H_2 and 5×10^{-6} for C_2H_6, which are similar in magnitude to the Jovian values. However, the

Saturnian spectrum shows evidence for further photochemical derivatives—C_3H_4 methyloacetylene and C_3H_8 propane—which also appear in the Titan spectrum. These higher-order photochemical products may also be related to the production of high-level haze layers, which may therefore relate to the rather pastel appearance of Saturn. Furthermore, the axial tilt of Saturn ($\sim 26°$) may cause some hemispheric asymmetries in the composition through the changes in incident solar energy and the resulting photochemistry.

The ultraviolet spectra of Jupiter and Saturn show important differences. Ammonia features, while evident in the Jovian spectrum, are not apparent in the Saturn observations made by the IUE observatory (Figure 2). While C_2H_2 is seen in each spectrum, the abundance is significantly greater on Saturn (Owen et al 1980); phosphine is also more dominant on this planet. Both of these features are consistent with the infrared observations. On Saturn, the region between 1900 and 2300 Å needs very careful analysis. It is possible that the shape of the albedo curve for this spectral interval is constrained by the phosphine absorption. A further unresolved feature occurs at 1600 Å. Gladstone et al (1982) suggest that it may be explained by the presence of diacetylene (C_4H_2), while Winkelstein et al (1983) suggest

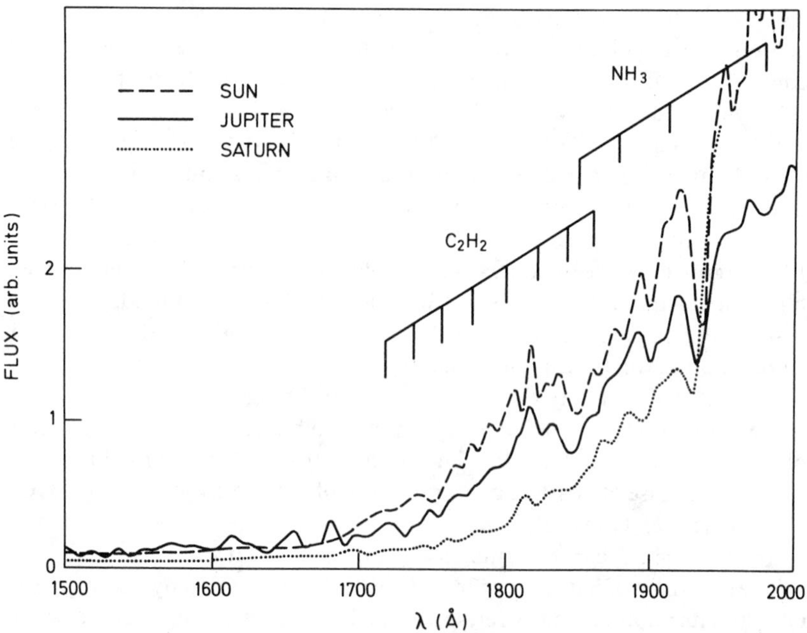

Figure 2 Spectra of the Sun, Jupiter, and Saturn between 1500 and 2000 Å. The Jupiter and Saturn spectra have been observed at Vilspa with IUE (Owen et al 1980).

the presence of water vapor, but this may be in the optical path near the rings rather than in the atmosphere of Saturn.

The observations of PH_3, GeH_4, and CO in the atmosphere of Jupiter and PH_3 on Saturn are not in agreement with models of thermochemical equilibrium (Prinn & Owen 1976). For example, one would expect phosphorus to be in the form of PH_3 only in regions where the temperature is greater than 800 K, in the deep, unobservable portions of the atmosphere. Barshay & Lewis (1978) suggest that PH_3 should not be observed, since it is expected to react with H_2O below a temperature of 2000 K. The discovery of PH_3 on Jupiter and Saturn at 2, 5, and 10 μm (Ridgway 1974, Larson & Fink 1977, Ridgway et al 1976, Tokunaga et al 1981) was entirely unexpected. Prinn & Lewis (1975) suggest that PH_3 was probably carried from deep atmospheric levels where it has been observed. For Jupiter, this corresponds to the 200–230 K level at 5 μm and the 130–145 K level at 9–10 μm. The time needed for the transportation would have to be short enough for the PH_3 to be observed at the top of the current before it has been completely oxidized with the available H_2O. Similar mechanisms could then account for GeH_4 and CO. As a consequence, the observations of these nonequilibrium species are further evidence of the dynamic, and particularly the convective, activity in the atmospheres of Jupiter and Saturn.

The abundance of H_2O in these atmospheres is affected by condensation processes. With higher spatial resolution, Hanel et al (1979) find concentrations of 5×10^{-6}, which is about five times higher than the previous estimates made from airplane measurements.

The abundance of sulfur and its vertical profile are also important for understanding the evolution and present state of these atmospheres. At depth, H_2S is expected to be the predominant sulfur compound, and the detection of its presence will require the direct measurements to be made by the Galileo probe later this decade. Certainly, there is still no direct evidence from existing observations of H_2S, which some authors still use in explaining the cloud colors. Furthermore, the analysis by Moore (1983) and Moore & Hunt (1983) suggests that H_2S may not be necessary in a model atmosphere to explain the spectral variations in the UV.

Atmospheric Structure

Both Jupiter and Saturn have internal heat sources and emit more energy than they receive from the Sun. The initial Earth-based observations of Aumann et al (1969) have now been refined by Hanel et al (1981a). With a global geometric albedo of 0.266 ± 0.013 and a phase integral of 1.25, derived from the Pioneer observations of Tomasko et al (1978), they estimate a Bond albedo of 0.333 ± 0.026. These values yield an effective brightness temperature of 124.4 ± 0.3 K and an energy balance of 1.67 ± 0.9.

For Jupiter, it is possible to account for this additional energy in terms of the gravitational contraction of the planet at a rate of about 1 mm yr^{-1} (Hubbard 1981). For Saturn, the situation may be slightly more complicated. Hanel et al (1982) estimate a Bond albedo of 0.33 ± 0.02, which produces an effective brightness temperature of $\sim 93.6 \pm 0.5$ K and an energy balance of 1.79 ± 0.1. The results shown in Table 2 summarize the heat balances at the time of the Voyager encounters. Hubbard (1981) suggests that this additional energy cannot be explained by simple cooling and contraction, and that this situation may also relate to the apparent helium depletion that has been detected. Stevenson & Salpeter (1977) suggest that a planet incorporating a mixture of hydrogen and helium has two types of energy—thermal and gravitational. If the mixing ratio of hydrogen to helium is constant throughout the mixture, then the two types of energy are released together in constant proportions. If, however, the mixing ratio changes by helium precipitating through the hydrogen, then an additional quantity of gravitational energy is released. This process may be happening on Saturn, since the planet has already cooled to the extent that helium is precipitating at the top of the metallic hydrogen zone. Stevenson & Salpeter (1977) suggest that this process is similar to rainfall. They suggest that the process began on Saturn some two billion years ago. Jupiter, in comparison, is more massive and has not yet reached the point in its cooling where the planet, at any level, is saturated with helium. The helium depletion of the Saturn atmosphere is observational evidence to support this hypothesis.

The internal heating of the massive atmosphere will have important consequences for atmospheric processes, since it provides an additional energy source for the respective weather systems.

Representative atmospheric profiles for Jupiter and Saturn are shown in Figure 3. For both planets, the tropospheric temperature structures are close to the adiabatic profile. Remote-sensing observations from Earth (see, for example, Orton 1981) and from spacecraft (Hanel et al 1979, Orton &

Table 2 Heat budget of the atmospheres of Jupiter and Saturn

	Saturn	Jupiter
Geometric albedo	0.232	0.274
Bond albedo	0.328	0.343
Absorbed solar energy (W)	1.04×10^{17}	5.02×10^{17}
Thermal emission (u)	1.87×10^{17}	8.37×10^{17}
Brightness temperatures (K)	93.6 ± 0.5	124.4 ± 0.3
Energy balance	1.79 ± 0.1	1.67 ± 0.9

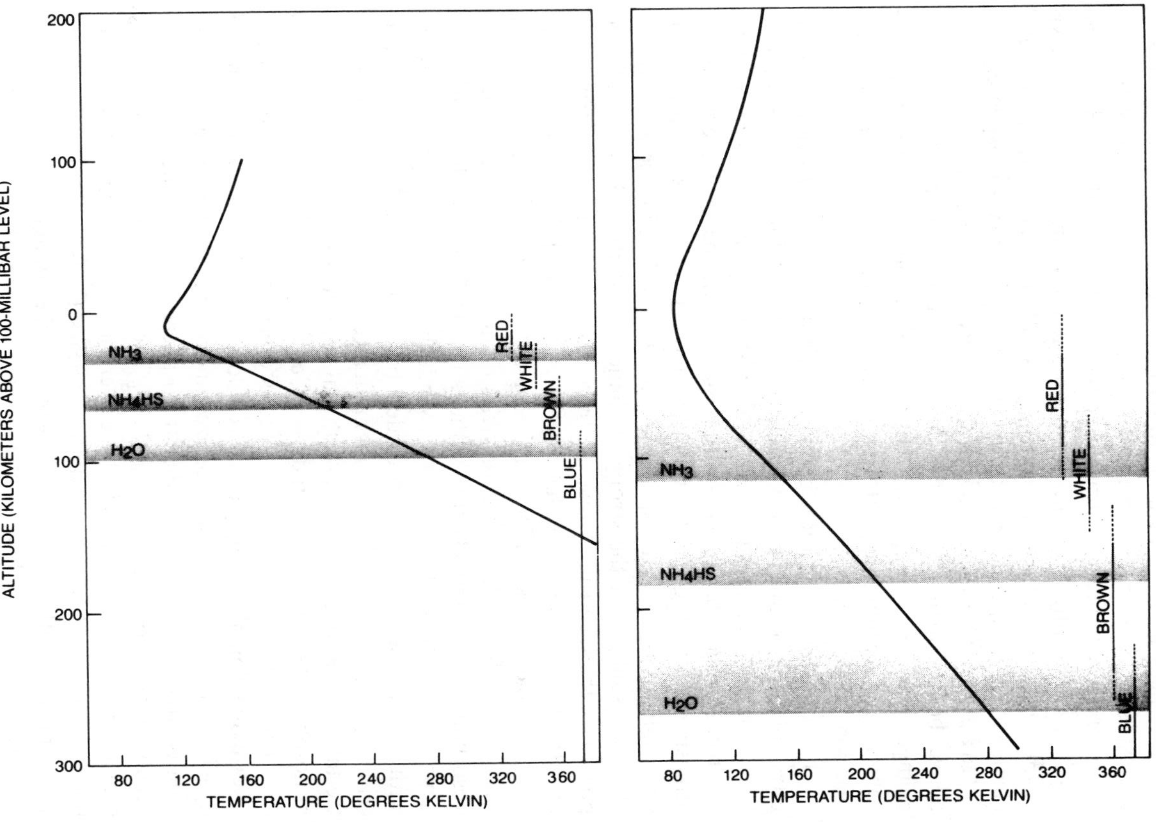

Figure 3 Atmospheric profiles for Jupiter (left) and Saturn (right) consistent with the available observations. The levels of the cloud layers expected to form in these atmospheres are indicated.

Ingersoll 1981) penetrate to pressure levels of ~1 bar. The lapse rate for the Jovian atmosphere is about 1.9 K km^{-1}, while for Saturn it is about 0.9 K km^{-1}. When these temperature variations are taken into account with atmospheres of near-solar composition, comparison with the saturation vapor pressure indicates that we may expect clouds of NH_3, NH_4SH, and H_2O at descending levels in the troposphere (Weidenschilling & Lewis 1973). Variations in these cloud structures will cause a corresponding local change in the temperature distribution of the planetary atmospheres.

On a global scale, the major differences between these atmospheres occur above the tropopause, in the less dense regions that are more sensitive to changes in solar radiation. We are primarily concerned with the stratospheric levels, where the photolysis of CH_4 to produce C_2H_2 is the major reaction (Atreya 1981).

The acetylene absorption band at ~1750 Å is clearly evident in the ultraviolet spectrum obtained by Owen et al (1980) for IUE observations. However, this feature is considerably stronger in the corresponding spectrum for Saturn, emphasizing a basic difference between the two upper atmospheres. It is generally thought that temperature structure in the stratospheres of the atmospheres is due to the heating created by the absorbed sunlight in the strong methane bands situated in the infrared portion of the spectrum. However, there may be additional contributions due to upper-atmosphere haze layers. Strobel (1973) suggested that hydrazine particles could form at these levels and act as nucleation agents. Prinn (1974) found that such particles of radius $\ll 1$ μm could form an absorbing layer of optical depth in the range of 0.2–0.25, with a tropopause temperature between 110 and 120 K. While this result is certainly consistent with the high-altitude haze provided by Axel (1972), Prinn's study is critically dependent upon the ammonia concentration at these levels. Certainly there is an extensive haze throughout the Jovian stratosphere and atmosphere, as the Voyager studies of Cook et al (1979b) have shown. Its composition is not known, but there is every reason to believe that the inward diffusion of the ring particles could be a familiar contribution to the opacity. With a variety of particle sizes, and different setting times, this may then create distinct layers in the stratosphere, which is suggested in the occultation data of Eshleman et al (1979). The stratosphere of Saturn shows similar structure (Figure 3). The haze over the southern hemisphere during the time of the Voyager encounter appears to extend throughout the stratosphere (Smith et al 1981). However, unlike Jupiter, there is a marked seasonal effect of this haze, which, as Trafton (1978) noted, affected the spectroscopic observations of Saturn.

The temperatures at the 209 mbar level at the time of the Voyager encounter were found to be warmer in the southern hemisphere compared with the northern region (Hanel et al 1981b). There is an apparent seasonal effect caused by Saturn's obliquity of $\sim 26°$. As a consequence, there is an important pole-to-equator temperature gradient, which is most noticeable in the winter (northern) hemisphere during the period of the recent Voyager encounter (Hanel et al 1981b). Although the temperatures obtained by Hanel et al (1981b) are not on an isobaric surface, their results suggest a variation from the south pole to the north pole of 108 K to 102 K at 535 mbar, and ~ 90 K to 86 K at the 209 mbar level. There are some important latitudinal variations too. Hanel et al find the latitudinal structure from 60° to the pole is reproduced in each hemisphere, which implies a symmetry in the temperature field, and through the thermal wind equation, in the resulting polar winds too. This is consistent with the imaging studies (see Smith et al 1981, 1982). There is not an immediate correlation between the observed temperature structure of Saturn's atmosphere and its visible appearance. Certainly, the large temperature gradient that exists in the region 10–25°N does not correlate with any imaging observations or the derived wind fields.

On a global scale, Saturn possesses small horizontal temperature gradients of 1–2 K at the ~ 150 mbar level and has little evidence of longitudinal structure (Figure 4*A*). The largest longitudinal variations are in the regions 35–42°N, 50–57°N, and poleward of 65°N. The first two regions correspond to relatively weak, retrograde jets (Figure 5*B*), while images of these regions suggest the presence of convective cloud features that are morphologically different from other locations.

Jupiter does not show any significant pole-to-equator temperature gradient, since the planet has a negligible inclination of $\sim 3°$. Consequently, the temperature field is essentially symmetric and consistent with a temperature difference of <3 K between equator and pole. On a global scale, there is considerable structure at cloud top (~ 800 mbar) and near the tropopause (~ 150 mbar), which does have some relationship with the visible appearance of the planet (Figure 4*B*). The Great Red Spot is evident at a longitude of $\sim 120°$, where its top is slightly colder than the surrounding clouds, indicating the elevated nature of this long cloud feature. Considerable longitudinal structure is seen at $\sim 9°$N, where the plume features are located. As Hunt et al (1981a) and Conrath et al (1981) have shown, these convective structures cause perturbation to the temperature structure of the entire troposphere. Maps of the planets at 5 μm, where there is negligible atmospheric opacity (see, for example, Terrile et al 1979), provide the most distinct information on the vertical structure of the upper clouds of the Jovian atmosphere. Regions of the South Equatorial Belt and the North Equatorial Belt are the most transparent; brightness temperatures of ~ 270 K have been detected in these regions, compared with a

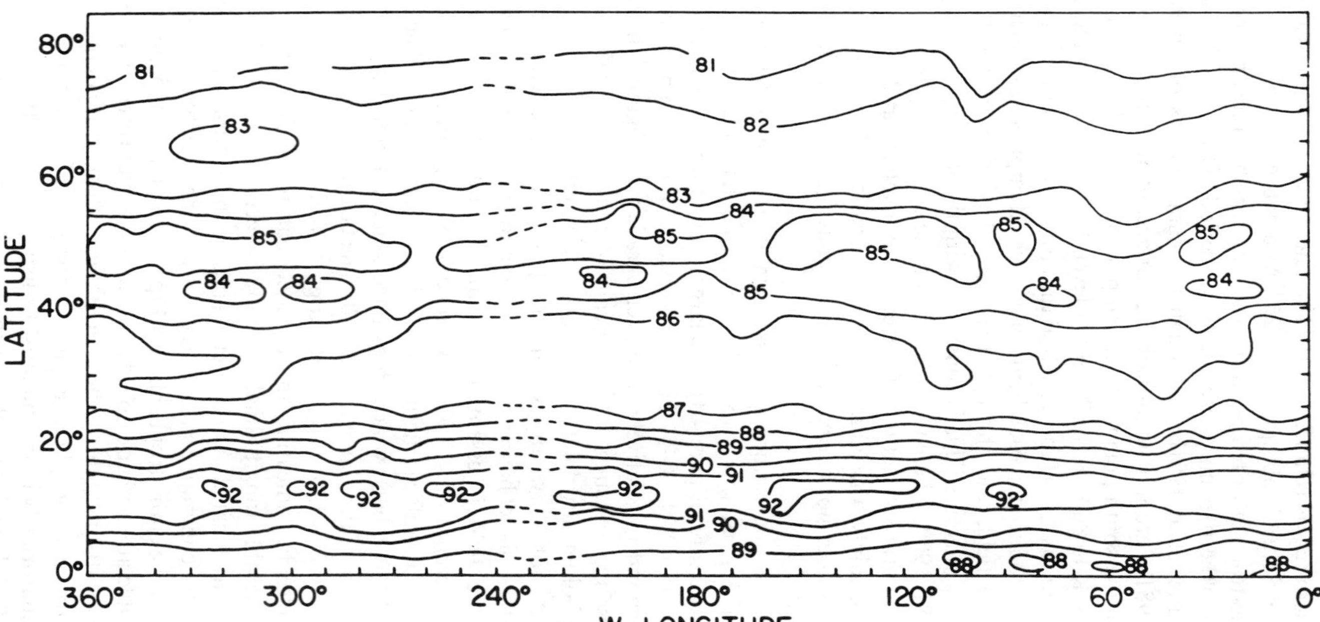

Figure 4A Map of Saturn brightness temperatures averaged within the spectral interval 330–400 cm^{-1}. Emission in this interval originates from a region centered near 150 mbar (Hanel et al 1982).

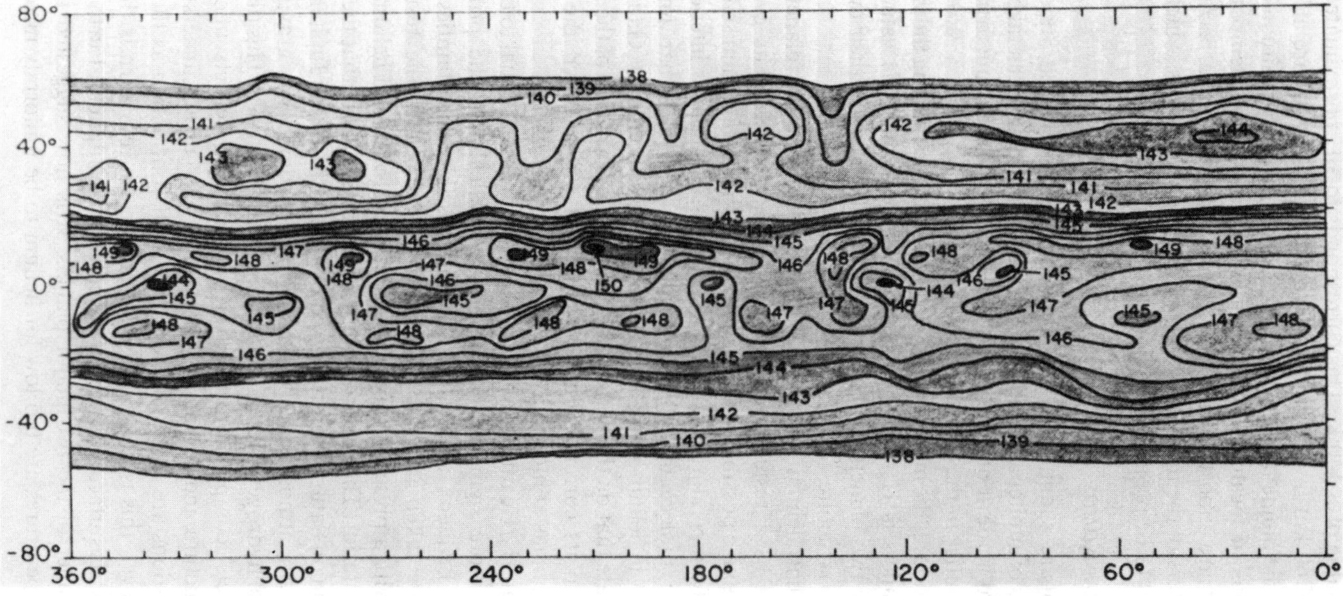

Figure 4B The Voyager 2 brightness temperature map of Jupiter at 226 cm^{-1}, corresponding to ~800 mbar (after Hanel et al 1979).

cloud-top value of ~140 K. This corresponds to an approximate altitude variation of 60 km in the emitting levels. Local hotspots are also found adjacent to the plume heads, where descending motions are situated. These small regions of localized thermal emission have strong temporal variations. They are important regions in the Jovian atmosphere, because they are the centers of the main emissions from the deeper levels, which enable these depths to be probed for important trace constituents.

Abundance Ratios

In Table 3 are summarized the abundance ratios of the atmospheric constituents of Jupiter and Saturn. Considerable care has been taken to use observations made at similar times and regions of the spectrum to minimize the uncertainties in the derived values.

Without doubt, the most important ratio for understanding the variation of these planets is H_2/He. We see from Table 3 that this value differs between the two planets, probably owing to differences in their evolution and internal structure.

The present helium abundances may differ from the planet's bulk composition as a result of helium differentiation during its evolution. Differentiation is possible, since helium and hydrogen are immiscible over the range of temperatures and pressures relevant to Jupiter's interior. Also, the metallic molecular hydrogen transition near 3 Mbar implies a discontinuity in helium abundance across the phase boundary. Gautier et al (1981) indicate that on the basis of the decay in the variation of the internal luminosity to its current value over the past 4.6×10^9 yr, the helium differentiation has at most only recently begun on Jupiter.

The Jovian H_2/He ratio (Table 3; Gautier et al 1981) is equal to the solar value and, more significantly, is slightly smaller than the primordial estimate of Lequeux et al (1979). If the possible uncertainties in the estimates are taken into account, it would seem that the results are consistent with a present uniform mixture of hydrogen and helium within the Jovian interior. The observation by Hanel et al (1981b) that the helium mass fraction on Saturn is only 11%, compared with 19% on Jupiter, is very significant. An atmospheric depletion implies significant gravitational separation of hydrogen and helium within Saturn's interior. This difference in the planets' composition and heat balance also reflects upon their internal structure and evolution. For Jupiter, the current understanding and observations are consistent with the small contraction as the planet cools. However, this is not the case for Saturn. Since Saturn is the more distant body, its surface and interior are colder than Jupiter's. Furthermore, it is the smaller of the two major planets. It is now suggested that the interior temperatures are too low for helium to be uniformly mixed with

Table 3 Abundance ratios of Jupiter and Saturn

Ratio	Spectral range	Jupiter	Saturn	Primordial nebula	Sun	Earth
H_2/H_2+He	thermal radiation	0.897 ± 0.03	0.94 ± 0.03	0.871 ± 0.02	0.89	—
C/H	scattering model	$(2–3) \times 10^{-3}$	—	—	—	—
	1–2 μm	8×10^{-4}	1.15×10^{-3}	—	$4.7^{+1.2}_{-1.0} \times 10^{-4}$	—
	thermal radiation	9.7×10^{-4}	—	—	—	—
D/H	HD/H_2 (visible)	$(5.1 \pm 0.7) \times 10^{-5}$	$(5.5 \pm 2.9) \times 10^{-5}$	2.5×10^{-5}	—	—
CH_3D/H_2	5 μm	$(2.5–5) \times 10^{-7}$	—	—	—	—
	10 μm	$(2–5) \times 10^{-7}$	—	—	—	—
$^{12}C/^{13}C$	1.1 μm	89^{+12}_{-10}	89^{+25}_{-18}	—	89 ± 5	89 ± 4
	~ 3 μm	160^{+40}_{-55}	—	—	—	—
$^{15}N/^{14}N$	10 μm	$(3.7 \pm 1.5) \times 10^{-3}$	—	—	—	3.7×10^{-3}

hydrogen throughout the metallic zone. Instead, the helium may be condensing at the top of this region, with motion of the "raindrops" of helium falling toward the center, where the gravitational energy is then turned into heat. It is estimated that this process started about 2 billion years ago when temperatures first dropped to the helium condensation point. For Jupiter, this situation can only have been reached recently. This hypothesis, which requires the Saturn helium to be concentrated toward the center of the planet, is consistent with the observations. Information on the stable isotopes $^{12}C/^{13}C$ and $^{15}N/^{14}N$ provides information on the chemical evolution of the Galaxy. Combes & Encrenaz (1979) estimate the $^{12}C/^{13}C$ ratios for Jupiter and Saturn to be 89^{+12}_{-10} and 89^{+25}_{-18} respectively, which are in good agreement with the solar value. Encrenaz et al (1980) identified the presence of $^{15}NH_3$ in the Jovian atmosphere and estimate a value of 0.0037 ± 0.0015 for the $^{15}N/^{14}N$ ratio. This is in good agreement with the terrestrial value.

It is possible to estimate the C/H ratio from measurements in both the near- and far-infrared portions of the spectrum (Tables 1 and 3). A detailed discussion is given by Wallace & Hunten (1978) on the possible sources of error associated with the various line formation methods used to correct for the scattering effects that contaminate the spectral lines. Encrenaz & Combes (1981) show that there are basically two classes of results. The Jovian C/H value, derived from visible and near-infrared data (which use scattering models), is 2 to 5 times the solar ratio. Estimates of an enrichment by less than a factor of 2 are obtained by methods using the thermal spectrum without scattering models. There is also some variation in the solar values. Encrenaz & Combes (1981) suggest that the most accurate value is that of Pagel (1977) and Lambert (1978) of 4.7×10^{-4}. From this estimate, it would seem that the C/H ratio is greater for Jupiter than for the Sun (Table 3), and the analysis of Voyager IRIS observations by Gautier et al (1982) suggests that the enrichment is by a factor of 2.07 ± 0.24. Using the values of Fink & Larson (1979) for Saturn, one finds that there is a carbon enrichment for this major planet too (Table 3).

The D/H ratio of these major planets is also important because of its astrophysical implications. Reeves et al (1973) argue that "big bang" nucleosynthesis is the only viable production mechanism, and that nuclear burning to produce ^3He is an efficient loss mechanism. It is therefore thought that the Jovian D/H ratio is indicative of the primordial value, but, as for the C/H value, the measurements are still controversial. There are two methods available. The HD molecule can be used in the visible and associated with the H_2 or CH_4 measurements in the same spectral range. Alternatively, the CH_3D molecule can be observed in the thermal infrared, and the ratio computed from the CH_3D/H_2 values from model studies.

Combes & Encrenaz (1979) derive a value of $D/H < 2.3 \times 10^{-5}$, which implies no deuterium enrichment on Jupiter. Their method is more accurate than that used by Trauger et al (1973), since it avoids the use of the H_2 quadrupole lines, which are difficult to measure.

In the thermal infrared, estimates of the CH_3D/H_2 ratio have been obtained in the 5- and 10-μm regions. However, Encrenaz & Combes (1981) have shown that there are significant problems in interpreting these data due to the possible variations in cloud structure, lack of knowledge of the spectral properties of the clouds, and uncertainties in the strongly temperature-dependent fractionation of CH_3D into its components. For example, Kunde et al (1982) derive a value for the CH_3D/H_2 ratio of 2.5×10^{-7} at 5 μm, but a value of 5×10^{-7} in the 8–9 μm region. This emphasizes the care that must be taken in choosing the spectral region for the determination of these abundance ratios.

Kunde et al (1982) have now improved the determination of the D/H ratio for Jupiter through the use of more accurate values of the mole fractions of CH_3D and CH_4, from which they estimate a value of $D/H = 3.6^{+1.0}_{-1.4} \times 10^{-5}$. This value is consistent with estimates from He^3 gas meteorites and the solar wind. If the Jovian value is representative of the bulk planet, then this apparent agreement with the meteoritic and solar wind values implies that Jupiter has retained its protosolar value without fractionation during accretion. Assuming this value to be representative of the solar nebula and using Audouze & Tinsley's (1978) model of the evolution of the Galaxy to correct for the evolution of the D/H ratio, Kunde et al (1982) estimated a value of $(5.5–9) \times 10^{-5}$ for the primordial D/H ratio. For this value, they took $(1.8–2.4) \times 10^{-31}$ gm cm^{-3} as an upper limit to the present-day baryon density, which according to Wagoner (1973) is consistent with an open universe in "big bang" models, provided that the mass of the universe is dominated by baryons.

Estimates of the D/H ratio for Saturn are still in an elementary state. Brault & Smith (1980) suggest a value of $D/H \sim 5.8 \times 10^{-5}$, using the $P_4(1)$ line of HD and the $S_4(1)$ of H_2. However, more extensive analyses with Voyager IRIS data have yet to be completed.

Courtin et al (1981) have recently derived the C^{12}/C^{13} ratio on Jupiter from the analysis of the ν_4 band of CH_4 in the spectra obtained from the Voyager 1 IRIS experiment. It is found to be $1.8^{+0.4}_{-0.6}$ times the terrestrial value of 89, namely 160^{+40}_{-55}. While this value disagrees with previous determinations for Jupiter, Courtin et al indicate that it is consistent with the hypothesis of solar system formations in which heavy neutron irradiation from the early Sun would have produced a ^{13}C enrichment of the inner planets and meteorites.

The presence of PH_3 is direct evidence of convective activity in these planetary atmospheres. Encrenaz et al (1980), Fink & Larson (1979), and Beer & Taylor (1979) estimate that the P/H value is depleted by a factor of 4 relative to the solar value. On Saturn, this ratio would appear to be enriched relative to the solar value. Tokunaga et al (1981) suggest an enrichment value of at least a factor of 3, while Larson et al (1980) suggest a factor of 2 from their 5 μm observations.

As Larson et al (1980) suggest, the reaction of PH_3 with H_2O may be slower than previously thought (Sill 1976). Furthermore, according to Strobel (1977), photodissociation of PH_3 is expected to be inhibited by the presence of gaseous NH_3. This situation is expected for Saturn and should account for the differences between the PH_3 abundances of the two planets (Tables 1 and 3).

Without doubt, NH_3 is one of the most important molecules in these atmospheres. It follows the saturation law below the level of minimum temperature on both major planets. Above this level, ammonia would follow a hydrostatic law in absence of photodissociation. For Jupiter, NH_3 is strongly depleted in the upper atmosphere. This information is found in the rotational band of NH_3 (40–110 μm) and in the 10 μm NH_3 band, where no thermal emission appears in the center of the NH_3 emission multiplets (see, for example, Goorvitch et al 1979, Gautier et al 1979, Marten et al 1980). Below the NH_3 cloud level at 145 K, the differing estimates of the NH_3/H_2 ratio from observations at various parts of the spectrum suggest the presence of nitrogen compound in this region. Combes & Encrenaz (1979) and Marten et al (1980) derive a N/H value depleted by a factor of 2 in the region. This suggests that nitrogen may be trapped as NH_4SH or NH_4OH cloud layers.

Information on the NH_3 distribution on Saturn is much more restricted. It has been observed in the visible region by Encrenaz et al (1974), but not in the near-infrared, where the bands are stronger (Owen et al 1977). Recent IUE observations in the 200-nm region do not show any pronounced features. These spectroscopic observations refer to the cloud-top region, so the effects of scattering particles may simply be complicating the spectral structure.

The current knowledge of the composition of Jupiter and Saturn is still too uncertain to specify any precise information on the internal structure of these planets. For Jupiter, it is believed that the enrichment in helium, deuterium, and carbon is moderate, and not sufficient to imply an inhomogeneous interior to the planet. More precise values are still required for the Saturn atmosphere.

However, the measurements of atmospheric composition may not necessarily determine the bulk composition of the planet. There are several possible separation processes that could give the interior a composition different from that of the atmosphere. For example, the planet could retain an original rocky core, and processes to separate helium in the interior have also been suggested. However, a knowledge of the atmospheric composition is a constraint on the bulk properties. In the future, we can gain further knowledge of the vertical distribution of the composition of Jupiter from the Galileo probe experiment.

In connection with this basic problem of the development of these planetary atmospheres is the origin of the colors. This matter is still strongly debated, and will not be resolved until more precise compositional measurements are available. The observations of lightning (Cook et al 1979a) provide a further energy source in the photochemical cycles that may involve CH_4, NH_3, H_2S, and the hydrocarbons that result from their reactions. Although H_2S has frequently been suggested as a coloring agent, it has yet to be detected. The importance of lightning is that it is localized and penetrates beneath the cloud that would otherwise absorb the incident solar ultraviolet energy. The study by Prinn (1970) and Sill (1976) indicates the importance of H_2S as a coloring agent. On the other hand, Sagan (1971) maintains that organic molecules are involved. However, two important constraints on this issue are the apparent lack of spectral contrast between red regions and neighboring white cloud areas in the Jovian atmosphere, and the apparently abrupt change between colors that is noticeable at the boundary between red/white regions in the neighborhood of the Great Red Spot. We note further that some of these "colored" clouds have lifetimes of decades. This is clearly a constraint on the energy source that may be related to the production of the colors. Solar UV energy may be too uniform, lightning too random, and charged particle bombardment too confined to special latitude regions. This may suggest a complicated involvement between the atmospheric motions and the chemistry.

Meteorology

For more than 300 years, observations of large-scale cloud features have provided the basic information on the gross characteristics of the atmospheres of Jupiter and Saturn (see, for example, Peek 1958, Smith & Hunt 1976, Alexander 1962). The visible appearance of Jupiter is one of alternating cloud bands of differing colors, separated by jet streams. Superimposed upon these cloud systems are large-scale features, such as the Great Red Spot and the three white ovals, which appear to have lifetimes varying from decades to centuries. Saturn is in many ways similar to Jupiter. Although the banded structure is clearly seen, the presence of haze layers above the main clouds seems to obscure the evidence of the larger-scale spots, which have now been observed at high resolution during the recent Voyager flybys (Smith et al 1981, 1982).

Unlike the meteorological systems of the terrestrial atmospheres, the weather systems of these planets are not solely driven by differential solar heating. We have seen that both planets have strong internal heat sources. Consequently, the meteorologies of these planets are influenced by two energy sources and by strong rotation. All the cloud velocities on Jupiter are referenced to the system III period of 9 h 55 min 29.711 s, and on Saturn to the system III period of 10 h 39.9 min $\pm$0.5 min.

On a large scale, there is little, if any, pole-to-equator energy transfer at the level of the visible clouds, which is the major difference between Earth and the giant planets. The Pioneer 11 measurements of Ingersoll et al (1976) have shown that the difference between the equator and pole temperature is not more than 3 K. At a latitude of $\pm 45°$, the belt zone structure breaks down in the Jovian atmosphere.

The temperature contrasts between the belts and zones are also small (see Hanel et al 1979), with contrasts of only 1–3 K at both cloud-top and tropopause levels. However, the location of the maximum contrast does vary significantly, and between the Pioneer 10/11 flybys of 1973/74 (Gehrels 1976) and the Voyager encounters of 1979 (Smith et al 1979a,b) it has shifted hemispheres. The bright white zone initially at 12–24°S has become narrower by a factor of 2, while the zone at 18–30°N has increased in width by a similar amount during this time. Even between the Voyager encounters, considerable changes were noticeable around the Great Red Spot (Smith et al 1979b).

By image-processing techniques, such as IPIPS (Hunt et al 1983), McIDAS (Limaye et al 1982), and that at JPL (Jepsen et al 1980), it has been possible for the first time to quantitatively analyze the Voyager images. Measurements of cloud winds have been obtained by tracking individual cloud elements between specific frames. The estimated errors in the zonal velocity are ± 2 m s^{-1}, and in the meridional velocity ± 1 m s^{-1} (Ingersoll et al 1981). The analyses of the Jovian data described here are for observations during the period around 26–27 February 1979 for Voyager 1 and the period around 1–2 July 1979 for Voyager 2. This is only a fraction of the data set extending from January to August 1979.

In Table 4, we have compared the zonal profiles obtained from the Voyager data by Ingersoll et al (1981) with 80 years of Earth-based observations summarized by Smith & Hunt (1976). From comparisons of these tabulations, it is apparent that the latitudes of the zonal jet maxima have changed very little during this period of 80 years. This is in marked contrast to the visible appearance of the planet. Also, there is a marked north-south symmetry in the zonal jet structure, which clearly shows seven jet maxima in the latitude range of 0–45°. This is also in complete contrast to the visible appearance of the planet.

Table 4 Latitudes of zonal jet maxima

Name of current	Latitude/deg[a] I	II	III	IV	$\bar{u}/(\mathrm{m\ s^{-1}})$[b]
N. Polar Region	—	—	—	56.5	10
	—	—	—	51.0	−13
	—	—	—	47.5	20
N.N.N. Temp. Ct.	43	44.46	42.8–45.9	43.0	−4
N.N. Temp. Ct. A	36–40	35–41	37.3–40.6	39.0	19
N.N. Temp. Ct. B	35	c	35.1–35.8	35.0	−19
N. Temp. Ct. A	29–33	28–32	30.2–31.4	31.5	−31
N. Temp. Ct. C	23	c	23.8–24.2	23.0	138
N. Trop. Ct. A	14–22	14–21	15.5–19.6	17.5	−26
N. Equat. Ct.	3–10	4–8	6.6–9.6	7.0	102
Central Equat. Ct.	—	—	—	0.0	95
S. Equat. Ct.	3–10	6–8	5.8–7.6	7.0	137
S. Edge SEB	19	18–22	20.3–21.7	19.5	−61
N. Edge STB	27	26	25.2–26.2	26.5	47
S. Temp. Ct.	29	32–35	33.6–33.7	32.0	−25
	—	—	—	36.5	34
S.S. Temp. Ct.	38–45	39–45	38.8–41.3	39.5	1
S. Polar Region	—	—	—	49.0	−3
	—	—	—	52.5	33
	—	—	—	56.5	−6

[a] Columns I, II, III are from Smith & Hunt (1976) and cover the years 1898–1948, 1946–1964, 1962–1970, respectively. Column IV is from Voyager (Ingersoll et al 1981), and covers the first half of 1979.
[b] Magnitude of the zonal velocity $\bar{u}$ is from Voyager (Ingersoll et al 1981).
[c] The current was not observed during the time interval.

Indeed, we find that the temperature structure of the troposphere measured by Hanel et al (1979) has a close resemblance to the visible cloud markings. Consequently these albedo features are more associated with the radiative budget of the Jovian atmosphere than with the jet structures, so that this lack of correlation between the visible and dynamical properties of Jupiter (and Saturn) is a major constraint on the numerical models that are being constructed to quantitatively explain the measurements.

In Figure 5*A*, the zonal profiles of the Jovian cloud system at the time of the encounters are shown; the differences between the zonal profiles are considerably smaller than the measurement errors. These data may be used to assess the stability of the jets by computing the latitudinal gradient $\mathrm{d}\zeta/\mathrm{d}y$ of the absolute vorticity associated with the zonal wind profile. By definition,

$$\frac{\mathrm{d}\zeta}{\mathrm{d}y} = \beta - \bar{u}'' = \frac{2\Omega\cos\theta}{r} - \frac{\mathrm{d}^2\bar{u}}{\mathrm{d}y^2}, \quad (1)$$

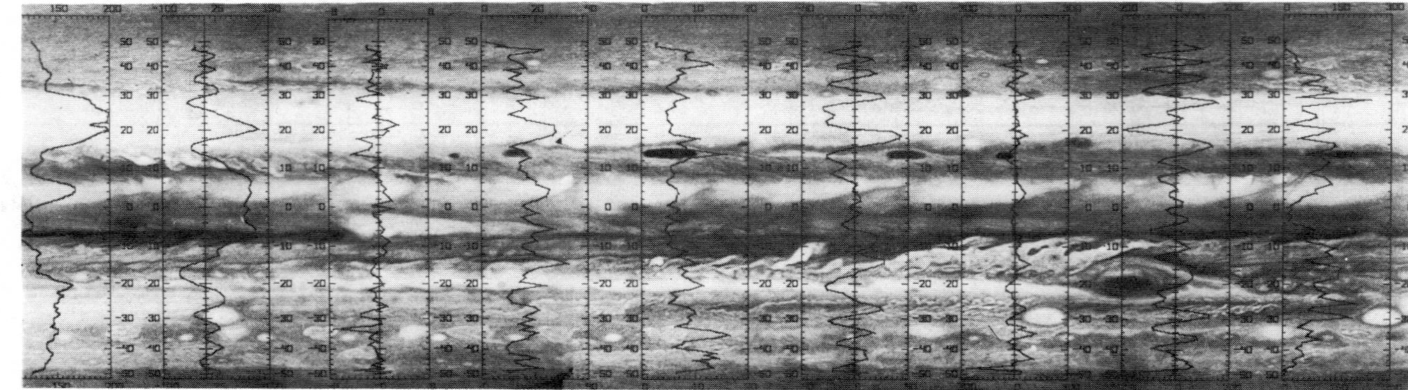

Figure 5A Zonal mean circulation parameters overlaid on a cylindrical projection image for rotation 357. From left to right the curves are normalized brightness, $\langle u \rangle$, $\langle v \rangle$, RMS deviation of $\langle u \rangle$, RMS deviation of $\langle v \rangle$, $d\langle u \rangle/dy$, $\langle u'v' \rangle$, $d^2\langle u \rangle/dy^2$, and the latitude cloud vector sampling density (after Limaye et al 1982).

where y is the northward component, β is the vertical component of vorticity coordinates, Ω is the planetary rotation rate, θ is the latitude, and $\bar{u}''$ is the curvature. Ingersoll et al (1981) have demonstrated that the barotropic stability condition $d^2\bar{u}/dy^2 < \beta$ is violated at the latitudes of the westward jets. It is apparent that $d^2\bar{u}/dy^2$ varies between -3 and $+2$ as a function of latitude. Earlier estimates by Ingersoll & Cuzzi (1969) using Earth-based data underestimated this parameter by a factor of 2. Numerical experiments (Rhines 1975, Williams 1979) with eddy mean flow interaction show that stratified and unstratified rotating fluids tend to relax to a state in which the flow is mainly zonal and $d^2\bar{u}/dy^2 < \beta$. The numerical experiments have been run with a variety of forcings and initial conditions,

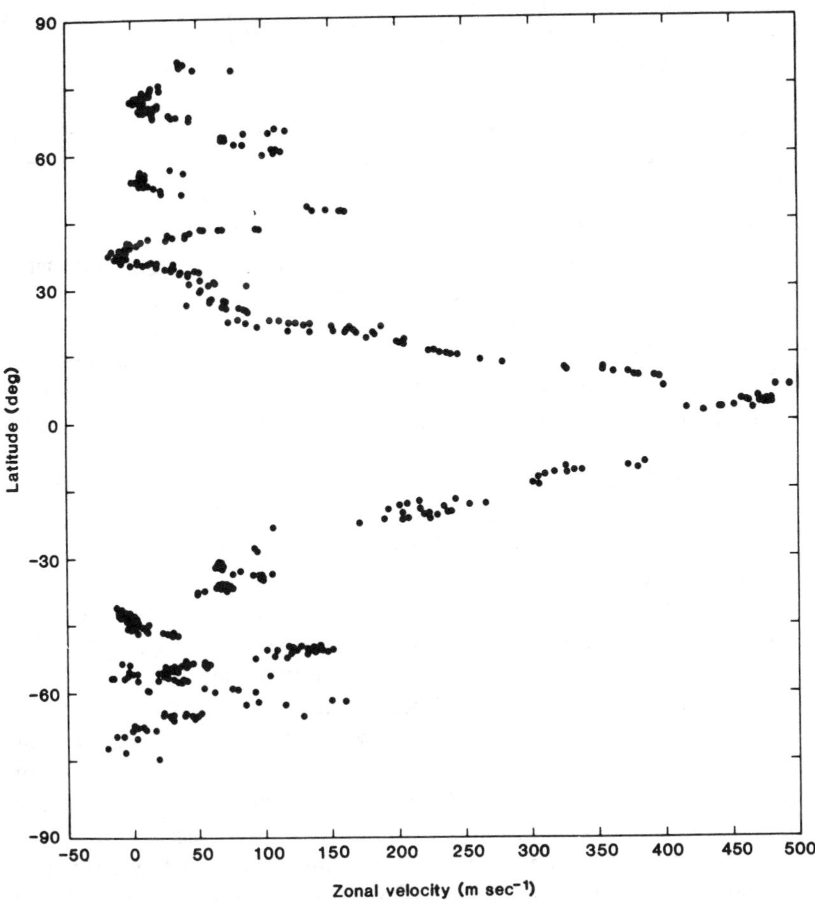

Figure 5B Zonal winds in the reference frame of Saturn's magnetic field (after Smith et al 1982).

including mechanical forcing and thermal (baroclinic) forcing. The computer flows seem to marginally satisfy the barotropic stability criterion $d^2\bar{u}/dy^2 < \beta$, but the question of whether a factor of two is significant requires further analysis.

The corresponding zonal profile for Saturn (Figure 5*B*) has some important differences. The equatorial winds are considerably stronger, with the zonal velocities reaching values of more than 500 m s^{-1}. There appear to be only three retrograde jets, symmetrically positioned in each hemisphere, which also appear to be barotropically unstable (Smith et al 1981). In spite of the axial tilt of the planet, the zonal wind profile, which extends to $\sim \pm 80°$ latitude, is symmetric about the equator. This suggests that the influence of the external solar energy must be small in comparison with the internal heating effects, which are probably spherically symmetric.

We can obtain insight into the possible driving mechanisms for these observed motions by measuring the cloud winds through the application of the image-processing systems.

At each point *i*, the velocities (u_i, v_i) may be measured, together with the quantities $\bar{u}' = v_i - \bar{u}$, $\bar{v}' = v_i - \bar{v}$, which are the deviations from the zonal mean quantities. From data sets of several thousand individual measurements, Beebe et al (1980) and Ingersoll et al (1981) have found that the eddy momentum flux variation with latitude, $\overline{u'v'}$, is positively correlated with $d\bar{u}/dy$ for both Voyager 1 and 2 data sets. This situation occurs for the entire global data set, and would suggest that the main motions are being driven by the conversion of eddy kinetic energy into zonal mean kinetic energy, as in the Earth's atmosphere (Holton 1973). However, Müller (1982) has found the measurement accuracy to be sensitive to navigational errors, while Limaye et al (1982) and Sromovsky et al (1982) have found that the statistical interpretation of these results is strongly affected by the sampling strategy used in the image-processing analyses. They found that significant sampling biases could produce a false positive correlation between the eddy momentum transports and the meridional shear of the zonal wind component. Their more careful examination of the data suggests that there is no significant correlation from the existing data set. However, physically, one must expect this transfer of energy to be taking place.

The rate of conversion $\{k'\bar{k}\}$ of eddy kinetic energy into zonal mean kinetic energy of 1.5–3 W m^{-2} for a layer 2.5 bar deep, estimated by Ingersoll et al (1981), is then an overestimate. The time constant for resupply of zonal mean kinetic energy is only 2–4 months, which is far too rapid when, according to Godfrey et al (1983), the eddy dissipation time is several Earth days. This problem still needs further detailed investigation.

From analysis of the kinetic energy spectra of the cloud winds, Müller et al (1983) find that the only significant spectral peak corresponds to the wave number 8 equatorial cloud feature. At extratropical latitudes, the spectra are remarkably flat, with energy generally decreasing by less than an order of magnitude between wave numbers 1 and 36. There is no hint of the steep spectral slope with the power-law exponent in the range -3 to -5, which is expected for two-dimensional or quasi-geostrophic flow (Charney 1971). The Jovian eddies are also highly anisotropic, being stretched in the zonal direction. This behavior can be contrasted with that of kinetic energy spectra of Earth, which are stretched zonally at wavelengths longer than those of the driving eddies, but are either meridionally stretched or isotropic at shorter wavelengths.

It would therefore seem that the dominant driving eddies of extratropical latitudes for Jupiter have wave numbers at least as great as 36. Saturn may behave in a similar manner.

Models of the Atmospheric Circulation

The major unresolved issue is the depth of the observed motions, since this is naturally the fundamental parameter for all models. There are currently two well-divided schools—the shallow modelers (Williams 1979) and the deep modelers (Ingersoll & Pollard 1982).

Williams has extended a terrestrial model to investigate the characteristics of the atmospheric circulations of the outer planets. His studies are performed for thin-layer β planes and rotating planets. In the barotropic model, vorticity is introduced by small-scale disturbances and is removed by the eddy viscosity. In the baroclinic model, the eddies arise spontaneously as instabilities that draw their energy from the mean equator-to-pole temperature gradient.

The suggestion of a correlation between $\overline{u'v'}$ and $d\bar{u}/dy$ in the Voyager measurements is one point of agreement with the model. However, the uncertainty in the precise magnitude of this energy transfer prevents a more detailed comparison of the planet's thermomechanical cycle.

There are some basic problems in the structure of the Williams model, which suggest that it is not a true representation of Jupiter and/or Saturn. The circulation in the model is driven by the Sun with a solid lower boundary, which completely decouples the atmosphere from the interior. In this sense, it resembles a general circulation model of the Earth. However, unlike the Earth's atmosphere or ocean, a significant amount of heat is introduced at the lower boundary of the atmosphere of both Saturn and Jupiter. A further weakness of the Earth analogy is its inability to represent the adiabatic structure and large heat capacity of the interior levels, as Ingersoll & Porco (1978) indicated. Despite these limitations, the model does appear to emulate some of the basic characteristics of the flows. The cloud bands are thought to be ultralong baroclinic waves, symmetrically arranged in pairs of alternating high- and low-pressure systems. Williams (1979) further suggests that the blocking effect of the planetary wave propagation on quasi-geostrophic turbulence cascades determines the width and zonality of the bands. It would appear that the degree of zonality is higher in the absence of surface drag. At polar regions, there is no horizontal temperature gradient and, correspondingly, no baroclinicity. Consequently, the belt/zone structure disappears at a latitude where internal heating dominates the contribution from solar heating.

The alternative view, that the motions may take place in deep atmospheres that may actually be affected by the interior characteristics, is taken by Ingersoll & Pollard (1982), who suggest that any steady zonal motion would take the form of differentially rotating cylinders, concentric about the planetary axis of rotation. Certainly the motions are deep if we follow the analysis of Smith et al (1982), who propose that the observed Saturn wind profile may extend well beneath the base of the visible clouds. They suggest that for Saturn the level of no motion may be at a pressure level of 10^3–10^4 bar. The observed zonal jets will be manifestations of these cylindrical patterns. From this model, we may expect the profile to be symmetric about the equator, and the zonal motions to terminate at the latitude corresponding to the tangent to the metallic core of the planet. For Saturn, the profile does appear symmetric (Figure 5*B*), and the organized motions do appear to terminate at $\sim 80°$. For Jupiter, Ingersoll & Pollard (1982) find that the profile is, in general, symmetric, although the deviations are more closely related with the Great Red Spot ($-18°$ to $-26°$) and the high-speed westerly jet in the north (18° to 24°). These deviations may be significant.

The analyses by Ingersoll & Pollard (1982) demonstrate that the zonal flow of the atmosphere could exist in the interior, but they do not show that it *must* exist. This will require a fully developed numerical model of convection in a rotating system. However, there must be some interaction between the internal processes and the atmosphere that makes the "star-like" interior of the planet assist in the production of apparently Earth-like weather systems.

Large-Scale Cloud Vortices

The Great Red Spot (GRS) has been the center of debates for many centuries, and the Voyager observations have provided some important observations that may assist in resolving its origin. This feature is not fixed but moves in an easterly direction relative to the main zonal flow at about 0.5° per day. At the time of the first encounter, small cloud vortices can be seen rotating in a counterclockwise manner around the GRS in a period of 6 days. However, a few months later, the growth of a large white cloud system to the east of the GRS forms a barrier to these cloud vortices.

The Voyager observations have shown that the GRS, the white ovals, and the small-scale spots at 41°S all possess similar meteorological features (Smith et al 1979a,b, Hanel et al 1979). Wind speeds of 110–120 m s^{-1} are observed near the edges of both features along their minor axes. Relative vorticity profiles reach a maximum of $6 \times 10^{-5}\ s^{-1}$. This is several times greater than the ambient $5 \times 10^{-5}\ s^{-1}$ of the meridional shear winds at the latitudes of these features. Their vorticities are in the range $(2–3) \times 10^{-5}\ s^{-1}$, with corresponding Rossby numbers for the flows within the GRS and White Oval (BC) of 0.36. Generally, the Rossby numbers within these features are much lower, indicating the geostrophy of the flow (Mitchell et al 1981). All these features rotate anticyclonically and are elevated relative to their surroundings (Conrath et al 1981). However, in contrast to the white ovals, the GRS possesses a large, quiescent interior region. These similarities strengthen the idea that all the features are of the same type, with the only difference being their individual size. The infrared observations of Hanel et al (1979) indicate that the GRS and the ovals have a cold region above the feature extending throughout the troposphere. This is consistent with a divergence at cloud-top level, although there is little evidence of organized flow in the GRS. Flasar et al (1981a) suggest that this upward motion may be driven by latent heat release in the water-vapor cloud region. Saturn's largest ovals are smaller than Jupiter's (Smith et al 1981).

However, to understand the origin of these features, it is necessary to examine the behavior of the white ovals that have been observed since their formation in 1939. Peek (1958) and Hunt & Beebe (1983) have shown that these features originally formed from a cloud system (zone) that stretched around the planet. In the past 40 years, the ovals have been contracting to their current size of 11,000 km × 5,000 km. It is very likely, therefore, that the GRS behaved in a similar way. Smith et al (1979a,b) have found that it is now only 24,000 km in length, compared with 46,000 km a century ago. The GRS is also contracting.

These observational characteristics are consistent with the numerical model of Williams (1979), which closely resembles the large-scale features of the Jovian atmosphere. His model predicts a large-scale circulation gyre in the position of the GRS, which corresponds to the warm anticyclonic core of a neutral baroclinic wave. Features of this type seem to appear naturally from the general circulation and may therefore account for all the large-scale features observed. However, their existence and the Voyager measurements are consistent with this type of driving mechanism. The solitary-wave theory suggested by Maxworthy et al (1978) may only apply to those special situations where the flow becomes primarily barotropic, and may account for the local interactions between cloud systems that are sometimes seen.

The lifetime of the vortices in the Jovian atmosphere is a further important problem since, unlike the terrestrial atmosphere, there are no solid surface features to constrain the flow. The radiative relaxation time is several years, so that with a cooling rate of about 10 K day^{-1} features will radiatively dissipate very slowly. However, Ingersoll & Cuong (1981) also suggest that the long-lived vortices maintain themselves against dissipation by absorbing small vortices produced by convection. They suggest that these large isolated vortices are modons, which, unlike solitons, tend to interact with one another.

The color of the GRS remains a major unresolved problem. The direct evidence is that there is upward motion in the spot. This supports the prediction of Prinn & Lewis (1975) that the color is due to the conversion of PH_3 into P_4, which condenses to form triclinic red phosphorus crystals. As a consequence, this explanation requires the GRS to extend more deeply into the atmosphere than the outer large-scale spots, such as the ovals.

Some Examples of Atmospheric Convection

It is certain that small-scale convection plays a major role in the meteorology of both Jupiter and Saturn. Examples of small-scale convective systems are evident in many of the Voyager images. However, in some situations large-scale organized convection may be seen at particular locations. Observations by the Voyager spacecraft imaging and IRIS instruments show an organized train of features moving in a westerly current at 9°N with a zonal speed of 100–120 m s^{-1} (Hunt et al 1981a). The region is characterized by a wave number 11–13 pattern, which was observed to fluctuate in its precise characteristics between the two encounters (Smith et al 1979a,b). Only a small number of plumes have active convective centers. Hunt et al (1981a) have shown that these cloud systems cause a perturbation to the temperature of the upper troposphere (Figure 6). The thermal structure of an individual plume supports the concept that the head is a region of strong upwelling while subsidence is occurring in the surrounding areas. Hunt et al (1981b) have measured the time variation of the change in the areas of the active plumes and estimated the divergences and vertical velocities associated with these features. The divergences are in the range of $(0.5\text{–}1.5) \times 10^{-5}\ s^{-1}$, and vertical velocities are in the range of 10–40 cm s^{-1} for the more active features.

The rapid development of convective activity is suggestive of an instability mechanism, while the global coherence implies that the same role is played by a planetary-scale wave system. A possible mechanism is wave

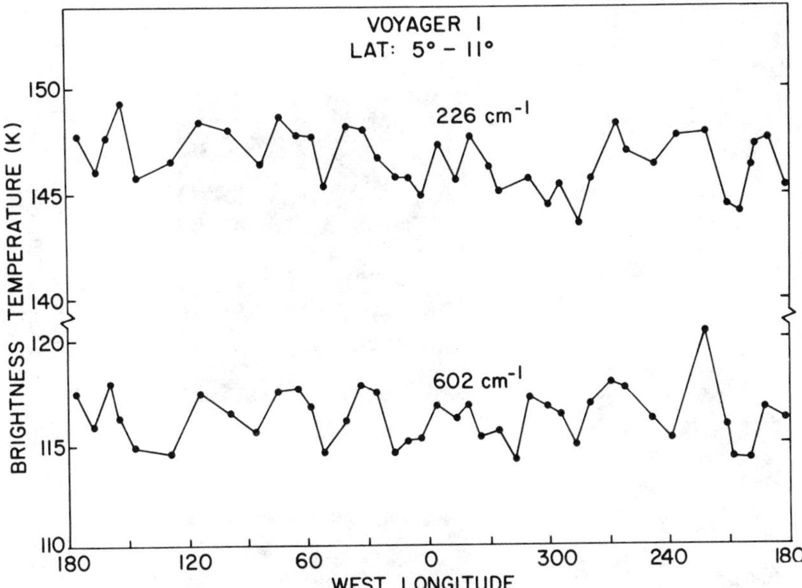

Figure 6 The brightness temperature at 226 cm^{-1} (cloud tops) and 602 cm^{-1} (tropopause) for the latitude range 5–11°, obtained by IRIS during the Voyager 1 encounter (after Hunt et al 1981a).

convective instability of the second kind (CISK) (Lindzen 1974), which is operative in the Intertropical Convergence Zone (ITCZ) of the tropical region of the Earth's atmosphere. This mechanism requires the presence of a finite-amplitude wave field with alternating lower-level regions of convergence and divergence along a moisture field capable of providing an energy power through latent heat release in the regions of upwelling that accompany convergence. The water vapor beneath the visible clouds may play an important role in developing the Jovian plumes. Furthermore, their presence on the northern edge of the equatorial region may be due to the presence of large-scale features in the southern hemisphere, such as the Great Red Spot and white ovals, that perturb the low-level convergence.

Plume-like features are not evident in the atmosphere of Saturn. However, evidence of small-scale convection is seen at many locations at high latitude, while the most organized systems are present in the strong easterly flow at 45°N. Hunt et al (1982) found that the divergences from these features were $(3\text{–}5) \times 10^{-5}\ s^{-1}$, which correspond to vertical velocities of $\sim 1\ m\ s^{-1}$. This is a stronger upward motion than is found in the Jovian atmosphere. The most unexpected system resembled a train of vortices formed behind a convective cloud tower (Figure 7; Godfrey et al

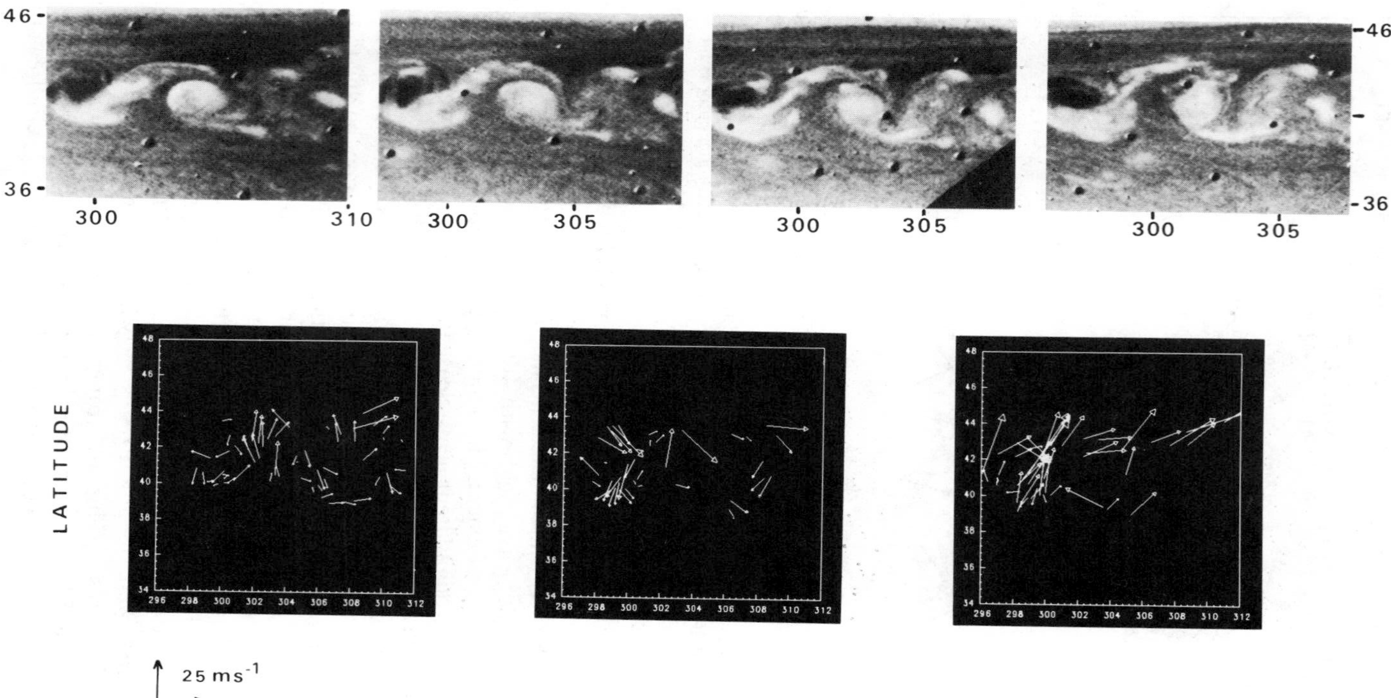

Figure 7 (*top*) The sequence of map projection green images of the midlatitudes of Saturn, showing the structure of the vortex street. From left to right, the images were obtained at 20 August 1981 14.24.47, 21 August 1981 01.02.23, 21 August 11.29.35, and 21 August 22.01.35. (*bottom*) The velocity measurements illustrating the flow in the vortex streets.

Table 5 A summary of the properties of vortex streets

Parameter	Symbol	Saturn	Gran Canaria (Berlin 1981)	Madeira (Chopra & Hubert 1965)	Predicted values
Diameter of obstacle	d	1.7×10^3 km	45 km	43 ± 3 km	—
Bandwidth of street	h	800 ± 120 km	55 ± 10 km	83 km	—
Longitudinal spacing	a	5×10^3 km	130–165 km	190 km	—
Undisturbed zonal flow	$\bar{u}_0$	(3.7–5.8) m s^{-1}	7.5–10 m s^{-1}	10 m s^{-1}	—
Zonal flow of eddies	U_e	(1.1–1.7) m s^{-1}	5 m s^{-1}	7.5 m s^{-1}	—
Length of vortex street	L	9×10^3 km	800 km	540 km	—
von Kármán ratio	h/a	0.16 ± 0.025	0.35–0.4	0.43	0.28–0.52
Eddy production rate	N	$(4.4–6.9) \times 10^{-7}$ s^{-1}	3.8×10^{-5} s^{-1}	3.95×10^{-5} s^{-1}	—
Eddy lifetime	$1/N$	(26–17) days	7.2 hr	7.15 hr	—
Vortex lifetime	t	(95–61) days	45 hr	20 hr	—
Stroudal number	S	0.2 (assumed)	0.22–0.17	0.17 ± 0.1	0.2 (cylinder)

1983). They show that all the characteristics of the flow resemble those of a Kármán vortex street, which can be seen formed in the wake of a solid object in a laboratory flow or behind an island in the Earth's atmosphere (Table 5). It would appear that the internal convective processes modulate the vertical extent of the cloud system that spawns the observed eddies.

In summary, there would seem to be many properties of Jovian and Saturnian cloud systems that may originate at depths greater than those of the clouds. The existence of these vortices and isolated cloud systems may prove to be special diagnostics that reveal how the atmospheres and interiors react. Our hope for understanding the dynamics of these atmospheres lies in finding a sufficient number of these diagnostics so that they will constrain the model and provide a unique representation of the observed behavior.

TITAN

Titan is the largest of the Saturnian satellites, and with a surface radius of 2575.0 ± 0.5 km (Lindal et al 1982) it is only slightly inferior to Ganymede in size. It is the only satellite with a substantial atmosphere, which is denser than that of any terrestrial planet except Venus.

With an escape velocity of ~ 2.5 km s^{-1} it is feasible that at the distance of the Saturnian system from the Sun, Titan might retain an atmosphere. The first indications of a possible atmosphere were reported in 1903 by Comas Solà (see Alexander 1962), who found that there was considerable limb-darkening, though the entire disk was brighter. Observations of this kind are difficult and frequently inconclusive. However, in the winter of 1943/44, Kuiper (1944) showed that methane was present in the Titan spectrum. It was not until Trafton (1972) detected the presence of hydrogen in the atmosphere that interest in Titan was renewed. Now, with the recent Voyager observations and improved Earth-based telescopic observations, our knowledge of this body has rapidly advanced.

Composition

In Table 6, we have summarized our current knowledge of the composition of Titan. Nitrogen, with a molecular weight of 28 amu (Tyler et al 1981), has been observed to be the major constituent of the atmosphere in EUV observations by Broadfoot et al (1981); methane, its derivatives, and the other detected gases probably make up about 1% of the atmosphere. Trafton (1981) has described in detail the pre-Voyager observations of Titan, which have now been substantially extended by the analyses of the IRIS measurements (see, for example, Kunde et al 1981). The presence of nitrogen as the major constituent makes Titan and the Earth unique in this respect. In addition, a very complete set of organic compounds have been

Table 6 Composition of the atmosphere of Titan

		Wave number (cm^{-1})	Approximate mole fraction
Nitrogen	N_2	EUV	0.94
Hydrogen	H_2	350	2×10^{-3}
Carbon-hydrogen			
Methane	CH_4	1304	3×10^{-2}
Acetylene	C_2H_2	729	2×10^{-6}
Ethylene	C_2H_4	950	4×10^{-7}
Ethane	C_2H_6	822	2×10^{-5}
Methyl-acetylene	C_3H_4	325, 633	3×10^{-8}
Propane	C_3H_8	748	2×10^{-5}
Diacetylene	C_4H_2	220, 628	10^{-8}–10^{-7}
Carbon-oxygen			
Carbon dioxide	CO_2	667	7×10^{-10}
Carbon monoxide	CO	—	5×10^{-3} [a]
Carbon-hydrogen-nitrogen			
Hydrogen cyanide	HCN	712	2×10^{-7}
Cyanoacetylene	HC_3N	500, 663	10^{-8}–10^{-7}
Carbon-nitrogen			
Cyanogen	C_2N_2	233	10^{-8}–10^{-7}

[a] Estimated value.

found at various levels of the Titan stratosphere (Table 6, Figure 8). These compounds originate by reactions of methane and nitrogen radicals in a predominantly nitrogen atmosphere. Although Titan is located at 9.35 AU, which reduces the solar flux by a factor of 87 compared with the Earth value, dissociation of significant quantities of methane still occurs due to Lyα radiation. Titan is at the edge of the Saturnian magnetosphere, and the boundary crosses the orbit of the satellite's path when appropriately influenced by the changing solar wind pressure. Consequently, there will be periods when the satellite will be outside the magnetospheric shield of Saturn. Even when it is inside this region, the satellite will be unprotected, since Titan does not possess a shielding magnetic field (Ness et al 1981). This causes energetic electrons and protons to ionize and dissociate the principal atmospheric constituents. Bremsstrahlung will extend the effective penetration depth of energetic particles down to an altitude of about 130 km. In addition to charged particles from Saturn's magnetosphere, Titan is probably also bombarded by galactic cosmic radiation consisting of protons in the GeV range, which may even penetrate to the surface. Kunde

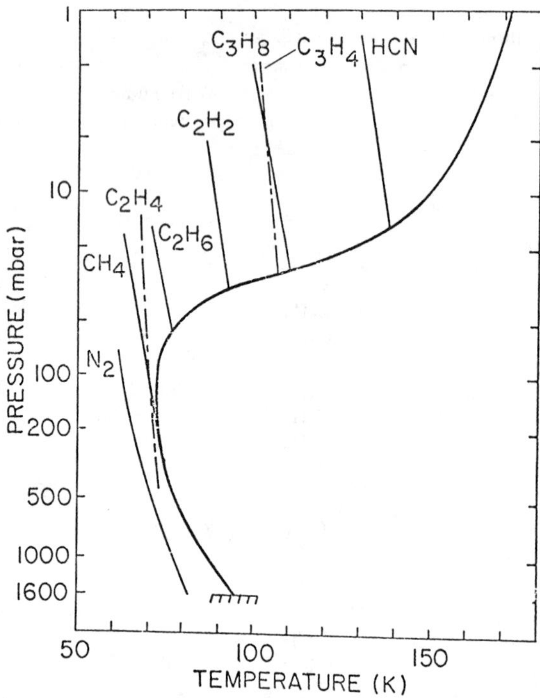

Figure 8 Vertical temperature profile of Titan's atmosphere and the condensation curves of N_2 and several hydrocarbons (after Maguire et al 1981).

et al (1981) suggest that particle ionization may then lead to neutral fragments and ions, so that the subsequent ion-molecule reactions involving methane, molecular hydrogen, and molecular nitrogen may produce the observed complex neutral hydrocarbons.

The wide variety of organic compounds with fairly high abundances of C_2H_2 and C_3H_8 indicate that the chemistry of Titan is significantly different from that of Jupiter and Saturn. Hanel et al (1981b) find that the Titan/Saturn ratio for C_2H_6 is ~ 4 and for C_2H_2 is ~ 150. Lovas et al (1979) find that the low abundance of H_2 ($\sim 0.2\%$) prevents the rapid recycling of methane and yields higher relative levels of the methane photolysis radicals (CH_2, CH_3, C_2H).

Although HCN has been observed in interstellar space (Snyder & Buhl 1971), the observations by Hanel et al (1981a) are the first in a planetary atmosphere. The detection of this organic compound is of considerable interest, since HCN is a key molecule in the synthesis of amino acids and bases present in nucleic acids. This does *not* suggest that the conditions on Titan are necessarily suitable for life. However, it does indicate that Titan may be "an Earth in deep freeze," and that the chemistry leading to the formation of organic molecules has occurred there even if the intense cold has prevented it from developing further. Strobel (1982) has suggested that HCN may be produced in significant amounts, both from electron bombardment and solar ultraviolet radiation.

The recent detection of CO_2 on Titan by Maguire et al (1982) was unexpected. This feature is clearly identified in emissions at 667 cm^{-1}, from which a mole fraction of 7×10^{-10} is determined. The authors suggest further that based upon the photochemical modeling, which is consistent with our current understanding of these atmospheres, a mole fraction of 5×10^{-3} for CO would be required. This value is consistent with the upper limit determined by Sandel et al (1982).

The origin of the CO_2 as the first oxygen-bearing molecule requires careful consideration. Since it has been detected in emission, it must therefore be in the stratosphere of Titan. This may suggest that the oxygen has been produced from impacting meteorites, as Strobel (1977) suggested for Jupiter. Owen (1982) has proposed an alternative hypothesis relating to the origin of the Titan atmosphere. He suggests that it may have been produced in part by the degassing of clathrate hydrates. He adds that the capture of CO, N_2, Ar, and CH_4 from the proto-Saturnian nebula by this mechanism would yield the currently observed atmosphere. The key observations that will assist in quantifying this hypothesis relate to the amount of Ar and Ne that may be currently present; such observations cannot be determined accurately by in situ measurements. However, the chemistry is also closely related to current surface and atmospheric conditions.

The atoms and molecules of hydrogen, which are produced by the photochemical reactions, easily escape from Titan because of the weak gravity, as McDonough & Brice (1973) first suggested. Hunten (1978) suggested that the hydrogen escape rate would be about 9×10^9 H_2 molecules cm^{-2} s^{-1}. Indeed, Lyα emission has been detected in the vicinity of Titan from the Orbiting Astronomical Observatory (OAO) (Barker et al 1980), from rocket observations (Weiser et al 1977), and from Voyager (Broadfoot et al 1981). Broadfoot et al find that the hydrogen is widely distributed throughout the Saturnian system and appears to form a cloud of uniform density encircling the planet between 8 and 25 R_S near the equatorial plane. Most of the hydrogen seems to be concentrated within 6 R_S, indicating a total volume of 2×10^{33} atoms cm^{-3} (which is a total content of 2×10^{34} hydrogen atoms). If the lifetime of a neutral hydrogen atom is 10^7 seconds, then the required supply rate is 2×10^{27} atoms per second. The torus, which may also contain some oxygen atoms, co-rotates with the magnetosphere of Saturn. It is still not certain if the hydrogen supply comes from Titan alone.

Atmospheric Structure, Clouds, and Motions

In Figure 8, we show the structure of the Titan atmosphere, which is consistent with the infrared observations of Hanel et al (1981b) and the occultation measurements of Tyler et al (1981). The profile has considerable similarity with the prediction of Hunten (1978). These observations span a pressure range from the surface 1.6 bar to 0.3 mbar, which is an altitude range of more than 200 km. Lindal et al (1982) find that near the surface, where the temperature is 93.8 ± 0.7 K, the temperature lapse rate is about 1.38 ± 0.1 K km^{-1}. They find an abrupt change in the temperature structure at an altitude of 3.5 km, which may mark either the boundary between a convective region with an adiabatic lapse rate near the surface and a higher stable region in radiative equilibrium, or the bottom of a methane haze layer. The spatial irregularities in the profile (Hinson & Tyler 1982) may suggest convective turbulence at these levels. The tropopause is situated at $\sim$43 km above the surface, where the temperature is 71.3 ± 0.5 K and the pressure is 140 mbar. Above the tropopause, the temperature increases with height throughout the stratosphere. The 1 mbar level corresponds to 160 ± 1 K near the evening terminator and 170 ± 15 K near the morning terminator. We see in Figure 8 that all the organic compounds were detected in the stratosphere of Titan, indicating the large opacity that is created by the tropospheric layers.

Since Titan has an effective temperature of $\sim$86 K, the observed surface temperature of $\sim$94 K indicates the presence of a small greenhouse effect. Samuelson et al (1981) find that this effect results from about 10% of the incident solar flux being absorbed by the surface layers. This model also requires about 11% argon to be present near the surface.

Flasar et al (1981b) have investigated the meridional temperature contrasts, which are shown in Figure 9 for 530, 200, and 1304 cm^{-1}. The variations at 200 cm^{-1} are small and in the northern hemisphere are $\sim$1 K higher than those at southern latitudes. The principle source of opacity at this spectral interval is thought to be methane clouds, which Flasar et al suggest are located in the neighborhood of the tropopause. At 530 cm^{-1}, the effective emission level is close to the ground (Samuelson et al 1981), suggesting that there is a real temperature variation with latitude. The major opacity is from the pressure-induced H_2 absorption, while the contribution from clouds is thought to be small. The thermal variation with latitude is approximately symmetric about the equator. The slightly warmer daytime temperatures found in the northern latitudes compared with the south are consistent with the north/south opacity difference inferred at 200 cm^{-1} by Flasar et al.

The observations at 1304 cm^{-1} relate to stratospheric levels, where the poles are apparently 20 K colder than the equatorial region (Figure 9). There is a hemispheric asymmetry, since the southern hemisphere is about 3 K warmer at midlatitudes than the equivalent region in the north.

Flasar (1982) suggests that the absorbed thermal contrasts at the 1 mbar level result from a similar situation of cyclostrophic winds that also occur in the Venus stratosphere (Leovy 1973). This requires the meridional gradient in the geopotential to be dynamically balanced, i.e. by the cyclostrophic zonal wind. Flasar (1982) has suggested that cyclostrophic winds of 50–100 m s^{-1} at the 1 mbar level and the absence of any longitudinal structure may ensure that the zonally symmetric flows and barotropic eddies are sufficient

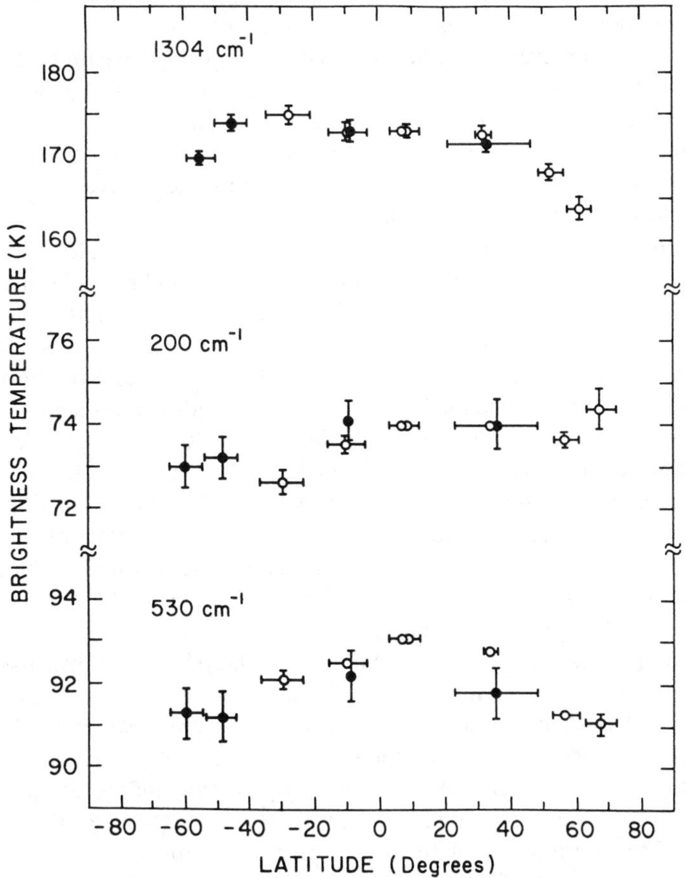

Figure 9 Latitudinal distributions of brightness temperatures at 200, 530, and 1304 cm^{-1} for Titan, observed by the IRIS instrument on Voyager 1. Daytime data shown as ●; nighttime as ○ (after Flasar et al 1981b).

to maintain the atmosphere with excess angular momentum. Like Venus, the upper atmosphere of Titan is super-rotating, since at the 1 mbar level the period is 2 days compared with the rotation period of 15 days. However, the stratosphere of Titan differs from that of Venus in its absence of diurnal effects and in the seasonal changes that are apparently present.

The visible appearance of Titan shows only a limited amount of detail and certainly insufficient contrast to enable winds to be determined by tracking cloud elements. At the time of the Voyager encounters, the north polar cap had a dark hood, while the equatorial region was marked by a distinct change of contrast between the hemispheres. In the troposphere, methane clouds probably form (Samuelson et al 1981) and may even rain or sleet onto the surface. Precise information on clouds at these levels cannot yet be obtained. The clouds and hazes at higher levels have been examined at a variety of wavelengths. Smith & Tomasko (1982) have found that the polarization observations are sensitive to the microphysical properties of the particles near $\tau = 0.5$. They find at the highest levels hazes of dark orange particles with radius $a = 0.1$ μm at stratospheric levels, and suggestions that these particles may increase in size at greater depths. A possible real refractive index is $n_r = 1.7$.

High-resolution images of the limit at high phase angles reveal an optically thin layer of haze $\sim$50 km thick and $\sim$100 km above the main aerosol layer (Smith et al 1981). The haze layer extends from the top of the north polar hood entirely around the illuminated hemisphere, continuing some distance beyond the terminator in both hemispheres. Broadfoot et al (1981) also found evidence for a set of discrete absorbing layers, which extended more than 500 km above the surface. Rages & Pollack (1982) found the average particle size of the detached haze to be about 0.3 μm; in addition, they found that at its thickest part, which occurs at an altitude of 350 km, the haze contains about 0.2 particles cm^{-3}. The total optical depth is $\sim$0.015 for latitudes less than 45°. At high northern latitudes, the top of the detached haze layer drops by $\sim$30 km, while near the north pole it thickens to merge with the main aerosol layer.

Color images (see, for example, Smith et al 1981) show a reddish disk with a readily discernible interhemispheric asymmetry. The northern hemisphere is darker and redder, and is surrounded by a still darker north polar hood. The interhemispheric boundary lies in Titan's orbital plane, and this symmetry is almost certainly determined by the satellite's rotation. At the time of the Voyager encounters, there was no evidence for a southern polar hood. The coloration itself, like that which occurs on Jupiter and Saturn, may be the result of a complicated mixture of cloud physics, photochemistry, and dynamics. However, the north/south asymmetry in the brightness of Titan does relate to the dynamical properties of the satellite.

Lockwood (1977) has found that between 1972 and 1976, Titan's disk-integrated brightness increased by 9% in the blue and about 5.5% in the yellow to a maximum during 1976–77 and then decreased in 1978. This variation (with a smaller variation) was also found to occur on Neptune (Lockwood & Thompson 1979, Suess & Lockwood 1980). These variations were thought to be related to solar variability, since the photochemical reactions in the upper atmosphere produce submicron aerosol particles, which then affect the reflection properties of the bodies.

Sromovsky et al (1981) suggested that for Titan the interhemispheric contrast may be the response to seasonal solar-heating variations, which result from the inclined spin axis of the satellite. The contrast significantly lags the solar forcing by almost exactly one season (90°, ~7 yr), suggesting that its production involves the atmosphere well below the visible layers. Sromovsky et al (1981) showed that this contrast has a significant effect on Titan's disk-integrated brightness as seen from the Earth, and they suggested that this probably accounts for most of the long-term variation, with the solar UV variations accounting for the remainder. Their prediction of the changes in the brightness for the remainder of this decade (Figure 10) will require careful observations from the Space Telescope, in addition to the continuation of the Earth-based photometric studies.

Discussion

In this article, I have reviewed our current knowledge of the atmospheres of Jupiter, Saturn, and Titan, which has been significantly advanced by the recent Voyager flyby measurements, by IUE, and by significant developments in image-processing methods. A major objective for studying planetary atmospheres is to provide information on all the planets in order to understand the origin of our solar system. There is no doubt that the variations in the H/He ratio between the planets and the Sun, and the other compositional differences, reflect upon their initial developments and place major constraints upon cosmological theories. A further reason for studying planetary atmospheres is to gain insight into the effects that are obscured and complicated by various mechanisms peculiar to the Earth's meteorology. The study of planetary atmospheres also provides a means of testing our understanding of physical processes in an environment different from that in which we have developed our initial ideas. The meteorologies of Jupiter, Saturn, and Titan offer a great challenge. Understanding the origin and evolution of the atmosphere of Titan, and its relationship with the development of the Earth's atmosphere, is a major problem for further analysis.

The past ten years have seen the renaissance period in planetary science. We have become used to regular space missions and a constant wealth of

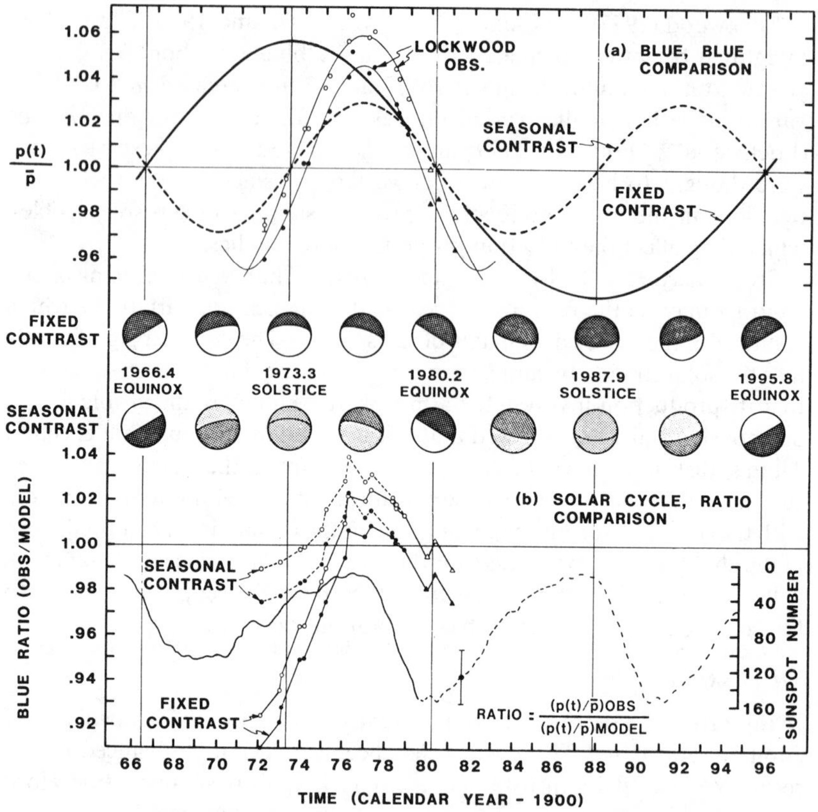

Figure 10 (*a*) Relative variations of Titan's blue albedo according to fixed contrast (heavy solid line) and seasonal contrast (heavy dashed line) models; (*b*) comparison of the solar cycle of smoothed sunspot number with the ratio of observed-to-predicted relative variations in Titan's geometric albedo (after Sromovsky et al 1981).

new observations. This situation has now changed, since we may have to wait until the end of this decade before the Galileo spacecraft reaches Jupiter. However, Space Telescope observations are certain to provide challenging data to extend the myopic view from the flyby space missions. Advances in telescopic facilities will also enable further advances to be made. But we have a major investment already in the huge volumes of spacecraft data that have recently been obtained, which, with careful analysis, will enable us to put into perspective the basic properties of the atmospheres of these major planets.

ACKNOWLEDGMENTS

It is a pleasure to thank my colleagues who have contributed to my understanding, which has enabled me to write this review. Discussions with Reta Beebe, Barney Conrath, Peter Gierasch, Andy Ingersoll, Conway Leovy, Toby Owen, Brad Smith, Ed Stone and Vern Suomi, and my Imperial College colleagues Peter Müller, Vivien Moore, and David Godfrey, are gratefully acknowledged. This research is supported by the Science and Engineering Research Council (UK).

Literature Cited

Alexander, A. F. 1962. *The Planet Saturn*. London: Faber & Faber
Atreya, S. K. 1981. *Adv. Space Res.* 1:8–26
Audouze, J., Tinsley, B. M. 1974. *Astrophys. J.* 192:187–96
Aumann, H. H., Gillespie, C. M., Low, F. J. 1969. *Astrophys. J. Lett.* 157:L69–72
Axel, L. 1972. *Astrophys. J.* 173:451–68
Barker, E., Cazes, S., Emerich, C., Vidal-Madjer, A., Owen, T. 1980. *Astrophys. J.* 242:383–94
Bar-Nun, A. 1979. *Icarus* 38:180–91
Barocki, W., Bar-Nun, A., Scarf, F. L., Cook, A. F., Hunt, G. E. 1983. *Icarus*. In press
Barshay, S. T., Lewis, J. S. 1978. *Icarus* 33:593–611
Beebe, R. F., Ingersoll, A. P., Hunt, G. E., Mitchell, J. L., Müller, J.-P. 1980. *Geophys. Res. Lett.* 7:1–41
Beer, R., Taylor, F. W. 1979. *Icarus* 40:189–92
Berlin, P. 1981. *ESA Bull.* 25:16–19
Brault, J. W., Smith, W. H. 1980. *Astrophys. J. Lett.* 235:L177–78
Broadfoot, A. L., Sandel, B. R., Shemansky, D. E., Holberg, J. B., Smith, G. R., et al. 1981. *Science* 212:206–11
Cameron, A. G. W. 1973. *Space Sci. Rev.* 15:121–35
Carlson, R., Judge, D. L. 1974. *J. Geophys. Res.* 79:3623–33
Charney, J. 1971. *J. Atmos. Sci.* 28:1087–95
Chopra, K., Hubert, L. F. 1965. *J. Atmos. Sci.* 22:652–57
Combes, M., Encrenaz, T. 1979. *Icarus* 39:1–27
Conrath, B. J., Flasar, F. M., Pirraglia, J. A., Gierasch, P. J., Hunt, G. E. 1981. *J. Geophys. Res.* 86:8769–77
Cook, A. F., Duxbury, T. C., Hunt, G. E. 1979a. *Nature* 280:794–95
Cook, A. F., Duxbury, T. C., Hunt, G. E. 1979b. *Nature* 280:780–93
Courtin, R., Gautier, D., Marten, A. 1981. *Bull. Am. Astron. Soc.* 13:722
Encrenaz, T., Combes, M. 1981. *Proc. IAU Conf. No. 96*, pp. 1–31
Encrenaz, T., Owen, T., Woodman, J. H. 1974. *Astron. Astrophys.* 37:49–57
Encrenaz, T., Combes, M., Zeau, Y. 1980. *Astron. Astrophys.* 84:148–61
Eshleman, V. R., Tyler, G. L., Wood, G. E., Lindal, G. F., Anderson, J. D., et al. 1979. *Science* 204:976–78
Fink, U., Larson, H. P. 1979. *Astrophys. J.* 233:1021–40
Flasar, F. M. 1982. In preparation
Flasar, F. M., Conrath, B. J., Pirraglia, J. A., Clark, P., French, R., Gierasch, P. J. 1981a. *J. Geophys. Res.* 86:8759–67
Flasar, F. M., Samuelson, R. E., Conrath, B. J. 1981b. *Nature* 292:693–98
Gautier, D., Grossman, K. 1972. *J. Atmos. Sci.* 29:788–92
Gautier, D., Marten, A., Baluteau, J. P., Lacombe, A. 1979. *Icarus* 37:214–27
Gautier, D., Conrath, B. J., Flasar, M., Hanel, R., Kunde, V., Chedin, A., Scott, J. 1981. *J. Geophys. Res.* 86:8713–20
Gautier, D., Bezard, B., Marten, A., Baluteau, J. P., Scott, N., Chedin, A., Kunde, V., Hanel, R. 1982. *Astrophys. J.* 257:901–12
Gehrels, T., ed. 1976. *Jupiter*, pp. 531–63. Tucson: Univ. Ariz. Press
Gladstone, C., Randall, R., Yung, Y. L. 1982. In preparation
Godfrey, D., Hunt, G. E., Suomi, V. E. 1983. *Geophys. Res. Lett.* In press
Goorvitch, D., Emikson, E. F., Simpson, J. P., Tokunaga, A. 1979. *Icarus* 40:75–86
Hanel, R., Conrath, B., Flasar, M., Herath, L., Kunde, V., et al. 1979. *Science* 206:952–56
Hanel, R. A., Conrath, B. J., Herath, L. W., Kunde, V. G., Pirraglia, J. A., et al. 1981a. *J. Geophys. Res.* 86:8705–12
Hanel, R., Conrath, B., Flasar, F. M., Kunde, V., Maguire, W., et al. 1981b. *Science* 212:192–200
Hanel, R., Conrath, B., Flasar, F. M., Kunde, V., Maguire, W., et al. 1982. *Science* 215:544–48

Hinson, D. P., Tyler, G. 1982. *Icarus*. Submitted for publication

Holton, J. R. 1973. *Introduction to Dynamical Meteorology*. New York: Academic

Hubbard, W. B. 1981. *Philos. Trans. R. Soc. London Ser. A* 303:315–26

Hunt, G. E. 1978. *Proc. Symp. Planet. Atmos.*, pp. 105–10. Ottawa: R. Soc. Canada

Hunt, G. E., Beebe, R. F. 1983. In preparation

Hunt, G. E., Conrath, B. J., Pirraglia, J. A. 1981a. *J. Geophys. Res.* 86:8777–83

Hunt, G. E., Müller, J.-P., Gee, P. 1981b. *Nature* 295:491–94

Hunt, G. E., et al. 1983. *Trans. IEEE*. In press

Hunten, D. 1978. In *The Saturn System*, ed. D. Hunten, D. Morrison, pp. 127–40. *NASA CP 2068*

Ingersoll, A. P., Cuong, P. G. 1981. *J. Atmos. Sci.* 38:2067–76

Ingersoll, A. P., Cuzzi, J. 1969. *J. Atmos. Sci.* 26:981–85

Ingersoll, A. P., Pollard, D. 1982. *Icarus* 52:62–80

Ingersoll, A. P., Porco, C. C. 1978. *Icarus* 35:27–43

Ingersoll, A. P., Münch, G., Neugebauer, G., Orton, G. S. 1976. See Gehrels 1976, pp. 197–205

Ingersoll, A. P., Beebe, R. F., Mitchell, J. L., Garneau, G. W., Yagi, G. M., Müller, J.-P. 1981. *J. Geophys. Res.* 86:8733–44

Jepsen, P., et al. 1980. *J. Br. Interplanet. Soc.* 33:315–20

Kuiper, G. 1944. *Astrophys. J.* 100:378–91

Kunde, V., Aikin, A. C., Hanel, R. A., Jennings, D. E., Maguire, W. C., Samuelson, R. E. 1981. *Nature* 292:686–88

Kunde, V., Hanel, R. A., Maguire, W. C., Gautier, D., Bjoraker, G. L. 1982. In preparation

Lambert, D. L. 1978. *Mon. Not. R. Astron. Soc.* 182:249–71

Larson, H. P. 1980. *Ann. Rev. Astron. Astrophys.* 18:43–76

Larson, H. P., Fink, U. 1977. *Bull. Am. Astron. Soc.* 9:515–16

Larson, H. P., Fink, U., Smith, H. A., Davis, D. S. 1980. *Astrophys. J.* 240:327–37

Lecacheaux, J., de Bergh, C., Combes, M., Maillard, J. P. 1976. *Astron. Astrophys.* 53:29–33

Leovy, C. 1973. *J. Atmos. Sci.* 30:1218–20

Lequeux, J., Peimbert, M., Rayo, J. F., Serrano, A., Torres-Peimbert, S. 1979. *Astron. Astrophys.* 80:155–66

Limaye, S., et al. 1982. *J. Atmos. Sci.* 39:1413–32

Lindal, G., Eshleman, V. R., Tyler, G. L. 1982. In preparation

Lindzen, R. S. 1974. *J. Atmos. Sci.* 31:156–79

Lockwood, G. W. 1977. *Icarus* 32:413–30

Lockwood, G. W., Thompson, D. T. 1979. *Nature* 280:43–48

Lovas, F. J., Snyder, L. E., Johnson, D. R. 1979. *Astrophys. J. Suppl.* 41:451–80

Maguire, W. C., Hanel, R. A., Jennings, D. E., Kunde, V. G., Samuelson, R. E. 1981. *Nature* 292:683–86

Maguire, W. C., Hanel, R., Jennings, D., Samuelson, R., Aikin, A., Yung, Y. 1982. In preparation

Maillard, J. P., Combes, M., Encrenaz, T., Lecacheux, J. 1973. *Astron. Astrophys.* 25:219–32

Marten, A., Courtin, R., Gautier, D., Lacombe, A. 1980. *Icarus* 41:410–22

Maxworthy, T., Redekopp, L. G., Weidman, P. D. 1978. *Icarus* 33:388–409

McDonough, T., Brice, N. R. 1973. *Icarus* 20:136–45

Mitchell, J. L., Beebe, R. F., Ingersoll, A. P., Garneau, G. W. 1981. *J. Geophys Res.* 86:8751–59

Moore, V. 1983. PhD thesis. Univ. London

Moore, V., Hunt, G. E. 1983. *Geophys. Res. Lett.* In press

Müller, J.-P. 1982. PhD thesis. Univ. London

Müller, J.-P., Hunt, G. E., Gorley, R. 1983. *Geophys. Res. Lett.* In press

Ness, N. F., Acuna, M. H., Lepping, R. P., Connerney, J. E., Behannon, K. W., et al. 1981. *Science* 212:211–17

Orton, G. S. 1981. *Proc. IAU Symp. No. 96*, pp. 35–57

Orton, G. S., Ingersoll, A. P. 1981. *J. Geophys. Res.* 85:5871–83

Owen, T. 1982. In preparation

Owen, T., McKellar, A. R. W., Encrenaz, T., Lecacheux, J., de Bergh, C., Maillard, J. P. 1977. *Astron. Astrophys.* 54:291–301

Owen, T., Caldwell, J., Rivolo, A., Moore, V., Lane, A., Sagan, C., Hunt, G. E., Ponnamperuma, C. 1980. *Astrophys. J. Lett.* 236:L39–42

Pagel, B. E. J. 1977. *Symp. Origin Distribution Elem. 2nd, Paris IAGC*

Peek, B. M. 1958. *The Planet Jupiter*. London: Faber & Faber

Prinn, R. G. 1970. *Icarus* 13:424–36

Prinn, R. G. 1974. *Bull. Am. Astron. Soc.* 6:375–76

Prinn, R. G., Lewis, J. S. 1975. *Science* 190:274–76

Prinn, R. G., Owen, T. 1976. See Gehrels 1976, pp. 319–71

Rages, K., Pollack, J. B. 1982. *Icarus*. In press

Reeves, H. J., Audouze, P., Fowler, W. A., Schramm, D. N. 1973. *Astrophys. J.* 179:909–30

Rhines, P. C. 1975. *J. Fluid Mech.* 69:417–43

Ridgway, S. T. 1974. *Astrophys. J. Lett.* 187:L41–43

Ridgway, S. T., Wallace, L., Smith, G. R. 1976. *Astrophys. J.* 207:1002–10

Sagan, C. 1971. *Space Sci. Rev.* 11:73–121

Samuelson, R. E., Hanel, R. A., Kunde, V. G., Maguire, W. C. 1981. *Nature* 292:688–93
Sandel, B. R., Shemansky, D. E., Broadfoot, A. L., Holberg, J. B., Smith, G. R., et al. 1982. *Science* 215:548–53
Sill, G. T. 1976. See Gehrels 1976, pp. 372–83
Smith, B. A., Hunt, G. E. 1976. See Gehrels 1976, pp. 564–85
Smith, B. A., Soderblom, L. A., Johnson, T. V., Ingersoll, A. P., Collins, S. A., et al. 1979a. *Science* 204:951–72
Smith, B. A., Beebe, R., Boyce, J., Briggs, G., Carr, M., et al. 1979b. *Science* 206:927–50
Smith, B. A., Soderblom, L., Beebe, R., Boyce, J., Briggs, G., et al. 1981. *Science* 212:163–91
Smith, B. A., Soderblom, L., Bateson, R., Bridges, P., Inge, J., et al. 1982. *Science* 215:504–37
Smith, P. H., Tomasko, M. 1982. *Icarus*. In press
Snyder, L. E., Buhl, D. 1971. *Chem. Phys. Lett.* 163:L47–51
Sromovsky, L., Suomi, V., Pollack, J. B., Krauss, R. J., Limaye, S. S., et al. 1981. *Nature* 292:698–702
Sromovsky, L., et al. 1982. *J. Atmos. Sci.* 39:1433–45
Stevenson, D. J., Salpeter, E. E. 1977. *Astrophys. J. Suppl.* 35:239–61
Strobel, D. F. 1973. *J. Atmos. Sci.* 30:489–98
Strobel, D. F. 1977. *Astrophys. J. Lett.* 214:L97–99
Strobel, D. F. 1982. *Geophys. Res. Lett.* In press
Suess, S. T., Lockwood, G. W. 1980. *Sol. Phys.* 68:393–409
Terrile, R. J., Capps, R. W., Becklin, E. E., Cruikshank, D. P. 1979. *Science* 206:995–96
Tokunaga, A., Caldwell, J., Gillett, F. C., Nott, I. G. 1979. *Icarus* 39:46–53
Tokunaga, A., Dinerstein, H. L., Lester, D. F., Rank, D. M. 1981. *Icarus* 48:283–89
Tomasko, M., West, R. A., Castillo, N. D. 1978. *Icarus* 33:558–92
Trafton, L. M. 1967. *Astrophys. J.* 147:765–81
Trafton, L. M. 1972. *Astrophys. J.* 175:285–91
Trafton, L. M. 1978. *The Saturn System*, pp. 31–53. NASA
Trafton, L. M. 1981. *Rev. Geophys. Space Sci.* 19:43–89
Trauger, J. T., Roesler, F. L., Carleton, N. P., Traub, W. 1973. *Astrophys. J. Lett.* 184:L137–41
Tyler, G. L., Eshleman, V. R., Anderson, J. D., Levy, G. S., Lindal, G. F., et al. 1981. *Science* 212:201–5
Wagoner, R. V. 1973. *Astrophys. J.* 179:343–51
Wallace, L., Hunten, D. M. 1978. *Rev. Geophys. Space Sci.* 16:289–319
Weidenschilling, S. J., Lewis, J. S. 1973. *Icarus* 20:465–76
Weiser, H., Vitz, R., Moose, H. W. 1977. *Science* 197:755–57
Williams, G. P. 1979. *J. Atmos. Sci.* 36:932–68
Winkelstein, P., Caldwell, J., Combes, M., Encrenaz, T., Hunt, G., Moore, V. 1983. *Icarus*. In press

Ann. Rev. Earth Planet. Sci. 1983. 11 : 461–94

ASTEROID AND COMET BOMBARDMENT OF THE EARTH[1]

Eugene M. Shoemaker
US Geological Survey, Flagstaff, Arizona 86001

INTRODUCTION

Two classes of solid bodies large enough to be detected by telescopes occur in orbits that overlap that of the Earth. These bodies are the Earth-crossing asteroids and comet nuclei. Although their orbits only rarely intersect the Earth's, the probabilities of their collision with the Earth are nevertheless finite and calculable. Systematic telescopic surveys carried out over the past two decades show that the flux of asteroids and comet nuclei in the Earth's neighborhood is sufficiently high that the effects of occasional collisions should be recognizable in the geological record. During these same two decades, an intensive international search for ancient impact structures has gone forward. The actual rate of bombardment of the Earth during the last half-billion years has been found to be roughly consistent with the present rate predicted from astronomical observations. Within a factor of about two, the average rate of bombardment of the Earth during the last half-billion years also appears to be consistent with the average rate of bombardment of the Moon over the last 3.3 billion years.

Spectacular new lines of study have developed in recent years leading to the recognition of rare large impact events that produce geochemical anomalies on a global scale. The possible effects of these large impacts on the Earth's biota have become the subject of vigorous debate. In this paper, I first review the astronomical and geologic evidence concerning the history of bombardment and then discuss the physical effects of large impacts, as they may apply to both the inorganic and organic worlds.

POPULATIONS OF EARTH-CROSSING BODIES

Earth-Crossing Asteroids

The term Earth-crossing asteroid is used here to designate a body of asteroidal appearance at the telescope whose orbit occasionally intersects the orbit of the Earth as a result of distant perturbations by the planets. Such intersections are possible under the condition that the perihelion (point nearest the sun) on the asteroid's orbit lies closer to the sun than the aphelion (point farthest from the sun) on the Earth's orbit. Under this condition, if the asteroid's perihelion were aligned on the same side of the sun with the Earth's aphelion and if the two orbits were coplanar, the orbits would overlap and there would be two points of intersection. In general, the major axes of the two orbits are not so aligned and the orbits are not coplanar. However, distant planetary perturbations cause the major axes of both the asteroid's orbit and the Earth's orbit to precess. During precession, the radius vectors from the sun to the nodes of the asteroid's orbit (the two points on the orbit that lie on the line of intersection of the asteroid's orbit with the Earth's orbit) oscillate between the asteroid's perihelion distance and aphelion distance. If there is continuous overlap of the two orbits, there will be four angles at which the two orbits *must* intersect during a 360° rotation of the major axis of the asteroid's orbit relative to the line of the nodes. If overlap is lost during part of the precession cycle there may be fewer or, in some cases, no intersections.

Distant planetary perturbations also cause secular changes in the eccentricity of both the Earth's orbit and the asteroid's orbit. Hence, the degree of orbital overlap varies as a result of variation in the orbital eccentricities as well as a result of precession. As a consequence, asteroid orbits that now overlap the Earth's orbit can lose overlap, and conversely, there are asteroids whose orbits are not now overlapping that are Earth crossers. In some cases, the combined effects of precession and secular variation of eccentricity lead to 8 or more intersections with the Earth's orbit in a complete cycle of rotation of the major axis of the asteroid's orbit. A semianalytical theory for accurate calculation of the secular perturbations of Earth-crossing asteroid orbits has been developed only in the last 15 years (Williams 1969), and its general application to Earth crossers has been carried out only in the last 5 years (Williams 1979, Shoemaker et al 1979).

The first Earth-crossing asteroid to be discovered was 887 Alinda, a moderately bright ~ 5 km diameter body found by M. Wolf at Heidelberg, Germany, in 1918. Its present perihelion distance is 1.15 astronomical units (AU); it was not recognized as an Earth crosser until a study of its orbital dynamics was carried out by Marsden (1970). The second Earth crosser to be discovered was 1221 Amor, found by E. Delporte at Uccle, Belgium, in 1932. This tiny body, about 1 km in diameter, has a perihelion distance of 1.08 AU and was not recognized as an Earth crosser until studied by Williams (1979). Asteroids that approach the Earth but whose orbits currently do not overlap the Earth's are now generally referred to as Amor asteroids. About half the known Amor asteroids, with perihelion distances $\leqslant 1.3$ AU, have been found to be Earth crossers, chiefly by application of William's secular-perturbation theory. Another asteroid found in 1932, this time by K. Reinmuth at Heidelberg, has an orbit that strongly overlaps the orbit of the Earth. This asteroid is 1862 Apollo, and its name is applied to the class of asteroids with orbits larger than Earth's and with current perihelion distances $\leqslant 1.017$ AU (the Earth's aphelion distance). In 1976, asteroid 2062 Aten was discovered by E. F. Helin at Palomar Mountain, California; this object is on an orbit smaller than Earth's, but it overlaps Earth's orbit at aphelion. Three more asteroids have since been recognized that belong to the Aten class.

A total of 49 Earth-crossing asteroids, comprising 4 Atens, 30 Apollos, and 15 Earth-crossing Amors, have been discovered through mid-1982 (Table 1). During the past decade, the average rate of discovery has been about 3 per year. Most of the Earth crossers are so small and intrinsically faint that they are only discovered when near the Earth. The absolute visual magnitude of 35 of the known Earth crossers is equal to or brighter than 18 (Table 1). On the basis of the discovery rate in various systematic surveys of the sky, the population of Earth crossers to absolute visual magnitude 18 is estimated at ~ 1300, including ~ 100 Atens, 700 ± 300 Apollos, and ~ 500 Earth-crossing Amors (Helin & Shoemaker 1979, Shoemaker et al 1979). Hence, the discovery of Earth-crossing asteroids to absolute visual magnitude 18 is estimated to be only 2 to 3% complete as of mid-1982. The impending launch of the Infrared Astronomical Satellite and the construction of a telescope at the Steward Observatory, Arizona, specially designed for the search for Earth-approaching asteroids may greatly accelerate their discovery in the next few years, however.

As they are observed merely as points of light at the telescope, our information about the sizes and other characteristics of Earth-crossing asteroids is derived entirely from analysis of the radiation reflected and emitted from them. Albedos of the asteroids can be determined from the ratio of reflected sunlight to emitted thermal radiation (Morrison 1977, Morrison & Lebofsky 1979). Cross-sectional areas and mean diameters can then be obtained from the absolute magnitudes. Albedos can also be estimated from measurements of polarization of the reflected sunlight (Zellner et al 1974, Dollfus & Zellner 1979) and from the dependence of apparent magnitude upon the angle between the vectors from the asteroid to the Earth and to the sun (Bowell & Lumme 1979).

When determinations of albedo are combined with broad- or narrow-band measurements of color in the visible and near-infrared parts of the spectrum, the asteroids can be classified into several broad groups or types (Bowell et al 1978, Zellner 1979). Asteroids designated *C-type* have visual geometric albedos in the range 0.02 to 0.065 and relatively neutral colors in the visible and near infrared; *S-type* asteroids have geometric albedos in the range 0.065 to 0.23 and reddish colors. Narrow-band spectrophotometry permits comparison of the reflectance spectra of the asteroids with meteorites. Moreover, the absorption bands of discrete mineral phases commonly can be recognized in the near-infrared (Gaffey & McCord 1979). C-type asteroids are generally believed to be compositionally similar to carbonaceous meteorites, whereas S-type asteroids are thought to be somewhat similar to ordinary chondritic meteorites, although only a few good matches of spectral reflectance are found between S-asteroids and ordinary chondrites. The majority of well-observed Earth crossers appear to be referable to the C or S types, but a significant proportion are found to have unusual colors and are not definitely assignable to any major spectrophotometric type (Shoemaker et al 1979).

At absolute visual magnitude 18, C-type asteroids have diameters in the range 1.3 to 2.3 km, and S-type in the range 0.7 to 1.3 km. Most of the Earth crossers observed by means of photoelectric photometry are S-type or have colors that probably are correlated with moderately high albedo, but the bulk of the volume for the observed objects is probably contained in a smaller number of low-albedo, relatively large, C-type objects (Shoemaker et al 1979). Because they are very faint and difficult to observe by photoelectric photometry, small C-type Earth crossers almost certainly are underrepresented in the present statistics.

The probability of collision with the Earth for a body whose orbit strongly overlaps Earth's can be calculated by methods first derived by Öpik (1951) and further developed by Wetherill (1967, 1968) and Shoemaker et al (1979). For orbits with shallow overlap or which overlap only part of the time, use must be made of Williams's (1969) perturbation theory, or of extensive numerical integration of the orbits to obtain the terms required to calculate the collision probability. The mean probability of collision with the Earth for Earth-crossing asteroids was found by Shoemaker et al (1979) to be $\sim 2.5 \times 10^{-9}$ yr^{-1}. When multiplied by the estimated population, this yields a collision rate of about 3.2 per million years for asteroids brighter than absolute magnitude 18. The impact speed may also be obtained from the terms used in the calculation of impact probability. The rms impact speed of Earth-crossing asteroids, weighted

Table 1 Earth-crossing asteroids discovered through May, 1982

Number	Name	Class	a^a (AU)	e^a	i^a (deg.)	q (AU)	Q (AU)	$V(1,0)$	Diameter (km)	References
1566	Icarus	Apollo	1.08	0.827	23.0	0.19	1.97	16.8	1.4	1
2212	Hephaistos	Apollo	2.16	0.835	11.9	0.36	3.97	14.4	(8.7)	2
	1974 MA	Apollo	1.78	0.762	37.8	0.42	3.13	14.2	~6	2
2101	Adonis	Apollo	1.87	0.764	1.4	0.44	3.30	18.8	~1	2
2340	Hathor	Aten	0.84	0.450	5.9	0.46	1.22	20.7	(0.2)	3
2100	Ra-Shalom	Aten	0.83	0.437	15.8	0.47	1.20	16.4	3.4	4
	1954 XA	Aten	0.78	0.345	3.9	0.51	1.05	19.2	~0.6	2
1864	Daedalus	Apollo	1.46	0.615	22.1	0.56	2.36	15.1	(3.3)	5
1865	Cerberus	Apollo	1.08	0.467	16.1	0.58	1.58	16.7	~2	2
	Hermes (1937 UB)	Apollo	1.64	0.624	6.2	0.62	2.66	17.3	~1	2
1981	Midas	Apollo	1.78	0.650	39.8	0.62	2.93	17.2	~2	2
	1981 VA	Apollo	2.46	0.744	22.0	0.63	4.29	17	~2	2
2201	1947 XC	Apollo	2.17	0.712	2.5	0.63	3.72	15.9	~3	2
1862	Apollo	Apollo	1.47	0.560	6.4	0.65	2.29	16.2	1.4	6
	1979 XB	Apollo	2.26	0.713	24.9	0.65	3.88	19.2	~0.6	2
2063	Bacchus	Apollo	1.08	0.349	9.4	0.70	1.45	17.9	~1	2
1685	Toro	Apollo	1.37	0.436	9.4	0.77	1.96	14.2	{4.7[b] {3.8[c]	3 7
2062	Aten	Aten	0.97	0.182	18.9	0.79	1.14	17.6	0.9	8
2135	Aristaeus	Apollo	1.60	0.503	23.0	0.79	2.40	18.4	~1	2
	1982 HR	Apollo	1.21	0.332	2.7	0.82	1.60	19	~0.5	
2329	Orthos	Apollo	2.40	0.658	24.4	0.82	3.99	15.5	~3	2
	6743P-L	Apollo	1.62	0.493	7.3	0.82	2.42	17.6	~1	2
1620	Geographos	Apollo	1.24	0.335	13.3	0.83	1.66	15.9	2.0	1
	1959 LM	Apollo	1.34	0.379	3.3	0.83	1.85	15.2	~3	2
	1950 DA	Apollo	1.68	0.502	12.1	0.84	2.53	16.1	~3	2

Table 1 (*continued*)

Number	Name	Class	a^a (AU)	e^a	i^a (deg.)	q (AU)	Q (AU)	$V(1,0)$	Diameter (km)	References
1866	Sisyphus	Apollo[d]	1.89	0.540	41.1	0.87	2.92	13.8	~10	2
	1973 NA	Apollo	2.46	0.642	68.1	0.88	4.04	15.3	~6	2
	1978 CA	Apollo	1.12	0.215	26.1	0.88	1.37	18.1	1.9	4
1863	Antinous	Apollo	2.26	0.606	18.4	0.89	3.63	15.7	~3	2
	1982 BB	Apollo	1.41	0.355	21.0	0.91	1.91	15	~4	
2102	Tantalus	Apollo	1.29	0.298	64.0	0.91	1.67	16.7	~3	2
	6344P-L	Apollo	2.58	0.635	4.6	0.94	4.21	22.3	~0.2	2
	1982 DB	Apollo	1.48	0.356	1.4	0.95	2.01	19	~1	
	1979 VA	Apollo	2.64	0.627	2.8	0.98	4.29	16.6	(3.2)	2
	1978 DA	Amor	2.48	0.587	15.6	1.02	3.93	18.0	0.9	4
	1980 PA	Amor	1.93	0.459	2.2	1.04	2.81	18.7	~1	2
2061	Anza	Amor	2.26	0.537	3.7	1.05	3.48	17.3	(2.3)	2
1915	Quetzalcoatl	Amor	2.52	0.583	20.5	1.05	3.99	18.6	0.4	7
	1980 AA	Amor	1.89	0.444	4.2	1.05	2.73	19.7	~0.5	2
1917	Cuyo	Amor	2.15	0.505	24.0	1.06	3.23	15.7	~3	2
1943	Anteros	Amor	1.43	0.256	8.7	1.06	1.80	15.7	2.0	7
	1981 QB	Amor	2.24	0.519	37.2	1.08	3.40	16	~3	2
1221	Amor	Amor	1.92	0.436	11.9	1.08	2.76	18.3	~1	2
	1980 WF	Amor	2.23	0.514	6.4	1.08	3.38	18.7	~1	2
	1982 DV	Amor	2.03	0.457	5.9	1.10	2.96	16	~3	
1580	Betulia	Amor	2.20	0.490	52.0	1.12	3.27	14.3	7.4	9
2202	Pele	Amor	2.29	0.510	8.8	1.12	3.46	17.7	~1	2
1627	Ivar	Amor	1.86	0.397	8.4	1.12	2.60	13.4	6.2	7
887	Alinda	Amor	2.52	0.544	9.1	1.15	3.88	14.1	4.7	3

Explanation of headings:

Number A permanent number is assigned to an Earth-crossing asteroid after its positions have been measured on two appearances near opposition (the direction opposite from the sun) and an accurate orbit has been determined.

Name For asteroids that have been assigned a permanent number, a name is proposed by the discoverer and adopted under the rules of the International Astronomical Union. At the time of discovery, a provisional designation is assigned (e.g. 1974 MA) which indicates the year (1974), half month (M), and sequence within the half month (A) in which the asteroid was reported to the Minor Planet Center. This designation is used until the asteroid is numbered, except for Hermes, which was named in violation of the rules. Asteroids discovered in the Palomar-Leiden survey for faint minor planets (van Houten et al 1970) received a special provisional designation, indicated by the suffix P-L.

Class Orbital class, generally named for the first member recognized (see text).

a Semimajor axis of orbit in astronomical units. The astronomical unit (AU) is the length of the semimajor axis of the Earth's orbit.

e Eccentricity of orbit.

i Inclination of orbit.

q Perihelion distance in astronomical units.

Q Aphelion distance in astronomical units.

$V(1,0)$ Absolute visual magnitude (mean magnitude as observed through a yellow filter and reduced to 1 AU distance from the Earth, 1 AU distance from the sun, and a position directly opposite from the sun). An increase of one magnitude corresponds to a decrease in brightness by a factor of $(100)^{1/5}$. Magnitudes for numbered asteroids are chiefly from Bowell et al (1979).

Diameter Diameter of a circular area equivalent to the mean cross-sectional area of the asteroid. Diameters based on measured albedos are reported to one decimal place without parentheses; diameters based on albedos inferred from broad-band color ratios are shown in parentheses; diameters based on albedos assumed without any photometric evidence are shown as approximate values.

References Index number for reference or source from which diameters are taken, as follows: 1. Zellner et al 1974; 2. Wetherill & Shoemaker 1982; 3. Bowell et al 1979; 4. Lebofsky et al 1979; 5. Bowell et al 1978; 6. Lebofsky et al 1981; 7. G. J. Veeder, personal communication, 1982; 8. Morrison et al 1976; 9. Lebofsky et al 1978.

Footnotes

[a] Orbital elements are from Ephemerides of Minor Planets (Akademia Nauk) and the Minor Planet Circulars.

[b] Diameter based on determination of albedo by polarimetry.

[c] Diameter based on determination of albedo by infrared radiometry.

[d] The orbit of Sisyphus generally does not intersect Earth's orbit during precession of the axis of the asteroid's orbit, owing to change of e and loss of overlap of the orbits.

according to probability of collision, was found to be 20.1 km s^{-1} (Shoemaker et al 1979).

The Earth-crossing asteroid swarm is depleted by collision with the Earth and with the other planets and by ejection of asteroids from the solar system as a result of successive close encounters with the planets. Typical dynamical lifetimes for these bodies are only about 3×10^7 yr (Wetherill & Williams 1968, Wetherill 1976). The swarm would be quickly diminished if the losses were not balanced by injection of new asteroids into Earth-crossing orbits. Rough consistency between the present estimated cratering rate and the geologic record of impact back to ~0.5 Gyr suggests that the population of Earth crossers is approximately in equilibrium. Analysis of the lunar-cratering record, however, suggests that the average flux of impacting bodies during the last half-billion years may have been about twice as high as the average flux over the past 3.3 Gyr (Shoemaker et al 1979). Some Earth-crossing asteroids are certainly derived from the main asteroid belt and are injected into Earth-crossing orbits by a combination of the effects of dynamical resonances and close encounters with Mars (Wetherill 1979). Others may be extinct, very short period comets (Öpik 1963, Marsden 1971, Sekanina 1971, Wetherill 1976, Degewij & Tedesco 1982). If a residue or core of rocky material is left when all the ices have sublimed from a comet like P/Encke, which is on an orbit like that of some Earth-crossing asteroids, the object would be recognized at the telescope as an Earth-crossing asteroid. Coarse rocky material may also form a lag deposit on the surface of an ablating comet nucleus, which might shut off observable cometary activity and thus produce an asteroid. If only about one in ten to a hundred comets like P/Encke evolve into objects of asteroidal appearance, the supply would appear to be adequate to maintain the Earth-crossing asteroid population (cf Wetherill 1979, Shoemaker et al 1979).

Comets

More than 10^{12} comet nuclei are estimated to reside in a spherical cloud, over a light year in diameter, that surrounds the sun (Weissman 1982a). Repeated perturbation of this distant cloud of comets by passing stars produces a small, but fairly steady, flux of comets in the region of the planets (Oort 1950, Weissman 1980). Most comets arrive in the Earth's neighborhood on extremely eccentric orbits, with periods ranging from thousands to millions of years; a substantial fraction of these are perturbed by the gravitational attraction of the giant planets into trajectories that allow them to escape from the solar system. Successive perturbations by the giant planets also lead to capture of about 0.01% of the long-period comets into orbits with periods less than 20 years. More than 500 long-period comets have been observed over the past few centuries, most of which passed inside the orbit of the Earth (Marsden 1979); somewhat more than 100 comets with periods $\leqslant 20$ years have also been discovered (Marsden 1979), but only a minor fraction of these cross the Earth's orbit.

Our information about the physical characteristics of comet nuclei is derived chiefly from observation of the gases and entrained dust that are liberated as the comets approach the sun. The nuclei are rotating solid bodies (Sekanina 1981, Whipple 1982), evidently composed chiefly of H_2O ice and embedded rocky particles (Whipple 1950, 1951, Delsemme 1982). An extended dusty atmosphere produced by insolation generally obscures the nucleus of the comet when it is close enough to the Earth for detailed photometric and radiometric observation. For this reason, present estimates of the sizes of comet nuclei are very uncertain. Observations of the amounts of gas and dust released, the acceleration of the comets by the jet effect of the released material (Whipple 1978), and photometric observations of comet nuclei when they are distant from the sun (Roemer 1966) suggest that the nuclei range in diameter from less than one kilometer to several tens of kilometers. Detection of P/Encke by radar indicates that the diameter of the nucleus of this comet is in the range of ~ 0.4 to 4.0 km (Kamoun et al 1981). Shoemaker (1981) and Shoemaker & Wolfe (1982) estimate that impacts of comet nuclei, dominantly on long-period orbits, may account for as much as $\sim 30\%$ of the recent production of impact craters larger than 10 km diameter on Earth. Weissman (1982b), on the other hand, estimates the cratering rate by comet impact at only $\sim 10\%$ of the total rate. This disagreement is due chiefly to differences in the evaluation of sizes of the long-period comet nuclei.

PRODUCTION OF IMPACT CRATERS ON THE CONTINENTS

Crater Scaling

If a colliding body is sufficiently large, the most easily recognized effect of its encounter with the Earth generally is an impact crater. The size of the crater produced depends primarily upon the mass, density, shock compressibility, and velocity of the impacting body, and upon the density, shock equation of state, and strength of the target rocks. The functional relationship between these variables and the size and shape of the crater has been the subject of fairly intensive theoretical and experimental investigation for the last two decades (Roddy et al 1977, Melosh 1980).

Impact cratering is such a sufficiently complicated process that we cannot yet predict with high accuracy and confidence the consequences of impact on land of the Earth-crossing bodies discovered at the telescope. The most reliable predictions, at present, are obtained from the use of state-of-the-art, large, finite difference computer codes, in which the model includes all of the above variables and whatever stratification or structure of the target may be appropriate. The target and the projectile are divided into a large number of cells in a rectangular array, and the history of a specific impact event is carried forward in small increments of time for each cell on the basis of the physics of the propagation of the shock and rarefaction waves generated by the impact. Even with the use of special very high speed computers, the calculation of a single case of impact is time-consuming and expensive. To simplify the calculation and reduce the expense, generally only cases of symmetrical projectiles impacting at vertical incidence have been investigated.

A number of code calculations, for the most part with progressively greater sophistication, have been run to simulate the impact event that produced Meteor Crater, Arizona. In this case, the properties of the meteorite and the target rocks, the structure of the target, and the resulting crater and associated structural deformation are all well known, but the mass, shape, and velocity of the projectile are unknown. Among the calculations published to date, one carried out by Roddy et al (1980) employs the most advanced code. The degree to which this code adequately represents the physical processes in cratering has been checked by numerous calculations of large explosion craters and comparison of the results with highly instrumental field experiments.

For impact of a single projectile at vertical incidence, Roddy et al (1980) estimate that the kinetic energy of the meteorite required to form Meteor Crater was $\sim$15 megatons TNT equivalent. (One megaton TNT is equivalent to 4.19×10^{22} ergs.) Their calculations suggest that the volume of craters about the size of Meteor Crater formed in rock is more nearly proportional to the kinetic energy of the projectile, as found by Gault (1973) for small experimental impact craters in rock, than it is to the momentum of the projectile, as predicted by Öpik (1958) and found by Schmidt (1980) for small impact craters in sand. If the rms impact velocity of 20 km s^{-1} obtained for Earth-crossing asteroids is adopted for the projectile that formed Meteor Crater, a spherical asteroid of meteoritic iron about 42 m in diameter has the kinetic energy required to form the initial 1.16 km diameter crater.

General scaling rules derived from dimensional analysis (Holsapple & Schmidt 1982) are a useful guide in extrapolating from the code calculations for Meteor Crater to obtain dimensions of craters produced by impact of Earth-crossing bodies $\sim$0.5 to 10 km diameter, comparable to those discovered at the telescope. Terrestrial impact craters larger than about 3 to 4 km diameter differ in an important way from Meteor Crater, however, and account must be taken of this difference. In these larger craters, the rock walls of the initial craters have collapsed and slumped toward the centers of the craters. Similar collapse has occurred in nearly all craters on the Moon larger than 25 km diameter. From measurement of the cumulative widths of the terraces formed by the slumped blocks along the walls, the diameter of the 90 km diameter lunar crater Copernicus is estimated to have been increased more than 35% by wall collapse (Shoemaker 1962). A conservative value of 30% enlargement by collapse will be used here for estimation of the final crater diameter (cf Shoemaker et al 1979). To calculate the size of projectile required to form a 10 km diameter crater on Earth, for example, an initial crater only $(1/1.3) \times 10$ km $= 7.69$ km diameter should be compared with Meteor Crater.

For two initial craters of similar form produced by projectiles with the same impact velocity (e.g. 20 km s^{-1}) in target rocks of the same strength, the ratio of the diameters of the projectiles that formed them is given by the following equation based on the Holsapple-Schmidt scaling rules (Wetherill & Shoemaker 1982):

$$d_2/d_1 = (\delta_1/\delta_2)^{(1+\beta-\gamma)/(3-\alpha)}(\rho_2/\rho_1)^{(1+\gamma)/(3-\alpha)}(D_2/D_1)^{3/(3-\alpha)}, \tag{1}$$

where d_1 and d_2 are the diameters and δ_1 and δ_2 the densities of the two projectiles, ρ_1 and ρ_2 are the densities of the target rocks at the sites of the two craters, D_1 and D_2 are the corresponding crater diameters, and α, β, and γ are scaling constants. Setting $d_1 = 42$ m and $\delta_1 = 7.86$ g cm^{-3} for the Meteor Crater projectile, $\rho_1 = 2.3$ g cm^{-3} for the mean density of the rocks and $D_1 = 1.16$ km for the diameter of Meteor Crater, $\delta_2 = 2.38$ g cm^{-3} for the effective density of an impacting S-type asteroid (Shoemaker et al 1979), $\rho_2 = 2.6$ g cm^{-3} for average rocks on the continental shields, $D_2 = 7.69$ km as the diameter of the initial crater that collapses to 10 km diameter, $\alpha = 0.39$ as a representative value intermediate between theoretical limits found by Holsapple & Schmidt (1982), $\beta = 0.11$, which is consistent with the value preferred by Schmidt (1980), and $\gamma = 0.06$, which is consistent with impact experiments in basalt targets, a diameter $d_1 = 0.63$ km is found as a representative diameter for an asteroid that produces a 10-km crater by impact on a continent.

An entirely independent estimate of the projectile diameter and energy can be made by use of an empirical scaling relationship based on experimental explosion craters (Nordyke 1961, 1962, Shoemaker et al 1963, Nordyke 1977, Shoemaker 1977, Shoemaker et al 1979). As modified by Shoemaker & Wolfe (1982) to account for differences in density of the target rocks, the diameter, D_t, of a terrestrial impact crater can be estimated from

$$D_t = c_f K_n (W \rho_a/\rho_t)^{1/3.4}, \tag{2}$$

where c_f is the crater collapse factor (nominally 1 for craters $\lesssim 3$ km diameter and 1.3 for craters $\gtrsim 4$ km in diameter), $K_n = 0.074$ km kilotons$^{-1/3.4}$ is an empirical constant derived from the diameter and explosive yield for the Jangle U nuclear crater, $\rho_a = 1.8$ g cm^{-3} is the estimated density of the alluvium at the Jangle U site in Yucca Flat, Nevada, ρ_t is the mean density of the target rocks, and $W = \pi d^3 \delta v^2/(12 \times 4.19 \times 10^{10})$ kilotons TNT equivalent is the kinetic energy of a projectile of diameter d, density δ, and velocity v, all measured in cgs units. Inversion of Equation (2) to obtain the kinetic energy of the Meteor Crater projectile gives an estimated energy of 15 megatons TNT equivalent, in agreement with the code calculation of Roddy et al (1980).

The success of Equation (2) rests on the choice of the 78 m diameter Jangle U crater for determination of the scaling coefficient K_n. Jangle U was selected because the initial energy density in the rocks shocked by nuclear explosion is comparable with that produced by asteroid impact (Shoemaker 1963), the explosion was at a scaled depth appropriate for comparison of explosion craters with most impact craters (cf Holsapple 1980), and the alluvium in Yucca Flat has a low strength, as required for scaling by Equation (2) from an experimental crater less than 100 m diameter with impact craters larger than 1 km that are formed in comparatively strong rock. The suitability of the strength properties of the Yucca Flat alluvium for scaling from small nuclear craters to large impact craters was judged on the basis of similarity of geometry of deformation of the material in the walls and floor of nuclear craters formed in the alluvium to that observed in larger impact craters (Shoemaker 1963). Applying Equation (2) to estimate the energy required to form a 10 km diameter impact crater, we get 1.04×10^4 megatons TNT; an asteroid with this kinetic energy and a density of 2.38 g cm^{-3} traveling at 20 km s^{-1} has a diameter of 0.57 km. When appropriate allowance is made for the uncertainties in all the scaling constants used, the $\sim 10\%$ difference between this diameter and the one obtained by means of scaling with Equation (1) should not be regarded as significant.

A source of some further uncertainty about the diameters of craters produced by asteroid and comet impact arises from effects of variation of impact angle. Code calculations of oblique impacts of asteroids are sufficiently difficult that none have yet been published, although codes exist and have been used to study oblique impact of small projectiles at much lower speeds, primarily in metal targets. Gault (1973) found that the diameters of small experimental craters formed by relatively low-speed impact in strong rock were approximately proportional to $(\sin i)^{2/3}$, where i is the elevation angle of impact. This relationship suggests that craters produced by asteroid impact at a modal elevation angle near 45° might have diameters 30% smaller than craters produced at vertical incidence.

It is not at all clear, however, that the Gault relationship applies to hypervelocity impact craters with diameters near 10 km, because coupling of the projectile to the target and the relative dependence of crater dimensions on rock strength and gravity vary with projectile velocity and energy. The principal result of changing impact angle in large hypervelocity impacts probably is to change the effective scaled depth of penetration of the projectile about in proportion to sin *i*. The net effect of this change on the diameter of the crater may be similar to the effect due to change of scaled depth of burst of explosion craters (Nordyke 1961, 1962).

The scaled depth of burst of the Jangle U crater, 5.2 m $kt^{-1/3.4}$ (kt = kiloton), is apparently equivalent to the effective depth of penetration for average S-type asteroids impacting in average continental rocks at vertical incidence. The dependence of crater diameter on depth of burst found for explosion craters is nearly linear at this scaled depth (Nordyke 1962), and the maximum decrease in diameter of the crater as the scaled depth decreases to 0 is only about 9.5%. A functional dependence of crater diameter on impact angle might therefore be written as

$$D_i = [1 - 0.095(1 - \sin i)]D_{90}, \quad (3)$$

where D_i is the mean diameter of a crater formed at impact angle i and D_{90} is the diameter of the crater formed by impact at vertical incidence; at a modal angle of about 45°, the mean diameter of the crater would be only ~3% smaller than the diameter produced at vertical incidence. If the above analysis is approximately correct, neglect of the impact angle in the calculation of mean crater diameter generally introduces an error considerably smaller than the error due to uncertainty about the crater scaling constants. As i becomes very small, the most evident effect is that the crater produced will be noticeably elongate in the direction of the trajectory, as is probably illustrated by the lunar craters Messier and Messier A. However, the frequency distribution of incidence angle is such that very few cases of impact at i approaching 0° are expected (Shoemaker 1962).

Present Cratering Rate by Asteroid and Comet Impact

In order to obtain the total production of craters larger than some specified diameter, the frequency distribution of size or energy of the impacting bodies or, alternatively, the frequency distribution of the craters produced must be known. The size distribution of the discovered Earth-crossing asteroids listed in Table 1 is strongly biased by observational selection. Large, bright objects are more easily discovered than small, faint ones. Wetherill & Shoemaker (1982) estimate that discovery may be roughly 20% complete for objects ⩾10 km diameter, whereas it is only 1 to 2% complete for objects ⩾1 km diameter. The discovery of bright Earth-crossing comets may be somewhat more complete than is the case for the brightest Earth-crossing asteroids (Shoemaker & Wolfe 1982), but we are hampered by a lack of accurate information on the actual sizes of the comet nuclei. The distribution of nuclear magnitudes of short-period comets suggests that the distribution of cumulative frequency, N_d, has the form $N_d \propto d^{-\lambda}$, where d is the nucleus diameter and the size index, λ, is close to 2 (Shoemaker & Wolfe 1982). A nearly identical size index is indicated by the distribution of absolute magnitudes of faint main-belt asteroids (van Houten et al 1970). The size distribution of craters on the lunar maria, on the other hand, suggests that the size index of objects larger than ~ 100 m that have struck the Moon over the last 3.3 Gyr has been significantly less than 2.

The density of craters with diameters $\geqslant 10$ km on 3.3 Gyr old lunar mare surfaces is about half that predicted from a steady flux of Earth-crossing bodies equal to the present flux (Shoemaker et al 1979); the apparent discrepancy is no greater than the possible error of estimation of the present cratering rate. It is reasonable to suppose, therefore, that the larger solid objects that impacted the Moon in the last 3.3 Gyr are related in origin to the present Earth-crossing bodies, and that the size and energy distributions of these objects were similar to the size and energy distributions of the present population of Earth crossers.

The size distribution of craters on the lunar maria between 3 and 100 km diameter is well represented by a power function with a size index of 1.7 (Shoemaker et al 1963). Craters at the small end of this size range have not collapsed, whereas all the largest craters have been enlarged by collapse. Correcting for collapse, in order to obtain the diameter distribution of the initial craters, yields a size index for initial crater diameters of 1.84. As $D_t \propto W^{1/3.4}$ (Equation 2) and $d \propto W^{1/3}$, the size index for impacting bodies of the same density and specific kinetic energy is given by $\lambda = 1.84 \times 3/3.4 = 1.62$. This low value of λ indicates that the ratio of large bodies to small bodies that struck the Moon in the last 3.3 Gyr is much higher than expected from the magnitude distributions of comet nuclei or small main-belt asteroids (cf Shoemaker et al 1963). The high proportion of large objects may be due to relatively frequent evolution of large short-period comets of the Encke-type into Earth-crossing asteroids, whereas the mass of most small short-period comets may usually be dissipated, and relatively few become asteroidal objects with long dynamical lifetimes.

Of the 3.2 Earth-crossing asteroids equal to or brighter than absolute magnitude 18 estimated to strike the Earth each million years, suppose that half are C-type and half S-type. The mean diameter of a magnitude 18 asteroid is estimated to be 0.89 km for S types and 1.73 km for C types (Shoemaker et al 1979). If the size index for both types is 1.62 and we use 0.57 km as the mean diameter required for an impacting S-type body to produce a 10-km crater, then, on average, $1.6 \times (0.89/0.57)^{1.62} = 3.3$ S-type asteroids that are capable of producing craters $\geqslant 10$ km diameter strike the Earth per million years. From Equation (2), using an estimated effective density of 1.7 g cm^{-3} (Shoemaker et al 1979), the diameter of an average C-type asteroid that will produce a 10-km crater is 0.61 km. Thus, on average, $1.6 \times (1.73/0.61)^{1.62} = 8.7$ C-type asteroids that are capable of producing craters $\geqslant 10$ km diameter strike the Earth each million years. An average of about 1/3 of the 12 total S- and C-type asteroids actually hit the continents each million years, on average, where they produce about 4 craters 10 km diameter or larger.

The derived cratering rate is sensitive to the fraction of C-type asteroids, but the minimum and maximum possible cratering rates, corresponding respectively to no C-type asteroids and all C types, do not deviate by more than $\pm 45\%$ from the rate derived for the case of one-half C types. Accounting for comet impact, the total cratering rate should be increased by $\sim 10\%$ to $\sim 30\%$ over that obtained from asteroid impact alone. Assuming that the errors are distributed lognormally, the log probable error (log *PE*) of the estimated crater production rate may be roughly evaluated as follows: log *PE* of the estimated Earth-crossing asteroid population $\simeq 0.18$; log *PE* due to uncertainty about crater diameters obtained from the scaling formula used $\simeq 0.18$; log *PE* due to uncertainty about the fraction of C-type asteroids $\simeq 0.18$. From the rms of the logs, the final cratering rate has a probable error of the order of a factor of 2. A systematic error may reside in the mean bulk densities assigned to C- and S-type asteroids. In the calculation of asteroid bulk densities by Shoemaker et al (1979), allowance was made for 24% porosity produced by impact brecciation. If all the asteroids were compact rather than fragmental, the final calculated cratering rate would be increased by 14%.

Few telescopic observations have been made of Earth-crossing bodies smaller than about 0.5 km diameter, but an estimate of the energy-frequency distribution of these smaller bodies can be made with the aid of the size-frequency distribution of the lunar craters. The estimated rate of collision with the Earth of bodies in the energy range of 1 kiloton to 10^5 megatons TNT equivalent is illustrated in Figure 1. The frequency of collision at 1.04×10^4 megatons TNT equivalent energy corresponds to the production rate of 10-km craters estimated above. In the energy range from 1 kiloton to 10,000 megatons, the extrapolated frequencies are from a smoothed crater-size distribution based on the observed crater-size distribution in Mare Cognitum, determined from Ranger VII television images of the Moon (Trask 1966). The preferred crater diameter scaling relationship is taken to be $D \propto W^{1/3.4}$, but extrapolated frequencies based on $D \propto W^{1/3.3}$ and $D \propto W^{1/3.5}$ are also shown for comparison. A corrected

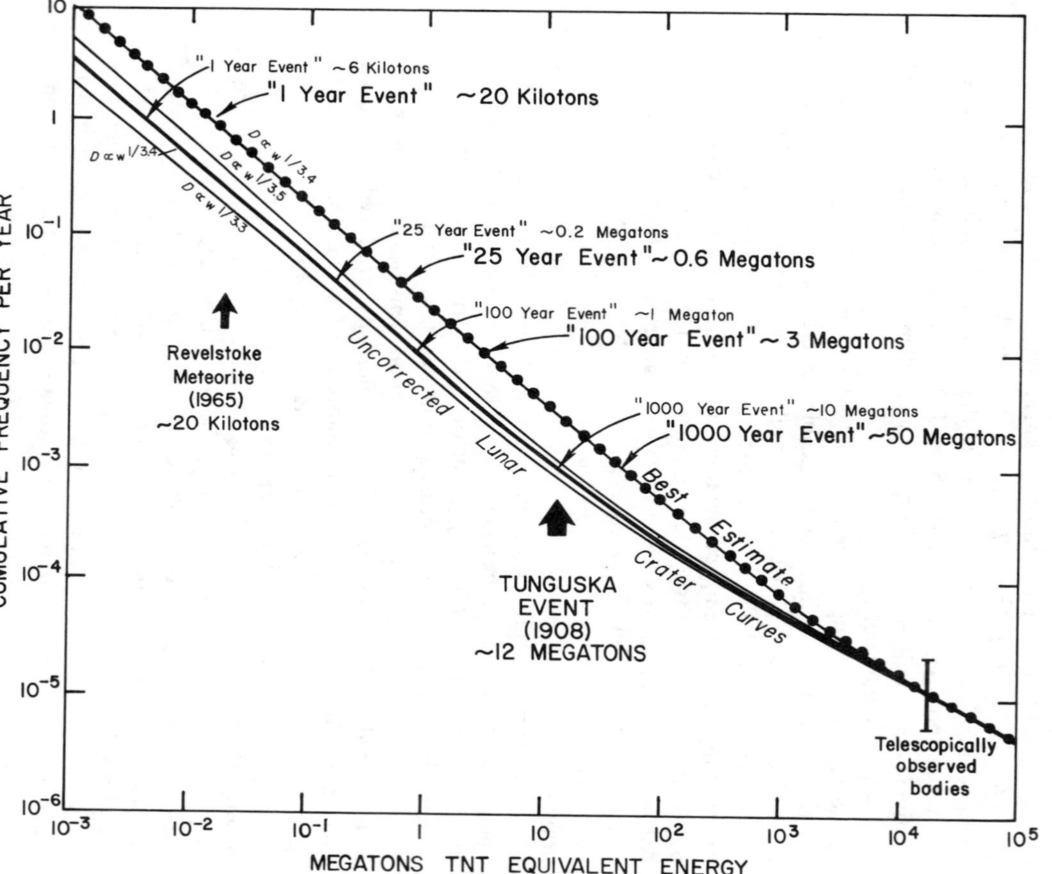

Figure 1 Estimated cumulative frequency distribution of kinetic energy of bodies colliding with the Earth.

frequency distribution, labeled *Best Estimate*, is based on the assumption that two thirds of the 10 km diameter craters counted by Trask (1966) predate the mare lavas that form the surface of Mare Cognitum, as suggested by studies of craters on the lunar maria by Neukum et al (1975).

Historic Collisions

The most energetic, historically well-documented encounter with the Earth of an Earth-crossing body produced a great meteoric fireball over the Podkamennaya-Tunguska River region of Siberia on the morning of June 30, 1908 (Krinov 1966). Traveling from southeast to northwest, the meteor passed nearly over the town of Kirensk; the endpoint of the trajectory was about 60 km northwest of the remote trading post of Vanovara, over a very sparsely inhabited part of the Siberian tiaga. The meteor was observed from distances as great as 600–1000 km from the endpoint; the atmospheric shock was audible at still greater distances. Trees were knocked down at distances up to 40 km from the endpoint, and circumstantial evidence suggests that dry timber was ignited by thermal radiation from the fireball at distances up to 15 km from the endpoint. Intensive investigation by expeditions from the Soviet Academy of Sciences carried out over many decades has shown that the Tunguska bolide disintegrated in the atmosphere; it deposited most of its kinetic energy at an estimated altitude of $\sim$8.5 km (Ben-Menahem 1975). Only microscopic spheres of glass and magnetite, formed by ablation, reached the ground (Florensky 1963).

Long-period atmospheric gravity and acoustic waves excited by the atmospheric shock were well recorded on weather-station barographs in Siberia and southern England; the passage of these waves in both the direct and reverse paths was recorded on a barograph at Potsdam, Germany. Coupling of the air wave to the ground near the endpoint produced seismic waves detected at Irkutsk, Tashkent, Tiflis, and Jena; local coupling of the air wave to the ground as it passed over some seismic stations was also detected. From the English barograms published by Whipple (1930) it is possible to estimate the kinetic energy of Tunguska bolide. Detailed analysis by Hunt et al (1960) indicates an energy of 12.5 megatons TNT equivalent. Scaling from large nuclear airbursts, Shoemaker (1977) independently found an energy of $\sim$12 megatons TNT; amplitudes of the air waves recorded on barographs in Siberia are consistent with this energy. However, a more recent analysis of the propagation of the air wave by ReVelle (personal communication, 1981) yields an estimated energy of $\sim$30 megatons.

On the basis of a thorough study of the seismic records, Ben-Menahem (1975) estimated the energy at 12.5 ± 2.5 megatons. The distance to which trees were felled near the endpoint is consistent with a $\sim$13-megaton explosion at an altitude of ~8 km (Ben-Menahem 1975), and, scaling from the effects of nuclear weapons (Glasstone 1964), the ignition of wood by the thermal radiation also suggests an energy of the order of 10 megatons.

Turco et al (1982) have recently reinterpreted Ben-Menahem's work on the seismic records and suggested an energy of $\gtrsim 670$ megatons. As the bolide was certainly stopped in the atmosphere, the near and far record of the air waves and the extent of felling and ignition of trees show unequivocally that the estimate of Turco et al is $\gtrsim 20$ times too high.

From Figure 1, it may be seen that the "best estimate" of the frequency of a 12-megaton encounter with the Earth is about once every 300 $\overset{\times}{\div}$ 2 years; a 30-megaton encounter occurs about every 700 $\overset{\times}{\div}$ 2 years. There is a ~12 to ~40% chance that a 12-megaton encounter will occur in an interval of 75 years (the approximate time elapsed since 1908) and a ~5 to ~20% chance that a 30-megaton event will occur during this interval. On the basis of the predicted frequency, estimates of the energy of the Tunguska event in the range of 10–15 megatons appear somewhat more likely than ReVelle's estimate of 30 megatons. There is no more than a 1.5% chance of encounter of a 670-megaton bolide in an interval of 75 years.

The physical characteristics of the Tunguska object remain unsolved, as only ablation products have been recovered. At a typical encounter velocity of 20 km s^{-1}, a 12-megaton stony asteroid with a density of 2.4 g cm^{-3} has a diameter of 60 m. A stony body this small will not normally survive passage through the atmosphere. Almost any body of rock tens of meters across can be expected to be riven with structural flaws, especially fragments produced by collisions in the asteroid belt. Because of their low strength, nearly all stony bodies less than ~150 m diameter are sheared apart by aerodynamic stress in the lower atmosphere (cf Passey & Melosh 1980). Depending on the strength and structure of the object, a cascade of fragmentation may occur, and the body can be pulverized, ablated, and essentially stopped in the air. Thus, contrary to a common misconception, there is no mystery about the failure of the Tunguska bolide to reach the ground and produce a crater. The fact that it was stopped in the air does not provide any clues about the object other than that it was not a large iron meteorite or a pallasite (stony iron). Failure of numerous Soviet expeditions to find macroscopic meteorite fragments on the ground, on the other hand, does suggest that the bolide was composed of very weak or friable stony material or, perhaps, very small rocky particles embedded in ices.

Whipple (1930) suggested that the Tunguska object may have been a comet, and this concept appears to have gained general acceptance among informed Soviet investigators (Krinov 1966). Kresák (1978) has shown that the approximate trajectory of the meteor was consistent with the hypothesis that the Tunguska bolide was a fragment of Comet Encke. The strongest evidence that the Tunguska object was a comet comes from the distribution of very fine particles in the high atmosphere on the night following the encounter; an enormous, bright noctilucent cloud extended westward from Siberia to Ireland ($\sim$6000 km) and as far south as 45°N latitude (Krinov 1966, Fesenkov 1962, 1966). The distribution of the cloud corresponds approximately to the pattern expected for encounter of a possible cometary tail associated with the bolide. Turco et al (1982), on the other hand, have pointed out that normal upper-atmospheric circulation could have spread dust, ice particles, and NO_x over much of the region where bright skies were observed on the night of June 30, 1908. Some of the observed optical effects may have been due to airglow from chemical reactions of NO_x produced in the atmosphere by the Tunguska fireball.

In February, 1947, another brilliant meteoric fireball was observed over Siberia, this time in the eastern coastal region of Sikhote-Aline (Krinov 1966). The fireball was due to atmospheric entry of an iron meteorite that broke up at an altitude of about 6 km and yielded more than 380 fragments that struck the ground. A strewn field of more than 100 craters larger than 0.5 m diameter was formed; the largest crater was 26.5 m in diameter. In four expeditions of the Soviet Academy of Sciences, more than 23 tons of meteorite fragments were collected; a total mass of 70 tons is estimated to have fallen. On the basis of an orbit computed by Fesenkov (1951), the encounter velocity was $\sim 14 \text{ km s}^{-1}$ and the kinetic energy of the bolide was $\sim$1.7 kilotons TNT equivalent. From Figure 1, about 7 $\overset{\times}{\div}$ 2 objects with this energy encounter the Earth each year, of which $\sim$2 $\overset{\times}{\div}$ 2 enter the atmosphere over the continents. However, only $\sim$6% of the meteorites recovered from observed falls are irons (Mason 1962), and only one quarter to one half of the bolides producing very bright fireballs apparently are strong enough to yield recoverable meteorites (Ceplecha & McCrosky 1976, Wetherill & ReVelle 1981). The predicted fall rate on the continents of iron meteorites with the energy of the Sikhote-Aline bolide, therefore, is once per 10 to 80 years, consistent with the record of one such fall in the time since civilization over most of the world has become scientifically observant. Perhaps one or more such falls occurred in this century that were not recorded.

On March 31, 1965, a brilliant fireball seen from distances as far away as 750–800 km flashed over the sky of southern British Columbia. Tiny fragments, aggregating about 1 g, of carbonaceous meteorite were later found on the snow near Revelstoke (Folinsbee et al 1967), and a rich collection of magnetite and silicate glass spherules was obtained from airborne collectors after the air mass penetrated by the meteor had passed over the United States (Carr 1970). A sensitive microbarograph array in operation at that time by the National Bureau of Standards at Boulder,

Colorado, recorded relatively long-period acoustic waves excited by the passage of the Revelstoke bolide (Shoemaker & Lowery 1967). Analysis of the air-wave record shows that the energy of the initial Revelstoke object was ~20 kilotons TNT equivalent. At an assumed modal velocity of 20 km s^{-1}, the indicated initial mass of the body was ~500 tons. An object with this or greater energy encounters the Earth about once a year (Figure 1), and one will penetrate the atmosphere over North America about once every 25 $\overset{\times}{\div}$ 2 years. The Revelstoke fireball was the most energetic meteor detected over North America by the Boulder microbarograph array, which was operated for less than 20 years. The largest event of probable meteoric origin observed anywhere over the globe in this time period had an energy of 0.5 megatons and was located in the South Atlantic; the occurrence of this event in a ~20-year time interval is also consistent with the frequency predicted from the Best Estimate curve of Figure 1.

Small Impact Craters Formed by Iron Meteorites

Although the atmosphere stops most stony bodies less than ~150 m diameter (energy less than ~200 megatons), smaller bodies of strong, high-density iron can reach the ground and form craters. If the frequency with which crater-forming iron meteorites fall on the continents is about one to several times per century, as suggested above, numerous small young impact craters must have been produced. Most of these would not be large enough to survive in easily recognizable form for more than a few centuries, but craters in semiarid to arid regions and most craters larger than several hundred meters diameter can survive for thousands or tens of thousands of years.

Prehistoric impact craters with associated iron or stony-iron meteorites are known from 13 localities around the world (Table 2). As expected, most have been found in semiarid and arid regions. The largest is the 1.2 km diameter Meteor Crater of Arizona, which was formed in the late Pleistocene between ~20,000 and ~30,000 years ago (Shoemaker 1963, Shoemaker & Kieffer 1974). Assuming that the fraction of iron objects of any given energy is 0.015 to 0.03, as estimated above, the frequency of encounter with the Earth of an iron asteroid with an energy of ~15 megatons can be estimated from Figure 1 at about once per 6300 to 50,000 years. On this basis, a feature like Meteor Crater would be formed on the continents once every ~20,000 to ~150,000 years. The lower bound of the predicted recurrence interval is comparable with the age of Meteor Crater.

Physical studies of asteroids indicate that the proportion among telescopically observable objects that are neither C- nor S-type is 10% (Zellner 1979), of which half or more may be iron objects. Using a fraction of 0.05 irons, a production rate on the continents of one impact crater the size

Table 2 Prehistoric impact craters with associated meteorites

Locality	Number of craters	Diameter of largest crater (m)	Age	References
Haviland, Kansas, USA	1	14	Holocene ?	Nininger & Figgins 1933
Dalgaranga, W.A., Australia	1	26	Holocene ?	Nininger & Huss 1960
Sobolev, Siberia, USSR	1	51	Holocene ?	Khryanina & Ivanov 1977
Morasko, Poland	8	60	Holocene ?	Pokrzywnicki 1964
Campo del Cielo, Argentina	20	90	Holocene ?	Cassidy et al 1965
Wabar, Saudi Arabia	2	91	6400 ± 2500 yr	Philby 1933, Storzer & Wagner 1977
Kaalijarv, Estonia	7	100	Holocene	Krinov 1961
Henbury, N. T., Australia	14	150	4200 ± 1900 yr	Milton 1968, Storzer & Wagner 1977
Odessa, Texas, USA	5	168	$\sim 25{,}000$ yr	Evans 1961
Boxhole, N.T., Australia	1	175	Holocene ?	Madigan 1937
Monturaqui, Chile[a]	1	455	Late Pleistocene ?	Sanchez & Cassidy 1966
Wolf Creek, W. A., Australia	1	875	Pleistocene	Fudali 1979
Meteor Crater, Arizona, USA	1	1200	$25{,}000 \pm 5000$ yr	Shoemaker & Kieffer 1974

[a] All meteoritic material discovered at Monturaqui, Chile, is completely oxidized, but the nature of the oxide shows that the meteorite was an iron. Many specimens of unoxidized iron meteorites reported from other localities in Chile may have been transported from this site by the prehistoric Indians (Sanchez & Cassidy 1966).

Table 3 Small impact craters unaccompanied by discovered meteorites

Crater	Diameter (km)	Impact metamorphosed rocks	References
Aouelloul, Mauritania[a]	0.4	Yes	Chao et al 1966 Fudali & Cassidy 1972
Temimichat, Mauritania[a]	0.7	No	Fudali & Cassidy 1972
Pretoria Salt Pan, South Africa	1.0	No	Wagner 1922 Milton & Naeser 1971
Lonar Lake, India	1.7	Yes	Fudali et al 1980
Tenoumer, Mauritania[a]	1.8	Yes	French et al 1970 Fudali & Cassidy 1972 Fudali 1974
Roter Kamm, southwest Africa	2.4	Yes?	Fudali 1973

[a] The three craters in Mauritania may have been formed simultaneously $(3.1 \pm 0.3) \times 10^6$ years ago by fragments produced from aerodynamic breakup of a single initial projectile (Dietz et al 1969, Fudali 1976).

of Meteor Crater every 23,000 $\overset{\times}{\div}$ 2 years is obtained, which is indistinguishable from the age of Meteor Crater. On the same basis, about 2 $\overset{\times}{\div}$ 2 craters $\geqslant$750 m diameter should have been formed on the continents in the last 23,000 years; two have been discovered (Table 2), but the smallest (Wolf Creek) may be older than 23,000 years. A total of 8 craters $\geqslant$100 m diameter with associated iron meteorites have been found (Table 2), but roughly 1000 should have been formed in the last 20,000 years. Completeness of discovery clearly drops very rapidly for impact craters below $\sim$0.5 km diameter.

In addition to the craters with associated iron meteorites, 6 other well-defined young craters of impact or probable impact origin are known that are smaller than 3 km diameter (Table 3). Again, most of these have been discovered in arid regions. No meteorites have been found at these craters, but projectiles of unusually high strength or density probably are required to form craters of this size; iron or stony-iron bodies are the most likely candidates. About 50 $\overset{\times}{\div}$ 2 craters $\geqslant$1 km diameter probably were formed on the continents by impact of iron meteorites in the last million years. Hence, $\sim$10 times as many craters as are shown in Table 3 may remain to be discovered.

Young Large Impact Craters

As 4 $\overset{\times}{\div}$ 2 craters $\geqslant$10 km diameter are predicted to be formed on the continents per 10^6 years, there is an 85% chance that at least one in this size range was actually produced in the last million years. Craters 10 km in diameter are large enough to survive in recognizable form in most parts of the world for times of the order of a million years. However, we may reasonably expect that any meteorite fragments that may have been associated with them have long since disappeared by weathering and erosion. Identification of impact origin, therefore, must rest on the structure of the craters (Shoemaker & Eggleton 1961, Dence et al 1977) and on the various products of shock metamorphism (French & Short 1968).

The best known young large impact crater contains Lake Bosumtwi, the sacred lake of the Ashanti tribe of Ghana. It is 10 km in diameter, and the fission-track age of the Bosumtwi impact glass is 1.04 ± 0.11 Myr (Storzer & Wagner 1977). The crater is the source of the contemporaneous Ivory Coast strewn field of tektites (Schnetzler et al 1967, Kolbe et al 1967). The impact origin of the Bosumtwi crater was suspected by Maclaren (1931) and confirmed by the discovery of coesite in the strongly shocked ejecta (Littler et al 1962).

A ~ 7 km diameter crater, referred to as Zhamanshin and located north of the Aral Sea in Kazakhstan, was recently recognized as an impact crater by Florensky (1975, 1977). A fission-track age of 1.07 ± 0.05 Myr was obtained from impact glass at Zhamanshin (Storzer & Wagner 1977). The crater appears to be the source of the nearby irghizite (wet tektite) strewn field (Florensky 1975, King & Arndt 1977).

Another large young impact crater in the Soviet Union, containing Lake Elgygytgyn (located near the Arctic coast of eastern Siberia), is 18 km diameter. The impact origin of this remote crater now appears to be well documented (Gurov et al 1978, 1979); a K/Ar age of 3.5 ± 0.5 Myr was reported for impact glass from the site. Assuming that the size index of large terrestrial craters is 1.84, the production rate of craters $\geqslant 18$ km on the continents is $(1.4 \overset{\times}{\div} 2) \times 10^{-6}$ yr^{-1}, and there is a 99% probability that at least one 18-km crater will be formed in a time interval of 3.5 Myr. The Elgygytgyn crater fulfills this prediction, but it is also likely that one or two other craters or eroded impact structures of this size and approximate age also exist.

Large, million-year-old craters can be deeply eroded or filled by sediments in humid regions or completely obliterated in areas of glaciation and on the continental shelves. Moreover, the geology of much of the land surface of the Earth remains to be mapped in detail. Hence, we should not expect that all the young continental impact structures $\geqslant 10$ km diameter have been recognized. A case in point is the source of the 700,000-year-old Australasian tektites. Their origin as terrestrial impactites derived from continental rocks is well established on chemical, isotopic, and mineralogical grounds (Chao et al 1962, Walter 1965, Taylor 1973, King 1977). Dietz (1977) postulated that the Australasian tektites were derived from

Elgygytgyn, and Taylor & McLennan (1979) argued for Zhamanshin as the source. Both craters appear to be eliminated as candidates on the basis of their ages. Judging from the modest strewn field of tektites associated with the 10-km Bosumtwi crater, Zhamanshin is also too small to be a likely source for the voluminous Australasian tektites. Apparently at least one large ($>$10 km diameter?) 700,000-year-old impact structure remains to be found somewhere on the continents, probably on the Asian land mass (cf Taylor & McLennan 1979).

Ancient Phanerozoic Impact Structures

When the impact record on the continents is examined for features older than a few million years, most of the craters are found to have been lost by erosion or by burial beneath sediments. The oldest known impact craters that preserve remnants of their original topography are the 14.7 Myr old Ries crater (27 km diameter) and nearby Steinheim Basin (3.4 km) in the Schwabian Alb of southern Germany (Pohl et al 1977, Reiff 1977). Some ancient craters, such as the Holleford crater (Cambrian) and the Brent crater (Ordovician), both in Ontario, and the Flynn Creek crater (Devonian) in Tennessee, were formed on the floors of shallow epicontinental seas and were buried beneath marine sediments (Dence & Guy-Bray 1972, Roddy 1977). Most older impact sites, however, can only be recognized by the structural disturbance of the bedrock and, where preserved, associated shock-metamorphosed rocks. In crystalline bedrock, the most easily recognized structural feature is a lens of breccia that was formed beneath the original crater floor. In stratified sedimentary targets, a highly deformed central structural uplift surrounded by a circular structural moat of down-dropped rock is commonly present. This distinctive structural pattern is the product of collapse of the initial crater. Very large structures ($>$50 km) can be considerably more complex (Dence et al 1977).

The record of impact structures older than a few million years essentially begins at craters or structures with diameters greater than 2 to 3 km. Fewer than 10 ancient features whose impact origin is firmly determined are smaller than 3 km, but many are close to this size. This limit corresponds approximately to the size of craters formed by the smallest stony asteroids ($\sim$100 to $\sim$150 m diameter) expected to survive passage through the atmosphere. The frequency of crater production probably jumps by a factor of $\sim$10 across the $\sim$3-km threshold.

About 70 craters and structures of probable impact origin that are larger than 3 km diameter have been identified throughout the world (Dence 1972, Robertson & Grieve 1975, Masaitis 1975, Grieve & Robertson 1979). Over 90% have been found in the United States, Canada, the Soviet Union, western Europe, and Australia, which constitute the major portion of the world most thoroughly investigated by geologists. It is also the part most thoroughly surveyed for impact craters. By far the largest number of impact structures occurs on the geologically stable platform or shield areas of the continents, called cratons, where ancient impact features tend to be preserved longest and where they are usually most easily recognized. Only the impact record from the cratons is sufficiently complete for meaningful comparison with the present cratering rate predicted from astronomical observations. Even on the cratons, however, most impact structures less than 10 km diameter evidently have been lost by erosion or have been so deeply eroded as to preclude recognition. The mean lifetime of 10-km impact structures against erosion has been estimated by Grieve & Robertson as 300 Myr.

Converting the estimated present cratering rate by asteroid impact given for the whole Earth, the rate per unit area is $(2.4 \stackrel{\times}{\div} 2) \times 10^{-14}\ \mathrm{km}^{-2}\ \mathrm{yr}^{-1}$ craters $\geqslant$10 km diameter. Adding cratering by comet impact increases this rate by $\sim$10–30%. Grieve & Dence (1979) found that the production over Phanerozoic time (the last $\sim$600 Myr) of craters $\geqslant$20 km diameter was $(0.36 \pm 0.1) \times 10^{-14}\ \mathrm{km}^{-2}\ \mathrm{yr}^{-1}$ on the North American craton and $(0.33 \pm 0.2) \times 10^{-14}\ \mathrm{km}^{-2}\ \mathrm{yr}^{-1}$ on the European craton. Averaging these rates and assuming a size index of 1.84 for the craters produced, the corresponding rate for craters $\geqslant$10 km is $(1.24 \pm 0.36) \times 10^{-14}\ \mathrm{km}^{-2}\ \mathrm{yr}^{-1}$. This rate lies just within the estimated error bar of the estimated present rate for asteroid impact, but it is of interest to inquire whether the result of Grieve & Dence might be biased by erosional loss of some craters $\geqslant$20 km diameter, particularly in the hearts of the Precambrian shields.

The most complete record of impact structures probably is preserved where the Precambrian rocks of the cratons are thinly veneered with Phanerozoic platform sediments (Grieve & Robertson 1979). Such an area is found in the Mississippi lowland of the United States. There, from a much smaller set of known structures, Shoemaker (1977) found a 10-km cratering rate of $(2.2 \pm 1.1) \times 10^{-14}\ \mathrm{km}^{-2}\ \mathrm{yr}^{-1}$, which is in excellent agreement with the estimated present rate. Within the errors of estimation, it appears that the cratering rate on the continents, averaged over intervals of the order of 10^8 years, has been relatively steady over Phanerozoic time.

Large Precambrian Impact Structures

Extrapolating to larger crater sizes with the size index of 1.84, the estimated present cratering rate on the continents is $6 \stackrel{\times}{\div} 2$ per 100 million years for 100 km diameter craters and $3 \stackrel{\times}{\div} 2$ per 100 million years for 140-km craters. Judging from the postmare lunar craters, the size distribution of craters $\geqslant$3 km diameter probably follows a simple power law up to about 100 km, but it appears to steepen at sizes above 100 km. As the area of the continents has not increased more than about 25% during the last 500 million years and as the cratering rate appears to have been nearly steady, about 25 $\overset{\times}{\div}$ 2 craters $\geqslant$100 km diameter should have been formed on the continents in this time.

Only one 100 km diameter impact structure of Phanerozoic age (Popigai in the USSR) and two structures approaching 100 km (Puchezh-Katunk in the USSR and Manicouagan in Quebec) have so far been recognized (Grieve & Robertson 1979). One reason so few have been found may be that the entire crust responds isostatically when craters of this size are formed, and the initial relief is largely lost through quasi-viscous flow of the mantle and lower crustal rocks. Other geologic processes, such as volcanism, may be triggered by these large impacts, and entire impact structures may be covered by lava or the evidence bearing on their origin may be obscured by these secondary geologic effects.

As many as 5 to 10 impact structures $\geqslant$140 km diameter probably were formed on the continents during the Phanerozoic; none have yet been found. Over the preceding 2.5 billion years, 5 to 10 additional impact structures this size probably were formed. Two 140 km diameter Precambrian structures of probable impact origin have been recognized: the Sudbury Basin of Ontario and the Vredefort Dome in South Africa (Grieve & Robertson 1979). The impact origin of the 1840$\pm$150 Gyr Sudbury Basin is well established from the shock metamorphism of rock fragments in a thick sequence of strongly shocked fallback material that is preserved in the center of the basin (French 1967, 1968). This material rests on a thick eruptive body of norite. Part of the eruptive material may be deep-seated magma whose formation was triggered by the impact, but considerable impact melt derived from the continental crust evidently is mixed with the norite (Hamilton 1970).

Although challenged by Nicolaysen et al (1963), the impact origin of the 1970$\pm$100 Gyr Vredefort Dome is strongly indicated by its structure (Wilshire & Howard 1968), by the occurrence and orientation of shatter cones (Manton 1965), and by abundant veins of microbreccia called pseudotachylyte (Shand 1916) that locally contain coesite and stishovite, the high-pressure polymorphs of SiO_2 (Martini 1978), and various shock-metamorphic features such as planar dislocations and lamellae in quartz (Carter 1965, Lilly 1981) and the crystallographic configuration of feldspars in a granite at the center of the dome (Aitken & Gold 1968). Dietz (1963) suggested that the great Bushveld complex of strataform igneous rock, which lies north of the Vredefort Dome, might have been formed by an eruption of a sheet of basalt in response to another large Precambrian impact event. On the basis of a structural interpretation of the Bushveld, Hamilton (1970) elaborated on Dietz's suggestion and proposed that the Bushveld eruptive sheet occupies the ring moats of three overlapping impact structures, each of which is roughly comparable to or larger than the Vredefort structure. Hamilton supposed that these structures, including the Vredefort, were all formed simultaneously.

To account for such a multiple impact event, he suggested that the impacting bodies may have been fragments of a comet that was broken up in the atmosphere or by gravitational disruption. As shown by Emiliani et al (1981), neither a comet nucleus nor an asteroid of the required size could be broken up and separated any significant distance by aerodynamic forces. If the initial body were sufficiently elongate or even multiple to begin with, on the other hand, it could be readily broken by tidal forces and the parts separated sufficiently to produce separate impact structures. At least two fairly rapidly rotating Earth-approaching asteroids, 433 Eros and 1620 Geographos, are sufficiently elongate that they are scarcely bound by their own gravity and would be readily broken and separated by tidal forces on near-encounter with the Earth. They may be multiple bodies to start with. Hence, a comet is not necessarily the most likely parent of multiple projectiles that may strike the Earth almost simultaneously. It is entirely plausible, and even moderately probable, that a multiple impact event comparable to that postulated by Hamilton for the Bushveld has occurred in the last two billion years. In fact, a pair of impact craters at Clearwater Lakes, Quebec, is an example of a multiple impact on a somewhat smaller scale that occurred about 290 million years ago (Grieve & Robertson 1979).

Whether the Bushveld structural system was actually formed in the manner suggested by Hamilton is a separate question, however. As all three impact structures posited by him are as large or larger than the Vredefort, the rocks comprising the central uplifts of these structures should have been brought up from depths comparable to the depth of origin of the granite in the central dome of Vredefort. Much of the largest central uplift in the Bushveld area (Marble Hall uplift) appears to be composed of upper-crustal sedimentary and volcanic rocks. If this is correct, the presence of these rocks may pose a serious difficulty for Hamilton's model. Clearly, much further work is needed to test his intriguing hypothesis.

EFFECTS OF BOMBARDMENT ON TERRESTRIAL LIFE

Asteroids and comet nuclei larger than 1 km diameter will unequivocally extinguish life locally wherever they strike the Earth, but the area affected by each impact crater and its surrounding ejecta blanket is too small to be of much consequence to biological evolution, even for impacting bodies up to tens of kilometers diameter. As the size of the crater increases, however, a threshold is reached where the material ejected into the high atmosphere has a strong transient effect on global climate. Discovery by Alvarez et al (1980) of anomalous abundances of iridium and other elements in a claystone layer at the Cretaceous-Tertiary boundary suggests that impact of a fairly large asteroid or comet is associated in time with a major extinction of living species. This anomalous claystone layer, which has been recognized from about 40 localities around the world (Alvarez et al 1982, Pillmore et al 1982), appears to record the fallout of highly shocked material that was heavily contaminated by the constituents of the impacting body. Typical thicknesses of the claystone are 1–2 cm. Alvarez et al (1980) estimated that an impacting asteroid on the order of 10 km diameter and with an iridium abundance comparable to that of common stony meteorites would be required to supply the amount of iridium distributed globally in the claystone.

Suspension in the stratosphere of a few tenths of a gram of micron- to submicron-size dust per square centimeter would almost totally block sunlight from the Earth's surface (Gerstl & Zardecki 1982). Toon et al (1982) have estimated that the settling time of most of the stratospheric dust would be of the order of several months to half a year. Under these conditions, only the warming of the air by the heat stored in the oceanic thermosphere would prevent the troposphere from becoming nearly isothermal. In a short time, the mean surface temperatures over land areas would drop by several tens of degrees C.

A somewhat longer-lived transient climatic effect may be produced by large impacts in the deep ocean. Enough water can be injected into the high atmosphere to replace most of the stratospheric air with water vapor. Emiliani et al (1981) have suggested that after the dust and excess water have precipitated and the atmosphere clears, the remaining water vapor would lead to a greenhouse effect that would raise surface temperatures by more than 10°C above the preexisting ambient conditions. This type of transient effect probably persists for times of the order of years, as the equilibrium composition of the stratosphere would be reestablished only by the slow processes of photochemical dissociation of the H_2O and diffusion across the tropopause.

Still other biologically deleterious effects of large impacts may be associated with large quantities of NO_x produced in the high-temperature atmospheric shock wave (Park 1978, Park & Menees 1978) and possibly with contamination of the atmosphere by the constituents of the projectile itself (cf Hsü 1980).

The threshold size of impacting bodies at which biologically significant perturbations of the atmosphere and oceanic thermosphere are produced is very poorly understood. If the abrupt extinction of marine plankton and other species at the end of the Cretaceous is related to impact of a body large enough to produce the observed noble-metal anomaly in the boundary claystone, then projectile diameters on the order of 10 km are evidently above the threshold. On the basis of the estimated Phanerozoic cratering rate and the size distribution of postmare craters on the Moon, 10 km diameter projectiles struck the Earth with a mean frequency of once every $\sim 0.5 \times 10^8$ years during the Phanerozoic. From the long-term lunar impact record, the mean frequency of Earth impact of 10-km bodies back to 3.3 Gyr was about once per 10^8 years.

The threshold size of projectiles that might cause mass extinction of species in certain environments may be considerably smaller than 10 km diameter. At 5 km diameter, the frequency of impact is about four times as great as at 10 km. It is entirely possible that dozens of impact-related ecological jolts severe enough to be reflected in the paleontologic record were delivered to the biosphere during the Phanerozoic. The test of this possibility will require close scrutiny of the stratigraphic and paleontologic evidence.

Bodies larger than 10 km diameter can also hit the Earth. For example, the Earth-approaching asteroid 433 Eros has a 20% chance of colliding in the next 400 million years (Wetherill & Shoemaker 1982); the mean diameter of Eros is ~ 20 km. Over the span of Precambrian time back to 3.3 Gyr it is likely that several asteroids and comet nuclei in the 20–30 km diameter range struck the Earth. Whether the climatic perturbations produced by these giant impacts were much greater than those produced by 10-km bodies is not clear. It may be that the climatic perturbations were limited by saturation and that any material introduced into the atmosphere exceeding a certain limit simply fell out again rather quickly. On the other hand, prompt effects, such as heating of the atmosphere due to compression under the initial load of material, probably had no upper bound.

The lunar-crater record shows that the cratering rate at 3.9 Gyr was about 25 times higher than at present, and that between 3.9 and 3.3 Gyr it decayed approximately exponentially with a half-life near 10^8 years. A dozen or more objects larger than 20 km diameter may have struck the Earth in the few hundred million years after deposition of the earliest-recorded Precambrian sediments of the Isua complex, Greenland, at 3.8 Gyr. At still earlier times, the bombardment was even more intense. Whatever primitive organisms may have existed then probably were subjected to frequently repeated environmental insults on a global scale.

Literature Cited

Aitken, F. K., Gold, D. P. 1968. The structural state of potash feldspar—a possible criterion for meteorite impact. See French & Short 1968, pp. 519–30

Alvarez, L. W., Alvarez, W., Asaro, F., Michel, H. V. 1980. Extraterrestrial cause for the Cretaceous-Tertiary extinction. *Science* 208:1095–1108

Alvarez, W., Asaro, F., Michel, H. V., Alvarez, L. W. 1982. Major impacts and their geological consequences. *Geol. Soc. Am. Ann. Meet., 95th*, pp. 431–32 (Abstr.)

Ben-Menahem, A. 1975. Source parameters of the Siberian explosion of June 30, 1908, from analysis and synthesis of seismic signals at four stations. *Phys. Earth Planet. Inter.* 11:1–35

Bowell, E., Lumme, K. 1979. Colorimetry and magnitudes of asteroids. In *Asteroids*, ed. T. Gehrels, pp. 132–69. Tucson: Univ. Ariz. Press. 1181 pp.

Bowell, E., Chapman, C. R., Gradie, J. C., Morrison, D., Zellner, B. 1978. Taxonomy of asteroids. *Icarus* 35:313–35

Bowell, E., Gehrels, T., Zellner, B. 1979. Magnitudes, colors, types and adopted diameters of the asteroids. In *Asteroids*, ed. T. Gehrels, pp. 1108–29. Tucson: Univ. Ariz. Press. 1181 pp.

Carr, M. H. 1970. Atmospheric collection of debris from the Revelstoke and Allende fireballs. *Geochim. Cosmochim. Acta* 34:689–700

Carter, N. L. 1965. Basal quartz deformation lamellae—a criterion for recognition of impactites. *Am. J. Sci.* 263:786–806

Cassidy, W. A., Villar, L. M., Bunch, T. E., Kohman, T. P., Milton, D. J. 1965. Meteorites and craters of Campo del Cielo, Argentina. *Science* 149:1044–64

Ceplecha, Z., McCrosky, R. E. 1976. Fireball end heights: a diagnostic for the structure of meteoric material. *J. Geophys. Res.* 81:6257–75

Chao, E. C. T., Adler, I., Dwornik, E. J., Littler, J. 1962. Metallic spherules in tektites from Isabella, the Philippine Islands. *Science* 135:97–98

Chao, E. C. T., Dwornik, E. J., Merrill, C. W. 1966. Nickel-iron spherules from Aouelloul glass. *Science* 154:759–65

Degewij, J., Tedesco, E. F. 1982. Do comets evolve into asteroids? Evidence from physical studies. In *Comets*, ed. L. Wilkening, pp. 665–95. Tucson: Univ. Ariz. Press. 766 pp.

Delsemme, A. H. 1982. Chemical composition of cometary nuclei. In *Comets*, ed. L. Wilkening, pp. 85–130. Tucson: Univ. Ariz. Press. 766 pp.

Dence, M. R. 1972. The nature and significance of terrestrial impact structures. *Proc. 24th Int. Geol. Congr., Montreal, Sect. 15*, pp. 77–89

Dence, M. R., Guy-Bray, J. V. 1972. Some astroblemes, craters and cryptovolcanic structures in Ontario and Quebec. *24th Int. Geol. Congr. Guideb., Field Excursion A65*. 61 pp.

Dence, M. R., Grieve, R. A. F., Robertson, P. B. 1977. Terrestrial impact structures: Principal characteristics and energy considerations. See Roddy et al 1977, pp. 247–75

Dietz, R. S. 1963. Vredefort Ring-Bushveld complex impact event and lunar maria. *Geol. Soc. Am. Spec. Pap. No. 73*, p. 35 (Abstr.)

Dietz, R. S. 1977. Elgygytgyn Crater, Siberia: a probable source of Australasian tektite field. *Meteoritics* 12:145–57

Dietz, R. S., Fudali, R., Cassidy, W. 1969. Richat and Semsiyat Domes (Mauritania): not astroblemes. *Bull. Geol. Soc. Am.* 80:1367–72

Dollfus, A., Zellner, B. 1979. Optical polarimetry of asteroids and laboratory samples. In *Asteroids*, ed. T. Gehrels, pp. 170–83. Tucson: Univ. Ariz. Press. 1181 pp.

Emiliani, C., Kraus, E. B., Shoemaker, E. M. 1981. Sudden death at the end of the Mesozoic. *Earth Planet. Sci. Lett.* 55:317–34

Evans, G. L. 1961. Investigations at the Odessa meteor craters. *Proc. Geophys. Lab./Lawrence Radiat. Lab. Cratering Symp.*, ed. M. D. Nordyke, pp. D1–11. *Rep. UCRL-6438*. Livermore, Calif: Lawrence Radiat. Lab.

Fesenkov, V. G. 1951. The orbit of the Sikhote-Aline meteorite. *Meteoritika* 9:27–31

Fesenkov, V. G. 1962. On the cometary nature of the Tunguska meteorite. *Sov. Astron. AJ* 5:441–51

Fesenkov, V. G. 1966. A study of the Tunguska meteorite fall. *Sov. Astron. AJ* 10:195–213

Florensky, K. P. 1963. The problem of the cosmic dust and the present state of studies on the Tunguska meteorite. *Geochimiya* 3L:284–96

Florensky, P. V. 1975. The Zhamanshin meteorite crater (The Northern Near-Aral) and its tektites and impactites. *Prob. Akad. Nauk Isvestzya* 10:73–86

Florensky, P. V. 1977. The meteoritic crater Zhamanshin (Northern Aral Region USSR) and its tektites and impactites. *Chem. Erde* 36:83–95

Folinsbee, R. E., Douglas, J. A. V., Maxwell, J. A. 1967. Revelstoke, a new Type I carbonaceous chondrite. *Geochim. Cosmochim. Acta* 31:1625–35

French, B. M. 1967. Sudbury structure, Ontario—some evidence for origin by meteorite impact. *Science* 156:1094–98

French, B. M. 1968. Sudbury structure, Ontario—Some petrographic evidence for an origin by meteorite impact. See French & Short 1968, pp. 383–412

French, B. M., Short, N. M., eds. 1968. *Shock Metamorphism of Natural Materials.* Baltimore, Md: Mono Book Corp. 646 pp.

French, B. M., Hartung, J. B., Short, N. M., Dietz, R. S. 1970. Tenoumer Crater, Mauritania: age and petrologic evidence for origin by meteorite impact. *J. Geophys. Res.* 75:4396–4406

Fudali, R. F. 1973. Roter Kamm: evidence for an impact origin. *Meteoritics* 8:245–57

Fudali, R. F. 1974. Genesis of the melt rocks at Tenoumer Crater, Mauritania. *J. Geophys. Res.* 79:2115–21

Fudali, R. F. 1976. Investigation of a new stony meteorite from Mauritania with some additional data on its find site: Aouelloul Crater. *Earth Planet. Sci. Lett.* 30:262–68

Fudali, R. F. 1979. Gravity investigation of Wolf Creek Crater, western Australia. *J. Geol.* 87:55–67

Fudali, R. F., Cassidy, W. A. 1972. Gravity reconnaissance at three Mauritanian craters of explosive origin. *Meteoritics* 7:51–70

Fudali, R. F., Milton, D. J., Fredriksson, K., Dube, A. 1980. Morphology of Lonar Crater, India: comparisons and implications. *The Moon and the Planets* 23:493–515

Gaffey, M. J., McCord, T. B. 1979. Mineralogical and petrological characterizations of asteroid surface materials. In *Asteroids*, ed. T. Gehrels, pp. 688–723. Tucson: Univ. Ariz. Press. 1181 pp.

Gault, D. E. 1973. Displaced mass, depth, diameter, and effects of oblique trajectories for impact craters formed in dense crystalline rocks. *The Moon* 6:32–44

Gerstl, S. A. W., Zardecki, A. 1982. The extinction of life due to stratospheric dust. *Nature*. In press

Glasstone, A., ed. 1964. *The Effects of Nuclear Weapons.* Washington DC: US At. Energy Comm. 730 pp.

Grieve, R. A. F., Dence, M. R. 1979. The terrestrial cratering record. II. The crater production rate. *Icarus* 38:230–42

Grieve, R. A. F., Robertson, P. B. 1979. The terrestrial cratering record. I. Current status of observations. *Icarus* 38:212–29

Gurov, E. P., Valter, A. A., Gurova, E. P., Serebrennikov, A. I. 1978. Elgygytgyn meteorite explosion crater in Chucotka. *Dokl. Akad. Nauk SSSR* 240:1407–10 (In Russian)

Gurov, E. P., Valter, A. A., Gurova, E. P., Kotlovskaya, F. I. 1979. Elgygytgyn Impact Crater, Chucotka: shock metamorphism of volcanic rocks. *Lunar Planet. Sci.* 10:479–81

Hamilton, W. B. 1970. Bushveld complex—product of impacts? *Geol. Soc. S. Afr. Spec. Publ. I*, pp. 367–79

Helin, E. F., Shoemaker, E. M. 1979. The Palomar planet-crossing asteroid survey, 1973–1978. *Icarus* 40:321–28

Holsapple, K. A. 1980. The equivalent depth of burst for impact cratering. *Proc. Lunar Planet. Sci. Conf., 11th*, pp. 2379–401

Holsapple, K. A., Schmidt, R. M. 1982. On the scaling of crater dimensions: 2. Impact processes. *J. Geophys. Res.* 87:1849–70

Hsü, K. 1980. Terrestrial catastrophe caused by cometary impact at the end of the Cretaceous. *Nature* 285:201–3

Hunt, J. N., Palmer, R., Penny, W. 1960. Atmospheric waves caused by large explosions. *Philos. Trans. R. Soc. London Ser. A* 252:275–315

Kamoun, P. G., Campbell, D. B., Ostro, S. J., Pettengill, G. H., Shapiro, I. I. 1981. Comet Encke: Radar detection of nucleus. *Science* 216:293–95

Khryanina, L. P., Ivanov, O. P. 1977. Structure of meteorite craters and astroblemes. *Dokl. Acad. Nauk SSSR* 233:457–60 (In Russian)

King, E. A. 1977. The origin of tektites: a brief review. *Am. Sci.* 65:212–18

King, E. A., Arndt, J. 1977. Water content of Russian tektites. *Nature* 269:48–49

Kolbe, P., Pinson, W. H., Saul, J. M., Miller, E. W. 1967. Rb-Sr study on country rocks of the Bosumtwi Crater, Ghana. *Geochim. Cosmochim. Acta* 31:869–75

Kresák, L. 1978. The Tunguska object—part of comet Encke? *Bull. Astron. Inst. Czech.* 29:129–34

Krinov, E. L. 1961. The Kaaliljarv meteorite craters on Saarema Island, Estonian SSR. *Am. J. Sci.* 259:430–40

Krinov, E. L. 1966. *Giant Meteorites.* Oxford: Pergamon. 397 pp.

Lebofsky, L. A., Veeder, G. L., Lebofsky, M. J., Matson, D. L. 1978. Visual and radiometric photometry of 1580 Betulia. *Icarus* 35:336–43

Lebofsky, L. A., Lebofsky, M. J., Rieke, G. H. 1979. Radiometry and surface properties of Apollo, Amor, and Aten asteroids. *Astron. J.* 84:885–88

Lebofsky, L. A., Veeder, G. J., Rieke, G. H., Lebofsky, M. J., Matson, D. L., Kowal, C., Wynn-Williams, C. G., Becklin, E. E. 1981.

The albedo and diameter of 1862 Apollo. *Icarus* 48:335–38

Lilly, P. A. 1981. Shock metamorphism in the Vredefort collar: evidence for internal shock sources. *J. Geophys. Res.* 86:10689–700

Littler, J., Fahey, J. J., Dietz, R. S., Chao, E. C. T. 1962. Coesite from the Lake Bosumtwi Crater. *Geol. Soc. Am. Spec. Pap. No. 68*, p. 218 (Abstr.)

Maclaren, M. 1931. Lake Bosumtwi, Ashanti. *Geogr. J.* 78:270–76

Madigan, C. T. 1937. The Boxhole crater and the Huckitta Meteorite (Central Australia). *Trans. Proc. R. Soc. South Aust.* 61:187–90

Manton, W. I. 1965. The orientation and origin of shatter cones in the Vredefort ring. *Ann. NY Acad. Sci.* 123:1017–49

Marsden, B. G. 1970. On the relationship between comets and minor planets. *Astron. J.* 75:206–17

Marsden, B. G. 1971. Evolution of comets into asteroids. In *Physical Studies of Minor Planets*, ed. T. Gehrels, pp. 413–21. *NASA SP-267*. Washington DC: US GPO. 687 pp.

Marsden, B. G. 1979. *Catalogue of Cometary Orbits*. Cambridge, Mass: Smithsonian Astrophys. Obs. 88 pp.

Martini, J. E. J. 1978. Coesite and stishovite in the Vredefort dome, South Africa. *Nature* 272:715–17

Masaitis, V. L. 1975. Astroblemes in the Soviet Union. *Sov. Geol.* 11:52–64

Mason, B. 1962. *Meteorites*. New York: John Wiley & Sons. 274 pp.

Melosh, H. J. 1980. Cratering mechanics—Observational, experimental, and theoretical. *Ann. Rev. Earth Planet. Sci.* 8:65–93

Milton, D. J. 1968. Structural geology of the Henbury meteorite craters, Northern Territory, Australia. *US Geol. Surv. Prof. Pap. 559-C*, pp. 1–17

Milton, D. J., Naeser, C. W. 1971. Evidence for an impact origin of the Pretoria Salt Pan, South Africa. *Nature Phys. Sci.* 229:211–12

Morrison, D. 1977. Asteroid sizes and albedos. *Icarus* 31:185–220

Morrison, D., Lebofsky, L. A. 1979. Radiometry of asteroids. In *Asteroids*, ed. T. Gehrels, pp. 184–205. Tucson: Univ. Ariz. Press. 1181 pp.

Morrison, D., Gradie, J. C., Reike, G. H. 1976. Radiometric diameter and albedo of the remarkable asteroid 1976 AA. *Nature* 260:691

Neukum, G., König, B., Fechtig, H., Storzer, D. 1975. Cratering in the Earth-Moon system: consequences for age determination by crater counting. *Proc. Lunar Sci. Conf., 6th*, pp. 2597–2620

Nicolaysen, L. O., De Villiers, J. W., Burger, A. J. 1963. The origin of the Vredefort Dome structure in the light of new isotopic data. *Int. Union Geod. Geophys., Berkeley, Abstr.*

Nininger, H. H., Figgins, J. D. 1933. The excavation of a meteorite crater near Haviland, Kiowa County, Kansas. *Proc. Colo. Mus. Nat. Hist.* 12:9–15

Nininger, H. H., Huss, G. I. 1960. The unique meteorite crater at Dalgaranga, Western Australia. *Mineral. Mag.* 32:619–39

Nordyke, M. D. 1961. Nuclear craters and preliminary theory of the mechanics of explosive crater formation. *J. Geophys. Res.* 66:3439–59

Nordyke, M. D. 1962. An analysis of cratering data from desert alluvium. *J. Geophys. Res.* 67:1965–74

Nordyke, M. D. 1977. Nuclear cratering experiments: United States and Soviet Union. See Roddy et al 1977, pp. 103–24

Oort, J. H. 1950. The structure of the cloud of comets surrounding the solar system, and a hypothesis concerning its origin. *Bull. Astron. Inst. Neth.* 11:91–110

Öpik, E. J. 1951. Collision probabilities with the planets and the distribution of interplanetary matter. *Proc. R. Ir. Acad. Sect. A* 54:165–99

Öpik, E. J. 1958. Meteor impact on solid surface. *Irish Astron. J.* 5:14–23

Öpik, E. J. 1963. The stray bodies in the solar system. Part 1. Survival of cometary nuclei and the asteroids. *Adv. Astron. Astrophys.* 2:219–62

Park, C. 1978. Nitric oxide production by Tunguska meteor. *Acta Astron.* 5:523–42

Park, C., Menees, G. P. 1978. Odd nitrogen production by meteoroids. *J. Geophys. Res.* 83:4029–35

Passey, Q. R., Melosh, H. J. 1980. Effects of atmospheric breakup on crater field formation. *Icarus* 42:211–33

Philby, H. St. J. B. 1933. *The Empty Quarter*. New York: Henry Holt & Co., pp. 157–80

Pillmore, C. L., Tschudy, R. H., Orth, C. J., Gilmore, J. S., Knight, J. D. 1982. Iridium abundance anomalies at the palynological Cretaceous/Tertiary boundary in coal beds of the Raton Formation, Raton basin, New Mexico and Colorado. *Geol. Soc. Am. Ann. Meet., 95th*, p. 588 (Abstr.)

Pohl, J., Stoffler, D., Gall, H., Ernstson, K. 1977. The Ries impact crater. See Roddy et al 1977, pp. 343–404

Pokrzywnicki, J. 1964. Meteorites of Poland, 6-Meteorites of Morasko. *Stud. Geol. Pol.* 15:49–70

Reiff, W. 1977. The Steinheim Basin—An impact structure. See Roddy et al 1977, pp. 309–20

Robertson, P. B., Grieve, R. A. 1975. Impact structures in Canada: Their recognition and characteristics. *J. R. Astron. Soc. Can.* 69:1–21

Roddy, D. J. 1977. Preimpact conditions and cratering processes at the Flynn Creek Crater, Tennessee. See Roddy et al 1977, pp. 277–308

Roddy, D. J., Pepin, R. O., Merrill, R. B., eds. 1977. *Impact and Explosion Cratering: Planetary and Terrestrial Implications.* New York: Pergamon. 1301 pp.

Roddy, D. J., Schuster, S. H., Kreyenhagen, K. N., Orphal, D. L. 1980. Computer code simulations of the formation of Meteor Crater, Arizona: Calculations MC-1 and MC-2. *Proc. Lunar Planet. Sci. Conf., 11th,* pp. 2275–308

Roemer, E. 1966. The dimensions of cometary nuclei. *Mem. Royal Sci. Soc. Liege,* 5th Ser. 12:23–28

Sanchez, J., Cassidy, W. 1966. A previously undescribed meteorite crater in Chile. *J. Geophys. Res.* 20:4891–95

Schmidt, R. M. 1980. Meteor Crater: Energy of formation—implications of centrifuge scaling. *Proc. Lunar Planet. Sci. Conf., 11th,* pp. 2099–2128

Schnetzler, C. C., Philpotts, J. A., Thomas, H. H. 1967. Rare earth and barium abundances in Ivory Coast tektites and rocks from the Bosumtwi Crater area, Ghana. *Geochim. Cosmochim. Acta* 31:1987–93

Sekanina, Z. 1971. A core-mantle model for cometary nuclei and asteroids of possible cometary origin. In *Physical Studies of Minor Planets,* ed. T. Gehrels, pp. 423–26. *NASA SP-267.* Washington DC: US GPO. 687 pp.

Sekanina, Z. 1981. Rotation and precession of cometary nuclei. *Ann. Rev. Earth Planet. Sci.* 9:113–45

Shand, S. J. 1916. The pseudotachylyte of Parijs (Orange Free State) and its relation to "trap-shotten gneiss" and "flinty crush-rock". *Q. J. Geol. Soc. London* 72:198–221

Shoemaker, E. M. 1962. Interpretation of lunar craters. In *Physics and Astronomy of the Moon,* ed. Z. Kopal, pp. 283–359. New York: Academic. 538 pp.

Shoemaker, E. M. 1963. Impact mechanics at Meteor Crater, Arizona. In *The Moon, Meteorites, and Comets,* ed. B. M. Middlehurst, G. P. Kuiper, 4:301–36. Chicago/London: Univ. Chicago Press. 810 pp.

Shoemaker, E. M. 1977. Astronomically observable crater-forming projectiles. See Roddy et al 1977, pp. 617–28

Shoemaker, E. M. 1981. The collision of solid bodies. In *The New Solar System,* ed. J. K. Beatty, B. O'Leary, A. Chaikin, 4:33–44. Cambridge, Mass: Sky Publ. Corp. 224 pp.

Shoemaker, E. M., Eggleton, R. E. 1961. Terrestrial features of impact origin. In *Proc. Geophys. Lab/Lawrence Radiat. Lab. Cratering Symp.,* ed. M. D. Nordyke, pp. A1–27. Livermore, Calif: Lawrence Radiat. Lab.

Shoemaker, E. M., Kieffer, S. W. 1974. *Guidebook to the Geology of Meteor Crater, Arizona.* Tempe: Ariz. State Univ. Cent. Meteorite Stud. 66 pp.

Shoemaker, E. M., Lowery, C. J. 1967. Airwaves associated with large fireballs and the frequency distribution of energy of large meteoroids. *Meteoritics* 3:123–24

Shoemaker, E. M., Wolfe, R. F. 1982. Cratering time scales for the Galilean satellites. In *The Satellites of Jupiter,* ed. D. Morrison, pp. 277–339. Tucson: Univ. Ariz. Press. 972 pp.

Shoemaker, E. M., Hackman, R. J., Eggleton, R. E. 1963. Interplanetary correlation of geologic time. *Adv. Astronaut. Sci.* 8:70–89

Shoemaker, E. M., Williams, J. G., Helin, E. F., Wolfe, R. F. 1979. Earth-crossing asteroids: Orbital classes, collision rates with Earth, and origin. In *Asteroids,* ed. T. Gehrels, pp. 253–82. Tucson: Univ. Ariz. Press. 1181 pp.

Storzer, D., Wagner, G. A. 1977. Fission track dating of meteorite impacts. *Meteoritics* 12:368–69

Taylor, S. R. 1973. Tektites: a Post-Apollo View. *Earth Sci. Rev.* 9:101–23

Taylor, S. R., McLennan, S. M. 1979. Chemical relationships among irghizites, zhamanshinites, Australasian tektites and Henbury impact glasses. *Geochim. Cosmochim. Acta* 43:1551–65

Toon, O. B., Pollack, J. B., Ackerman, T. P., Turco, R. P., McCay, C. P., Liu, M. S. 1982. Evolution of an impact generated dust cloud and its effects on the atmosphere. In *Geological Implications of Impacts of Large Asteroids and Comets on the Earth, Geol. Soc. Am. Spec. Pap.,* ed. L. T. Silver, T. H. Schultz, pp. 187–200

Trask, N. J. 1966. Size and spatial distribution of craters estimated from the Ranger photographs. *Jet Propulsion Lab. Tech. Rep. 32-800,* pp. 252–63

Turco, R. P., Toon, O. B., Park, C., Whitten, R. C., Pollack, J. B., Noerdlinger, P. 1982. An analysis of the physical, chemical, optical and historical impacts of the 1908 Tunguska Meteor fall. *J. Geophys. Res.* In press

van Houten, C. J., van Houten-Groeneveld, I., Herget, P., Gehrels, T. 1970. The Palomar-Leiden survey of faint minor planets. *Astron. Astrophys. Suppl.* 2:339–448

Wagner, P. A. 1922. The Pretoria salt-pan—a

soda caldera. *S. Afr. Dept. Mines Geol. Surv. Mem. 20.* 136 pp.

Walter, L. S. 1965. Coesite discovered in tektites. *Science* 147 : 1029–32

Weissman, P. R. 1980. Stellar perturbations of the cometary cloud. *Nature* 288 : 242–43

Weissman, P. R. 1982a. Dynamical history of the Oort cloud. In *Comets*, ed. L. Wilkening, pp. 637–58. Tucson: Univ. Ariz. Press. 766 pp.

Weissman, P. R. 1982b. Terrestrial impact rates for long and short-period comets. In *Geological Implications of Impacts of Large Asteroids and Comets on the Earth, Geol. Soc. Am. Spec. Pap.*, ed. L. T. Silver, T. H. Schultz, pp. 15–24

Wetherill, G. W. 1967. Collisions in the asteroid belt. *J. Geophys. Res.* 72 : 2429–44

Wetherill, G. W. 1968. Relationships betwen orbits and sources of chondritic meteorites. In *Meteorite Research*, ed. P. Millmann, pp. 573–89. Dordrecht: Reidel. 942 pp.

Wetherill, G. W. 1976. Where do the meteorites come from? A re-evaluation of the Earth-crossing Apollo objects as sources of chondritic meteorites. *Geochim. Cosmochim. Acta* 40 : 1297–1317

Wetherill, G. W. 1979. Steady-state populations of Apollo-Amor objects. *Icarus* 37 : 96–112

Wetherill, G. W., ReVelle, D. O. 1981. Which fireballs are meteorites? A study of the Prairie Network photographic meteor data. *Icarus* 48 : 308–28

Wetherill, G. W., Shoemaker, E. M. 1982. Collision of astronomically observable bodies with the Earth. In *Geological Implications of Impacts of Large Asteroids and Comets on the Earth, Geol. Soc. Am. Spec. Pap.*, ed. L. T. Silver, T. H. Schultz, pp. 1–13

Wetherill, G. W., Williams, J. G. 1968. Evaluation of the Apollo asteroids as sources of stone meteorites. *J. Geophys. Res.* 73 : 635–48

Whipple, F. J. W. 1930. The great Siberian meteor and the waves, seismic and aerial, which it produced. *R. Meteorol. Soc. J.* 56 : 287–304

Whipple, F. L. 1950. A comet model. I. Acceleration of Comet Encke. *Astrophys. J.* 111 : 374–94

Whipple, F. L. 1951. A comet model. II. Physical relations for comets and meteors. *Astrophys. J.* 113 : 464–74

Whipple, F. L. 1978. Comets. In *Cosmic Dust*, ed. J. A. M. McDonnell, pp. 1–73. New York: John Wiley & Sons. 693 pp.

Whipple, F. L. 1982. The rotation of comet nuclei. In *Comets*, ed. L. Wilkening, pp. 227–50. Tucson: Univ. Ariz. Press. 766 pp.

Williams, J. G. 1969. *Secular perturbations in the solar system.* PhD thesis. Univ. Calif., Los Angeles. 270 pp.

Williams, J. G. 1979. Classification of planet-crossing asteroids. *Lunar Sci. X*, p. 1349 (Abstr.)

Wilshire, H. G., Howard, K. A. 1968. Structural pattern in central uplifts of cryptoexplosion structures as typified by Sierra Madera. *Science* 162 : 258–61

Zellner, B. 1979. Asteroid taxonomy and the distribution of the compositional types. In *Asteroids*, ed. T. Gehrels, pp. 783–806. Tucson: Univ. Ariz. Press. 1181 pp.

Zellner, B., Gehrels, T., Gradie, J. 1974. Minor planets and related objects. XVI. Polarimetric parameters. *Astron. J.* 79 : 1100–10

SUBJECT INDEX

K

L

M

N

O

P

CUMULATIVE INDEXES

CONTRIBUTING AUTHORS VOLUMES 1–11

CHAPTER TITLES VOLUMES 1–11

TECTONOPHYSICS AND REGIONAL GEOLOGY

A NONPROFIT SCIENTIFIC PUBLISHER
Annual Reviews Inc.
4139 EL CAMINO WAY • PALO ALTO, CA 94306 USA • (415) 493-4400

Please list the volumes you wish to order by volume number. If you wish a standing order (the latest volume sent to you automatically each year), indicate volume number to begin order. Volumes not yet published will be shipped in month and year indicated. All prices subject to change without notice. Prepayment required from individuals. Telephone orders charged to VISA, MasterCard, American Express, welcomed.

ANNUAL REVIEW SERIES

		Prices Postpaid per volume USA/elsewhere	Regular Order Please send: Vol. number	Standing Order Begin with: Vol. number
Annual Review of ANTHROPOLOGY				
Vols. 1-10	(1972-1981)	$20.00/$21.00		
Vol. 11	(1982)	$22.00/$25.00		
Vol. 12	(1983)	$27.00/$30.00		
Vol. 13	(avail. Oct. 1984)	$27.00/$30.00	Vol(s). ______	Vol. ______
Annual Review of ASTRONOMY AND ASTROPHYSICS				
Vols. 1-19	(1963-1981)	$20.00/$21.00		
Vol. 20	(1982)	$22.00/$25.00		
Vol. 21	(1983)	$44.00/$47.00		
Vol. 22	(avail. Sept. 1984)	$44.00/$47.00	Vol(s). ______	Vol. ______
Annual Review of BIOCHEMISTRY				
Vols. 29-50	(1960-1981)	$21.00/$22.00		
Vol. 51	(1982)	$23.00/$26.00		
Vol. 52	(1983)	$29.00/$32.00		
Vol. 53	(avail. July 1984)	$29.00/$32.00	Vol(s). ______	Vol. ______
Annual Review of BIOPHYSICS AND BIOENGINEERING				
Vols. 1-10	(1972-1981)	$20.00/$21.00		
Vol. 11	(1982)	$22.00/$25.00		
Vol. 12	(1983)	$47.00/$50.00		
Vol. 13	(avail. June 1984)	$47.00/$50.00	Vol(s). ______	Vol. ______
Annual Review of EARTH AND PLANETARY SCIENCES				
Vols. 1-9	(1973-1981)	$20.00/$21.00		
Vol. 10	(1982)	$22.00/$25.00		
Vol. 11	(1983)	$44.00/$47.00		
Vol. 12	(avail. May 1984)	$44.00/$47.00	Vol(s). ______	Vol. ______
Annual Review of ECOLOGY AND SYSTEMATICS				
Vols. 1-12	(1970-1981)	$20.00/$21.00		
Vol. 13	(1982)	$22.00/$25.00		
Vol. 14	(1983)	$27.00/$30.00		
Vol. 15	(avail. Nov. 1984)	$27.00/$30.00	Vol(s). ______	Vol. ______

SEE ORDERING INFORMATION ON PAGE 4.

		Prices Postpaid per volume USA/elsewhere	Regular Order Please send: Vol. number	Standing Order Begin with: Vol. number
Annual Review of ENERGY				
Vols. 1-6	(1976-1981)	$20.00/$21.00		
Vol. 7	(1982)	$22.00/$25.00		
Vol. 8	(1983)	$56.00/$59.00		
Vol. 9	(avail. Oct. 1984)	$56.00/$59.00	Vol(s). ______	Vol. ______
Annual Review of ENTOMOLOGY				
Vols. 7-16, 18-26	(1962-1971; 1973-1981)	$20.00/$21.00		
Vol. 27	(1982)	$22.00/$25.00		
Vol. 28	(1983)	$27.00/$30.00		
Vol. 29	(avail. Jan. 1984)	$27.00/$30.00	Vol(s). ______	Vol. ______
Annual Review of FLUID MECHANICS				
Vols. 1-13	(1969-1981)	$20.00/$21.00		
Vol. 14	(1982)	$22.00/$25.00		
Vol. 15	(1983)	$28.00/$31.00		
Vol. 16	(avail. Jan. 1984)	$28.00/$31.00	Vol(s). ______	Vol. ______
Annual Review of GENETICS				
Vols. 1-15	(1967-1981)	$20.00/$21.00		
Vol. 16	(1982)	$22.00/$25.00		
Vol. 17	(1983)	$27.00/$30.00		
Vol. 18	(avail. Dec. 1984)	$27.00/$30.00	Vol(s). ______	Vol. ______
Annual Review of IMMUNOLOGY				
Vol. 1	(1983)	$27.00/$30.00		
Vol. 2	(avail. April 1984)	$27.00/$30.00	Vol(s). ______	Vol. ______
Annual Review of MATERIALS SCIENCE				
Vols. 1-11	(1971-1981)	$20.00/$21.00		
Vol. 12	(1982)	$22.00/$25.00		
Vol. 13	(1983)	$64.00/$67.00		
Vol. 14	(avail. Aug. 1984)	$64.00/$67.00	Vol(s). ______	Vol. ______
Annual Review of MEDICINE: Selected Topics in the Clinical Sciences				
Vols. 1-3, 5-15	(1950-1952; 1954-1964)	$20.00/$21.00		
Vols. 17-32	(1966-1981)	$20.00/$21.00		
Vol. 33	(1982)	$22.00/$25.00		
Vol. 34	(1983)	$27.00/$30.00		
Vol. 35	(avail. April 1984)	$27.00/$30.00	Vol(s). ______	Vol. ______
Annual Review of MICROBIOLOGY				
Vols. 17-35	(1963-1981)	$20.00/$21.00		
Vol. 36	(1982)	$22.00/$25.00		
Vol. 37	(1983)	$27.00/$30.00		
Vol. 38	(avail. Oct. 1984)	$27.00/$30.00	Vol(s). ______	Vol. ______
Annual Review of NEUROSCIENCE				
Vols. 1-4	(1978-1981)	$20.00/$21.00		
Vol. 5	(1982)	$22.00/$25.00		
Vol. 6	(1983)	$27.00/$30.00		
Vol. 7	(avail. March 1984)	$27.00/$30.00	Vol(s). ______	Vol. ______
Annual Review of NUCLEAR AND PARTICLE SCIENCE				
Vols. 12-31	(1962-1981)	$22.50/$23.50		
Vol. 32	(1982)	$25.00/$28.00		
Vol. 33	(1983)	$30.00/$33.00		
Vol. 34	(avail. Dec. 1984)	$30.00/$33.00	Vol(s). ______	Vol. ______

SEE ORDERING INFORMATION ON PAGE 4.